Statistics

Concepts and Applications

Statistics

Concepts and Applications

David R. Anderson
University of Cincinnati

Dennis J. Sweeney
University of Cincinnati

Thomas A. Williams
Rochester Institute of Technology

West Publishing Company
St. Paul   New York   Los Angeles   San Francisco

Complete Teaching Package to Accompany

STATISTICS: CONCEPTS AND APPLICATIONS

Anderson/Sweeney/Williams

SOLUTIONS MANUAL
Solutions for All Problems

STUDENT STUDY GUIDE
Chapter Summaries, Problems, Multiple Choice Questions

TEST BANK
Multiple Choice and Short Answer Questions, Problems

DEMONSTRATION PROBLEMS AND LECTURE NOTES
Transparencies

MICROCOMPUTER STATISTICAL PACKAGE

Editorial Production Services: Cobb Dunlop Production Services, Inc.
Composition: Science Press
Cover Design: Delor Erickson

COPYRIGHT © 1986 by WEST PUBLISHING COMPANY
50 West Kellogg Boulevard
P.O. Box 64526
St. Paul, MN 55164-1003

Printed in the United States of America

Library of Congress Cataloging-in-Publication Data
Anderson, David Ray, 1941–
   Statistics: concepts and applications.

   Bibliography: p.
   Includes index.
   1. Mathematical statistics.   I. Sweeney, Dennis J.
II. Williams, Thomas Arthur, 1944–   .   III. Title.
QA276.A599   1986   519.5   85-26325
ISBN 0-314-93146-5

2nd Reprint—1988

Contents

List of Statistics in the News and Statistics in Practice

■Statistics in the News
■Statistics in Practice

Preface

The purpose of this book is to provide students from a wide variety of academic programs with a sound conceptual introduction to the field of statistics. The text is applications oriented, and has been written with the needs of the nonmathematician in mind. The mathematical prerequisite is a course in college algebra.

APPLICATIONS AND METHODOLOGY INTEGRATED

Applications of the statistical methodology are an integral part of the organization and presentation of the material. Statistical techniques are introduced in conjunction with problem scenarios where the techniques have been successfully applied. The discussion and development of each technique is thus centered around an application setting, with the statistical results providing information helpful in solving the underlying problem. Examples, which are numbered and set off in the text, introduce additional application settings and provide the student with worked out illustrations of the use of each statistical technique.

Although the book is applications oriented, we have taken care to provide a sound methodological development. Throughout the text we have utilized notation that is generally accepted for the topic being covered. Thus, students will find that the text provides good preparation for the study of more advanced statistical material. A bibliography that should prove useful as a guide to further study has been included as an appendix. The primary features of the text are as follows:

LEARNING OBJECTIVES

Each chapter begins with a statement of learning objectives under the heading "What You Will Learn in This Chapter." This list contains the concepts the student will be expected to master and should help guide the student's study of the material.

STATISTICS IN THE NEWS

Each chapter then opens with a general interest news article that demonstrates a use of the statistical procedures that will be introduced in the chapter. These "Statistics in the News" applications are based on actual news articles appearing in magazines and newspapers such as *Time, USA Today, The New York Times,* and so on. The "Statistics in the News" is a shortened version of the original article and specifically focuses on the use of statistics as reported in the publication. Topical selection such as cost of a college education, marriage statistics, pay differentials for men and women,

TABLE 1

An Overview of "Statistics in the News" Articles

Chapter	Chapter Topic	"Statistics in the News" Topic	Source
1	Introduction	Marriage	*USA Today*
2	Descriptive Statistics I	Favorite Soft Drink	*Time*
3	Descriptive Statistics II	The Dream of Home-ownership	*U.S. News & World Report*
4	Two Variables	Employee Honesty	*Personnel Psychology*
5	Probability	Probability and Your Future	*How You Rate*
6	Probability Distributions	Joe Dimaggio and Pete Rose: Baseball's Greatest	*Sports Illustrated*
7	Normal Distribution	IQ Scores	*How You Rate*
8	Sampling and Sampling Distributions	Financial Aid for College Students	*Cincinnati Enquirer*
9	Population Mean	Average Annual Cost of a College Education	*New York Times*
10	Population Proportion	Miss America Pageant	*USA Today*
11	Two Populations	Pay Differentials for Men and Women	*Monthly Labor Review*
12	Population Variance	Japanese Product Quality	*Fortune*
13	Chi Square Tests	Public Opinion Survey About War	*Tampa Tribune*
14	Analysis of Variance	Self-Esteem and Academic Achievement	*Journal of School Psychology*
15	Regression	Statistics and Professional Football	*Cincinnati Enquirer*
16	Multiple Regression	Children of Divorce	*Journal of R&D in Education*
17	Nonparametric Statistics	Families	*Journal of Marriage and the Family*

the Miss America pageant, professional sports, and so on have been made in an attempt to capture the student's interest and indicate that the statistical procedures he or she is about to learn have some interesting applications. The list of the "Statistics in the News" topics and the source of the articles are shown in Table 1.

CHAPTER PEDAGOGY

Each chapter introduces statistical methodology in the context of problem scenarios and examples which demonstrate the use of the methodology in a wide variety of general interest applications. Problems are provided after each section to enable the student to check his or her progress. Answers to the even numbered problems are provided at the back of the book. Each chapter concludes with a summary that reviews the key concepts and topics that have been introduced in the chapter. A "Statistics in Practice" section then describes a practical application of the material. A glossary follows with review definitions of the statistical terms found in the chapter, and a key formula section itemizes the important equations that the student should know how to apply. A review quiz is then included to reinforce the key concepts presented. Supplementary exercises, based on the material throughout the chapter, are presented to provide students with additional opportunities to practice applying the methodology presented. Where appropriate, computer printouts are provided; these printouts demonstrate how computer packages can be used to provide the statistical computations and summaries. Further discussion of some of these features is provided below.

STATISTICS IN PRACTICE ARTICLES

To further emphasize the applications of statistics, each chapter contains an article describing an actual organization and how it is now using the statistical methodology introduced in the chapter. In most instances, these articles have been supplied by practitioners; the industries represented by these organizations are varied and include health care, government, business, and entertainment. We feel these applications help motivate the student to learn the material and provide an appreciation for some of the ways statistical methods are used in practice. Table 2 contains a guide to the organizations and types of applications included.

REVIEW QUIZZES

Each chapter includes a review quiz consisting of true-false and multiple choice questions. Each review quiz provides the student with the opportunity to evaluate his or her progress after the chapter material has been covered. Answers for the review quiz questions are included at the back of the book.

TABLE 2

An Overview of Statistics in Practice Articles

Chapter	Chapter Topic	Organization	Application
1	Introduction	Kings Island Inc.	Consumer Profile Survey
2	Descriptive Statistics I	Colgate-Palmolive Company	Quality Assurance
3	Descriptive Statistics II	St. Luke's Hospital	Time in Hospice Program
4	Two Variables		Grading Practices of School Teachers
5	Probability	Morton Thiokol Inc.	Customer Service
6	Probability Distributions	Many Organizations	Acceptance Sampling
7	Normal Distribution	Burroughs Corporation	Credit Cards for Banking
8	Sampling and Sampling Distributions	United Media Enterprises	Public Opinion Surveys
9	Population Mean	Thriftway, Inc.	LIFO Inventory Estimation
10	Population Proportion	Harris Corporation	Testing for Defectives
11	Two Populations	California Elementary Schools	Private vs. Public Schools
12	Population Variance	U.S. General Accounting Office	Water Pollution Control
13	Chi Square Tests	United Way	Community Perceptions of Charities
14	Analysis of Variance	Burke Marketing Services Inc.	Cereal Taste Test
15	Regression	Monsanto Company	Poultry Feed Additive
16	Multiple Regression	Champion International Corp.	Control of a Production Process
17	Nonparametric Statistics	West Shell Realtors	Real Estate Prices by Neighborhood

COMPUTER EXERCISES

Twelve chapters conclude with an exercise which has been designed to be solved using a statistical computer package. These exercises, which are available for assignment at the option of the instructor, contain larger data sets than would normally be processed with hand calculations. Students may treat the exercises as cases in which the computer results and their personal evaluation and judgment may combine to provide the desired solution and/or recommendation.

FLEXIBILITY

There is a significant amount of flexibility possible in selecting material to satisfy specific course needs. As an illustration, a possible outline for a two-quarter sequence in introductory statistics is given below:

Possible Two-Quarter Course Outline

First Quarter	Second Quarter
Introduction (Chapter 1)	Population Mean (Chapter 9)
Descriptive Statistics (Chapters 2 and 3)	Proportions (Chapter 10)
Two Variables (Chapter 4)	Two Populations (Chapter 11)
Probability (Chapter 5)	Variance (Chapter 12)
Probability Distributions (Chapter 6)	Chi-Square Tests (Chapter 13)
Normal Distribution (Chapter 7)	Analysis of Variance (Chapter 14)
Sampling and Sampling Distributions (Chapter 8)	Regression/Correlation (Chapter 15)

Other possibilities exist for such a course, depending upon the time available and the background of the students.

ANCILLARIES

Accompanying the text is a complete package of support materials. These include an instructor's manual, a student study guide, a computer software package, a bank of test questions, and Demonstration Problems and Lecture Notes (transparency masters of worked out demonstration problems not found in text). The instructor's manual, prepared by the authors, includes the completely worked solutions to all problems. The study guide provides an additional source of problems and explanations for the students. The test bank, prepared by W. Robert Stephenson, provides a series of multiple choice questions and problems that will aid in the preparation of exams. The computer software package, prepared by the authors, is available for solving problems, computer exercises, etc.

We believe that the applications orientation of the text, its special interest features of "Statistics in the News," "Statistics in Practice," the "Review Quiz," and "Computer Exercises," combined with the package of support materials, provide an ideal basis for introducing students with a wide variety of academic backgrounds and interests to statistics and its many areas of application.

ACKNOWLEDGMENTS

We owe a debt to many of our colleagues and friends for their helpful comments and suggestions during the development of this manuscript. Among these are:

Paul B. Berger	Boston University
Ben P. Bockstege	Broward Community College

John M. Burns	Mt. San Antonio College
Louis J. Cote	Purdue University
Carl Cuneo	Essex Community College
Shirley Dowdy	West Virginia University
Penelope Greene	Harvard University
Terry H. Hughes	Arizona State University
Robert L. Lacher	South Dakota State University
Stanley M. Lukawecki	Clemson University
David R. Lund	University of Wisconsin—Eau Claire
Robert Mee	University of South Alabama
Jeff Mock	Diablo Valley College
Alex Papadopoulos	University of North Carolina—Charlotte
Franklin D. Rich	Sam Houston State University
Arnold L. Schroeder	Long Beach City College
A. K. Shah	University of South Alabama
W. Robert Stephenson	Iowa State University
Bill Stines	North Carolina State University—Raleigh
Barbara Treadwell	Western Michigan University
David L. Turner	Utah State University
Vasant B. Waikar	Miami University

Our associates who supplied information for the "Statistics in Practice" applications made a major contribution to the text. They are:

Bill Mefford and Tom Russell, Kings Island, Inc.
William R. Fowle, Colgate Palmolive Company
Ricki O'Meara, St. Luke's Hospital
Michael Haskell, Morton Thiokol
Frank C. Garcia, Burroughs Corporation
Kenneth R. Sayers, Thriftway, Inc.
Richard A. Marshall, Harris Corporation
Art Foreman and Dale Ledman, U.S. General Accounting Office
Philip R. Tyler, United Way
Ron Tatham, Burke Marketing Services
James R. Ryland and Robert M. Schisla, Monsanto Company
Marion Williams and Bill Griggs, Champion International Corporation
Rodney Fightmaster, West Shell Realtors

We are also indebted to our editor, Mary C. Schiller, and others at West Publishing Company for their editorial counsel and support during the preparation of this text. Finally, we would like to express our appreciation to Phyllis Trosper for her typing and secretarial support.

David R. Anderson
Dennis J. Sweeney
Thomas A. Williams

February 1986

Statistics

Concepts and Applications

CHAPTER 1

Introduction

What You Will Learn in This Chapter

- the two interpretations of the term statistics

- the difference between a population and a sample

- the advantages of summarizing data

- the process of statistical inference and the role of probability

- an overview of various applications of statistical analysis

Contents

1.1 The Population and the Sample
1.2 Data Summarization
1.3 Statistical Inference and Probability
1.4 Some Illustrative Applications

Statistics in the News

THINKING OF MARRIAGE? STATISTICS MAY HELP YOU DECIDE

More than 90% of the people living in the United States will marry sometime during their lives. From youth, thoughts of marriage, decisions of whether to marry or not marry, adjusting to living with another person, decisions about whether to divorce or not divorce, and so on, preoccupy most of us. Lured by such notions as falling in love and living happily ever after, many couples look to marriage as the key to fulfillment and the achievement of a quality life. But, what do statistics have to tell us about marriage and the impact it has on people's lives?

The Gordon S. Black Corporation of Rochester, New York, sampled 1504 adults in order to investigate the quality of life of people living in the United States. The sample results showed that 65% of the married adults stated their lives were satisfying, 55% of the single adults stated their lives were satisfying, and 28% of the divorced adults stated their lives were satisfying. Additional statistics showed the following:

Median age of first marriage (men)	24.6 years
Median age of first marriage (women)	22.1 years
Percent of all marriages that end in divorce	49.2%
Average length of a marriage	9.4 years
Percent of divorces that involve children	55.4%

Based on "Measuring Our Quality of Life," *USA Today* (March 14, 1985).

Fine Art Studio

Most people participate in a wedding ceremony sometime during their lifetime.

These additional statistics reveal some of the risks of entering into marriage. Although more married people are satisfied with life, almost 50% of all marriages end in divorce. Since divorced people are the least satisfied with life it is clear that the decisions to marry or to divorce should not be taken lightly. The rewards of a successful marriage are great, but the consequences of failure are significant.

The next time you read a newspaper, look for items such as the following:

46% of the people surveyed believe that the President is doing a good job in foreign affairs.

The average selling price of a new house is $74,500.

The unemployment rate is 8.9%.

New car sales are up 2.4% over last year.

The numerical facts—or data—in these news items (46%, \$74,500, 8.9%, 2.4%) are commonly referred to as statistics. Several other statistics concerning marriage and divorce were given in "Statistics in the News." In everyday usage, then, the term *statistics* refers to numerical facts or data.

The field, or subject, of statistics involves much more than simply the calculation and presentation of numerical data. In a broad sense, the subject of statistics involves the study of how numerical facts or data are collected, how they are analyzed, and how they are interpreted.

In this chapter we provide an overview of the fundamental concepts associated with collecting, analyzing, interpreting, and presenting data. We begin by defining the terms *population* and *sample*. Then we briefly discuss methods for summarizing a set of data; the objective in using these methods is to present the data in a more convenient and easily interpreted form. Next we discuss the concept of statistical inference and its role in helping us to develop conclusions about a population based upon sample data. The chapter concludes with several illustrations that show the diversity of fields in which the techniques of statistical analysis have been successfully applied.

1.1 THE POPULATION AND THE SAMPLE

Let us consider the case of a major political party that would like to estimate the percentage of voters currently favoring its presidential candidate. How could the party develop such an estimate?

There are approximately 100 million voters in the United States. For the above problem, then, this group of approximately 100 million voters is called the population. In general, we define the population as follows:

Population

A population is the collection of all items of interest in a particular study.

In theory, every voter could be contacted and asked if he or she preferred the party's candidate.

We can see that attempting to contact 100 million voters is impractical from both a time and cost perspective. Instead, let us suppose that the political party selects a subset of 1500 voters believed to be representative of the 100 million voters. This subset of voters is referred to as a sample.

Sample

A sample is a portion of the population selected to represent the whole population.

Furthermore, suppose that of the 1500 voters actually contacted, 600 favor the party's candidate. Expressing this result as a percentage, we find that $(600/1500) \times 100\% =$

40% of the individuals in the sample favor the candidate. This percentage found for the sample could be used as an estimate of the percentage of all voters (the population) that favor the candidate.

Many situations have characteristics similar to those of the above illustration in that there exists a large group (individuals, voters, households, products, customers, etc.) about which information is being sought. Because of time, cost, or other considerations, however, data are collected only from a small portion of the group. As defined above, the larger group of items in a particular study is called the *population,* and the smaller group, the group actually contacted, is called the *sample.*

In "Statistics in the News," several statistics are presented for a sample of 1504 adults; for instance, 65% of the married adults sampled are satisfied with their lives, the average length of a marriage is 9.4 years, and so on. Based upon these data, we might conclude that for the population of all Americans, an estimate of the percentage of married adults that are satisfied with their lives is 65%; similarly, an estimate of the average length of a marriage ending in divorce is 9.4 years, and so on. This concept of using sample data in order to draw conclusions about a population is one of the most important concepts in statistics; it is called statistical inference. In Section 1.3 we discuss this important concept further.

1.2 DATA SUMMARIZATION

In many statistical studies we are interested in summarizing a set of data in order to present it in a more convenient or more easily interpreted form. For example, suppose

TABLE 1.1

Monthly Sales for 200 Salespersons

107	73	68	97	76	79	94	59	98	57
54	65	71	70	84	88	62	61	79	98
66	62	79	86	68	74	61	82	65	98
62	116	65	88	64	79	78	79	77	86
74	85	73	80	68	78	89	72	58	69
92	78	88	77	103	88	63	68	88	81
75	90	62	89	71	71	74	70	74	70
65	81	75	62	94	71	85	84	83	63
81	62	79	83	93	61	65	62	92	65
83	70	70	81	77	72	84	67	59	58
78	66	66	94	77	63	66	75	68	76
90	78	71	101	78	43	59	67	61	71
96	75	64	76	72	77	74	65	82	86
66	86	96	89	81	71	85	99	59	92
68	72	77	60	87	84	75	77	51	45
85	67	87	80	84	93	69	76	89	75
83	68	72	67	92	89	82	96	77	102
74	91	76	83	66	68	61	73	72	76
73	77	79	94	63	59	62	71	81	65
73	63	63	89	82	64	85	92	64	73

that a sales manager has monthly sales figures (number of units sold) for 200 of the firm's salespeople; these data are shown in Table 1.1. It is difficult to draw any conclusions about sales performance looking at the data in this form. Thus, in order to provide the manager with better information about sales performance, the data could be summarized and presented in tabular fashion; Table 1.2 shows this type of summary.

TABLE 1.2

Summary of Monthly Sales for 200 Salespersons

Monthly Sales Volume (units sold)	Number of Salespersons
40–49	2
50–59	10
60–69	52
70–79	65
80–89	44
90–99	22
100–109	4
110–119	1
Total	200

From Table 1.2 we see that the most frequent sales volumes occur in the interval 70–79. In addition, we see that $52 + 65 + 44 = 161$ salespersons sold between 60 and 89 units and that only 5 of the 200 salespersons (2.5%) were able to sell 100 units or more. A number of additional observations that are not apparent from Table 1.1 can be made with the data summarized and presented in the tabular form of Table 1.2.

As another means of summarizing the data we could add the sales volumes for each salesperson and divide by 200 to compute an average sales volume; doing so yields an average of 76 units per salesperson. While this average of 76 units provides a summarization of the data in a single numerical value, the tabular presentation in Table 1.2 provides more information about the variability in the data. Graphical approaches can also be used to summarize a data set. The study of methods for data summarization is referred to as *descriptive statistics*.

1.3 STATISTICAL INFERENCE AND PROBABILITY

Much of statistics is concerned with analyzing sample data in order to learn about characteristics of a population. In order to make our discussion more concrete, let us consider a situation in which a production manager has to decide whether or not to send a recently completed production run to the warehouse. Suppose that the run consists of 2000 items. Suppose also that the manager follows a policy of sending the entire run to the warehouse if no more than 3% of the items are defective; otherwise the items are reworked. In this situation the population consists of the 2000 items in the production run. The characteristic of interest is the percentage of items that are defective. Because of time and cost considerations the production manager has concluded that it is not practical to inspect every item in the population. Instead, the decision has been made to take a sample of 150 items. If more than 3% are defective in the sample, the batch will be reworked.

Let us suppose that the sample of 150 items is taken and that 3 defective items are found. An estimate of the percent defective in the population is $(3/150) \times 100\% = 2\%$. According to the production manager's policy, the entire batch would be accepted and shipped to the warehouse.

The process of estimating the percent defective in the population based on the percent defective in the sample is one example of the use of statistical inference in a decision-making context. Whenever we make a conclusion or inference about an entire population based on sample results, we have to recognize the following: Since the results are based on an analysis of only a small part of the population, they will not be exactly the same as if the entire population had been used. Hence it is desirable to provide some type of indication of how good the sample results are likely to be in terms of estimating the population characteristics. This is where probability plays an important role in statistical inference.

When statisticians make statements about the precision of their estimates, a measure of uncertainty is included. For instance, a statistician might state that the population percent defective is 2% with a possible error of ±.5%. Thus an interval estimate of 1.5% to 2.5% is provided. With the help of probability theory, the statistician can state how likely it is that the interval estimation procedure leads to an interval that contains the actual proportion defective in the population. By applying probability concepts to the analysis of data in a sample, we will learn how to provide estimates of the characteristics of a population, including probabilistic statements about the quality or precision of the estimates.

1.4 SOME ILLUSTRATIVE APPLICATIONS

In this section we present some applications that illustrate how statistics are currently being used in practice.

Medicine

Medical researchers have developed a new drug that is believed to provide a shorter recovery period than the drugs currently available for a particular illness. Under controlled laboratory conditions, a sample of patients with the illness was treated with the new drug. By recording the length of the recovery period for each patient, the researchers were able to estimate the average time it takes for a patient receiving the new drug to recover from the illness.

Education

Educational studies frequently address the question of how the abilities of current students compare to the abilities of previous students. Information from these studies provides educators with a better understanding of the trends in the educational system as well as a basis for evaluating new educational programs and curriculums. For example, a reading test was given to a sample of children completing the sixth grade. The fact that the average score on the exam was 5 points above the average score for a sample of sixth graders 2 years previously was a positive indicator of the benefits associated with a new reading curriculum.

Chemistry

Chemists work to develop a variety of products, including fly sprays and insect repellants. Any new chemical formulations for sprays and repellants are developed and tested under laboratory conditions. In such tests the insects are exposed to the chemicals, and data on survival rates and the resistances of the insects are recorded. A statistical analysis of these data is then used to evaluate the potential of the spray or repellant.

Psychology

Children in an orphanage were used as a sample in a study designed to evaluate the effect of love on the educational and emotional growth and development of young children. One group of children remained in the orphanage under normal everyday conditions and were given no special attention other than that generally found in the orphanage. A second group of children also lived in the orphanage under normal conditions but were given additional daily contact with a person who demonstrated love by providing one-on-one attention, listening, and caring. The educational and emotional growth of the two groups of children were studied over a 5-year period. The statistical summaries of the data showed remarkable differences in the educational and emotional growth of the two groups of students; for example, children exposed to the loving environment showed a much higher level of educational development and a much lower level of emotional problems.

Sociology

The effect of parenting styles on children raised in two different cultural settings was the topic of a sociology study. A sample of children was selected from each of the two cultural settings. Data on attitudes, openness, and emotional stability were collected on the parents and the children. An overall conclusion of the study was that in the cultural environment that contained more parental control and direction of children, the children were less independent as they grew into adulthood.

Business

In an attempt to measure consumer acceptance of a new product, a firm identified several test-market areas throughout the country and selected a sample of potential customers in each area. The individuals in the sample were asked to try the product for 30 days. After the trial period, the individuals were asked what they liked about the product, what they didn't like about the product, and whether or not they preferred the new product to the product they were currently using. In addition to learning about consumers' evaluations of the new product, statistical analysis of the sample data provided the firm with an estimate of the percentage of all potential customers who would prefer the new product.

Engineering

The engineering group of an aircraft manufacturer has been working on a design modification for a major component used in aircraft engines. In order to evaluate the proposed change in design for this component, prototypes of the new design were developed; engines containing the prototypes were then tested in a laboratory environment that simulated actual operating conditions. Of particular concern was the length of time the engine could be operated before the component in question failed to operate. A statistical analysis of the sample data collected showed that the new design had significant reliability problems as compared to the component currently in use.

Agriculture

Agricultural specialists are interested in providing improved growth conditions and productivity for grain products. A university agricultural research program resulted in the development of a new variety of corn. Sample plots of the corn were grown and tested. The data collected from the sample showed that the new corn provided a higher yield per acre and a higher degree of disease resistance.

Government

United States government officials continually make decisions, evaluate programs, and allocate budgets based on statistical studies and statistical information. In a recent study designed to evaluate the physical condition of bridges over 10 years old, a sample of 200 bridges was identified. The data collected showed that 120 of the bridges sampled (60%) did not meet current safety standards. Estimates of the costs to make the necessary repairs were also developed. Data from this study were instrumental in showing the legislative body the critical nature of the problem and in helping them develop funding projections to correct the deficiencies.

In the applications described, we can see the wide diversity of statistical applications and statistical studies. It is anticipated that the importance of statistical procedures will continue to grow in the future. In addition to the areas of applications cited, statistical inquiry is continuing to expand in fields such as law, archeology, economics, history, public policy and so on. Regardless of your field of study or career direction, you can anticipate that statistical studies and statistical data will continue to provide important information.

Summary

Although the term *statistics* as used in everyday practice usually refers to numerical facts or data, the field of statistics is much broader. It involves the design of studies as well as the collection, analysis, interpretation, and presentation of the data collected. Two key components of nearly all statistical studies are the population and the sample.

Statistical inference involves the process of drawing conclusions about the population and its characteristics based on information available from the sample. Probability plays an important role in statistical inference by enabling the statistician to make statements concerning the precision of the sample returns and the likelihood of error.

Because statistics has been successfully applied to so many diverse fields of study, nearly every college student is required to take a course in statistical methodology. A brief overview of typical areas of statistical applications is presented in Section 1.4.

Statistics in Practice

SAMPLING ATTITUDES OF AMUSEMENT PARK VISITORS

Kings Island Entertainment Center is a 1600-acre year-round recreational, sports, and shopping complex located in southwestern Ohio. As one of America's top amusement parks, Kings Island provides rides, entertainment, and other attractions, which draw nearly 3 million people annually.

The Kings Island research staff is responsible for designing questionnaires, selecting samples, conducting interviews, and analyzing data that provide information about attitudes, perceptions, and preferences of individuals who visit the Kings Island facilities. Such information guides operating policies and future plans for the park.

Samples of park visitors are taken throughout the day. As the visitors enter the park, interviewers ask sampled individuals to answer a short questionnaire about themselves and why they came to the park. Examples of data collected include home zip code, distance traveled to the park, type of admission ticket (group sales, season pass, or regular ticket), group size, respondent's age, and so on.

Methods from descriptive statistics are used to summarize the sample results. Each piece of statistical data collected is of interest to someone in the organization. For example, the home zip code provides an indicator of how each market area is doing in drawing visitors to the park. A wide variety of plans, strategies, and decisions are based on the information provided by the samples of park visitors.

Glossary

Population—The collection of all items (individuals, households, products, customers, etc.) of interest in a particular study.

Sample—A portion of the population selected to represent the whole population. Data collected from the sample provide the basis for making inferences about the characteristics of the population.

Descriptive statistics—The study of methods for data summarization.

Review Quiz

TRUE/FALSE

1. A population is the collection of all items of interest in a particular study.
2. A sample need not be representative of the population in order to be useful in drawing inferences.
3. The study of methods for data summarization is referred to as prescriptive statistics.
4. The use of probability enables the statistician to make statements concerning the precision of statistical inferences.

MULTIPLE CHOICE

5. When data are collected for only a portion of the elements of interest, we are using a
   a. population
   b. sample
   c. statistical inference
   d. summary
6. Descriptive statistics is that branch of statistics concerned with
   a. arriving at a conclusion for the population based upon sample information
   b. the summarization and presentation of data
   c. statistical inference
7. Statistical inference is that branch of statistics concerned with
   a. arriving at a conclusion for the population based upon sample information
   b. the summarization and presentation of data
8. In a recent study based upon an inspection of 100 homes in Central City, 60 homes were found to violate one or more city codes. Based upon this information, the city manager released a statement that 60% of Central City's 500 homes were in violation of city codes. The manager's statement is an example of
   a. descriptive statistics
   b. statistical inference
9. Refer to question 8. The manager's statement that 60% of Central City's 500 homes are in violation of city codes is
   a. exactly correct
   b. only an approximation, since it is based upon sample information
   c. obviously wrong, since it is based upon a study of only 100 homes
10. The statement "Based on previous experience, we expect to sell $50,000 worth of snow removal equipment this winter" is an example of
    a. descriptive statistics
    b. statistical inference

Supplementary Exercises

1. Discuss the difference between the concept of *statistics* as numerical facts or data and the concept of *statistics* as a discipline or field of study.

2. A sample of 10 grocery stores in Montgomery, Alabama, on a particular day revealed that the average price per pound for hamburger was $1.60.
   a. What is the population of interest in this study?
   b. In the statistical inference process, we would like to estimate the average price per pound for hamburger for all grocery stores in Montgomery, Alabama. Suggest a value for such an estimate.

3. In a recent study of causes of death in males 60 years of age and older, a sample of 120 showed that 48 men had died due to some form of heart disease.
   a. Compute a descriptive statistic that describes the proportion of males 60 years of age or older that die because of some form of heart disease.
   b. Discuss the role of statistical inference in the context of this example.

4. A study of 250 households in Cedar Bluff showed that a household produced an average of 4 pounds of garbage per day.
   a. What is the sample in this study?
   b. What might be the population of interest in this study?
   c. Discuss the role of statistical inference in the context of this example.

5. Select a daily newspaper and observe the uses of statistical data.
   a. Note any articles that contain statistical information.
   b. Note any advertisements that base their appeal on statistical information.

6. In December, 1984, the *Journal of the American Medical Association* reported the results of a 7-year study that found that women whose mothers took the drug DES during pregnancy are twice as likely to develop tissue abnormalities that might lead to cancer as compared to women whose mothers did not take the drug.
   a. This study involved the comparison of two populations. What were the populations involved?
   b. In the sample of 3980 women who were exposed to DES before birth, 63 developed the tissue abnormalities. Compute a descriptive statistic that would report the number of women out of 1000 who were found to have the tissue abnormality.
   c. For every 1000 women who were not exposed to DES before birth, what is a rough estimate of the number who would develop the tissue abnormality?

7. A firm is interested in testing the advertising effectiveness of a new television commercial. As part of the test, the commercial is shown on a 6:00 P.M. local news program in Denver, Colorado. Two days later a market research firm conducts a telephone survey to obtain information on recall rates (percentage of viewers who recall seeing the commercial) and impressions of the commercial.
   a. What is the population for this study?
   b. What is the sample for this study?

8. The quality control department of a large manufacturing firm is responsible for maintaining product specifications for a variety of production line operations.
   a. List some of the information that the quality control department might want in order to determine whether or not product specifications are being met.

b. Why would the firm be interested in sampling concepts from the area of statistics?

9. Comment on the problem of misleading statistical data that may result in each of the following situations.

   a. In order to estimate the support for a particular political candidate, a pollster visits a major shopping center from 10:00 A.M. to 3:00 P.M. and interviews shoppers.

   b. In order to determine the favorite type of vacation for families in the United States, a study by a Florida tourism promoter reports interviews with 50 out-of-state families visiting Orlando, Florida.

   c. A door-to-door interviewer is instructed to open each interview with the words, "Hello, I am conducting a survey for H&G Soap Products Company. Do you like using H&G products?"

10. A sample of midterm grades for five students showed the following results: 72, 65, 82, 90, 76. Which of the following statements are correct and which should be challenged as being too generalized?

    a. The average midterm grade for the sample of five students is 77.

    b. The average midterm grade for all students who took the exam is 77.

    c. An estimate of the average midterm grade for all students who took the exam is 77.

    d. More than half of the students who take this exam will score between 70 and 90.

    e. If five other students are included in the sample, their grades will be between 65 and 90.

11. A survey of starting salaries for 1985 college graduates was conducted in the summer of 1985. The survey reported an average annual starting salary of $18,800. The survey result was based on a nationwide sample of 400 college graduates who had accepted job offers during the spring of 1985.

    a. What is the population in this study?

    b. What would be a good estimate of the average starting salary for the population?

CHAPTER 2

Descriptive Statistics I: Tabular and Graphical Methods

What You Will Learn in This Chapter

- how the measurement process leads to a classification of data as either nominal, ordinal, interval, and ratio, or, alternatively, as either qualitative or quantitative

- how to construct and interpret summarization procedures for qualitative data such as:
    frequency distributions
    relative frequency distributions
    bar graphs and pie charts

- how to construct and interpret summarization procedures for quantitative data such as:
    frequency and relative frequency distributions
    cumulative frequency and cumulative relative frequency distributions
    histograms
    frequency polygons and ogives

- how to use stem-and-leaf displays to summarize a data set quickly

Contents

Statistics in the News

THE BATTLE OF THE SOFT DRINKS

Soft-drink lovers spent $25 billion on their favorite beverages last year. Competition for a share of the market is intense; new products are proliferating and much money is being spent on advertising campaigns. Including diet and caffeine-free drinks, consumers may now choose from 235 brands; 7 years ago 152 brands were available.

A graphical summary in the figure below shows the market leader is Coke. Pepsi, 7-Up, Dr. Pepper, and Tab round out the top five, with RC Cola dropping to the number 9 position in sales. The most promising newcomer appears to be Diet Coke, which is challenging the current leaders, Tab and Diet Pepsi, in the diet soft-drink market. Coke was slow to introduce Diet Coke because the company feared that putting its famous Coke name on another product would reduce the sales of its flagship brand. However, two-thirds of Diet Coke sales are currently coming from new soda drinkers or from drinkers of competing brands.

The future promises a continued growth of the soft-drink market, with manufacturers introducing even more new products in an attempt to

Fifteen little leaguers select fifteen different brands of soft drink as their personal favorites.

capture a share of this expanding market. However, it is doubtful that the market can continue growing at its current rate. By one estimate, if the growth rate of the past 35 years continues over the next 35 years, by the year 2017 every American will down 1900 bottles of soda annually. That is more than 5 soft drinks per day.

Based on "A Hot Fight Over Cold Drinks," *Time,* May 16, 1983. By early 1985, Diet Coke had edged 7-Up for the number 3 position in the battle of the soft drinks.

The purpose of this chapter is to introduce several tabular and graphical procedures that are commonly used to summarize data. We begin with a discussion of how the process of measuring an observed phenomenon leads to data that can be classified as being qualitative or quantitative. Then we present tabular and graphical techniques for summarizing qualitative and quantitative data. An illustration of a graphical summary of data concerning soft-drink market shares is contained in the figure at the end of "Statistics in the News."

The chapter concludes with an introduction to exploratory data analysis, a set of techniques that rely upon simple arithmetic and easy-to-use graphical presentations to summarize data.

2.1 CLASSIFICATION OF DATA

Data are obtained through the process of *measurement*. The rules that govern how we assign labels or numerical values to entities determine what is called the level of measurement. There are four basic levels of measurement: nominal, ordinal, interval, and ratio.

Nominal Measurement

The level of measurement is *nominal* when the measure assigned to an item is a label used to identify the item.

EXAMPLE 2.1

In a survey, individuals identified their political affiliations as Democrat, Republican, or Independent. Since the measure that was assigned (Democrat, Republican, or Independent) is just a label corresponding to a distinct political category, this is an example of nominal measurement.

• • •

Ordinal Measurement

The level of measurement is *ordinal* when the measures assigned permit the items to be ordered with respect to some criterion.

EXAMPLE 2.2

One common classification of automobiles is based upon size; that is, cars can be classified as compact, intermediate, or full size. Compact, intermediate, and full size are the measures assigned to the cars. Note that this type of measurement implies much more than just a label for automobiles. The cars can be ordered according to the criterion of size. Compact cars are smaller than intermediate cars, and intermediate cars are smaller than full-size cars. Nothing is said or implied, however, about how much smaller compact cars are compared to intermediate cars or how much smaller intermediate cars are compared to full-size cars.

• • •

In the examples of nominal and ordinal measurement, the measures assigned were not numbers. However, numbers can also be used as nominal and ordinal measures.

EXAMPLE 2.3

The numbers on baseball player uniforms are nominal measures—that is, labels used to identify the players. These numbers are useful for scorekeeping because they enable fans and others to recognize the players easily. It makes no sense to perform ordinary arithmetical operations, such as addition and subtraction, on nominal data, even though the measures assigned are numbers.

• • •

EXAMPLE 2.4

McNicolas High School assigns a class rank to each student based upon grade-point averages. The student with the highest grade-point average is ranked number 1, the student with the second highest is ranked number 2, and so on. These class ranks are ordinal data. As with nominal data, ordinary arithmetical operations with ordinal data make no sense.

• • •

Interval Measurement

The level of measurement is *interval* when there is a fixed numerical unit of measurement and each measure assigned is expressed as a quantity of those units.

EXAMPLE 2.5

The measurement of temperature using a thermometer is an example of interval measurement. The fixed unit of measurement is the degree, and each measure assigned specifies a quantity of degrees. Note that addition and subtraction make sense with interval data. For instance, the increase in temperature between a reading of 35° and 40° is the same as the increase in temperature between 85° and 90°. In each case the increase is 5°.

• • •

While distances between interval measures are comparable, one cannot talk meaningfully about ratios with interval data. For instance, it does not make sense to say that 80°F is twice as hot as 40°F because the amount of heat represented by 0°F is not referring to a condition of no heat. Also, on the Celsius scale a temperature of 40°F is 4.44°C, and 80°F is 26.67°C; the higher temperature on the Celsius scale is no longer twice the lower temperature because the location of the zero point is different (32°F corresponds to 0°C).

The fact that interval-level measurement does not have an inherently determined zero point means that we cannot make statements concerning ratios and proportions. In order to make such statements, we must have ratio-level measurement.

Ratio Measurement

The level of measurement is *ratio* when there is a fixed unit of measure and the zero point is inherently defined on the scale of measurement. Ratio comparisons can be made with these measures.

EXAMPLE 2.6

When we measure physical distances such as the length and width of a piece of paper, a value of zero denotes the absence of any distance. As a result, we can conclude that a piece of paper 10 inches in length is twice as long as a piece of paper that is 5 inches long. Such conclusions could not be drawn based upon interval-level measurements.

• • •

In Table 2.1 we provide a summary of the four levels of measurement just discussed. Since the types of data available correspond to these four levels of measurement, we can classify data as being either nominal data, ordinal data, interval data, or ratio data.

TABLE 2.1

Summary of Four Types of Measurement

Type of Measurement	Characteristics	Example
Nominal	• Measure assigned is a label.	• Measure is used to identify political affiliation: Democrat, Republican, Independent.
Ordinal	• Measure assigned permits ordering of items.	• Measurement of size of car: compact, intermediate, full size. • Class rank for high school students.
Interval	• Measures defined in terms of fixed and equal units. • Items can be ordered.	• Measurement of temperature—e.g., 30°, 60°, etc. *Note:* 60° *is not* twice as warm as 30°.
Ratio	• Zero point is inherently defined by the measurement scale. • Measures defined in terms of fixed and equal units. • Items can be ordered.	• Measurement of physical distance; e.g., 5 feet, 10 feet; 5 meters, 10 meters. *Note:* 10 feet *is* twice as long as 5 feet, and 10 meters is twice as long as 5 meters.

Qualitative and Quantitative Data

A distinguishing characteristic of nominal and ordinal data is that ordinary mathematical operations such as addition and subtraction do not lead to meaningful results. On the other hand, interval and ratio data share the distinguishing characteristic that the measures assigned are quantities defined in terms of fixed and equal units. Thus a somewhat simpler classification of data types arises if we choose to look at the data in terms of whether or not the data values are based upon a fixed unit of measurement. *Quantitative data* are data for which the measure assigned is based upon a fixed unit of measurement. *Qualitative data* are data for which the measure assigned is not based upon a fixed unit of measurement.

With this classification scheme, qualitative data is data obtained from nominal- and ordinal-level measurement; quantitative data is data obtained from interval- and ratio-level measurement. We will use this simpler classification scheme whenever it is necessary to distinguish between the types of data available. Perhaps the most important distinction is that with quantitative data, simple statistical calculations such as computing averages make sense.

EXERCISES

1. Based upon their score on a mathematics achievement test, each student in the seventh grade was assigned to one of three mathematics programs: remedial, regular, or enriched. Do the labels assigned represent ordinal or nominal data? Explain.

2. A state agency classified workers as either professional, white-collar, or blue-collar. What type of measurement does this classification represent? Explain.

3. A large city parking garage uses colors to distinguish different parking areas in order to make it easier for customers to locate their cars. If the colors used are red, green, blue, and yellow, what type of data does this represent? Explain.

4. What is the difference between interval and ratio data?

5. How would you classify the measurement 29,028 feet, the height of Mt. Everst? Explain.

6. Shoe widths are designated by the letters A, B, C, D, and E. What type of data do the labels assigned represent? Explain.

7. Cloud shapes are described using the terms cirrus, stratus, and cumulus. What type of measurement does this classification represent? Explain.

8. In the game of golf, the following measurement scheme could be used to represent a player's score: 0 for a par, $+1$ for a score of 1 over par, $+2$ for a score of 2 over par, -1 for a score of 1 under par, -2 for a score of 2 under par, and so on. A golfer obtained the following scores for 9 holes: 0, $+1$, -1, 0, 0, $+2$, 0, $+1$, 0. What type of data does this represent? Explain.

2.2 SUMMARIZING QUALITATIVE DATA

Frequency Distribution

We begin our discussion of how tabular and graphical methods can be used to summarize qualitative data with the definition of a *frequency distribution.*

Frequency Distribution

A *frequency distribution* is a tabular summary of a set of data showing the frequency (or number) of items in each of several nonoverlapping classes.

The objective in developing a frequency distribution is to provide insights about the data that cannot be quickly obtained if we look only at the original data.

EXAMPLE 2.7

In order to help select the next meeting site for their national convention, 80 members of an organization were asked to indicate their preference from among the following four cities: Miami, New Orleans, New York, and San Francisco. Table 2.2 shows the data, consisting of the 80 choices made. To develop a frequency distribution for this data set, we must first count the number of data items associated with each city. This can be accomplished by preparing a tally sheet from the original data, like the one shown in Table 2.3. Simply summing the tallies leads us to the frequency distribution shown in Table 2.4.

• • •

TABLE 2.2

Data Showing City Preference for a Sample of 80 Members

New York	San Francisco	New Orleans	New York	New Orleans
New Orleans	San Francisco	San Francisco	Miami	New Orleans
San Francisco	Miami	San Francisco	San Francisco	San Francisco
Miami	New York	New York	San Francisco	New Orleans
San Francisco	New Orleans	San Francisco	Miami	San Francisco
Miami	Miami	New Orleans	New Orleans	San Francisco
New Orleans	New Orleans	San Francisco	New Orleans	New Orleans
San Francisco	San Francisco	New York	San Francisco	New York
New Orleans	Miami	New York	San Francisco	New Orleans
New York	New Orleans	San Francisco	San Francisco	New York
San Francisco	San Francisco	San Francisco	Miami	San Francisco
Miami	Miami	New Orleans	New Orleans	New Orleans
New York	San Francisco	New York	New Orleans	New Orleans
San Francisco	San Francisco	Miami	New York	San Francisco
Miami	Miami	New Orleans	Miami	San Francisco
New Orleans	Miami	Miami	New Orleans	New Orleans

TABLE 2.3

Tally Sheet of City Preference for the National Convention

City	Tally
Miami	N̄ N̄ N̄ I
New Orleans	N̄ N̄ N̄ N̄ IIII
New York	N̄ N̄ II
San Francisco	N̄ N̄ N̄ N̄ N̄ III

The frequency distribution shown in Table 2.4 provides us with a better understanding of the member preferences described in Example 2.7 than does the original data set shown in Table 2.2. We can now see at a glance that 16 of the members surveyed prefer Miami, 24 prefer New Orleans, 12 prefer New York, and 28 prefer San Francisco. Of course this information was also contained in Table 2.2, but it is much easier to grasp quickly when the data have been systematically organized (as in a frequency distribution).

TABLE 2.4

Frequency Distribution of City Preference for the National Convention

City	Frequency
Miami	16
New Orleans	24
New York	12
San Francisco	28
Total	80

Relative Frequency Distribution

A frequency distribution shows the number (frequency) of data items in each of several nonoverlapping classes. However, often we are interested in knowing the fraction, or proportion, of the data items that fall within each class. The *relative frequency* of a class is simply the fraction, or proportion, of the total number of data items belonging to the class. For a data set having a total of n observations, or items, the relative frequency of each class is given by

$$\text{Relative Frequency of a Class} = \frac{\text{Frequency of the Class}}{n} \qquad (2.1)$$

EXAMPLE 2.7 (continued)

Using (2.1) we can compute a *relative frequency distribution* for the data presented in Table 2.2. It is shown in Table 2.5. The relative frequency distribution enables us to make percentage statements about the data. For example, 16 of the 80 members surveyed, or 20%, prefer Miami; 24 out of 80, or 30%, prefer New Orleans; 12 out of 80, or 15%, prefer New York; and 28 out of 80, or 35%, prefer San Francisco. Note that the sum of the relative frequencies for all classes is 1.0.

• • •

TABLE 2.5

Relative Frequency Distribution of City Preference for the National Convention

City	Relative Frequency
Miami	.20
New Orleans	.30
New York	.15
San Francisco	.35
Total	1.00

Bar Graphs and Pie Charts

A *bar graph* is a graphical device for depicting the information presented in a frequency distribution. On the horizontal axis of the graph we specify a fixed distance to represent the width of a bar. There is one bar for each class of the data, and each bar is separated from the others to enhance the visual presentation. The frequency of each class is indicated on the vertical axis.

EXAMPLE 2.7 (continued)

In Figure 2.1 we show a bar graph for the frequency distribution in Table 2.4; it is just a graphical representation of the frequency distribution. In Figure 2.2 we present another bar graph for this data; it differs from Figure 2.1 in that the scale on the vertical axis has been changed—it starts at a frequency of 10. Clearly, the visual impression given by Figure 2.2 is not the same as the visual impression given by Figure 2.1. In Figure 2.2, it appears that the number of people preferring New Orleans is at least twice the number preferring Miami. However, it is actually only 1½ times as many. The change in the vertical axis has tended to distort comparisons between classes.

• • •

Figures 2.1 and 2.2 show us that by varying the scale on the vertical axis, we can change the visual impression created. When developing bar graphs, we need to exercise care so that the bar graph presents a fair picture of the data. One way to ensure this is

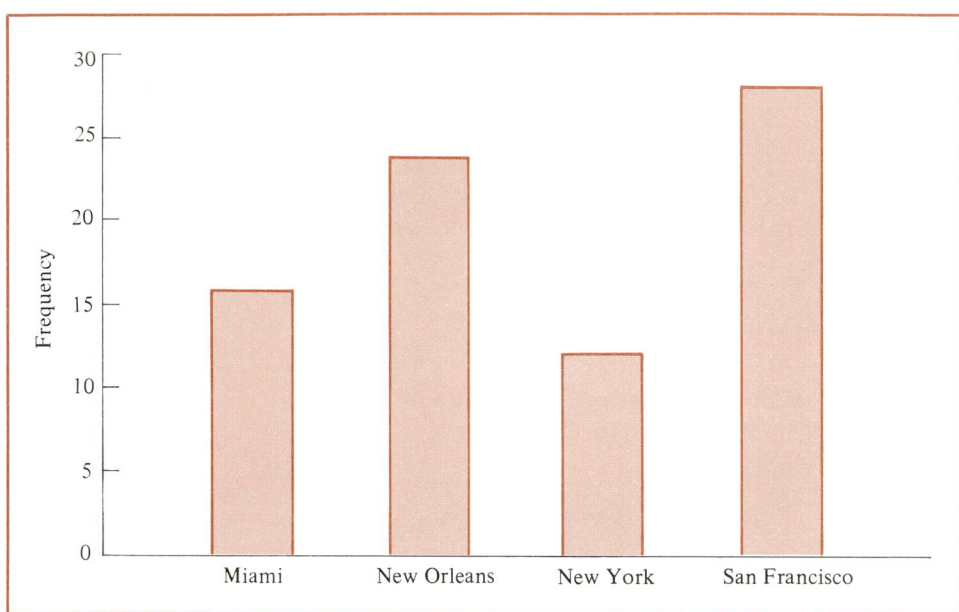

FIGURE 2.1
Initial Bar Graph for Example 2.7

to require that the area of each bar divided by the total area of all the bars be equal to the relative frequency of the class. To satisfy this requirement, the following two guidelines can be adopted:

1. Choose the same width for each bar.
2. Start the vertical axis at a frequency of 0.

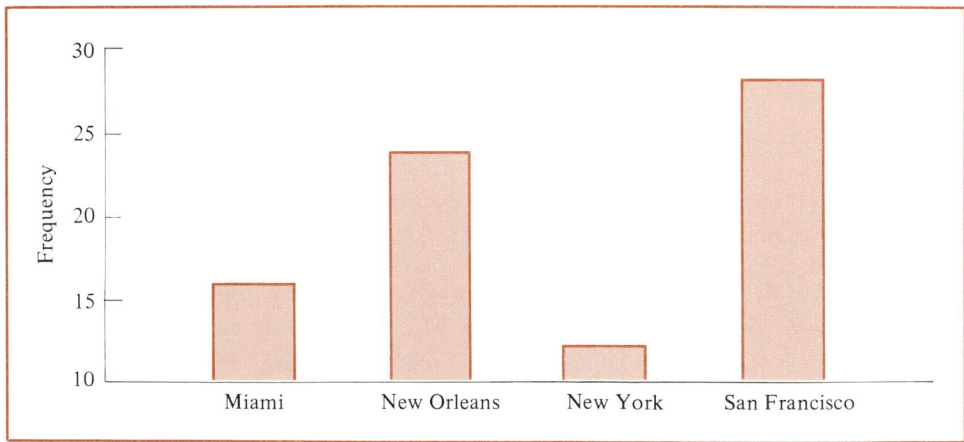

FIGURE 2.2
Revised Bar Graph for Example 2.7

Both the bar graphs in Figures 2.1 and 2.2 follow the first guideline. The bar graph in Figure 2.1 also follows the second guideline, but the one in Figure 2.2 does not and, as a result, gives a biased impression.

A *pie chart* is a commonly used pictorial device for presenting relative frequencies. To draw a pie chart, we draw a circle; then, the relative frequency distribution is used to subdivide the circle into sectors, or parts, that correspond to the relative frequency for each class.

EXAMPLE 2.7 (continued)

In Figure 2.3 we show a pie chart based upon the relative frequency distribution shown in Table 2.5. Since there are 360° in a circle, we used a protractor to break the circle into four parts; each part corresponds to the relative frequency of one of the classes. For example, to identify the sector corresponding to Miami, we identified a part of the circle that corresponds to 20% of 360°, or .20(360) = 72°. Note that regardless of the size of the circle, the pie chart always presents the same visual perception, since the sectors we draw correspond to the relative frequencies.

• • •

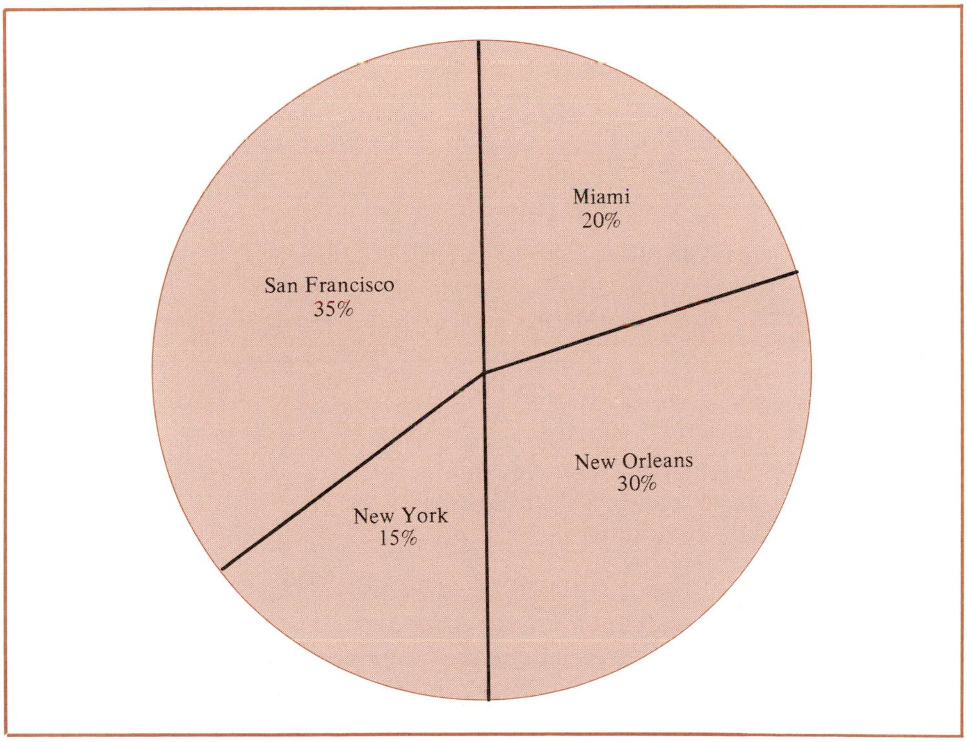

FIGURE 2.3
Pie Chart for Example 2.6

EXERCISES

9. Consumers were asked to rate the taste of a new diet drink as being poor (P), good (G), or excellent (E). The following data were obtained.

G	P	G	E	G
G	E	P	G	G
G	G	E	P	E
E	G	P	G	G
P	G	G	E	E

    Summarize the data by constructing
    a. a frequency distribution
    b. a relative frequency distribution
    c. a bar graph
    d. a pie chart

10. Voters were asked in an exit poll to state how they voted: Democrat (D), Republican (R), or Independent (I). The following data were obtained.

R	R	D	R	D	R	I	R	D	R
D	D	I	I	R	R	R	D	R	D
R	I	D	R	D	R	D	R	R	I

    Summarize the data by constructing
    a. a frequency distribution
    b. a relative frequency distribution
    c. a bar graph
    d. a pie chart

11. Freshmen entering the college of science at Eastern University were asked to indicate their preferred major. The following data were obtained:

Major	Chemistry	Physics	Biology	Geology
Number	112	33	65	21

    Summarize the data by constructing
    a. a frequency distribution
    b. a bar graph
    c. a pie chart

12. Employees at Electronics Associates are on a flextime system; under this system, the employees can begin their working day at 7:00, 7:30, 8:00, 8:30, or 9:00 A.M. The following data represent a sample of the starting times selected by the employees.

7:00	8:30	9:00	8:00	7:30	7:30	8:30	8:30	7:30	7:00
8:30	8:30	8:00	8:00	7:30	8:30	7:00	9:00	8:30	8:00

Summarize the data by constructing
a. a frequency distribution
b. a relative frequency distribution
c. a bar graph
d. a pie chart

13. A national restaurant chain provides a card on each table which contains questions regarding the customers' opinions about their meal, the service, and so on. One question asks the customer to rate the quality of the service as poor, below average, average, above average, or outstanding. The following data represent the results obtained for one restaurant in the chain.

Poor: 7
Below average: 14
Average: 33
Above average: 67
Outstanding: 19

Summarize the data by constructing
a. a frequency distribution
b. a bar graph
c. a pie chart

14. The application form for an automobile loan at a local bank asks applicants to indicate the length of payment period desired; the possible choices are 24, 36, 48, or 60 months. The following data were obtained for the loan applications received the previous week.

48 48 36 60 24 48 48 36 48 48
60 36 48 48 24 48 48 60 24 36
36 48 48 36 60

Summarize the data by constructing
a. a frequency distribution
b. a relative frequency distribution
c. a bar graph
d. a pie chart

2.3 SUMMARIZING QUANTITATIVE DATA: TABULAR METHODS

Frequency Distribution

Frequency distributions are also commonly used for summarizing quantitative data. Let us consider the data presented in Example 2.8.

EXAMPLE 2.8

A mathematics achievement test consisting of 150 questions was given to 50 sixth-grade students. The following data show the number of questions answered correctly by each student.

112	72	69	97	107
73	92	76	86	73
126	128	118	127	124
82	104	132	134	83
92	108	96	100	92
115	76	91	102	81
95	141	81	80	106
84	119	113	98	75
68	98	115	106	95
100	85	94	106	119

We could treat each possible test score as a separate class and use the procedures described in Section 2.2 to develop a frequency distribution. But, since there are so many different scores, such a procedure would not help to summarize the data (there would be over 40 separate classes). With quantitative data, it is customary to group data values that are close together in the same classes.

Suppose we group the test-score data into 8 nonoverlapping classes, with each test score belonging to one and only one class. Table 2.6 contains a tally sheet and a frequency distribution with 8 classes. We see that Table 2.6 provides a good summary of the data. From the frequency distribution we note the following:

1. The most frequently occurring scores are in the interval 95–104, with 10 of the students, or 20%, getting this many correct answers.
2. Only 5 of the 50 students, or 10%, answered 74 or fewer of the questions correctly.
3. Approximately half of the students (26 of 50) answered 75–104 of the questions correctly.
4. Only 6 of the 50 students, or 12%, answered at least 125 of the questions correctly.

Other relevant observations are possible. We see that the potential value of a frequency distribution is that it provides some insights about the entire set of data that cannot be obtained easily by viewing the data in their original unorganized form.

•   •   •

In constructing frequency distributions for quantitative data, there are no hard-and-fast rules. The objective is to present the data in a tabular format so that it reveals existing patterns without losing too much of the individual differences in the data. Important considerations are choosing the proper width for each class, determining the number of classes, and choosing class limits. The following guidelines are often recommended for making appropriate choices.

TABLE 2.6

**Tally Sheet and Frequency
Distribution for Data in Example 2.8**

Tally Sheet

Number of Questions Answered Correctly	Tally
65–74	|||||
75–84	||||| ||||
85–94	||||| ||
95–104	||||| |||||
105–114	||||| ||
115–124	|||||	
125–134	|||||
135–144		

Frequency Distribution

Number of Questions Answered Correctly	Frequency
65–74	5
75–84	9
85–94	7
95–104	10
105–114	7
115–124	6
125–134	5
135–144	1
Total	50

Guidelines for Selection of Class Width and Number of Classes

1. Use classes of equal width.
2. Use between 5 and 20 classes for the frequency distribution.

There is obviously a close relationship between class width and the number of classes; larger class widths mean fewer classes, and vice versa. Once a class width has been chosen, the following expression can be used to compute the approximate number of classes needed.

$$\text{Approximate number of Classes} = \frac{\text{Largest Data Value} - \text{Smallest Data Value}}{\text{Class Width}} \tag{2.2}$$

The recommended number of classes is obtained by rounding up the value obtained using (2.2).

EXAMPLE 2.8 (continued)

For the math achievement scores of Example 2.7, the lowest value is 68 and highest value is 141. Using (2.2) with 10 as a convenient choice for class width yields an approximate number of classes equal to $(141 - 68)/10 = 7.3$. Rounding up, we find that 8 classes (which is between 5 and 20) are recommended for the frequency distribution.

• • •

In practice, we determine the appropriate class width and number of classes by trial and error. A convenient class width is chosen, (2.2) is used to find the corresponding approximate number of classes needed, and this value is rounded up to determine the recommended number of classes needed. If the two given guidelines are satisfied, a frequency distribution is constructed. If, in the judgment of the analyst, the resulting frequency distribution reveals the patterns in the data without losing too many of the individual differences, the frequency distribution is used. If not, another class width is chosen and the process is repeated.

Table 2.7 shows frequency distributions using 2 and 4 classes. We see that using 2 classes provides too much grouping of the data at the expense of showing its dispersion. Even with 4 classes, the amount of detail provided is somewhat inadequate. Thus, faced with a choice of 2, 4, or 8 classes, most people would elect to use 8 classes. This is not meant to imply that 8 intervals is "the" correct choice. Clearly, frequency distributions

TABLE 2.7

Frequency Distributions for Example 2.8 Using Different Number of Classes

Frequency Distribution: 2 Classes

Number of Questions Answered Correctly	Frequency
65–104	31
105–144	19
Total	50

Frequency Distribution: 4 Classes

Number of Questions Answered Correctly	Frequency
65–84	14
85–104	17
105–124	13
125–144	6
Total	50

with more than 8 intervals could be constructed. A subjective decision must be made as to which frequency distribution best describes the observed data.

Apparent and Real Class Limits

The apparent class limits are the upper and lower class limits, as shown in the frequency distribution. For instance, in Table 2.6, the apparent lower limit for the first class is 65 and the apparent upper limit is 74. For that example the apparent class limits and real class limits are the same, since each test score must fall within the apparent limits of one of the classes. In other cases, however, the data summarized in the frequency distribution has been rounded. In such cases, the rounded data will always fall within one of the intervals defined by the apparent class limits. However, before rounding, some of the actual data values may fall between the apparent upper limit of one class and the apparent lower limit of the next class. In such cases, the real and apparent class limits differ. As an illustration, consider the following example.

EXAMPLE 2.9

A doctor's office has collected data on the waiting times for patients who arrive at the office with a request for emergency service. The following data were collected over a 1-month period. For convenience, the waiting times shown were rounded to the nearest whole minute.

$$2, 5, 10, 12, 4, 4, 5, 17, 11, 8, 9, 8, 12, 21, 6, 8, 7, 13, 18, 3$$

The following frequency distribution was developed for these data.

Waiting Time (Minutes)	Frequency
0–4	4
5–9	8
10–14	5
15–19	2
20–24	1
Total	20

Consider the interval 10–14. The lower limit for this interval (10) and the upper limit (14) are referred to as the *apparent limits*. However, since the data values were originally rounded to the nearest minute, any actual times observed that are greater than or equal to 9.5 minutes but less than 14.5 minutes are included in this interval. Thus, 9.5 and 14.5 are the *real* limits for this interval.

• • •

In general, when we round data, the smallest difference possible between the values that we round to is referred to as the *unit difference*. For example, if we were measuring the weights of individuals and our measurements were rounded to the nearest pound, the unit difference would be 1 pound. Thus if the apparent limits for a weight class are 140 and 180, the real limits are 139.5 and 180.5; in this case the class width is $180.5 - 139.5 = 41$ pounds. If the weights were measured to the nearest tenth of a pound, the unit difference is .1 and the real class limits are 139.95 and 180.05. Thus the following relationship can be established.

Determining the Real Class Limits for Rounded Data

$$\frac{\text{Real Lower}}{\text{Limit}} = \frac{\text{Apparent Lower}}{\text{Limit}} - \tfrac{1}{2}(\text{Unit Difference}) \tag{2.3}$$

$$\frac{\text{Real Upper}}{\text{Limit}} = \frac{\text{Apparent Upper}}{\text{Limit}} + \tfrac{1}{2}(\text{Unit Difference}) \tag{2.4}$$

We see that in order to determine the real limits in the class, we must decide upon the accuracy represented in the data.

Relative Frequency Distribution

We define the relative frequency distribution for quantitative data in the same manner that we did for qualitative data. First, recall that the relative frequency is simply the fraction or proportion of the total number of items belonging to the class. That is, for a data set having n observations,

$$\text{Relative Frequency of a Class} = \frac{\text{Frequency of the Class}}{n}$$

Hence, a relative frequency distribution is a tabular summary of a set of data showing the relative frequency in each class.

EXAMPLE 2.8 (continued)

The relative frequency distribution for the data of Example 2.8 based upon 8 class intervals is as follows.

Number of Questions Answered Correctly	Relative Frequency
65–74	.10
75–84	.18
85–94	.14
95–104	.20
105–114	.14
115–124	.12
125–134	.10
135–144	.02
Total	1.00

We see that 2% of the students answered at least 135 questions correctly, 10% answered between 125 and 134 questions correctly, and so on.

• • •

Cumulative Frequency and Cumulative Relative Frequency Distributions

One of the variations of the basic frequency distribution, called the *cumulative frequency distribution,* provides additional information concerning a data set. The cumulative frequency distribution contains the same number of classes as the frequency distribution. However, the cumulative frequency distribution shows the total number of data items with value less than or equal to the real upper limit for the class.

EXAMPLE 2.9 (continued)

The frequency distribution and the cumulative frequency distribution for the data presented in Example 2.9 are shown in the accompanying table.

Frequency Distribution		Cumulative Frequency Distribution	
Waiting Time	Frequency	Waiting Time	Cumulative Frequency
0–4	4	Less than or equal to 4.5	4
5–9	8	Less than or equal to 9.5	12
10–14	5	Less than or equal to 14.5	17
15–19	2	Less than or equal to 19.5	19
20–24	1	Less than or equal to 24.5	20
	Total 20		

To understand how the cumulative frequency distribution is constructed, consider the interval 10–14 in the frequency distribution. Recall that the real upper limit for this interval is 14.5, since the original data were rounded to the nearest minute. To determine the number of data values less than or equal to 14.5, we simply sum the frequencies for the intervals 0–4, 5–9, and 10–14; thus we obtain $4 + 8 + 5 = 17$. Hence the value that we enter in the cumulative frequency distribution corresponding to a waiting time of less than or equal to 14.5 minutes is 17.

• • •

As a final point we note that a cumulative relative frequency distribution shows the *fraction* of data items with values less than or equal to the upper real class limit.

EXAMPLE 2.9 (continued)

The cumulative relative frequency distribution for Example 2.9 is as shown below.

Waiting Time	Cumulative Relative Frequency
less than or equal to 4.5	.20
less than or equal to 9.5	.60
less than or equal to 14.5	.85
less than or equal to 19.5	.95
less than or equal to 24.5	1.00

From the cumulative relative frequencies, we see that 100% of the waiting times are less than or equal to 24.5 minutes, 95% are less than or equal to 19.5 minutes, and so on.

• • •

EXERCISES

15. The given data show the number of automobiles arriving at a toll booth during 20 intervals, each of 10 minutes duration.

26	26	38	24
32	22	15	33
19	27	21	28
16	20	34	24
27	30	31	33

Summarize the data by constructing
a. a frequency distribution
b. a relative frequency distribution
c. a cumulative frequency distribution
d. a cumulative relative frequency distribution

16. A psychologist asked 25 of her patients to take a written examination designed to measure the patient's depression level. The following data were obtained (higher scores indicate a higher level of depression).

75	68	33	69	26
62	77	54	61	96
87	57	61	56	79
78	67	78	68	75
28	89	61	51	41

Summarize the data by constructing
a. a frequency distribution
b. a relative frequency distribution
c. a cumulative relative frequency distribution

17. A survey of 250 company employees asks each employee to indicate the one-way mileage (rounded to the nearest mile) from his or her home to work. A partial relative frequency distribution is shown:

Miles	Relative Frequency
0–4	.10
5–9	.22
10–14	
15–19	.18
20–24	.16

   a. Complete the relative frequency distribution.
   b. Construct the frequency distribution for these data.

18. National Airlines accepts flight reservations by phone. Shown below are the call durations (in minutes) for a sample of 20 phone reservations. Construct the frequency and relative frequency distributions for the data.

2.1	4.8	5.5	10.4
3.3	3.5	4.8	5.8
5.3	5.5	2.8	3.6
5.9	6.6	7.8	10.5
7.5	6.0	4.5	4.8

19. The numbers of television sets sold by Globe TV Sales are shown below. Data show weekly sales for a 15-week period.

13	10	6	9	10
11	5	8	3	9
10	7	8	10	6

   Construct a frequency distribution and a relative frequency distribution for the data.

20. The given data are the bowling scores an individual obtained for the most recent 20 games:

160	170	181	156	176
148	198	179	162	150
162	156	179	178	151
157	154	179	148	156

   Summarize the data by constructing
   a. a frequency distribution
   b. a relative frequency distribution
   c. a cumulative frequency distribution
   d. a cumulative relative frequency distribution

21. The given data are the final examination scores for students enrolled in section 1 of college algebra.

$$
\begin{array}{ccccccccccc}
69 & 62 & 60 & 72 & 70 & 85 & 81 & 77 & 57 & 84 & 66 \\
48 & 70 & 92 & 81 & 98 & 51 & 75 & 62 & 73 & 73 & 83 \\
94 & 93 & 89 & 79 & 60 & 77 & 69 & 90 & 71 & 91 & 89
\end{array}
$$

Summarize the data by constructing
a. a frequency distribution
b. a relative frequency distribution
c. a cumulative frequency distribution
d. a cumulative relative frequency distribution.

2.4 SUMMARIZING QUANTITATIVE DATA: GRAPHICAL METHODS

Frequency, relative frequency, cumulative frequency, and cumulative relative frequency distributions have been used as tabular procedures for summarizing quantitative data. Graphical summaries can also be used—often with greater impact. The most commonly used graphical presentations of quantitative data sets are histograms, frequency polygons, and ogives.

When developing graphical presentations of quantitative data sets, the usual practice is to work with the real class limits. For example, suppose that we are working with data that could take on only integer values between 1 and 50, with corresponding class intervals of 1–5, 6–10, 11–15, and so on. In developing graphical presentations for such data, we assume the real limits to be .5–5.5, 5.5–10.5, 10.5–15.5, and so on.

Histogram

The most common form of graphical presentation of a frequency distribution is a *histogram.* A histogram is constructed by placing the real class limits on the horizontal axis of a graph and the frequencies on the vertical axis. Each class is shown on the graph by drawing a rectangle whose base is the class interval and whose height is the corresponding frequency for the class.* The base for each interval extends from the lower to the upper real limit of the interval.

EXAMPLE 2.10

Dinner checks for the Lakeside Restaurant were rounded to the nearest dollar and then used to develop the given frequency distribution.

---

*When class intervals must be unequal because of some particular feature of the data set, the method of constructing a histogram should be modified. References providing a discussion of the necessary modifications are given in Appendix A.

Check Amount	Frequency
20–29	6
30–39	15
40–49	33
50–59	24
60–69	12
70–79	10
Total	100

Since the data were rounded to the nearest dollar, the real class limits are 19.5–29.5, 29.5–39.5, and so on. Thus, when we draw the histogram, the base for the rectangle for the class interval 20–29 must begin at 19.5 and end at 29.5, the base for the rectangle for the class interval 30–39 must begin at 29.5 and end at 39.5, and so on. The histogram for this data set is shown in Figure 2.4.

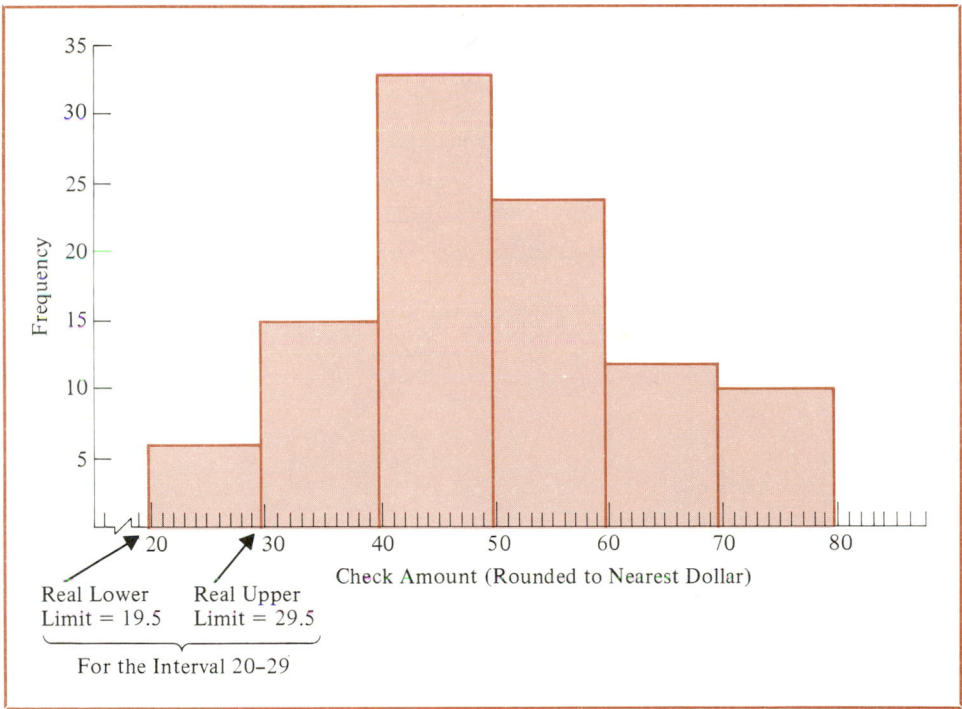

FIGURE 2.4
Histogram for Data Set of Example 2.10

• • •

Frequency Polygon

A *frequency polygon* provides an alternative to a histogram as a means of presenting a frequency distribution graphically. Again, the data values are placed on the horizontal axis and the frequencies on the vertical axis. However, instead of using rectangles, as with the histogram, we find the *class midpoints* on the horizontal axis and then plot points directly above the class midpoints at a height corresponding to the frequency of the class.

The definition of class midpoint is given.

Class Midpoint

The midpoint of any class interval is the average of the class limits (apparent or real).

Classes of zero frequency are added at each end of the frequency distribution so that the frequency polygon touches the horizontal axis at both ends of the graph. The frequency polygon is then formed by connecting the points corresponding to each class midpoint with straight lines.

EXAMPLE 2.10 (continued)

The frequency polygon for the frequency distribution of check amounts is shown in Figure 2.5. Note that the interval added at the lower end is 10–19 (with real limits of 9.5–19.5), and the interval added at the top end is 80–89 (with real limits of 79.5 and 89.5). As an illustration of determining a class midpoint, consider the interval 20–29.

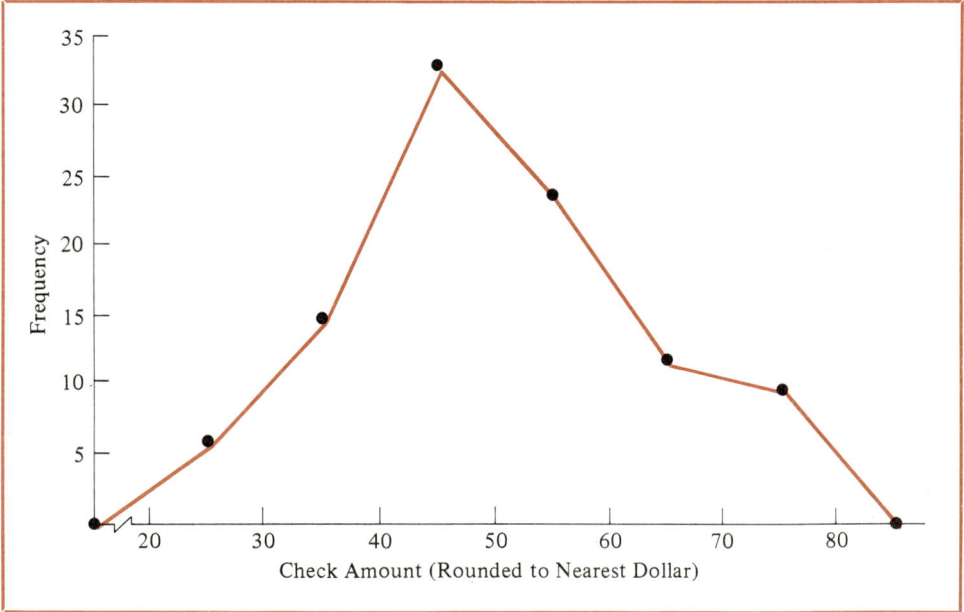

FIGURE 2.5
Frequency Polygon for Data Set of Example 2.10

$$\text{Class Midpoint} = \frac{20 + 29}{2} = 24.5$$

We get the same answer if the real class limits are used.

$$\text{Class Midpoint} = \frac{19.5 + 29.5}{2} = 24.5$$

Similarly, the class midpoint for the interval 30–39 is:

$$\text{Class Midpoint} = \frac{30 + 39}{2} = 34.5$$

The accompanying table shows the class midpoints for all the intervals with a class of zero frequency added on each end.

Check Amount	Apparent Lower Limit	Apparent Upper Limit	Class Midpoint
10–19	10	19	14.5
20–29	20	29	24.5
30–39	30	39	34.5
40–49	40	49	44.5
50–59	50	59	54.5
60–69	60	69	64.5
70–79	70	79	74.5
80–89	80	89	84.5

Note that each succeeding class midpoint is just 10 units—the class width—above the preceding class midpoint. Thus identifying the class midpoints for all classes is a simple matter once the midpoint for any one class has been determined. We used these class midpoints to draw the frequency polygon shown in Figure 2.5

• • •

While the histogram and frequency polygon in Figures 2.4 and 2.5 were based on the frequency distribution of dinner check amounts, they could have just as easily been based on the relative frequency distribution. The graphical presentations based on the relative frequency distribution would have been identical to Figures 2.4 and 2.5 with the exception that the vertical axis would have been measured in terms of relative frequency instead of actual frequency.

Ogive

A graph of the cumulative frequency distribution or cumulative relative frequency distribution is called an *ogive*. The data values are on the horizontal axis and cumulative frequencies are on the vertical axis. A point is plotted directly above each upper real limit at a height corresponding to the cumulative frequency at that upper real limit. One additional point is then plotted above the lower real limit for the first class at a height of zero. These points are then connected by straight line segments.

EXAMPLE 2.10 (continued)

The cumulative frequency distribution for the data of Example 2.10 is as follows.

Check Amount	Cumulative Frequency
Less than or equal to 29.5	6
Less than or equal to 39.5	21
Less than or equal to 49.5	54
Less than or equal to 59.5	78
Less than or equal to 69.5	90
Less than or equal to 79.5	100

To draw the ogive, we plot a point with height 100 at 79.5, a point with height 90 at 69.5, and so on. Note that for the class 20–29, we plot a point at 19.5, the lower real limit, at a height of zero. Figure 2.6 shows the ogive for this data set.

• • •

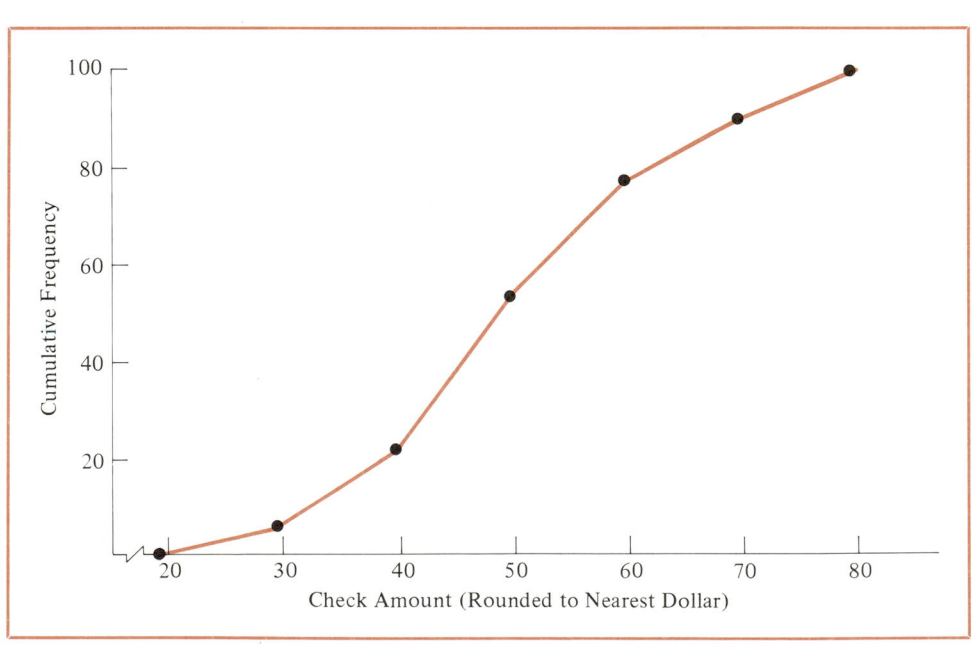

FIGURE 2.6
Ogive for Example 2.10

EXERCISES

22. RC Radio Corporation collects data on the monthly sales for its car telephone units. Over the past 12 months the sales have been as follows:

Month	Sales	Month	Sales
Jan	80	July	88
Feb	115	Aug	91
Mar	82	Sept	89
Apr	102	Oct	95
May	94	Nov	105
June	90	Dec	108

Summarize the data by constructing
a. a frequency distribution using the intervals 80–89, 90–99, and so on
b. a histogram
c. a frequency polygon
d. an ogive

23. Consider the following frequency distribution.

Interval	Frequency
10–19	12
20–29	15
30–39	24
40–49	18
50–59	10

a. Construct the histogram for this frequency distribution.
b. Draw the corresponding frequency polygon.
c. Draw the ogive for these data.

24. A vocabulary test given to 50 tenth-grade pupils resulted in the following frequency distribution.

Number Correct	Frequency
5–9	3
10–14	7
15–19	12
20–24	14
25–29	9
30–34	5

a. Construct the corresponding histogram.
b. Draw the frequency polygon.
c. Draw the ogive for these data.

25. The duration of 20 long-distance telephone calls is given (time recorded in minutes).

10.5	5.0	15.3	16.8	9.2
4.2	12.6	7.8	11.5	12.6
20.2	27.5	8.9	12.2	18.2
14.5	14.0	5.5	15.5	8.9

Summarize the data by constructing
a. a frequency distribution
b. a relative frequency distribution
c. a histogram
d. a frequency polygon
e. an ogive

26. A Better Business Bureau conducted a study of automobile repair shops in order to investigate price variations among shops. A test car was set up to have a specific front-end alignment problem; different shops were then asked to correct the problem. The following data shows the amounts charged (in dollars) by 25 different repair shops.

81	111	58	81	88	62	88	87	77	99
79	62	104	95	83	76	74	72	64	52
87	66	86	67	87					

Summarize the data by constructing
a. a frequency distribution
b. a relative frequency distribution
c. a histogram
d. an ogive

2.5 EXPLORATORY DATA ANALYSIS

The techniques of exploratory data analysis focus on how simple arithmetic and easy-to-draw pictures can be used to summarize data quickly. In this section we study how one of these techniques—referred to as a *stem-and-leaf display*—can be used to rank order data and provide an idea of the shape of the underlying distribution.

One simple method of displaying data involves arranging the data in ascending or descending order. This process, referred to as rank ordering the data, provides some degree of organization. However, such an approach provides little sense of the shape of the distribution of data values. A stem-and-leaf display is a device that provides a display of both rank order and shape simultaneously.

EXAMPLE 2.8 (continued)

In Example 2.8 we considered a set of data that showed the results of a mathematics achievement test (consisting of 150 questions) given to 50 sixth-grade students. The number of questions answered correctly by each student is repeated.

112	72	69	97	107
73	92	76	86	73
126	128	118	127	124
82	104	132	134	83
92	108	96	100	92
115	76	91	102	81
95	141	81	80	106
84	119	113	98	75
68	98	115	106	95
100	85	94	106	119

The first digits of each data item are arranged to the left of a vertical line. To the right of the vertical line we record the second digit for each item as we pass through the scores in the order they were recorded. The second digit for each item is placed on the horizontal line corresponding to its first digit:

```
 6 | 9  8
 7 | 2  3  6  3  6  5
 8 | 6  2  3  1  1  0  4  5
 9 | 7  2  2  6  2  1  5  8  8  5  4
10 | 7  4  8  0  2  6  6  0  6
11 | 2  8  5  9  3  5  9
12 | 6  8  7  4
13 | 2  4
14 | 1
```

Given this organization of the data, it is a simple matter to rank order the second digits on each horizontal line. Doing so leads to the following stem-and-leaf display of the data:

```
 6 | 8  9
 7 | 2  3  3  5  6  6
 8 | 0  1  1  2  3  4  5  6
 9 | 1  2  2  2  4  5  5  6  7  8  8
10 | 0  0  2  4  6  6  6  7  8
11 | 2  3  5  5  8  9  9
12 | 4  6  7  8
13 | 2  4
14 | 1
```

Each line in this display is referred to as a *stem,* and each piece of information on a stem is a *leaf.* For example, consider the first line

```
 6 | 8  9
```

The meaning attached to this line is that there are two data values in the data set whose first digit is 6: 68 and 69. Similarly, the second line,

$$7 \mid 2 \quad 3 \quad 3 \quad 5 \quad 6 \quad 6$$

specifies that there are six data values whose first digit is 7: 72, 73, 73, 75, 76, and 76. Thus we see that the data values in this stem-and-leaf display are separated into two parts. The label for each stem is the one- or two-digit first part of the number (that is, 6, 7, 8, 9, 10, 11, 12, 13, or 14)—to which we refer as the starting point—and the leaf is the single digit second part (that is, 0, 1, 2, . . . , 8, 9). The vertical line simply serves to separate the two parts of each number listed.

To focus on the shape indicated by the stem-and-leaf display, let us use a rectangle to depict the "length" of each stem. Doing so we obtain the following:

```
 6 |  8   9
 7 |  2   3   3   5   6   6
 8 |  0   1   1   2   3   4   5   6
 9 |  1   2   2   2   4   5   5   6   7   8   8
10 |  0   0   2   4   6   6   6   7   8
11 |  2   3   5   5   8   9   9
12 |  4   6   7   8
13 |  2   4
14 |  1
```

Rotating this page counterclockwise onto its side provides a picture of the data very similar to that provided by a histogram of the data with classes of 60–69, 70–79, 80–89, and so on.

Although the stem-and-leaf display may appear at first glance to offer little more information about the data set than that provided by a histogram, there are two primary advantages:

1. The stem-and-leaf is easier to construct; in fact, it is easier to construct such a display than to describe how to do it.
2. Within any interval, the stem-and-leaf display provides more information than the histogram, since the stem-and-leaf shows us the actual data values.

In the same way that there is no right number of classes in a frequency distribution or histogram, there is no right number of rows or stems in a stem-and-leaf display. If we believe that our original stem-and-leaf display has condensed the data too much, it is a simple matter to stretch the display by using two or more stems for each starting point. For example, to use two lines for each starting point, we place all data values ending in 0, 1, 2, 3, or 4 on one line and all values ending in 5, 6, 7, 8 and 9 on a second line.

•   •   •

EXAMPLE 2.8 (continued)

Consider the following stretched stem-and-leaf display:

```
 6*  |
 6·  | 8  9
 7*  | 2  3  3
 7·  | 5  6  6
 8*  | 0  1  1  2  3  4
 8·  | 5  6
 9*  | 1  2  2  2  4
 9·  | 5  5  6  7  8  8
10*  | 0  0  2  4
10·  | 6  6  6  7  8
11*  | 2  3
11·  | 5  5  8  9  9
12*  | 4
12·  | 6  7  8
13*  | 2  4
13·  |
14*  | 1
14·  |
```

The usual convention is to attach an asterisk (*) to each label that corresponds to leaves of 0–4 and a dot (·) to each label that corresponds to leaves of 5–9. This stretched stem-and-leaf display is similar to a histogram with intervals of 60–64, 65–69, 70–74, 75–79, and so on.

Although our illustration of a stem-and-leaf display shows leaves consisting of one digit, stem-and-leaf displays consisting of leaves with two digits are similarly displayed.

• • •

EXAMPLE 2.11

The following data show the number of hamburgers sold by a fast-food restaurant for each of 15 weeks.

1852	1644	1766	1888	1912
2044	1812	1790	1679	2008
1565	1852	1967	1954	1733

Following is a stem-and-leaf display of this data:

```
15 | 65
16 | 44  79
17 | 33  66  90
18 | 12  52  52  88
19 | 12  54  67
20 | 08  44
```

We see that each leaf consists of two digits and each two-digit leaf is separated by a space.

• • •

EXERCISES

27. A psychologist developed a new test of adult intelligence. The test was administered to 20 individuals, and the following data were obtained.

114	99	131	124	117
102	106	127	119	115
98	104	144	151	132
106	125	122	118	118

    Construct a stem-and-leaf display for these data.

28. The certification board of a national computer users group has been working on a test designed to measure competencies in computer center management. A preliminary version of the test, consisting of 150 multiple-choice questions, was given to 25 computer center managers. The number of questions each individual answered correctly is shown.

102	91	72	98	115
57	89	121	89	124
122	136	105	80	79
64	108	113	83	63
84	96	99	75	97

    Construct a stem-and-leaf display to summarize these data.

29. In a study of postmeningitic brain damage, a series of tests were administered to 50 subjects. The following data were obtained, where low scores represent considerable brain damage.

87	76	67	58	92	59	41	50	90	75
80	81	70	73	69	61	88	46	85	97
50	47	81	87	75	60	65	92	77	71
70	74	53	43	61	89	84	83	70	46
84	76	78	64	69	76	78	67	74	64

    Construct a stem-and-leaf display for these data.

30. The high temperature at 30 selected cities on December 15 resulted in the following data.

53	66	48	84	77	38	86	90	39	52
41	49	74	34	58	68	72	61	71	47
46	48	55	66	47	44	51	58	73	89

    Construct a stem-and-leaf display for these data.

Summary

A set of data, even if modest in size, is often difficult to interpret directly in the form in which it is gathered. Tabular and graphical procedures provide ways of organizing and summarizing the data so they are more easily interpreted. Frequency distributions, relative frequency distributions, cumulative frequency distributions, and cumulative relative frequency distributions were introduced as tabular methods of summarizing data. These distributions can also be presented graphically through the use of histograms, frequency polygons, and ogives. The chapter concluded with a brief introduction to exploratory data analysis. The purpose of the descriptive statistical procedures presented in this chapter is to provide tabular and graphical approaches that facilitate the interpretation of data.

Statistics in Practice

QUALITY ASSURANCE FOR DETERGENTS

The central business philosophy of the Colgate-Palmolive Company is expressed by their slogan, "Quality products since 1806." The Quality Assurance and Improvement Department within Colgate devotes full-time efforts toward achieving this goal. In this capacity, a variety of statistical techniques are used. Relative frequency distributions and graphical techniques such as histograms are some of the most useful tools for communicating data and ideas.

As an example of the use of statistical techniques at Colgate, consider the production of the familiar heavy-duty detergent used for home laundries. The regular-size carton of the detergent has a stated weight of 20 ounces. Of particular concern is the density of the detergent powder that is placed in the carton. Even with rigid quality control standards in the powder-production process, at times the powder varies in its weight per unit volume. For example, if the weight of the powder is on the heavy side (a high specific gravity), it will not take as much powder to reach the 20-ounce-per-carton weight limit. In this case, the company is faced with the problem of filling cartons with 20 ounces but having the carton appear slightly underfilled when it is opened by the user.

To reduce this problem and maintain the quality standards, the detergent powder is sampled periodically before being placed in the cartons. When the powder reaches an unacceptably high density or specific gravity, corrective action is taken to reduce the specific gravity of the powder before the filling operation is permitted to resume.

Repeated samples provide more and more data about the specific gravity of powder. At some point, various parties in the company are interested in knowing how the powder-production process is doing in terms of meeting density guidelines. Tabular and graphical summaries provide convenient ways to present the data. For example, the accompanying figure shows a histogram for the specific gravity of 150 samples taken over a 1-week period. Since it has been concluded that an undesirably high specific gravity occurs around .40, this summary shows that the operation is meeting its quality guidelines, with practically all the data showing values less than .40. Production management personnel should be pleased with the quality aspect of the powder product as indicated by this type of statistical summary.

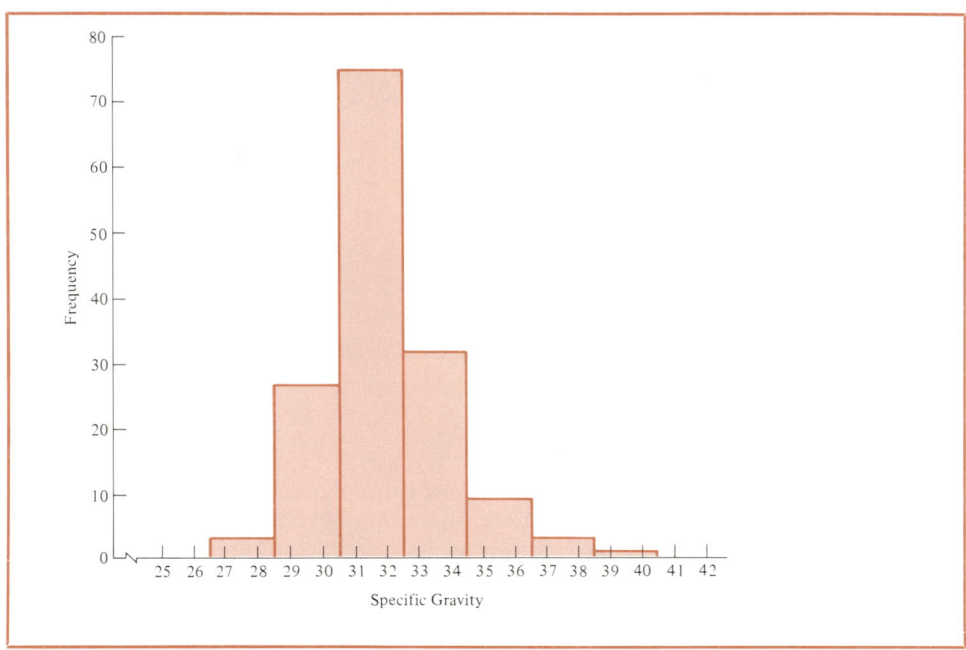

Glossary

Measurement—The process of assigning a label or numerical value to some observed phenomenon.

Nominal-level measurement—The measure assigned to an item is simply a label used for identification.

Ordinal-level measurement—The measure assigned permits the items to be rank ordered with respect to a criterion.

Interval-level measurement—The measure assigned conforms to a fixed numerical unit of measurement and each measure is expressed as a quantity of those units.

Ratio-level measurement—The measure assigned has all the properties of interval-level measurement plus the property that a zero point is inherently defined by the measurement scheme.

Quantitative data—Data obtained from interval- and ratio-level measurement.

Qualitative data—Data obtained from nominal- and ordinal-level measurement.

Frequency distribution—A tabular summary of a set of data showing the frequency (or number) of items in each of several nonoverlapping classes.

Relative frequency distribution—A tabular summary of a set of data showing the relative frequency—that is, the fraction or proportion—of the total number of items in each of several nonoverlapping classes.

Bar graph—A graphical device for depicting the information presented in a frequency distribution of qualitative data.

Pie chart—A pictorial device for presenting qualitative data summaries based upon subdividing a circle into sectors that correspond to the relative frequency for each class.

Apparent limits—The lower and upper limits actually shown for the classes of a frequency distribution.

Real limits—The lower and upper limits based upon the *actual values observed* before rounding.

Unit difference—The smallest difference possible between the rounded values.

Cumulative frequency distribution—A tabular summary of a set of data that shows the total number of data items with value less than or equal to the real upper limit for the class.

Histogram—A graphical presentation of a frequency distribution constructed by placing the class intervals on the horizontal axis of a graph and the frequencies on the vertical axis. Each class corresponds to a rectangle whose base is the real class interval and whose height is the class frequency.

Frequency polygon—A graphical presentation formed by connecting the points corresponding to each class midpoint and frequency with a straight line.

Class midpoint—The average of the lower and upper limits (apparent or real).

Ogive—A graph of the cumulative frequency distribution or cumulative relative frequency distribution.

Exploratory data analysis—The use of simple arithmetic and easy-to-draw pictures to look at data more effectively.

Stem-and-leaf display—An exploratory data analysis technique that simultaneously rank orders the data and provides insight into the shape of the underlying distribution.

Key Formulas

Relative Frequency of a Class

$$\frac{\text{Frequency of the Class}}{n} \tag{2.1}$$

Approximate number of Classes

$$\frac{\text{Largest data value} - \text{smallest data value}}{\text{Class Width}} \tag{2.2}$$

Real Lower Limit

$$\text{(Apparent lower limit)} - \frac{1}{2} \text{(unit difference)} \qquad (2.3)$$

Real Upper Limit

$$\text{(Apparent upper limit)} + \frac{1}{2} \text{(unit difference)} \qquad (2.4)$$

Review Quiz

TRUE/FALSE

1. The speed of an airplane in miles per hour is ordinal data.
2. A football player's number is nominal data.
3. Room numbers in a building are examples of interval data.
4. The place a person finishes in a golf tournament (first, second, third, etc.) is an example of ordinal data.
5. Ordinal and ratio data are classified as quantitative data.
6. Nominal and ordinal data are classified as qualitative data.
7. A frequency distribution is a graphical summary of a set of data.
8. A relative frequency distribution has twice as many classes as a frequency distribution.
9. In a bar graph the width of the bar is proportional to the number of items in the class.
10. The class midpoint is halfway between the apparent lower and apparent upper class limit.
11. A histogram is a graphical presentation of a frequency distribution.
12. A frequency polygon is a graphical presentation of a cumulative frequency distribution.
13. A stem-and-leaf display provides insight into the shape of the distribution of a set of data.

MULTIPLE CHOICE

14. Which of the following measures involve nominal data?
    a. the test score on an exam
    b. the number on a basketball player's jersey
    c. the speed of an automobile
    d. the class rank of a college student
15. Which of the following measures involve ordinal data?
    a. the score in a baseball game
    b. the place of a baseball team in the league standings

c. the height of a flagpole
d. the weight of a fish

16. A group of union members indicated what they felt was the most important issue in the upcoming labor management negotiations; these preferences are shown in the accompanying frequency distribution.

Issue	Frequency
Wages	22
Medical benefits	10
Retirement	15
Working conditions	13
Total	60

The relative frequency of the wages issue is closest to

a. .22
b. .15
c. .40
d. .60

17. Refer to the frequency distribution in question 16. Fringe benefits include medical benefits and retirement. What percentage of the union membership feels that fringe benefits are the most important issue?

a. 16.7%
b. 25%
c. 63%
d. 41.7%

18. Consider the following frequency distribution

Completion Time (min)	Frequency
8–10	5
11–13	8
14–16	15
17–19	12
20–22	7
23–25	3
Total	50

The real lower limit for the third class is

a. 13.5
b. 14
c. 15
d. 16

19. Refer again to the frequency distribution in question 18. The midpoint for the second class is
   a. 11
   b. 11.5
   c. 12
   d. 12.5

20. Consider the following stem-and-leaf display.

$$
\begin{array}{c|cccccc}
6 & 2 & 2 & 4 \\
7 & 3 & 6 & 6 & 7 & 8 & 9 \\
8 & 1 & 5 & 7 \\
9 & 0 & 1
\end{array}
$$

The data set consists of how many items?
   a. 14
   b. 18
   c.   6
   d. 20

Supplementary Exercises

31. Determine whether the following represent nominal, ordinal, interval, or ratio data. Explain.
   a. religious preference
   b. age (in years)
   c. student grade point average
   d. credit card numbers
   e. SAT math scores
   f. marital status (single, married, divorced)

32. The owner of a major league baseball team has been concerned that attendance has been falling. The following data are the attendance figures (in thousands) for the previous 20 home games.

17	33	14	21	23
27	15	15	29	17
12	29	16	22	24
22	24	18	22	19

Summarize the data by constructing
   a. a frequency distribution
   b. a relative frequency distribution
   c. a cumulative frequency distribution

33. Refer to the baseball attendance figures provided in exercise 32.
   a. Construct a histogram for these data.
   b. Draw the frequency polygon.
   c. Draw an ogive for these data.

34. Frequent airline travelers were asked to indicate the airline they believed offered the best overall service. The four choices were American Air(A), East Coast Air(E), Suncoast(S), and Great Western(W). The following data were obtained.

```
E  A  E  S  W  W  E  S  W  E
W  E  E  A  S  S  W  E  A  W
W  S  E  E  A  E  E  S  W  A
S  E  A  W  A  A  W  E  S  W
```

Summarize the data by constructing
a. a frequency distribution
b. a relative frequency distribution
c. a bar graph
d. a pie chart

35. The administrator of a large city hospital has been collecting data regarding the number of patients treated in the emergency room on weekends. The following data are the numbers treated for each of the previous 15 weeks.

```
154  177  164  145  110
214  131  122  180  191
172  148  157  174  160
```

Summarize the data by constructing
a. a frequency distribution
b. a relative frequency distribution
c. a cumulative relative frequency distribution
d. a histogram

36. A psychological test was designed to measure a subject's ability to anticipate the next item in a series while viewing or hearing its immediate predecessor. The number of items correctly anticipated was recorded.

```
21  13   8  10  17
15   6  23   3  19
12  17   8  12  14
```

Use appropriate tabular and graphical methods to summarize the data.

37. A nursery school offers programs for 4-year olds ranging from a 1-day-a-week program to a 5-day-a-week program. To help in planning, the school's director surveyed parents regarding the type of program they preferred. The following data, which represents the number of days, were obtained

```
3  3  1  2  2  4  4  2  3  3
3  5  3  3  2  2  1  5  3  3
2  4  5  3  3  3  4  2  2  4
```

Summarize the data by constructing
   a. a frequency distribution
   b. a relative frequency distribution
   c. a histogram
   d. a frequency polygon
38. Dinner check amounts for La Maison's French Restaurant are shown below:

42.65	36.12	52.90	44.26	52.00
34.10	39.86	29.40	48.75	82.00
38.40	44.50	79.80	74.45	71.81
46.62	56.12	63.00	63.06	59.42

   a. Construct frequency and relative frequency distributions for the data.
   b. Construct a cumulative relative frequency distribution for the data.
39. The following data represent quarterly sales volumes for 40 selected corporations.

17,864,000	15,065,000	42,200,000	13,523,000
49,747,000	20,510,000	5,520,000	7,985,000
3,624,000	11,556,000	1,855,000	9,023,000
3,804,000	5,933,000	23,900,000	6,145,000
9,232,000	2,979,000	1,059,000	42,789,000
5,143,000	33,380,000	20,779,000	6,145,000
2,141,000	17,768,000	18,017,000	42,800,000
5,090,000	41,626,000	12,003,000	6,840,000
3,669,000	37,738,000	40,765,000	21,946,000
13,614,000	39,914,000	7,846,000	25,837,000

   a. Construct a frequency distribution to summarize these data. Use a class width of $5,000,000.
   b. Develop a relative frequency distribution for the data.
   c. Construct a cumulative frequency distribution for the data.
   d. Construct a cumulative relative frequency distribution for the data.
40. Use the data in exercise 39 for the following.
   a. Construct a histogram as a graphical representation of the data.
   b. Construct a frequency polygon for the data.
   c. Construct an ogive for the data.
41. The given data show home mortgage loan amounts (in dollars) handled by a particular loan officer in a savings and loan company. Use a frequency distribution, relative frequency distribution, and histogram to help summarize these data.

20,000	38,500	33,000	27,500	34,000
12,500	25,999	43,200	37,500	36,200
25,200	30,900	23,800	28,400	13,000
31,000	33,500	25,400	33,500	29,200
39,000	38,100	30,500	45,500	30,500
52,000	40,500	51,600	42,500	44,800

42.   Morrison Communications, Inc. periodically reviews sales personnel performance records. One member of the sales force has had the following weekly sales volume (units sold) over the past quarter, or 13 weeks. Use a relative frequency distribution, a cumulative frequency distribution, and a frequency polygon to summarize these sales data.

$$13, 19, 20, 17, 21, 27, 9, 15, 22, 18, 18, 23, 20$$

43.   Given below are the closing prices at week's end for 40 common stocks.

$7\frac{1}{2}$	$16\frac{1}{4}$	$19\frac{3}{4}$	$7\frac{3}{8}$	$10\frac{1}{8}$
$5\frac{3}{4}$	$7\frac{7}{8}$	$6\frac{7}{8}$	$24\frac{5}{8}$	$17\frac{1}{8}$
$12\frac{3}{4}$	11	5	$24\frac{3}{4}$	$34\frac{3}{4}$
$14\frac{3}{4}$	$10\frac{3}{8}$	$35\frac{1}{8}$	$20\frac{1}{4}$	$12\frac{5}{8}$
42	$10\frac{3}{8}$	57	$19\frac{3}{4}$	28
$63\frac{7}{8}$	$10\frac{3}{4}$	$7\frac{7}{8}$	$17\frac{5}{8}$	48
$17\frac{1}{4}$	$9\frac{3}{4}$	44	$11\frac{1}{8}$	20
$41\frac{3}{4}$	$16\frac{5}{8}$	$21\frac{1}{4}$	$8\frac{5}{8}$	$16\frac{3}{8}$

   a.   Construct frequency and relative frequency distributions for these data.
   b.   Construct cumulative frequency and cumulative relative frequency distributions for these data.

44.   Use the data in exercise 43 for the following:
   a.   Construct a histogram for the data. Plot relative frequency on the vertical axis.
   b.   Construct a frequency polygon for the data.
   c.   Construct an ogive for the data.

45.   The grade point averages for 30 students majoring in economics are given.

2.21	3.01	2.68	2.68	2.74
2.60	1.76	2.77	2.46	2.49
2.89	2.19	3.11	2.93	2.38
2.76	2.93	2.55	2.10	2.41
3.53	3.22	2.34	3.30	2.59
2.18	2.87	2.71	2.80	2.63

   a.   Construct a relative frequency distribution for the data.
   b.   Construct a cumulative relative frequency distribution for the data.

46.   Use the data in exercise 45 for the following:
   a.   Construct a histogram for the data. Plot relative frequency on the vertical axis.
   b.   Construct an ogive for the data.

47.   Hospital records show the following number of days of hospitalization for 20 patients.

5	7	7	15
21	15	22	10
10	6	8	18
14	5	7	8
3	8	4	10

    a.   Construct frequency and relative frequency distributions for the data.
    b.   Construct a cumulative relative frequency distribution for the data.
    c.   Construct a histogram.

48. Points scored by the winning team in 25 A.C.C. (Atlantic Coast Conference) basketball games are shown.

86	79	74	72	91
82	64	75	72	74
63	80	78	95	82
86	77	73	69	72
81	85	92	62	90

    a.   Construct frequency and relative frequency distributions for the data.
    b.   Construct a cumulative relative frequency distribution for the data.
    c.   Construct a histogram.

49. The final examination scores in a section of calculus resulted in the following data.

56	77	84	82	42
61	44	95	98	84
93	62	96	78	88
58	62	79	85	89
89	97	53	76	75

Draw a stem-and-leaf display for these data.

50. A 150-question social-awareness test was given to a group of college freshmen. The following data show the number of questions answered correctly by each of the students.

121	114	94	136	144	126	98	103	118	127
135	97	119	117	122	138	142	141	102	105

Draw a stem-and-leaf display for these data.

51. The following data show total yardage accumulated over the football season for 20 receivers.

744	652	576	1112	971	451	1023	852	809	596
941	975	400	711	1174	1278	820	511	907	1251

Draw a stem-and-leaf display for these data.

Computer Exercise

Consolidated Foods, Inc. has opened several new grocery stores at a variety of locations over the past 2 years. One of the special services at these new stores is that customers may pay for their purchases using Visa or Mastercard credit cards as well as using either cash or an approved check. In order to better understand how customers are using the new payment feature, a sample of 100 customers was selected over a 1-week period. Data collected for each customer included how much was spent during the shopping trip and the method of payment. The data collected are shown in the accompanying table.

Amount Spent ($)	Method of Payment	Amount Spent ($)	Method of Payment
84.12	Check	86.34	Check
34.66	Credit card	20.23	Credit card
37.27	Credit card	108.70	Check
38.82	Credit card	45.36	Credit card
46.50	Credit card	83.31	Check
99.67	Check	64.45	Credit card
70.18	Check	54.33	Credit card
99.21	Check	16.78	Cash
138.42	Check	115.96	Check
93.68	Check	95.83	Check
120.89	Check	19.76	Cash
10.14	Cash	35.37	Cash
74.51	Check	111.98	Check
17.91	Check	103.95	Check
49.59	Check	90.40	Credit card
4.74	Cash	6.68	Cash
48.14	Cash	32.09	Credit card
65.67	Credit card	79.70	Credit card
89.66	Check	96.08	Credit card
96.40	Check	20.60	Cash
54.16	Credit card	78.81	Check
79.55	Check	123.62	Check
67.95	Check	125.01	Check
30.69	Cash	41.58	Credit card
151.89	Check	36.73	Credit card
130.41	Check	52.07	Credit card
98.80	Check	19.78	Cash
23.59	Cash	66.44	Check
104.67	Check	5.08	Cash
90.04	Check	50.15	Credit card
77.62	Check	114.42	Check
36.01	Cash	97.26	Credit card
88.17	Check	22.75	Cash
66.76	Credit card	53.63	Credit card
23.50	Cash	132.31	Check
127.34	Check	105.54	Check
26.02	Cash	66.09	Check
79.77	Check	62.24	Check
29.35	Check	97.93	Check
71.31	Credit card	10.57	Cash
43.57	Credit card	51.21	Credit card
76.18	Credit card	90.17	Check
59.38	Credit card	24.08	Credit card
72.99	Credit card	42.72	Cash

Continues next page

Amount Spent ($)	Method of Payment	Amount Spent ($)	Method of Payment
19.24	Cash	97.72	Check
80.20	Check	112.67	Check
55.79	Cash	14.30	Cash
134.27	Check	28.76	Credit card
64.68	Credit card	81.85	Check
75.54	Check	56.84	Credit card

QUESTIONS

1. Develop tabular and graphical summaries of the data that will be helpful in describing the amount spent and the method of payment.
2. Does there appear to be any relationship between the amount spent and the method of payment?
3. Use exploratory data analysis to help summarize the data.

CHAPTER 3

Descriptive Statistics II: Measures of Location and Dispersion

What You Will Learn In This Chapter

- how to compute and interpret the mean, median, and mode for a set of data

- what a percentile is and how it is computed

- how to compute and interpret measures of dispersion, such as the range, variance, standard deviation, and coefficient of variation

- how to compute descriptive statistics from grouped data

- Chebyshev's theorem and how to use it

- how to construct a 5-number summary and a box-and-whisker plot

Contents

Statistics in the News

THE AMERICAN DREAM: IS IT STILL AFFORDABLE?

Homeownership is the ultimate fulfillment of the American dream. It represents a way of life sought by most Americans. However, with the recent history of rapid inflation and high interest rates, the dream of ownership is beginning to fade for a significant portion of the population. The 1980 census showed that 64.4% of all households were homeowners. However, over the past few years this percentage has begun to decline.

In 1977, the median price of a home was $44,000; an average monthly principal-plus-interest payment was $277. By 1984, the median price of similar single-family home had risen to $74,200. The higher cost and higher interest rates had pushed the average monthly principal-plus-interest payment to $622. Without question, individuals in the middle- and lower-income brackets are beginning to be forced out of the housing market. If the trend continues, it is estimated that homeownership will be out of reach of all but the wealthiest 10% of American families.

The upward trend in housing costs affects not only families, builders, construction workers, building-materials suppliers, and realtors, but, as some suggest, it may begin to have an impact on our democracy itself. Anything can happen when a human dream goes unfulfilled. And the evidence is that the American dream of homeownership may be unfulfilled in the future.

Median resale prices in some major metropolitan areas are as follows:

A real estate representative prepares to show a house to prospective buyers.

Anaheim-Santa Ana	$135,100
San Francisco	$130,500
Los Angeles	$117,700
New York	$105,400
Boston	$ 95,600
Dallas-Ft. Worth	$ 86,500
Denver	$ 82,400
Chicago	$ 80,300
Milwaukee	$ 68,100
Memphis	$ 64,300
Atlanta	$ 63,900
St. Louis	$ 62,400
Kansas City	$ 59,600
Tampa	$ 58,200
Louisville	$ 49,300

Based on "Housing Prices in Key Cities," *U.S. News & World Report*, August 27, 1984.

In this chapter we continue our discussion of descriptive statistics by introducing several numerical measures of location and dispersion: for example, the mean, the median, the variance, and the standard deviation. We show how to compute these measures from the original data set and then from the grouped data available in a frequency distribution. Numerical measures that are computed for a population are called *population parameters;* when they are computed for a sample, they are called *sample statistics.*

As we showed in the "Statistics in the News" article on housing costs, numerical measures are an important part of many presentations; for example, in that article, median resale prices are reported as a measure of housing costs in various cities.

3.1 MEASURES OF LOCATION

Mean

Perhaps the most important numerical measure is the *mean,* or average value, of the data. The mean provides a good measure of central location for a data set. It is obtained by adding all the data values and dividing by the number of items. If the data set is a sample the mean is denoted by $\bar{x}$ (pronounced *x* bar); if the data set is a population the mean is denoted by the Greek letter μ (pronounced m$\bar{\text{u}}$).

EXAMPLE 3.1

A sample of five sections in the College of Science resulted in the following data regarding the number of students in each section.

$$46, 54, 42, 46, 32$$

We compute the sample mean for this data as follows:

$$\bar{x} = \frac{46 + 54 + 42 + 46 + 32}{5}$$

$$= \frac{220}{5} = 44$$

Thus for the five sections sampled, the mean number of students in a section is 44.

$$\bullet \quad \bullet \quad \bullet$$

In specifying general statistical formulas, it is customary to denote the value of the first data item by x_1, the value of the second data item by x_2, and so on. (The symbol x_1 is pronounced *x* sub 1, x_2 is pronounced *x* sub 2, and so on.) Using this notation, (3.1) gives a general formula for the sample mean.

Sample Mean

$$\bar{x} = \frac{x_1 + x_2 + \cdots + x_n}{n} = \frac{\sum_{i=1}^{n} x_i}{n} \qquad (3.1)$$

where n = number of items in the sample.

In this formula, the numerator denotes the sum of the data values starting with $i = 1$ and ending with $i = n$. That is,

$$\sum_{i=1}^{n} x_i = x_1 + x_2 + \cdots + x_n$$

The uppercase Greek letter Σ (sigma) is used as a summation sign.

EXAMPLE 3.1 (continued)

Using the notation x_1, x_2, x_3, x_4, x_5 to represent the number of students in each of the 5 sections sampled, we have:

$$x_1 = 46$$
$$x_2 = 54$$
$$x_3 = 42$$
$$x_4 = 46$$
$$x_5 = 32$$

Thus to compute the sample mean, we can write

$$\bar{x} = \frac{\sum_{i=1}^{n} x_i}{n} = \frac{\sum_{i=1}^{5} x_i}{5} = \frac{x_1 + x_2 + x_3 + x_4 + x_5}{5} = \frac{46 + 54 + 42 + 46 + 32}{5} = 44$$

$$\bullet \quad \bullet \quad \bullet$$

When we want to sum all the data items, we use the following abbreviated summation notation:

$$\sum x_i = \sum_{i=1}^{n} x_i$$

The starting and ending points for the summation, shown below and above Σ, respectively, are dropped in the abbreviated notation. It is understood that all the data values are to be included in the sum. A summary of the summation notation and operations used in this text is contained in Appendix C.

EXAMPLE 3.2

A survey of television-viewing habits among college students provided the following data on viewing time for each student in hours per week.

14, 9, 12, 4, 20, 26, 17, 15, 18, 15, 10, 6, 16, 15, 8, 5

Since there are 16 sample observations, $n = 16$. The sample mean is given by

$$\bar{x} = \frac{\Sigma\, x_i}{n} = \frac{14 + 9 + \cdots + 5}{16} = \frac{210}{16} = 13.125$$

Hence, the average time spent viewing television for the 16 students is 13.125 hours.

$$\bullet \quad \bullet \quad \bullet$$

Equation (3.1) shows how the mean is computed for a sample of n items. The formula for computing the mean of a population is similar, but we use different notation to indicate that we are dealing with the entire population. The number of items in the population is denoted by N, and, as we mentioned previously, the symbol for the population mean is μ.

Population Mean

$$\mu = \frac{\Sigma\, x_i}{N} \tag{3.2}$$

Median

The *median* is another statistical measure of central location for a set of data. The median for a set of data is that value falling in the middle when the data items are arranged in ascending order (rank ordered from smallest to largest). If there are an odd number of data items, the median is the middle item. If the number of data items is even, there is no single middle value. We follow the convention of defining the median to be the average of the middle two values in this case. For convenience this definition is restated below.

Median

If there is an odd number of items in the data set, the median is the value of the middle item when all items are arranged in ascending order.

If there is an even number of items in the data set, the median is the average value of the two middle items when all items are arranged in ascending order.

EXAMPLE 3.1 (continued)

Let us apply the above definition to compute the median for the sample of five section sizes sampled from the College of Science. Arranging the five data values in ascending order provides the following rank-ordered list:

$$32 \quad 42 \quad 46 \quad 46 \quad 54$$

Since $n = 5$ is odd, the median is the middle item in the above rank-ordered list. That is, the median corresponds to the third value, or 46. Even though there are two values of 46 for this data set, each value is treated as a separate item when we place the data in rank order and hence when we determine the median.

• • •

EXAMPLE 3.2 (continued)

A rank ordering of the viewing times for the 16 students produced the following list:

$$4 \quad 5 \quad 6 \quad 8 \quad 9 \quad 10 \quad 12 \quad 14 \quad 15 \quad 15 \quad 15 \quad 16 \quad 17 \quad 18 \quad 20 \quad 26$$

Median = 14.5

Since $n = 16$ is even, the median is the average value of the two middle items, that is, the eighth and ninth items. Hence the median is $(14 + 15)/2 = 14.5$.

• • •

While the mean is the most commonly used measure of central location, there are a number of situations in which the median is a better measure. The mean is influenced by extreme values in a data set, but the median is not.

EXAMPLE 3.3

A sample of five families in Herrold, Iowa, showed the following annual family incomes:

$$\$17,500 \qquad \$23,000 \qquad \$24,000 \qquad \$26,000 \qquad \$320,000$$

The median annual income is \$24,000 (the third item). The mean is given by

$$\bar{x} = \frac{\$17,500 + \$23,000 + \$24,000 + \$26,000 + \$320,000}{5} = \$82,100$$

Clearly the median is a better indication of typical incomes than the mean. The mean has been substantially influenced by the single family with a very large income. Indeed, the mean annual income for the sample is over three times the annual income for 4 of the 5 families.

• • •

Because of the kinds of situations represented by Example 3.3, the median is usually the preferred measure of central location when there are extreme values in the data. When conducting a study, we recommend presenting both the mean and the median. The user can then decide which is more appropriate.

Mode

A third statistical measure, the *mode,* is sometimes used as a measure of central location. The mode is defined as follows.

Mode

The mode of a set of data is the value that occurs with greatest frequency.

EXAMPLE 3.1 (continued)

Referring to the sample of five section sizes, we see that the only value that occurs more than once is 46. Since this value, occurring with a frequency of 2, has the greatest frequency in the data set, it is the mode.

• • •

EXAMPLE 3.2 (continued)

Referring to the 16 weekly television-viewing times for college students, we see that 15 hours of viewing time occurs with the greatest frequency. Thus 15 (occurring with a frequency of 3) is the mode.

• • •

Although there is a single value that occurs with the greatest frequency in both of these examples, situations can arise for which the greatest frequency occurs at two or more different values. The data set introduced in Example 2.8 of Chapter 2 gives the scores of 50 sixth-grade students on a mathematics achievement test. The score of 92 occurred 3 times and the score of 106 occurred 3 times. Thus both 92 and 106 are modes for this data set; we say that the data are *bimodal*. If a data set has more than two modes it is said to be *multimodal;* in such cases the mode is almost never reported since listing three or more modes would not do a very good job of describing a central location for the data. In the extreme case, every data value can be different, and an argument can be made that every observation is a mode. We take the position that in such cases the mode is not an appropriate measure of central location, and thus no value would be reported.

The type of data for which the mode is considered a good measure is qualitative data. For example, the data set introduced in Example 2.7 of Chapter 2 resulted in the following frequency distribution of preferences for convention cities.

City	Frequency
Miami	16
New Orleans	24
New York	12
San Francisco	28
Total	80

Hence the mode, or most frequently preferred city, is San Francisco. For this type of data it obviously makes no sense to speak of the mean or median. But the mode does provide a good indicator of what we are interested in, the city preferred by the greatest number of people.

As another illustration of the use of the mode as a measure of location for qualitative data, suppose that a manufacturer markets a product in three different package designs. Measuring frequency of purchase, the modal package design would be the one most frequently purchased. Management is obviously interested in knowing the mode, but it would not be possible, nor would it make any sense, to compute a mean or median for the data. When working with nominal data, the mode is the only measure of central location that is useful.

Percentiles

A *percentile* is a statistical measure that locates values in the data set that are not necessarily central locations. In addition to identifying locations, this measure provides information regarding how the data items are spread over the interval from the lowest to the highest values. Hence percentiles can also be viewed as measures of dispersion, or variability, in the data set. In large data sets that do not have numerous repeated values, the pth percentile is a value that divides the data set into two parts. Approximately p percent of the items take on values less than the pth percentile; approximately $(100 - p)$ percent of the items take on greater values.

Percentile

The pth percentile of a data set is a value such that *at least p* percent of the items take on this value or less and *at least* $(100 - p)$ percent of the items take on this value or more.

Admission test scores for colleges and universities are frequently reported in terms of percentiles. For instance, suppose an applicant has a raw score of 54 on the verbal portion of an admissions' test. It may not be readily apparent how this student performed relative to other students taking the same test. However, if the raw score of 54 corresponds to the 70th percentile, then we know that approximately 70% of the students had scores less than this individual and approximately 30% scored better.

The above definition is not very useful for calculating the pth percentile. We recommend the following procedure.

Calculating the pth Percentile

STEP 1   Arrange the data values in ascending order.

STEP 2   Compute an index i as follows:

$$i = \left(\frac{p}{100}\right)n$$

where p is the percentile of interest and n is the number of data values.

STEP 3   (a) If *i is not an integer,* the next integer value greater than i denotes the position of the pth percentile.

(b) If *i is an integer value,* the *p*th percentile is the average of the data values in positions *i* and *i* + 1.

EXAMPLE 3.2 (continued)

As an illustration of this procedure, let us determine the 90th percentile for the television-viewing times of Example 3.2.

STEP 1   Arrange the 16 data values in ascending order:

4   5   6   8   9   10   12   14   15   15   15   16   17   18   20   26

STEP 2   Compute *i* as follows:

$$i = \left(\frac{90}{100}\right) 16 = 14.4$$

STEP 3   Since *i* is not an integer, the position of the 90th percentile is the next integer value greater than 14.4, the fifteenth position. Thus the 90th percentile is 20.

As another illustration of this procedure, let us compute the 50th percentile for this data set. Applying step 2, we obtain

$$i = \left(\frac{50}{100}\right) 16 = 8$$

Since *i* is an integer, step 3(b) indicates that the 50th percentile is the average of the eighth and ninth data values; thus the 50th percentile is $(14 + 15)/2 = 14.5$. Similarly, we find that the 25th percentile is 8.5 and the 75th percentile is 16.5.

• • •

In the above example we saw that the 50th percentile is 14.5 hours of weekly television-viewing time. Note that this value is also the median for this data set. Recall that in the discussion of the median, we found the median divided the data into two equal parts. In terms of percentiles, the median is the 50th percentile. At times, the 25th percentile and/or the 75th percentile may be of particular interest. These two percentiles are referred to as the first and third quartiles, respectively. In conjunction with the median, they divide the data set into quarters.

EXERCISES

1. A hospital emergency room recorded the number of patients treated on Sunday for the past 16 weeks.

11, 14, 18, 14, 21, 17, 13, 21, 25, 19, 17, 13, 28, 13, 17, 18

Compute the mean, median, mode, and the 90th percentile for these data.

2. A bowler has the following scores for six games:

   182, 168, 184, 190, 170, 174

   Using these data as a sample, compute the following descriptive statistics.
   a. mean
   b. median
   c. mode
   d. 75th percentile

3. Monthly sales data for car telephone units for the RC Radio Corporation are:

   80, 115, 82, 102, 94, 90, 88, 91, 89, 95, 105, 108

   Compute the mean, median, and mode for monthly sales volumes.

4. A sample of 15 college seniors showed the following credit hours taken during the final term of the senior year:

   15, 21, 18, 16, 18, 21, 19, 15, 14, 18, 17, 20, 18, 15, 16

   What are the mean, median, and mode for credit hours? Compute and interpret the 70th percentile for these data.

5. The data given below show the number of automobiles arriving at a toll booth during 20 intervals, each of 10 minutes duration. Compute the mean, median, mode, first quartile, and third quartile for the data.

26	26	38	24
32	22	15	33
19	27	21	28
16	20	34	24
27	30	31	33

6. In automobile mileage and gasoline consumption testing, 13 automobiles were road tested for 300 miles in both city and country driving conditions. The following data were recorded for miles-per-gallon performance:

   City: 16.2, 16.7, 15.9, 14.4, 13.2, 15.3, 16.8, 16.0, 16.1, 15.3, 15.2, 15.3, 16.2.
   Country: 19.4, 20.6, 18.3, 18.6, 19.2, 17.4, 17.2, 18.6, 19.0, 21.1, 19.4, 18.5, 18.7.

   Use the mean, median, and mode to make a statement about the difference in performance for city and country driving.

7. A survey of television-viewing habits among college students provided the following data on viewing time in hours per week:

   14, 9, 12, 4, 20, 26, 17, 15, 18, 15, 10, 6, 16, 15, 8, 5

    a. Compute the mean, median, and mode.

    b. Compute the 10th and 80th percentiles.

3.2 MEASURES OF DISPERSION

When data are collected, it is often desirable to consider the variability or dispersion in the data values. For example, assume that an individual had the choice of using public transportation in a city to get to work or driving his or her own car. One consideration of interest would likely be the travel times associated with both modes of transportation.

Suppose that over a period of several months the individual used both modes of transportation about the same number of times and found that the mean time to get to work averaged around 30 minutes for both alternatives. At first glance, then, it would appear that both alternatives offer comparable service. However, before reaching any final conclusions, let us first consider the relative frequency histograms shown in

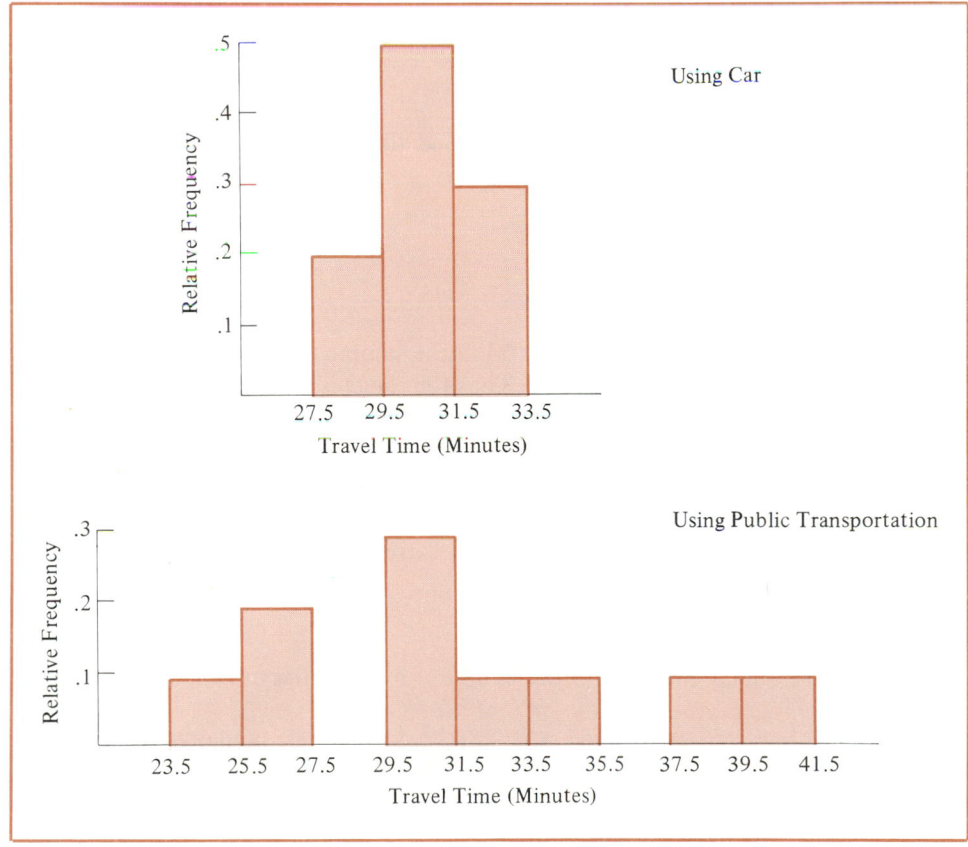

FIGURE 3.1
Relative Frequency Histograms for Travel Time

Figure 3.1. Although the mean travel time is roughly 30 minutes for both modes of transportation, do both alternatives possess the same degree of reliability in terms of getting to work on time? Note the variability or dispersion in the data. Which mode of transportation would you prefer?

For many people the variability exhibited in the times for the public transportation system would be a major concern. That is, to protect against arriving late, one would have to allow for the maximum possible travel times of approximately 42 minutes for public transportation. With a car, one would only need to allow for maximum times up to 34 minutes. Of even more concern, however, are the wide extremes (23.5 to 41.5 minutes) that must be expected when using public transportation. This illustration shows that although the average, or mean, travel time is an important consideration, the dispersion, or variability, in the travel times is at least as important and might in this case be an overriding consideration.

We turn now to a discussion of some commonly used numerical measures of the variability, or dispersion, in a set of data.

Range

The range is perhaps the simplest measure of variability to compute.

Range

The *range* for a set of data is the difference between the largest and smallest values.

EXAMPLE 3.1 (continued)

The largest section size was 54 in the sample taken at the College of Science and the smallest section size was 32. Thus the range is $54 - 32 = 22$ students.

• • •

While the range is the easiest of the statistical measures of dispersion to compute, it is not widely used because the range is based on only two of the items in the data set and thus is influenced too much by extreme values.

EXAMPLE 3.3 (continued)

The annual incomes for the sample of five families in Herrold, Iowa, are

$17,500 $23,000 $24,000 $26,000 $320,000

The range for this data set is $320,000 - 17,500 = $302,500$. Here the range is not a very good measure of the variability in the data. For four of the five families the differences in annual incomes are $8500 or less.

• • •

A measure sometimes used to overcome this sensitivity to extreme values is the *interquartile range*. This measure is the difference between the third and first quartiles.

EXAMPLE 3.3 (continued)

The first quartile for the annual incomes of the sample of five families from Herrold, Iowa, is $23,000. The third quartile is $26,000. The interquartile range for the data is

$$\$26,000 - \$23,000 = \$3000.$$

• • •

A better measure of variability that involves differences among all the items in a data set is the variance.

Variance

A key step in computing the variance involves the computation of the difference between each data value and the mean for the data set. The difference between each observation x_i and the mean ($\bar{x}$ for a sample, μ for a population) is called a *deviation about the mean*.

EXAMPLE 3.1 (continued)

Recall that the sample mean for the data set of Example 3.1 is a section size of 44 students. A summary of the data, including the computation of the deviations from the sample mean, is shown below.

Number of Students in Section (x_i)	Mean Section Size $\bar{x}$	Deviation from Sample Mean ($x_i - \bar{x}$)
46	44	2
54	44	10
42	44	−2
46	44	2
32	44	−12
	Total	0

Note: Sum of deviations about the mean is 0.

• • •

We might first think of summarizing the dispersion in a data set by computing the average deviation about the mean. However, a little reflection based upon a study of the above table would lead us to discard that idea—the sum of the deviations about the mean for the five sections sampled is equal to zero. This is true for any data set; that is

$$\Sigma \, (x_i - \overline{x}) = 0$$

The positive and negative deviations cancel each other, causing the average deviation for any data set to equal zero. Thus the average deviation cannot measure the variability in a data set since it will always be equal to $0/n = 0$.

One approach to preventing the positive and negative deviations from canceling out is to take the absolute value of each deviation. The average absolute deviation can then be computed as a measure of variability. While this measure is sometimes used, the most common approach to preventing the cancellation of deviations is to square them.

EXAMPLE 3.1 (continued)

The squared deviations and their sum are shown below.

Number of Students in Section (x_i)	Deviation from Sample Mean ($x_i - \overline{x}$)	Squared Deviation ($x_i - \overline{x}$)^2
46	2	4
54	10	100
42	−2	4
46	2	4
32	−12	144
Totals	0	256
	$\Sigma \, (x_i - \overline{x})$	$\Sigma \, (x_i - \overline{x})^2$

• • •

One measure of variability based on the squared deviations is the *average squared deviation*. If the data set involved is a population, the average of the squared deviations is called the *population variance*. The population variance is denoted by the Greek symbol σ^2 (pronounced sigma squared). Given a population of N items and using μ to represent the population mean, the definition of the population variance is given by Equation (3.3).

Population Variance

$$\sigma^2 = \frac{\Sigma \, (x_i - \mu)^2}{N} \tag{3.3}$$

In most statistical applications, the data set being analyzed is a sample, When we compute a measure of variability for a sample, we are often interested in using the sample statistic obtained as an estimate of the population parameter σ^2. At this point it might seem that the average of the squared deviations of the sample values from $\bar{x}$ would provide a good estimate of the population variance. However, statisticians have found that the average squared deviation for the sample has the undesirable feature of providing a biased estimate of the population variance σ^2; specifically, it tends to underestimate the population variance.

Although it is beyond the scope of this text, it can be shown that if the sum of the squared deviations in a sample is divided by $n - 1$, and not n, then the resulting sample statistic provides an unbiased estimate of the population variance. For this reason, the *sample variance* is not defined as the average squared deviation in the sample. Rather, it is denoted by s^2 and is defined as follows.

Sample Variance

$$s^2 = \frac{\Sigma(x_i - \bar{x})^2}{n - 1} \tag{3.4}$$

In later chapters we frequently use the sample variance as an estimate of the population variance. Because of this, we use the definition of s^2 provided by (3.4).

EXAMPLE 3.1 (continued)

Let us now compute the sample variance for the data on the number of students in the College of Science classes. Recall that for this data set, $\Sigma(x_i - \bar{x})^2 = 256$. Hence with $n - 1 = 4$, we obtain

$$s^2 = \frac{\Sigma(x_i - \bar{x})^2}{n - 1} = \frac{256}{4} = 64$$

$$\bullet \quad \bullet \quad \bullet$$

While admittedly it is difficult to obtain an intuitive feel for the meaning of the variance, we can note that larger variances could only be obtained from data sets with larger deviations about the mean and, therefore, more dispersion.

Before concluding this subsection on the use of variance as a measure of dispersion, we provide shortcut formulas for computing the variance of a data set. The shortcut formula for the sample variance is obtained through some algebraic manipulation of the numerator in (3.4).

Sample Variance (Shortcut Formula)

$$s^2 = \frac{\Sigma x_i^2 - (\Sigma x_i)^2/n}{n - 1} \tag{3.5}$$

The advantage of working with (3.5) over (3.4) is that the step of calculating and squaring each deviation is not necessary. Thus there are fewer calculations and fewer rounding errors. Formula (3.5) can be modified easily to compute a population variance. The appropriate formula is as shown.

Population Variance (Shortcut Formula)

$$\sigma^2 = \frac{\Sigma x_i^2 - (\Sigma x_i)^2/N}{N} \qquad (3.6)$$

EXAMPLE 3.2 (continued)

Recall that the sample mean for average weekly television-viewing time of a sample of college students is $\bar{x} = 13.125$. Let us use both (3.4) and (3.5) to compute the sample variance for this data set.

x_i	x_i^2	$(x_i - \bar{x})$	$(x_i - \bar{x})^2$
14	196	.875	.765625
9	81	−4.125	17.015625
12	144	−1.125	1.265625
4	16	−9.125	83.265625
20	400	6.875	47.265625
26	676	12.875	165.765625
17	289	3.875	15.015625
15	225	1.875	3.515625
18	324	4.875	23.765625
15	225	1.875	3.515625
10	100	−3.125	9.765625
6	36	−7.125	50.765625
16	256	2.875	8.265625
15	225	1.875	3.515625
8	64	−5.125	26.265625
5	25	−8.125	66.015625
Totals 210	3282	0.000	525.750000
Σx_i	Σx_i^2	$\Sigma(x_i - \bar{x})$	$\Sigma(x_i - \bar{x})^2$

Using (3.4) we obtain

$$s^2 = \frac{\Sigma (x_i - \bar{x})^2}{n - 1} = \frac{525.75}{15} = 35.05$$

Using (3.5) we obtain

$$s^2 = \frac{\Sigma x_i^2 - (\Sigma x_i)^2/n}{n - 1} = \frac{3282 - (210)^2/16}{15} = 35.05$$

We see from the example that both formulas produce the same result. However, additional work is required if we use (3.4), since each deviation and its square must be calculated.

• • •

Standard Deviation

The *standard deviation* of a data set is defined to be the positive square root of the variance. Following the notation we adopted for a sample variance and a population variance, we use s to denote the sample standard deviation and σ to denote the population standard deviation. The standard deviation is derived from the variance in the following manner:

Standard Deviation

$$\text{Sample Standard Deviation} = s = \sqrt{s^2} \qquad (3.7)$$

$$\text{Population Standard Deviation} = \sigma = \sqrt{\sigma^2} \qquad (3.8)$$

EXAMPLE 3.1 (continued)

Recall that the sample variance for the sample of section sizes at the College of Science is $s^2 = 64$. Thus the sample standard deviation is $s = \sqrt{64} = 8$.

• • •

Obviously, the standard deviation is also a measure of dispersion, since the square root of a larger variance will provide a larger standard deviation. However, the standard deviation is more often used as a measure of dispersion because it is in the same units as the data. For instance, in the example just considered, the variance of section sizes is 64 "students squared." The standard deviation is 8 students. The fact that variance is reported in units of the original data squared makes it difficult to obtain an intuitive feel for it as a measure of variability.

The standard deviation is also used when applying Chebyshev's theorem and in computing the coefficient of variation. Let us see how the computations are made and why.

Chebyshev's Theorem

Chebyshev's theorem provides a method for determining a lower bound on the percentage of items in a data set that lie within a specified number of standard deviations of the mean. To apply this theorem, we need know only the mean and standard deviation for the data set.

Chebyshev's Theorem

For any set of data and any value of k greater than or equal to 1, at least $1 - (1/k^2)$ of the data must be within $\pm k$ standard deviations of the mean.

In computing the standard deviation to apply Chebyshev's theorem, we use the formula for the population standard deviation because we are using the theorem to make a statement about only the data set itself, not to make an inference about some larger population. Thus the data set itself is the population of interest.

EXAMPLE 3.2 (continued)

Recall that the mean weekly television-viewing time for a sample of 16 college students was 13.125 hours. Suppose we would like to know what fraction of the weekly viewing times are within 2 standard deviations of the mean. From Chebyshev's theorem, we know this must be *at least* $1 - (1/k^2) = 1 - (1/2)^2 = \frac{3}{4}$ of the viewing times.

To find an interval containing at least $\frac{3}{4}$ of the viewing times, we must first find the population standard deviation for the data set. The population variance is

$$\sigma^2 = \frac{525.75}{16} = 32.859375$$

Thus the population standard deviation is

$$\sigma = \sqrt{\sigma^2} = \sqrt{32.859375} = 5.732$$

An interval guaranteed to contain at least 75% of the viewing times in this data set is given by

$$13.125 - 2(5.732) \quad \text{to} \quad 13.125 + 2(5.732)$$

or

$$1.661 \quad \text{to} \quad 24.589$$

A check of the original data set verifies this result. Fifteen of the 16 weekly viewing times (almost 94%) actually fall within the interval 1.661 to 24.589 hours.

• • •

In closing we note that Chebyshev's theorem provides a very conservative estimate of the number of data items in an interval. In later chapters we see that when knowledge is available concerning the probability distribution for the data, a better estimate can be obtained.

Coefficient of Variation

Often we are more interested in a relative measure of the variability in a data set than the absolute measure provided by the standard deviation or variance. For example, a standard deviation of 1 inch would be considered very large for a batch of motor-mount bolts used in automobiles. However, a standard deviation of 1 inch would be considered small for the length of a telephone pole. When the means for data sets differ greatly, we do not get an accurate picture of the relative variability in the two data sets by comparing the standard deviation. A measure of variability that can be used for such comparisons is the *coefficient of variation*. The formula for computing the coefficient of variation is given by (3.9).

Coefficient of Variation

$$\frac{\text{Standard Deviation}}{\text{Mean}} \times 100\% \qquad (3.9)$$

For sample data the coefficient of variation is $(s/\overline{x}) \times 100\%$, and for a population it is $(\sigma/\mu) \times 100\%$.

EXAMPLE 3.1 (continued)

Recall that for the sample of section sizes at the College of Science, the sample mean and sample standard deviation are 44 and 8, respectively. Thus for this data set the coefficient of variation is $(8/44) \times 100\% = 18.18\%$. In other words, we could say that the standard deviation of the sample is 18.18% of the value of the sample mean.

• • •

We caution that the coefficient of variation should be used only for data sets involving positive numbers. Otherwise negative or very small values could be obtained for the mean. This leads to meaningless values for the coefficient of variation.

EXERCISES

8. The number of freshmen majoring in computer science at Maine Institute of Technology for the past 5 years is:

    390, 380, 400, 410, 420

    Compute the sample mean, variance, and standard deviation for this data set.

9. In exercise 2 a bowler's scores for six games were as follows.

    182, 168, 184, 190, 170, 174

    Using these data as a sample, compute the following descriptive statistics.
    a. range
    b. variance
    c. standard deviation
    d. coefficient of variation

10. Compute the range, variance, and standard deviation for the sample of credit hours taken by college seniors in their final term. The data are as follows.

    15, 21, 18, 16, 18, 21, 19, 15, 14, 18, 17, 20, 18, 15, 16

11. Given are the yearly household incomes (in dollars) for 10 families in Grimes, Nebraska.

10,648	17,416
6,517	13,555
14,821	9,226
152,936	11,800
18,527	12,222

   a. Compute the range as a measure of variability.
   b. Compute the interquartile range as a measure of variability.
   c. Compute the standard deviation as a measure of variability.
   d. Which of the above measures do you feel is the best measure of variability in the data? Why?

12. A production department uses a sampling procedure to test the quality of newly produced items. The department employs the following decision rule at an inspection station: If a sample of 14 items has a variance of more than .01 for a certain characteristic, the production line must be shut down for repairs. Suppose the following data have just been collected:

3.43	3.45	3.43
3.48	3.52	3.50
3.39	3.48	3.41
3.38	3.49	3.45
3.51	3.50	

   Should the production line be shut down? Why or why not?

13. The following times were recorded by the quarter-mile and mile runners of a university track team (times are in minutes):

Quarter-Mile Times:	.92	.98	1.04	.90	.99
Mile Times:	4.52	4.35	4.60	4.70	4.50

   After viewing this sample of running times, one of the coaches commented that the quarter-milers turned in the more consistent times. Use the standard deviation and the coefficient of variation to summarize the variability in the data sets. Does the use of the coefficient of variation measure indicate that the coach's statement should be qualified? Why?

14. The monthly charges for credit card holders at Schip's Department Store have a population mean of $250 and a population standard deviation of $100. Use Chebyshev's theorem to answer the following questions:
   a. What can be said about the percentage of the card holders who will have monthly charges between $100 and $400?
   b. What can be said about the percentage of card holders who will have monthly charges between 0 and $500?
   c. Provide a range of credit card charges that will include at least 80% of all credit card customers.

15. During a recent football season, a major conference reported that the average attendance for its conference games was 45,000. The standard deviation in the attendance figure was $\sigma = 4,000$. Use Chebyshev's theorem for the following:

    a. Develop an interval that contains the attendance figures for at least 75% of the games.

    b. Develop an interval that contains the attendance figures for a least 8/9 of the games.

    c. The commissioner claims that at least 90% of the games had attendance between 29,000 and 61,000. Is this statement warranted given the information we have? Explain.

3.3 MEASURES OF LOCATION AND DISPERSION FOR GROUPED DATA

In most cases measures of central location and dispersion for a data set are computed using the individual data values. However, sometimes we are presented with data in grouped or frequency distribution form. This section describes how approximations of the mean, variance, and standard deviation can be made from the frequency distribution.

Mean

Recall that in order to compute the sample mean using the individual data values, we simply sum all the values and divide by n, the sample size. If the data are available only in frequency distribution form, we can approximate the sum of the data values by first finding an approximation for the sum of the data values for each class and then adding these for all classes.

To do this, we treat the midpoint of each class as if it were the mean of the items in the class. Let M_i denote the midpoint for class i and f_i denote the frequency of the class. Then an approximation to the sum of the items in class i is given by $f_i M_i$. Summing these values over all classes, we obtain $\Sigma f_i M_i$, which approximates the sum of all the data values.

Once this approximation of the sum of all the data values is obtained, an approximation of the mean is computed by dividing this sum by the total number of data items. The following formula can be used to compute the sample mean from grouped data.

Sample Mean for Grouped Data

$$\bar{x} = \frac{\Sigma f_i M_i}{n} \tag{3.10}$$

As we indicated at the beginning of this section, we do not expect the calculations based on the ungrouped data and grouped data to provide exactly the same numerical result. Thus $\bar{x}$ calculated using (3.10) is an approximation of $\bar{x}$ calculated when all sample values are known. The difference between these two values is known as *grouping error*. In practice it has been found that grouping errors are usually relatively small if 15 or more classes are used.

EXAMPLE 3.4

In Example 2.8 of Chapter 2 we analyzed a data set involving the scores obtained by each of 50 sixth graders on a mathematics achievement test. We developed a frequency distribution using the eight class intervals 65–74, 75–84, ..., 135–144. Table 3.1 shows a frequency distribution for this same data involving 16 classes. Note that we have added two columns to the frequency distribution. One is for the class midpoints and the other is for the approximation of the sum of data values in each class, $f_i M_i$. The sum of the items in the last column provides the approximation to the sum of the 50 data values. Using this sum we compute the mean for the grouped data as follows:

$$\bar{x} = \frac{\Sigma f_i M_i}{50} = \frac{4950}{50} = 99$$

Referring to the original data as shown in Example 2.8 and computing the sample

TABLE 3.1

Computing a Sample Mean from Grouped Data for Example 3.4

Number of Questions Answered Correctly	f_i	M_i	$f_i M_i$
65–69	2	67	134
70–74	3	72	216
75–79	3	77	231
80–84	6	82	492
85–89	2	87	174
90–94	5	92	460
95–99	6	97	582
100–104	4	102	408
105–109	5	107	535
110–114	2	112	224
115–119	5	117	585
120–124	1	122	122
125–129	3	127	381
130–134	2	132	264
135–139	0	137	0
140–145	1	142	142
Totals	50		4950
	n		$\Sigma f_i M_i$

mean using the ungrouped data, we find $\bar{x} = 98.92$. Thus the grouping error is $99.00 - 98.92 = 0.08$. Obviously our approximation is a good one in this case.

• • •

Variance

The approach to computing the variance for a set of grouped data is to use a slightly altered form of the shortcut formula for the variance as provided in (3.5). Since we no longer have the individual data values, we treat the class midpoint as being a representative value for the data items in each class. Then, just as we did with the sample mean calculation for grouped data, we weight each midpoint by the frequency of its corresponding class. Using this approach, (3.5) is modified as follows.

Sample Variance for Grouped Data

$$s^2 = \frac{\sum f_i M_i^2 - (\sum f_i M_i)^2/n}{n - 1} \tag{3.11}$$

TABLE 3.2

Computation of the Variance for Example 3.4 Using Grouped Data

Class	Frequency f_i	Midpoint M_i	$f_i M_i$	M_i^2	$f_i M_i^2$
65–69	2	67	134	4,489	8,978
70–74	3	72	216	5,184	15,552
75–79	3	77	231	5,929	17,787
80–84	6	82	492	6,724	40,344
85–89	2	87	174	7,569	15,138
90–94	5	92	460	8,464	42,320
95–99	6	97	582	9,409	56,454
100–104	4	102	408	10,404	41,616
105–109	5	107	535	11,449	57,245
110–114	2	112	224	12,544	25,088
115–119	5	117	585	13,689	68,445
120–124	1	122	122	14,884	14,884
125–129	3	127	381	16,129	48,387
130–134	2	132	264	17,424	34,848
135–139	0	137	0	18,769	0
140–144	1	142	142	20,164	20,164
Totals	50		4950		507,250

$$\sum f_i M_i = 4950 \qquad s^2 = \frac{507,250 - (4950)^2/50}{49}$$

$$= 351.02$$

EXAMPLE 3.4 (continued)

The calculation of the sample variance for the grouped data of Example 3.4 is shown in Table 3.2. We see that $s^2 = 351.02$.

• • •

Standard Deviation

The standard deviation computed from grouped data is simply the square root of the variance computed from grouped data. For the data of Example 3.4, the sample standard deviation computed from grouped data is $s = \sqrt{351.02} = 18.74$

Population Summaries—Grouped Data

Before closing this section on computing measures of location and dispersion from grouped data, we note that the formulas in this section were presented only for data sets constituting a sample. Population summary measures are computed in a similar manner. The grouped data formulas for a population mean and variance are as follows.

Population Mean for Grouped Data

$$\mu = \frac{\Sigma f_i M_i}{N} \tag{3.12}$$

Population Variance for Grouped Data

$$\sigma^2 = \frac{\Sigma f_i M_i - (\Sigma f_i M_i)^2 / N}{N} \tag{3.13}$$

EXERCISES

16. The following frequency distribution for the first examination in sociology was posted on the department bulletin board.

Examination Grade	Frequency
40–49	3
50–59	5
60–69	11
70–79	22
80–89	15
90–99	6
Total	62

Treating these data as a sample, compute the mean, variance, and standard deviation.

17. A service station has recorded the following frequency distribution for the number of gallons of gasoline (rounded to the nearest gallon) sold per car on a given day.

Gasoline (gallons)	Frequency
0–4	74
5–9	192
10–14	280
15–19	105
20–24	23
25–29	6
Total	680

Compute the standard measures of central location and dispersion for these grouped data. If the service station expects to service about 120 cars on a given day, what is an estimate of the total number of gallons of gasoline that will be sold?

18. Scores obtained by patients on a depression level test are summarized in the following frequency distribution:

Depression Level Score	Frequency
25–34	3
35–44	1
45–54	2
55–64	6
65–74	4
75–84	6
85–94	2
95–104	1
Total	25

Using the above grouped data, compute the folloiwng.
a. mean
b. variance
c. standard deviation

19. Consider the following frequency distribution for automobile repair costs (rounded to the nearest dollar) for an insurance company's minor claims category:

Repair Cost (dollars)	Frequency
0–99	10
100–199	28
200–299	60
300–399	70
400–499	52
Total	220

Compute the mean, variance, and standard deviation for these grouped data.

3.4 EXPLORATORY DATA ANALYSIS

In Chapter 2 we introduced exploratory data analysis. Recall that the focus of exploratory data analysis is on using simple arithmetic and easy-to-draw pictures to summarize data. In this section we continue our introduction of exploratory data analysis by considering 5-number summaries and box-and-whisker plots.

5-Number Summaries

In a 5-number summary, the following 5 numbers are used to summarize a data set:

1. Lowest value in the data set
2. First quartile (25th percentile), called the *lower hinge*
3. Median
4. Third quartile (75th percentile), called the *upper hinge*
5. Largest value in the data set

EXAMPLE 3.4 (continued)

In Table 3.3 we provide a stretched stem-and-leaf display for the data set involving scores of sixth graders on the mathematics achievement test. From this we see that the lowest value is 68 and the highest value is 141. To compute the lower hinge, or 25th percentile, we can use the approach presented in Section 3.1. Since the data values are already in ascending order in Table 3.3, we proceed directly to Step 2 as shown below.

$$i = \left(\frac{25}{100}\right) 50 = 12.5$$

Since i is not an integer, 13 denotes the position of the 25th percentile. The thirteenth item is 83, so we refer to 83 as the lower hinge. In a similar fashion we find that the median is 97.5 and the upper hinge (75th percentile) is 113. Hence, we can present the 5-number display shown.

| 68 | 83 | 97.5 | 113 | 141 |

To provide additional information we can add to this figure the distance between adjacent values.

| 68 | 83 | 97.5 | 113 | 141 |
| | 15 | 14.5 | 15.5 | 28 |

Further refinement of the 5-number display also shows the following differences: median − lowest value; upper hinge − lower hinge; highest value − median.

68	83	97.5	113	141
	15	14.5	15.5	28
		29.5	30	43.5

The values in this bottom row are called *half-ranges*, since each is the range for a data set consisting of one-half of the original data set. These three values are frequently referred to as the *lowspread*, *midspread*, and *highspread*, respectively.

• • •

TABLE 3.3

Stretched Stem-and-Leaf Display

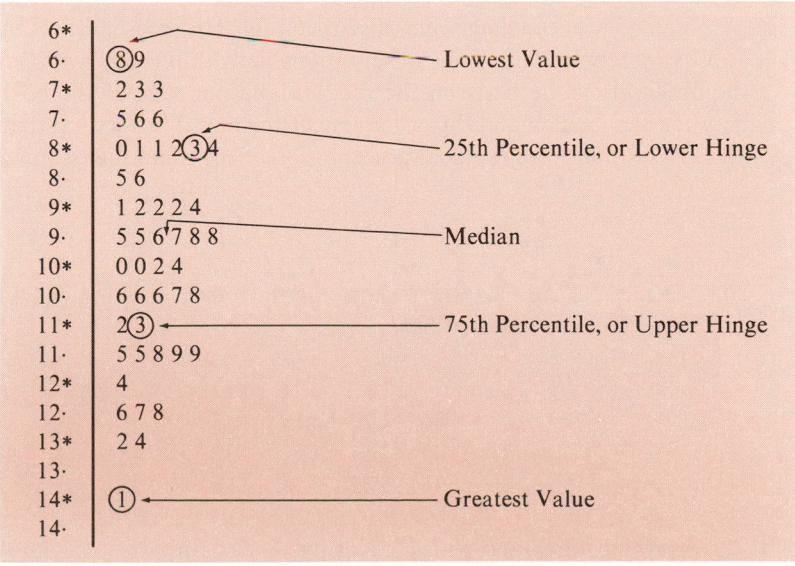

The advantage of the 5-number summary is that it provides information about the location of the underlying distribution (through the median and quartiles), the dispersion or spread in the data (through the range or half-ranges), and the shape (by comparing the distances shown).

Box-and-Whisker Plots

A *box-and-whisker plot* is a graphical presentation of a 5-number summary. In Figure 3.2 we illustrate a box-and-whisker plot for the 5-number summary of the data in Table 3.3. The dashed lines are known as *whiskers,* and in this plot we have shown them extending from each hinge to the lowest and highest values in the data set. A different perspective would be obtained if each whisker extended one midspread beyond each hinge. A complete discussion of the choice of the length of the midspread is, however, beyond the scope of this text.

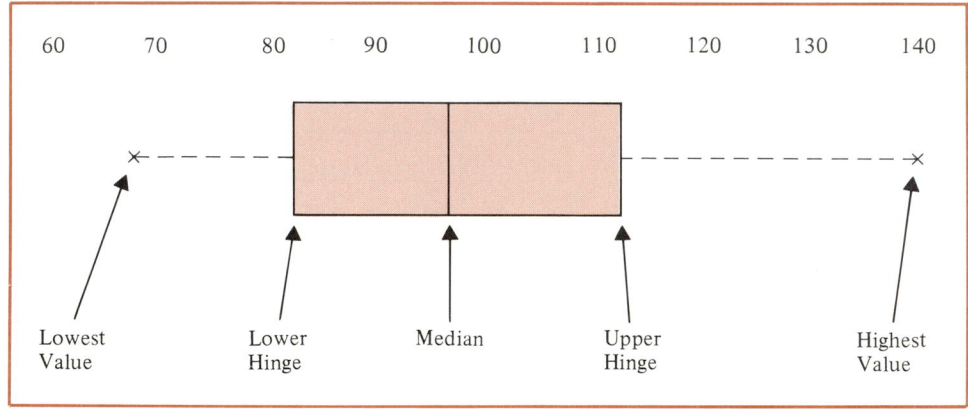

FIGURE 3.2
Box-and-Whisker Plot

Before we conclude our discussion of box-and-whisker plots, we note that approximately half of all the observations lie within the box. Moreover, one-fourth of the observations lie between the median and one side of the box and one-fourth lie between the median and the other side of the box. Thus we see that such plots provide an easy-to-interpret graphical presentation of the 5-number summary.

EXERCISES

20.   Exercise 27 in Chapter 2 showed a data set consisting of 20 intelligence scores. The data are repeated here.

114	99	131	124	117
102	106	127	119	115
98	104	144	151	132
106	125	122	118	118

Develop a box-and-whisker plot for these data.

21. Exercise 28 in Chapter 2 showed the following data set, consisting of the number of questions answered correctly on a 150-question examination in computer-center management.

102	91	72	98	115
57	89	121	89	124
122	136	105	80	79
64	108	113	83	63
84	96	99	75	97

Develop a box-and-whisker plot for these data.

22. Exercise 29 in Chapter 2 showed a data set consisting of scores on a test designed to measure the degree of postmeningitic brain damage. The data set is repeated below, where low scores represent considerable brain damage.

87	76	67	58	92	59	41	50	90	75
80	81	70	73	69	61	88	46	85	97
50	47	81	87	75	60	65	92	77	71
70	74	53	43	61	89	84	83	70	46
84	76	78	64	69	76	78	67	74	64

Develop a box-and-whisker plot for these data.

23. In exercise 30 in Chapter 2, the following data set showed the high temperature at 30 selected cities.

53	66	48	84	77	38	86	90	39	52
41	49	74	34	58	68	72	61	71	47
46	48	55	66	47	44	51	58	73	89

Develop a box-and-whisker plot for these data.

3.5 THE ROLE OF THE COMPUTER IN DESCRIPTIVE STATISTICS

In this section we describe the role of computers and computer software packages in statistical analysis by showing how a statistical package can be used to generate descriptive statistics for a data set. In future chapters we provide further illustrations showing how statistical computing systems can support the analysis and interpretation of data.

In the 1960s there were relatively few computerized statistical packages for analyzing data. Since that time, however, the situation has changed dramatically. Today the user has a choice of packages such as SAS (Statistical Analysis System), SPSS (Statistical Package for the Social Sciences), BMDP (UCLA Biomedical statistical package), and Minitab, to name just a few. In this text we use the Minitab system to illustrate the application of statistical computing systems.

Minitab is a general-purpose statistical computing system that can be used on a variety of mainframe or personal computers. It has been designed for users who have had little or no previous computer experience. Although it is very easy to use, it offers a great deal of power for data summarization and statistical analysis.

In the illustrations of Minitab, we describe its use in what is referred to as *interactive mode*. In this mode the user enters data and commands from a computer terminal or personal computer keyboard; Minitab carries out each user command as soon as it is given. In the figures containing the steps of a Minitab session, we show the computer responses from the Minitab system in black and the input by the user in color.

Minitab consists of a worksheet of rows and columns in which data are stored. The columns are denoted with labels c1, c2, and so on, unless the user elects to name the columns with specific labels. The rows of the worksheet correspond to the individual items or elements of the data set. That is, a separate row is used for each item. The Minitab system consists of about 150 commands, which can be used to analyze the data stored in the worksheet. To provide an illustration of how Minitab works, we use the data set originally presented in Example 2.8. Recall that this data showed the number of questions answered correctly by each of 50 students who took a 150-question mathematics achievement test. For convenience, this data set is shown in Table 3.4.

Referring to Figure 3.3, we assume that the user has loaded the Minitab system into the computer. The MTB > symbol appears on the terminal screen, which indicates that the Minitab system is waiting for a command from the user. The user then inputs read c1, indicating that the system is to take the data from the lines that follow and store the data in column 1 of the worksheet. Note then that the next line shows that Minitab's response is DATA>; thus the user enters the first data value. This process continues with one data value being entered per line. When all the data values have been input, the data entry process is concluded when the user responds to the request for more data with the command end. The data set now resides in the Minitab worksheet in column 1.

TABLE 3.4

Data Set Originally Provided in Example 2.8 Showing Mathematics Achievement Test Scores

112	72	69	97	107
73	92	76	86	73
126	128	118	127	124
82	104	132	134	83
92	108	96	100	92
115	76	91	102	81
95	141	81	80	106
84	119	113	98	75
68	98	115	106	95
100	85	94	106	119

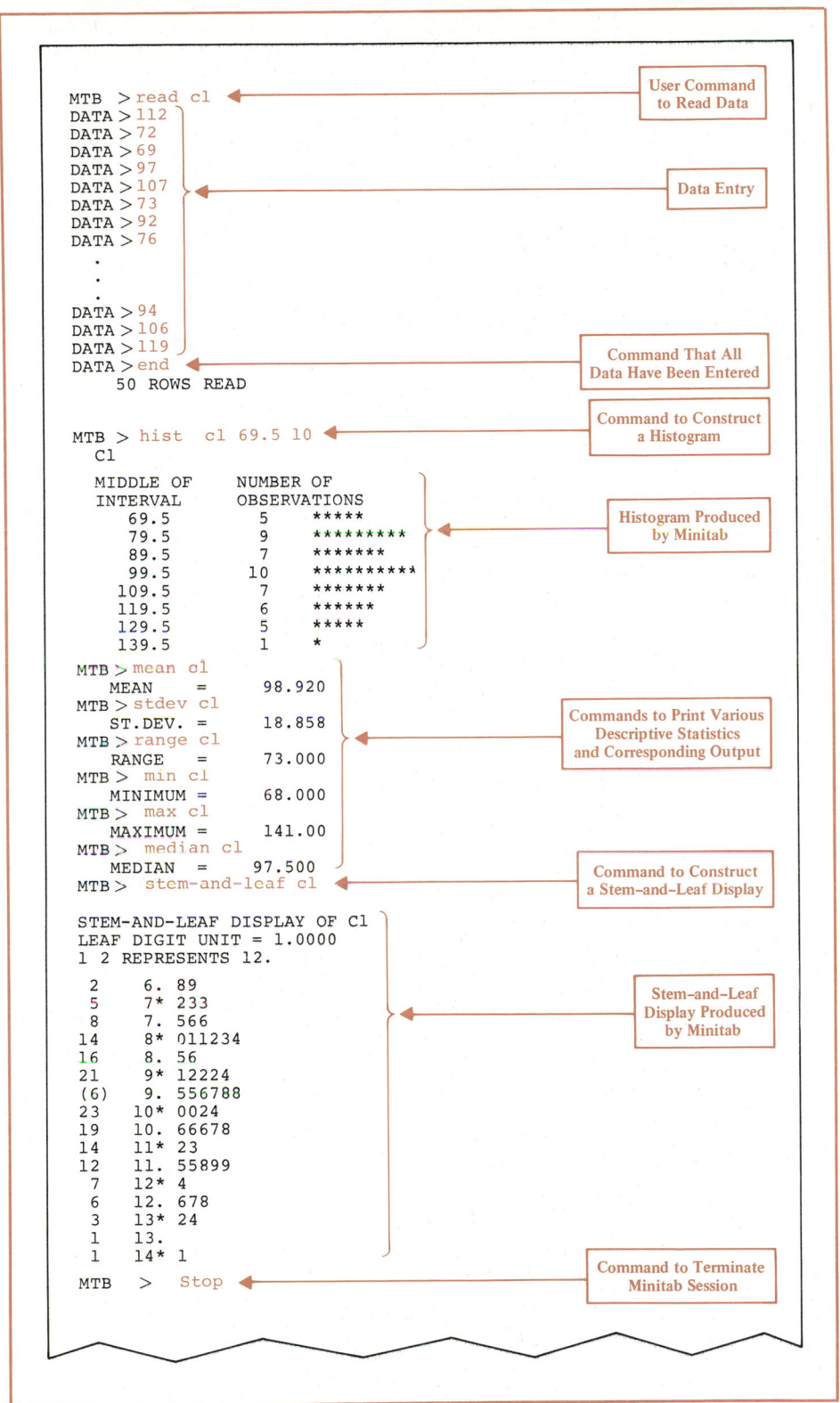

```
MTB  > read c1                    ← User Command
DATA > 112                          to Read Data
DATA > 72
DATA > 69
DATA > 97
DATA > 107                        ← Data Entry
DATA > 73
DATA > 92
DATA > 76
     .
     .
     .
DATA > 94
DATA > 106
DATA > 119                        ← Command That All
DATA > end                          Data Have Been Entered
      50 ROWS READ

MTB  > hist  c1 69.5 10           ← Command to Construct
      C1                            a Histogram

      MIDDLE OF       NUMBER OF
      INTERVAL        OBSERVATIONS
        69.5             5     *****
        79.5             9     *********     ← Histogram Produced
        89.5             7     *******          by Minitab
        99.5            10     **********
       109.5             7     *******
       119.5             6     ******
       129.5             5     *****
       139.5             1     *

MTB  > mean c1
      MEAN     =        98.920
MTB  > stdev c1
      ST.DEV. =         18.858    ← Commands to Print Various
MTB  > range c1                     Descriptive Statistics
      RANGE    =        73.000     and Corresponding Output
MTB  > min c1
      MINIMUM =         68.000
MTB  > max c1
      MAXIMUM =         141.00
MTB  > median c1
      MEDIAN  =         97.500    ← Command to Construct
MTB  > stem-and-leaf c1            a Stem-and-Leaf Display

STEM-AND-LEAF DISPLAY OF C1
LEAF DIGIT UNIT = 1.0000
1 2 REPRESENTS 12.

    2      6. 89
    5      7* 233
    8      7. 566
   14      8* 011234                ← Stem-and-Leaf
   16      8. 56                       Display Produced
   21      9* 12224                     by Minitab
   (6)     9. 556788
   23     10* 0024
   19     10. 66678
   14     11* 23
   12     11. 55899
    7     12* 4
    6     12. 678
    3     13* 24
    1     13.
    1     14* 1

MTB   >   Stop                    ← Command to Terminate
                                    Minitab Session
```

FIGURE 3.3
Computer Analysis Using Minitab

Now that the data is in the Minitab worksheet, the user can continue with the analysis of the data. The user command shown in Figure 3.3 is hist c1 69.5 10, which is a command for Minitab to construct a histogram from the data in column 1; the midpoint of the first class interval is 69.5 and the class width is 10. Note the output that is produced directly after giving this command. In this way the analyst can quickly look at the output, form some initial judgments, and continue. Data analysis done interactively provides quick computer response and is an important reason why statistical packages such as Minitab are so valuable to an analyst. The session with Minitab continues with some other commands for producing descriptive summaries of the data. In each case the names of the commands shown in Figure 3.3 are self-explanatory. Note that this session with Minitab concludes with the user requesting a stem-and-leaf display of the data.

Summary

In this chapter we introduced several statistical measures that can be used to describe the central location and dispersion within a data set. Unlike the tabular and graphical procedures for summarizing data, the measures introduced in this chapter summarize the data in terms of numerical values. When the numerical values obtained are for a sample, they are called sample statistics. When they are for a population, they are called population parameters. Some of the sample statistics and population parameters introduced are summarized below.

	Sample Statistic	Population Parameter
Mean	$\bar{x}$	μ
Standard deviation	s	σ
Variance	s^2	σ^2
Size	n	N

As measures of central location, we defined the mean, median, and mode for both sample and population data sets. Then the concept of a percentile was used to describe the location of other values in the data set. Next, we presented the range, interquartile range, variance, standard deviation, and coefficient of variation as statistical measures of variability or dispersion in a data set. Finally, we described how the mean; variance, and standard deviation could be computed for grouped data. However, we recommend using the measures based on the individual data values unless the grouped format is the only manner in which the data are available.

A brief discussion of two exploratory data analysis techniques that can be used to summarize data more effectively was included in Section 3.4. Specifically, we showed how to develop a 5-number summary and a box-and-whisker plot in order to provide simultaneous information about the location, dispersion, and shape of the underlying distribution. The chapter concluded with a discussion of the role of the computer in descriptive statistics. An interactive session with the software package Minitab was used to illustrate how statistical computing systems can support the analysis and interpretation of data.

Statistics in Practice

HEALTH CARE PROGRAMS

The Hospice Program at St. Luke's Hospital in St. Louis, Missouri, is based on the belief that people with advanced diseases can live and die among family and friends in familiar surroundings—free of pain, and alert. Hospice offers an integrated set of services, which care for people and their families when the doctors have said there is nothing more that can be done to cure the diseases. The goal of the Hospice Program is to enable dying persons to live the remainder of their lives at home among loved ones and in familiar surroundings as free from pain and other symptoms of terminal illness as possible.

In the coordination and administration of St. Luke's Hospice program, reports containing statistical summaries help the program's administrators better understand the ongoing operation of the program. Since patients admitted to the program are terminally ill, life expectancy and the corresponding length of stay in the program are expected to be relatively short. However, administrators who monitor the program and plan for the future in terms of number of beds, number of patients that can be admitted, and so on, need periodic information showing the number of days the patients stay in the Hospice Program.

A sample of 54 recent patients in the program provided the following descriptive statistics on length of stay:

Mean stay:	59.5	days
Median stay:	38	days
Modal stay:	4	days

Note that the mean number of days a patient is in the program is about 2 months (59.5 days). However, the median shows that half of the patients are in the program approximately 1 month (38 days) or less. The value of the mode shows that the most frequent length of stay is 4 days. Descriptive statistics such as these provide program administrators with information essential for decision making and planning, which will maximize the impact and contribution of the Hospice Program.

Glossary

Population parameter—A numerical value used as a summary measure for a population of data (e.g., the population mean, μ, and the population variance, σ^2).

Sample statistic—A numerical value used as a summary measure for a sample (e.g., the sample mean, $\bar{x}$, and the sample variance, s^2).

Mean—A measure of the central location of a data set. It is computed by summing all the values in the data set and dividing by the number of items.

Median—A measure of central location of a data set. It is the value which splits the data set into two equal groups—one with values greater than or equal to the median, and one with values less than or equal to the median.

Mode—A measure of central location of a data set, defined as the most frequently occurring data value.

Percentile—A value such that at least $p\%$ of the items in the data set are less than or equal to its value and at least $(100 - p)\%$ of the items are greater than or equal to it. The median is the 50th percentile, the first quartile is the 25th percentile, and the third quartile is the 75th percentile.

Range—A measure of dispersion for a data set, defined to be the difference between the highest and lowest values.

Interquartile range—A measure of dispersion for a data set, defined to be the difference between the third and first quartiles.

Variance—A measure of dispersion for a data set, found by summing the squared deviations of the data values about the mean and then dividing the total by N if the data set is a population or by $n - 1$ if the data set is from a sample.

Standard deviation—A measure of dispersion for a data set, found by taking the square root of the variance.

Chebyshev's theorem—A theorem which allows the use of knowledge of the standard deviation and mean to draw conclusions about the fraction of data items within k standard deviations of the mean.

Coefficient of variation—A measure of relative dispersion for a data set, found by dividing the standard deviation by the mean and multiplying by 100%.

Grouped data—Data available in class intervals as summarized by a frequency distribution. Individual values of the original data are not recorded.

5-number summary—An exploratory data analysis technique that uses the following 5 numbers to summarize the data set: lowest value, 25th percentile (lower hinge), median, 75th percentile (upper hinge), largest value.

Box-and-whisker plot—A visual presentation of a 5-number summary.

Key Formulas

Sample Mean

$$\bar{x} = \frac{\Sigma x_i}{n} \tag{3.1}$$

Population Mean

$$\mu = \frac{\Sigma x_i}{N} \tag{3.2}$$

Population Variance

$$\sigma^2 = \frac{\Sigma (x_i - \mu)^2}{N} \tag{3.3}$$

Sample Variance

$$s^2 = \frac{\Sigma(x_i - \overline{x})^2}{n - 1} \tag{3.4}$$

Sample Variance (Shortcut Formula)

$$s^2 = \frac{\Sigma x_i^2 - (\Sigma x_i)^2/n}{n - 1} \tag{3.5}$$

Population Variance (Shortcut Formula)

$$\sigma^2 = \frac{\Sigma x_i^2 - (\Sigma x_i)^2/N}{N} \tag{3.6}$$

Standard Deviation

$$\text{Sample Standard Deviation} = s = \sqrt{s^2} \tag{3.7}$$

$$\text{Population Standard Deviation} = \sigma = \sqrt{\sigma^2} \tag{3.8}$$

Coefficient of Variation

$$\left(\frac{\text{Standard Deviation}}{\text{Mean}}\right) \times 100\% \tag{3.9}$$

Sample Mean for Grouped Data

$$\overline{x} = \frac{\Sigma f_i M_i}{n} \tag{3.10}$$

Sample Variance for Grouped Data

$$s^2 = \frac{\Sigma f_i M_i^2 - (\Sigma f_i M_i)^2/n}{n - 1} \tag{3.11}$$

Population Mean for Grouped Data

$$\mu = \frac{\Sigma f_i M_i}{N} \tag{3.12}$$

Population Variance for Grouped Data

$$\sigma^2 = \frac{\Sigma f_i M_i^2 - (\Sigma f_i M_i)^2/N}{N} \tag{3.13}$$

Review Quiz

TRUE/FALSE

1. The mean for a set of data is found by adding all the data values and dividing by the number of items minus 1.
2. The mean and median can never be equal.
3. The median and the 50th percentile are the same.
4. The mode of a set of data is the value that occurs with greatest frequency.
5. The standard deviation is the average of the differences between the data values and the mean.
6. The variance is the square root of the standard deviation.
7. The interquartile range is the difference between the fourth and second quartiles.
8. The sample variance is the average of the squared deviations in a sample.
9. The coefficient of variation is a measure of the relative variability in a data set.
10. The highspread is the difference between the largest value in a data set and the median.

MULTIPLE CHOICE

11. The mean for the data 6, 9, 10, 12, 13 is closest to
    a. 9.5
    b. 9.9
    c. 10.5
    d. 11.0
12. The median for the data set in question 11 is
    a. 9.5
    b. 10
    c. 11
    d. 12
13. The standard deviation for the data set in question 11 is closest to
    a. 3
    b. 5
    c. 8
    d. 10
14. The range for the data set in question 11 is closest to
    a. 3
    b. 4
    c. 5
    d. 6
15. The coefficient of variation for the data set in question 11 is closest to
    a. 27%
    b. 2%
    c. 5%
    d. 20%

16. The midspread for the data set in question 11 is closest to
   a. 4
   b. 3
   c. 1
   d. 7

Supplementary Exercises

24. A sample of six recent home mortgage loans showed the following interest rates.

   12.5, 13.2, 11.2, 13.0, 12.0, 12.5

Compute the following descriptive statistics for the data set
   a. mean
   b. median
   c. mode
   d. 25th percentile
   e. range
   f. interquartile range
   g. variance
   h. standard deviation
   i. coefficient of variation

25. The following data show home mortgage loan amounts handled by a particular loan officer in a savings and loan association:

20,000	38,500	33,000	27,500	34,000
12,500	25,900	43,200	37,500	36,200
25,200	30,900	23,800	28,400	13,000
31,000	33,500	25,400	33,500	20,200
39,000	38,100	30,500	45,500	30,500
52,000	40,500	51,600	42,500	44,800

Find the mean, median, and mode for these data.

26. Calculate the variance, standard deviation, and range for the mortgage amounts shown in exercise 25. Conversion of the data to 1000s (e.g., 38,500 is listed as 38.5) may ease the burden of having to work with large numerical values.

27. Morrison Communications, Inc., periodically reviews sales personnel performance records. One member of the sales force has had the following weekly sales volume (units sold) over the past quarter, or 13 weeks:

   13, 19, 20, 17, 21, 27, 9, 15, 22, 18, 18, 23, 20

   a. Compute the mean, median, and mode.
   b. Compute the first and third quartiles.
   c. Compute the range and interquartile range.
   d. Compute the standard deviation and coefficient of variation.

28. A sample of ten stocks on the New York Stock Exchange shows the following price–earnings ratios:

$$9, 4, 6, 7, 3, 11, 4, 6, 4, 7$$

Using the above data, compute the mean, median, mode, range, variance, and standard deviation.

29. A sample of recent oil drilling locations shows oil found at the following depths (feet):

| 1500 | 1200 | 1600 | 1700 | 1500 | 2000 |

Compute the mean, median, mode, range, variance, and standard deviation for the drilling depth data.

30. The number of patients treated at the Morton Hospital emergency room per day are shown below. Data are from a random sample of 12 days.

$$45, 50, 36, 59, 28, 42, 55, 67, 33, 35, 40, 50$$

Compute the mean, median, mode, range, variance, and standard deviation for these data.

31. National Airlines accepts flight reservations by phone. Shown below are the durations (in minutes) for a sample of 20 phone reservations.

2.1	4.8	5.5	10.4
3.3	3.5	4.8	5.8
5.3	5.5	2.8	3.6
5.9	6.6	7.8	10.5
7.5	6.0	4.5	4.8

Using the following descriptive statistical methods, summarize these data

a. mean
b. median
c. mode
d. range
e. variance
f. standard deviation

32. Soft-drink purchases at the Wright Field concession stands show the following 1-day totals:

Drink	Units Purchased
Cola	4553
Diet cola	2125
Uncola	1850
Orange soda	1288
Root beer	1572

What is the mode for the above sample data?

33. Light bulbs manufactured by a well known electrical equipment firm are known to have a mean life of 800 hours, with a standard deviation of 100 hours.
    a. What percentage of the light bulbs will have a life of 600 to 1000 hours?
    b. What percentage of the light bulbs will have a life of 550 to 1050 hours?
    c. Provide an interval for light bulb life that will be true for at least 50% of the light bulbs.

34. Daily volume for the stock market over a 6-month period showed a mean of 40 million shares with a standard deviation of 7 million shares.
    a. What percentage of the days during the 6-month period showed volumes between 30 and 50 million shares?
    b. What percentage of the days during the 6-month period showed volumes between 20 and 60 million shares?
    c. Provide an interval for daily volume that must include at least two-thirds of the days.

35. A frequency distribution for the duration of 20 long-distance telephone calls (rounded to the nearest minute) is shown below:

Call Duration	Frequency
4–7	4
8–11	5
12–15	7
16–19	2
20–23	1
24–27	1
Total	20

Compute the mean, variance, and standard deviation for the above data.

36. Dinner check amounts at La Maison French Restaurant (round to the nearest dollar) have the following frequency distribution:

Dinner Check (dollars)	Frequency
25–34	2
35–44	6
45–54	4
55–64	4
65–74	2
75–84	2
Total	20

Compute the mean, variance, and standard deviation for the given data.

37. Automobiles traveling on the New York State Thruway are checked for speed by a state police radar system. A frequency distribution of speeds is shown.

Speed (Miles per Hour)	Frequency
40–44	10
45–49	40
50–54	150
55–59	175
60–64	75
65–69	15
70–74	7
75–79	3
Total	475

a. What is the mean speed of the automobiles traveling on the New York State Thruway?

b. Compute the variance and the standard deviation.

38. In exercise 49 of Chapter 2, we provided data which showed the final examination scores in a section of calculus. The data are repeated:

```
56  77  84  82  42
61  44  95  98  84
93  62  96  78  88
58  62  79  85  89
89  97  53  76  75
```

Develop a box-and-whisker plot for these data.

39. In exercise 50 of Chapter 2, we provided the results of a 150-question social-awareness test that was given to a group of first-year college students. The following data show the number of questions answered correctly by each of the students.

```
121  114   94  136  144  126   98  103  118  127
135   97  119  117  122  138  142  141  102  105
```

Develop a box-and-whisker plot for these data.

40. In exercise 51 in Chapter 2, we showed the following data set, consisting of total yardage accumulated over the football season for 20 receivers.

```
744  652  576  1112   971   451  1023  852  809   596
941  975  400   711  1174  1278   820  511  907  1251
```

Develop a box-and-whisker plot for these data.

Computer Exercise

A national association of nurses has sponsored a study to determine the job satisfaction of nurses employed in hospitals. As part of the study, 50 nurses were asked to indicate their degree of satisfaction in their work, in their pay, and in their opportunities for promotion. Each of the three aspects of satisfaction were measured on a scale of 0 to 100, with larger values indicating higher degrees of satisfaction. Data were also collected on the type of hospital where each nurse was employed. The hospital types considered in the study were investor-owned hospitals, Veterans Administration (VA) hospitals, and university hospitals. The following data were collected:

Satisfaction Scores			Type of Hospital	Satisfaction Scores			Type of Hospital
Work	Pay	Promotion		Work	Pay	Promotion	
71	49	58	VA	72	76	37	VA
84	53	63	University	71	25	74	Investor-owned
84	74	37	VA	69	47	16	University
87	66	49	University	90	56	23	University
72	59	79	University	84	28	62	Investor-owned
72	37	86	VA	86	37	59	VA
72	57	40	Investor-owned	70	38	54	Investor-owned
63	48	78	VA	86	72	72	VA
84	60	29	VA	87	51	57	Investor-owned
90	62	66	Investor-owned	77	90	51	University
73	56	55	VA	71	36	55	University
94	60	52	VA	75	53	92	University
84	42	66	Investor-owned	74	59	82	Investor-owned
85	56	64	Investor-owned	76	51	54	University
88	55	52	University	95	66	52	VA
74	70	51	University	89	66	62	Investor-owned
71	45	68	Investor-owned	85	57	67	Investor-owned
88	49	42	Investor-owned	65	42	68	VA
90	27	67	VA	82	37	54	VA
85	89	46	University	82	60	56	VA
79	59	41	University	89	80	64	University
72	60	45	Investor-owned	74	47	63	Investor-owned
88	36	47	Investor-owned	82	49	91	Investor-owned
77	60	75	Investor-owned	90	76	70	VA
64	43	61	Investor-owned	78	52	72	VA

QUESTIONS

1. Develop descriptive measures of location for each of the three job-satisfaction variables. What aspect of the job is the most satisfying for the nurses? What

appears to be the most critical issue, or issue of lowest satisfaction for nurses? Explain.

2. Develop descriptive measures of dispersion for each of the three measures of job satisfaction. Which measure of satisfaction appears to have the greatest difference of opinion among the nurses?

3. Develop summary information for the type of hospital variable.

4. Show how exploratory data analysis can help summarize these data.

CHAPTER 4

Descriptive Statistics III: Data Analysis Involving More Than One Variable

What You will Learn in This Chapter

- how to analyze data sets involving two or more variables

- how to compute and interpret the sample covariance and the sample correlation coefficient

- how to use regression analysis to develop an equation that estimates how two variables are linearly related

- how to use contingency tables to analyze data sets involving two or more qualitative variables

Contents

Statistics in the News

IN SEARCH OF AN HONEST EMPLOYEE

When a retail outlet is having problems with stolen cash or merchandise, the first thought may be that customer shoplifting has gotten out of control. However, recent studies show customer shoplifting ranks second to employee theft as the most significant cause of stolen cash and merchandise. It has been estimated that department stores are losing at least $1.35 billion a year to employees. In the 1970s roughly 2% of department store employees were detected in some form of theft. In 1982, some department stores reported as many as 10% of their employees had been caught in instances of theft.

One approach to curtailing employee theft is to screen job applicants carefully to eliminate those who have a high potential for theft. Currently, pencil-and-paper honesty tests are widely used; at least 5000 firms report that honesty tests are given as a routine part of their hiring process. These tests are used in retail businesses, financial institutions, warehouse operations, and other places where employees have easy access to cash and/or merchandise.

Researchers Paul Sackett and Michael Harris of the University of Illinois at Chicago have completed a study investigating the validity of pencil-and-paper tests in predicting employee honesty. One phase of their study compared pencil-and-paper honesty tests with polygraph tests. Each employee in the study received a score on honesty from a pencil-and-paper test and a second score on honesty based on a polygraph test. The correlation of the pencil-and-paper test scores with

Mark Antman, Stock, Boston

Honesty is essential for employees whose job involves handling cash transactions.

the polygraph test scores varied with the specific pencil-and-paper test used. However, the correlation coefficients reported ranged from .27 to .86, indicating that the pencil-and-paper test scores were positively related to the polygraph test scores.

The correlation analysis indicated that employees who score high on pencil-and-paper honesty tests also tend to score high on polygraph tests. While the study supports the use of pencil-and-paper honesty tests, the researchers noted that since the correlation is not perfect, some errors may be made in drawing conclusions about the honesty of individual applicants. They caution against communicating honesty test results to rejected applicants.

Based on "Honesty Testing for Personnel Selection: A Review and Critique," *Personnel Psychology* (1984).

In Chapters 2 and 3 we discussed the use of descriptive statistics for summarizing a set of data involving one variable (e.g., test scores or television-viewing time). In many situations, however, a data set consists of measurements on two or more variables of interest. One approach to computing summary measures on such a data set would be to apply the methods of Chapters 2 and 3 to each variable individually. However, such an approach would not recognize the important relationships that may exist among the variables. In cases where data have been collected on two or more variables, we are often more interested in describing how the variables are related to each other than in preparing summary statistics for the individual variables. For example, the "Statistics in the News" discusses a research study concerning the relationship between two variables; scores on pencil-and-paper honesty tests were found to be related to scores on polygraph tests.

EXAMPLE 4.1

In a recent study involving a sample of 64 cities in the United States, researchers recorded the fluoride level in the drinking water and the death rate due to cancer for each city. One approach to summarizing the data set would be to compute the sample mean and sample standard deviation for each of these variables (fluoride level and cancer death rate), along with appropriate graphical presentations such as histograms. However, given this data set, many of us would want to go one step further to see whether or not there is any evidence that the fluoride level in the drinking water is related to the rate of cancer deaths.

• • •

In Example 4.1 the objective is to determine if the two variables are related. In this chapter we provide a brief introduction to how *correlation analysis* can be used to determine the extent to which two variables are *linearly related*. Our analysis shows how to compute a measure of the strength of the relationship between two variables. In situations such as Example 4.1, where the analysis involves two variables, we say we have a *bivariate* data set. Anytime there are two or more variables involved, we have a *multivariate* data set. One question that arises when analyzing bivariate data is, Are the two variables related? The usual followup to that question is, If the two variables are related, what is the nature of the relationship?

EXAMPLE 4.2

A firm specializing in the computerized translation of technical and scientific documents from English to German conducted a study involving the number of words in a document and the time required to perform the translation. The objective of the study was to develop an equation that could be used to predict the translation time for a document given the number of words.

• • •

In Example 4.2 the objective is to utilize the relationship between the two variables to predict translation time. In this chapter we also introduce *regression analysis,* a statistical technique that can be used to develop a mathematical equation showing how two variables are related. The variable that is being predicted is called the *dependent* variable; the variable being used to predict the value of the dependent variable is called the *independent* variable. Thus in Example 4.2, the desire to predict the translation time would suggest making translation time the dependent variable; hence the number of words on a page would be the independent variable.

Although the previous examples involve quantitative data, there are many practical applications where the data set consists of several qualitative variables such as hair color, sex, occupation, and so on. In these situations, the use of contingency tables and crosstabulation analysis can provide insight into possible relationships. Section 4.3 provides an introduction to these methods.

4.1 CORRELATION ANALYSIS

In this section we consider the analysis of bivariate data where the objective is to develop a measure of the degree of linear association between the two variables, identified as x and y.

EXAMPLE 4.3

The management of a stereo and sound-equipment store located in San Francisco would like to investigate whether or not there is any relationship between the number of commercials (x) shown on Friday evening television and the resulting sales volume for Saturday (y), measured in hundreds of dollars. The following sample data were obtained.

x = number of commercials	2	5	1	3	4	1	5
y = sales volume (\$100s)	24	28	22	26	25	24	26

Note that each x_i value is paired with a y_i value; that is, when two commercials were shown the sales volume was \$2400 ($x_1 = 2$, $y_1 = 24$), when five commercials were

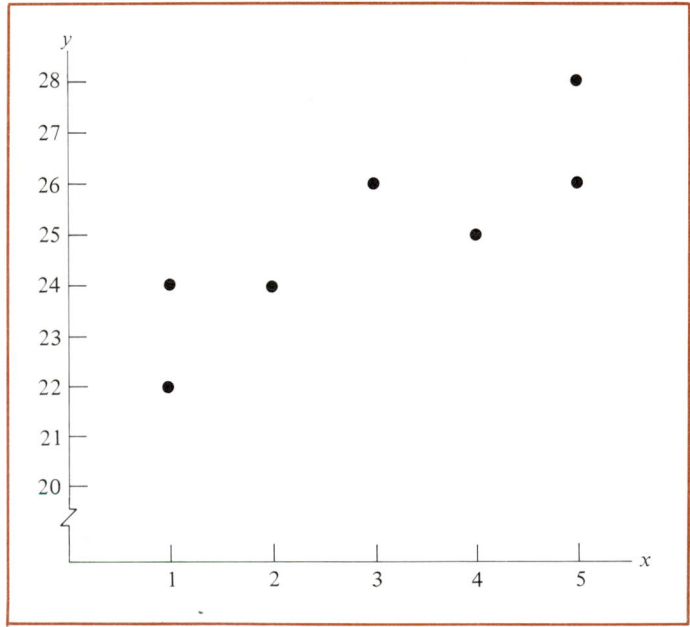

FIGURE 4.1
Scatter Diagram for Example 4.3

shown the sales volume was $2800 ($x_2 = 5$, $y_2 = 28$), and so on. In Figure 4.1 we show a graph of the data; this type of graph is referred to as a *scatter diagram*. In this case, the scatter diagram appears to indicate that there is an increasing linear relationship between x and y. To measure the degree of linear association between these two variables, we need to define a measure of linear association known as the *covariance*.

• • •

Covariance

Sample covariance is defined as follows.

Sample Covariance

$$s_{xy} = \frac{\Sigma(x_i - \bar{x})(y_i - \bar{y})}{n - 1} \qquad (4.1)$$

In this formula each x_i value is paired with a y_i value. We then sum the products obtained by multiplying the deviation of each x_i from its sample mean, $\bar{x}$, times the deviation of the corresponding y_i from its sample mean, $\bar{y}$; this sum is then divided by $n - 1$. Intuitively, this is very close to computing the average of the sum of the above products.

EXAMPLE 4.3 (continued)

To measure the strength of the linear relationship between the number of commercials (x) and the sales volume (y), we use (4.1) to compute the sample covariance. The following calculations illustrate the computation of $\Sigma(x_i - \bar{x})(y_i - \bar{y})$. Note that $\bar{x} = 21/7 = 3$ and $\bar{y} = 175/7 = 25$.

x_i	y_i	$x_i - \bar{x}$	$y_i - \bar{y}$	$(x_i - \bar{x})(y_i - \bar{y})$
2	24	−1	−1	1
5	28	2	3	6
1	22	−2	−3	6
3	26	0	1	0
4	25	1	0	0
1	24	−2	−1	2
5	26	2	1	2
Totals 21	175	0	0	17

Using (4.1) we obtain

$$s_{xy} = \frac{\Sigma(x_i - \bar{x})(y_i - \bar{y})}{n - 1} = \frac{17}{6} = 2.8333$$

• • •

The formula for computing the covariance of a population of size N is similar to (4.1), but we use different notation to indicate that we are dealing with the entire population.

Population Covariance

$$\sigma_{xy} = \frac{\Sigma(x_i - \mu_x)(y_i - \mu_y)}{N} \tag{4.2}$$

In (4.2) we use the notation μ_x for the population mean of the variable x and μ_y for the population mean of the variable y. The sample covariance, s_{xy} is an estimate of the population covariance, σ_{xy} based upon a sample of size n.

Interpretation of the Covariance

To aid in the interpretation of the *sample covariance,* consider Figure 4.2. It is the same as the scatter diagram of Figure 4.1 with a vertical line at $x = 3$ (the value of $\bar{x}$) and a horizontal line at $y = 25$ (the value of $\bar{y}$). Four quadrants have been identified on the graph. Points that fall in quadrant I correspond to x_i values greater than $\bar{x}$ and y_i values greater than $\bar{y}$; points that fall in quadrant II correspond to x_i values less than $\bar{x}$ and y_i values greater than $\bar{y}$, and so on. Thus the value of $(x_i - \bar{x})(y_i - \bar{y})$ must be positive for points located in quadrant I, negative for points located in quadrant II,

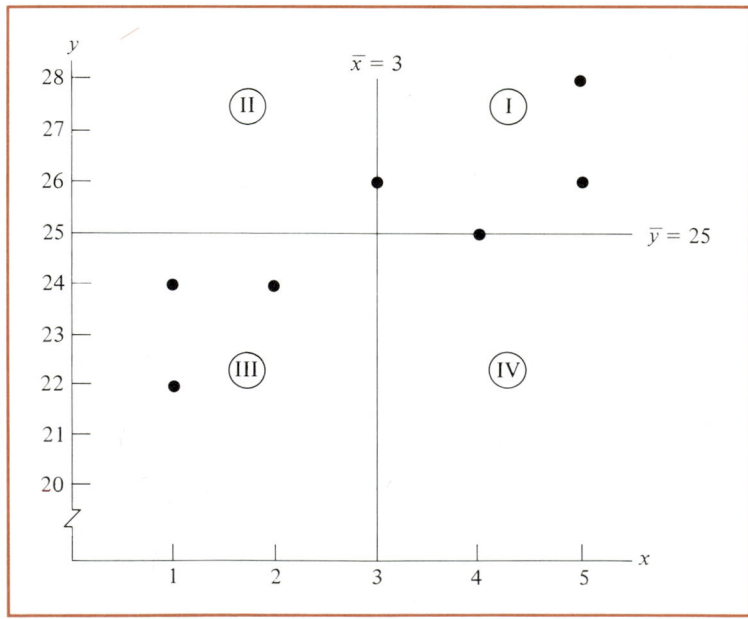

FIGURE 4.2
Quadrants I, II, III, and IV for Example 4.3

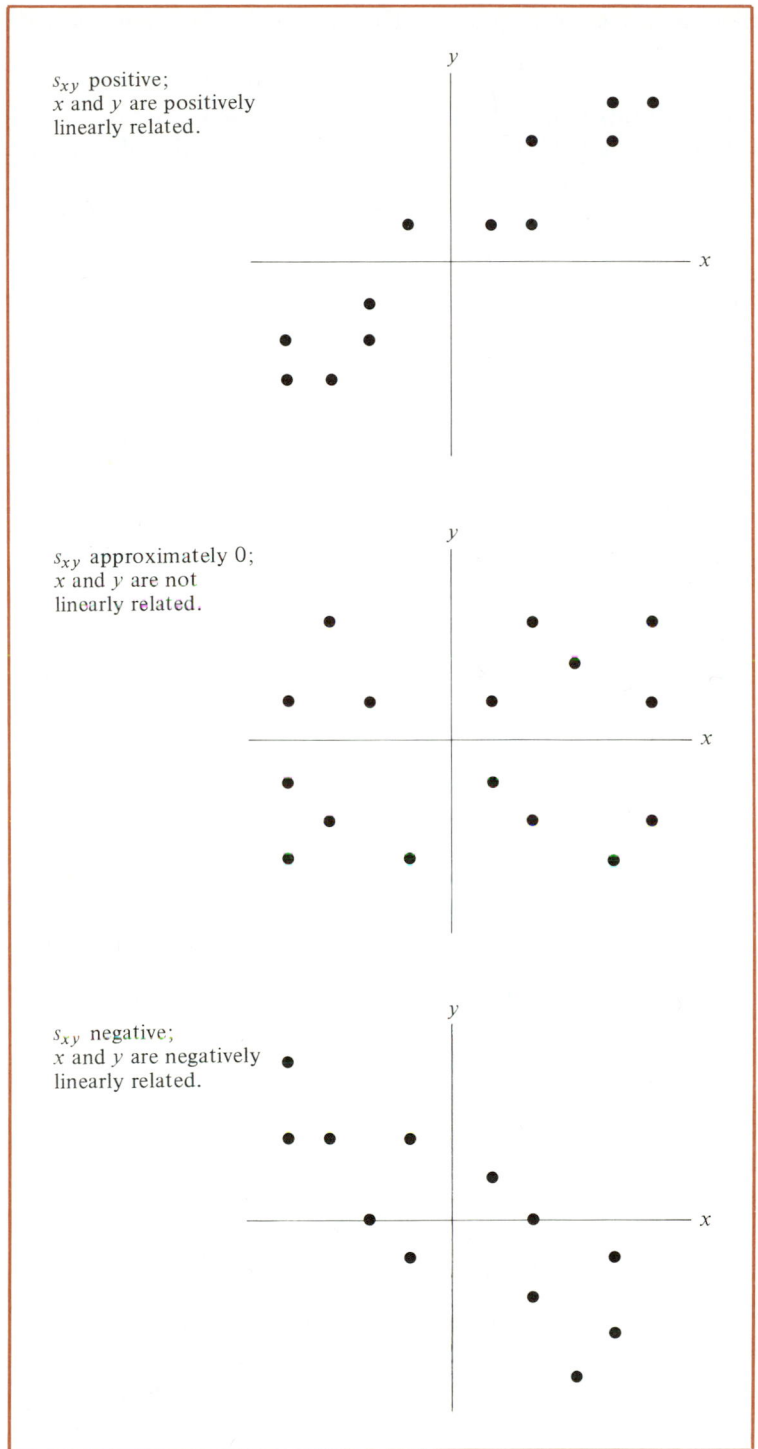

FIGURE 4.3
Interpretation of Sample Covariance

positive for points located in quadrant III and negative for points located in quadrant IV.

If the value of s_{xy} is positive, the points that have had the greatest effect on s_{xy} must lie in quadrants I and/or III, and hence a positive value for s_{xy} is indicative of a positive linear association between x and y; that is, as the value of x increases, the value of y increases. If the value of s_{xy} is negative, however, the points that have had the greatest effect on s_{xy} lie in quadrants II and/or IV, and hence a negative value for s_{xy} is indicative of a negative linear association between x and y; that is, as the value of x increases, the value of y decreases. Finally, if the points are evenly distributed across all four quadrants, the value of s_{xy} will be close to 0, indicating no linear association between x and y. Figure 4.3 shows the values of s_{xy} that can be expected with these three different types of scatter diagrams.

From the previous discussion it might appear that a large positive value for the covariance is indicative of a strong positive linear relationship and that a large negative value is indicative of a strong negative linear relationship. However, one problem with using covariance as a measure of the strength of the linear relationship is that it is difficult to say what is a large or small value for the covariance. This problem is made even worse when we consider the fact that the value we obtain for the covariance depends upon the unit of measurement. For example, suppose we were interested in the relationship between height (x) and weight (y) for individuals. If height is measured in inches, we will get much larger numerical values for ($x_i - \overline{x}$) than if it is measured in feet. Thus we would obtain larger values for $\Sigma(x_i - \overline{x})(y_i - \overline{y})$—and hence a larger covariance—when, in fact, there is no difference in the relationship. A measure of relationship that avoids this difficulty is the *correlation coefficient*.

Correlation Coefficient

For sample data, the Pearson Product Moment correlation coefficient is defined as follows.

Pearson Product Moment Correlation Coefficient—Sample Data

$$r = \frac{s_{xy}}{s_x s_y} \tag{4.3}$$

where

s_{xy} = sample covariance

s_x = sample standard deviation of x

s_y = sample standard deviation of y

Equation (4.3) shows that the Pearson Product Moment correlation coefficient for sample data (commonly referred to more simply as the sample correlation coefficient) is computed by dividing the sample covariance by the product of the standard deviation of x and the standard deviation of y. Before we consider further interpretation of the sample correlation coefficient, let us consider the use of (4.3) for a sample problem.

EXAMPLE 4.3 (continued)

Using the data presented in Example 4.3, we can compute the sample correlation coefficient:

$$s_x = \sqrt{\frac{\Sigma(x_i - \bar{x})^2}{n - 1}} = \sqrt{\frac{18}{6}} = 1.7321$$

$$s_y = \sqrt{\frac{\Sigma(y_i - \bar{y})^2}{n - 1}} = \sqrt{\frac{22}{6}} = 1.9149$$

and since $s_{xy} = 2.8333$, we have

$$r = \frac{s_{xy}}{s_x \, s_y} = \frac{2.8333}{(1.7321)(1.9149)} = .854$$

$$\bullet \quad \bullet \quad \bullet$$

When using a calculator to compute the sample correlation coefficient, the following formula is preferred. Since the computation of each deviation $x_i - \bar{x}$ and $y_i - \bar{y}$ is not necessary, less round-off error is introduced.

Pearson Product Moment Correlation Coefficient-Sample Data Alternate Formula

$$r = \frac{\Sigma x_i y_i - (\Sigma x_i \, \Sigma y_i)/n}{\sqrt{\Sigma x_i^2 - (\Sigma x_i)^2/n} \, \sqrt{\Sigma y_i^2 - (\Sigma y_i)^2/n}} \tag{4.4}$$

Algebraically, equations (4.3) and (4.4) are equivalent.

EXAMPLE 4.3 (continued)

Let us use (4.4) to recompute the value of the sample correlation coefficient.

	x_i	y_i	$x_i y_i$	x_i^2	y_i^2
	2	24	48	4	576
	5	28	140	25	784
	1	22	22	1	484
	3	26	78	9	676
	4	25	100	16	625
	1	24	24	1	576
	5	26	130	25	676
Totals	21	175	542	81	4397

Using (4.4) we obtain:

$$r = \frac{542 - (21)(175)/7}{\sqrt{81 - (21)^2/7}\,\sqrt{4397 - (175)^2/7}} = \frac{17}{19.8997} = .854$$

Thus we see that the value obtained for r using (4.4) is the same (to three decimal places) as the value obtained using (4.3).

The formula for computing the correlation coefficient of a population, denoted by the Greek letter ρ (rho, pronounced row), is as follows.

Pearson Product Moment Correlation Coefficient—Population Data

$$\rho = \frac{\sigma_{xy}}{\sigma_x \sigma_y} \tag{4.5}$$

where

σ_{xy} = population covariance

σ_x = population standard deviation for x

σ_y = population standard deviation for y

Thus we think of r as an estimate of ρ based upon a sample of size n.

Interpretation of the Correlation Coefficient

First let us consider a simple example which illustrates the concept of perfect positive linear association.

EXAMPLE 4.4

Consider the following bivariate data set consisting of $n = 3$ pairs of points and the associated scatter diagram:

x_i	y_i
1	10
2	30
3	50

Scatter Diagram

The straight line drawn through each of the three points shows that there is a perfect linear relationship between the two variables x and y. The computation of r for this data is as follows:

	x_i	y_i	$x_i y_i$	x_i^2	y_i^2
	1	10	10	1	100
	2	30	60	4	900
	3	50	150	9	2500
Totals	6	90	220	14	3500

$$r = \frac{\Sigma x_i y_i - (\Sigma x_i \Sigma y_i)/n}{\sqrt{\Sigma x_i^2 - (\Sigma x_i)^2/n} \, \sqrt{\Sigma y_i^2 - (\Sigma y_i)^2/n}}$$

$$= \frac{220 - (6)(90)/3}{\sqrt{14 - (6)^2/3} \, \sqrt{3500 - (90)^2/3}} = \frac{40}{40} = 1$$

Thus we see that the value of the sample correlation coefficient for this data set is 1.

$\bullet \quad \bullet \quad \bullet$

In general, it can be shown that if all the points in a data set fall on a straight line having positive slope, then the value of the sample correlation coefficient is $+1$; that is, a sample correlation coefficient of $+1$ corresponds to a perfect positive linear association between x and y. Moreover, if the points in the data set fall on a straight line having negative slope, the value of the sample correlation coefficient is -1; ie, a sample correlation coefficient of -1 corresponds to a perfect negative linear association between x and y.

Let us now suppose that for a certain data set there is a positive linear association between x and y but that the relationship is not perfect. The value of r will be less than $+1$, indicating that the points in the scatter diagram do not all fall on a straight line. As the points in a data set deviate more and more from a perfect positive linear association, the value of r becomes smaller and smaller. A value of r equal to 0 indicates no linear relationship between x and y and values of r near zero indicate a weak relationship.

EXAMPLE 4.3 (continued)

Recall that for the data set in this example, $r = +.854$. Since $r = +.854$, we conclude that there appears to be a positive linear association between the number of commercials and Saturday sales volume. More specifically, an increase in the number of commercials is associated with an increase in sales volume.

$\bullet \quad \bullet \quad \bullet$

We have stated that values of r near $+1$ indicate a strong linear association between two variables and values of r near zero indicate little or no linear association between the variables. But we must be careful not to conclude that a value of r near zero means there is no relationship between the variables. Example 4.5 shows that we can have cases where $r = 0$ and there is no linear relationship; in fact, in this case, there is a perfect curvilinear relationship between the variables.

EXAMPLE 4.5

Consider the following sample data and the associated scatter diagram.

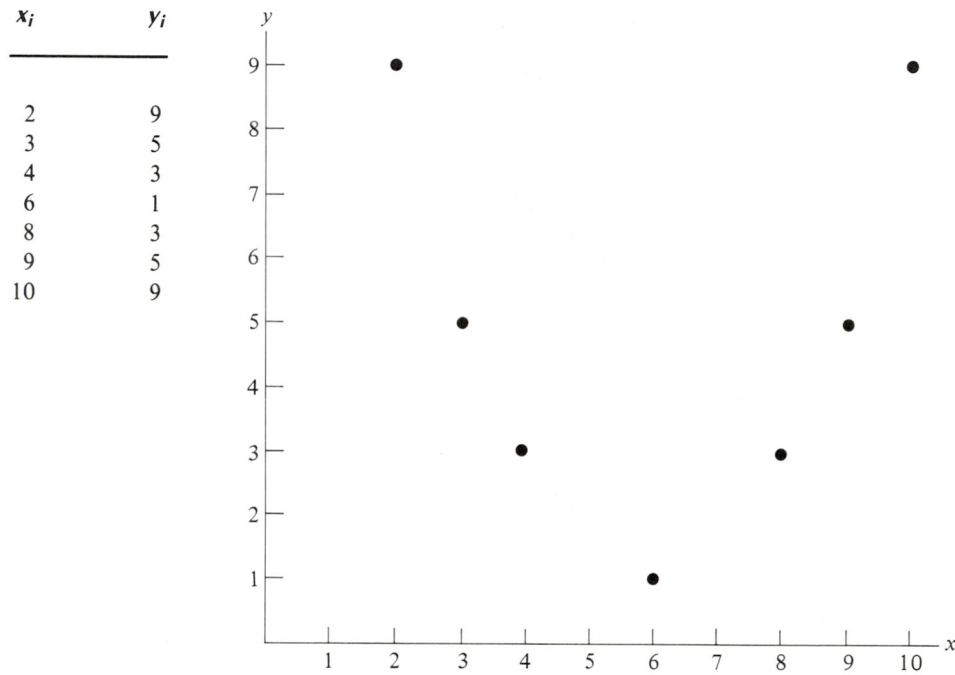

x_i	y_i
2	9
3	5
4	3
6	1
8	3
9	5
10	9

The computation of r is as follows:

	x_i	y_i	x_iy_i	x_i^2	y_i^2
	2	9	18	4	81
	3	5	15	9	25
	4	3	12	16	9
	6	1	6	36	1
	8	3	24	64	9
	9	5	45	81	25
	10	9	90	100	81
Totals	42	35	210	310	231

Using (4.4) we obtain:

$$r = \frac{210 - (42)(35)/7}{\sqrt{310 - (42)^2/7}\ \sqrt{231 - (35)^2/7}} = \frac{0}{56.9912} = 0$$

Thus $r = 0$ and we conclude that there is no linear relationship between x and y. Note, however, that there is an obvious curvilinear relationship between x and y as can be seen from looking at Figure 4.4, where we have drawn a line through the points in the scatter diagram.

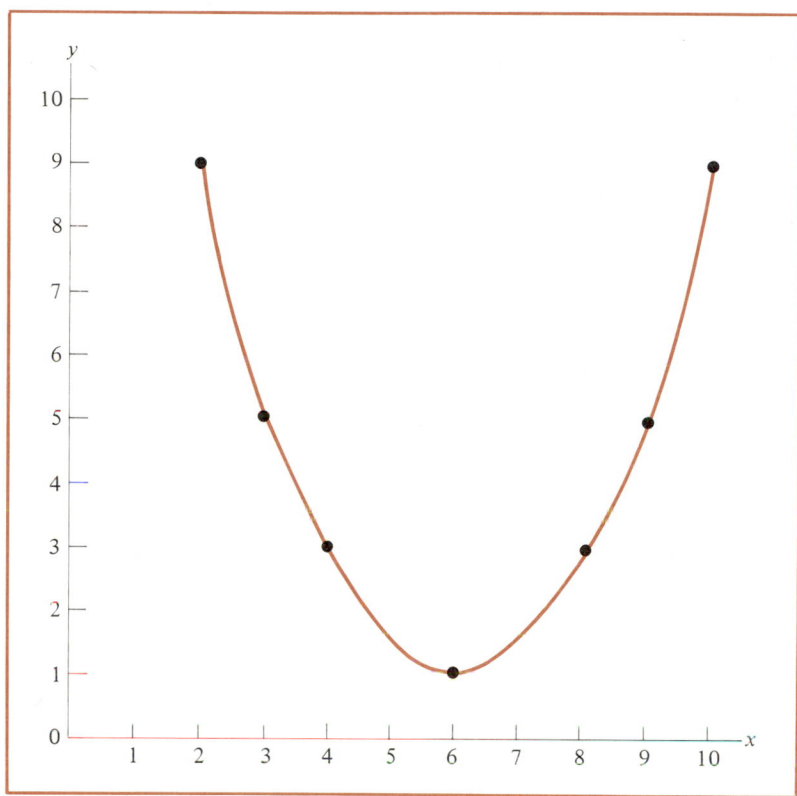

FIGURE 4.4
Even Though $r = 0$, There Is a Perfect Curvilinear Relationship for the Data of Example 4.4

• • •

To reiterate, Example 4.5 illustrates an important concept regarding the proper interpretation of the sample correlation coefficient. The sample correlation coefficient measures only the degree of *linear association* between the two variables. A value of r equal to 0 cannot be interpreted as implying that there is no relationship between the two variables. One should always look at the associated scatter diagram as well as the value of the sample correlation coefficient when attempting to determine if and how two variables are related.

In closing this section we caution that while a correlation coefficient near ± 1 does imply a strong linear relationship between two variables, it does not imply a cause-effect relationship. For instance, it has been noted that as women's hemlines are raised, stock prices go up. There is a positive correlation, but it would be folly to infer a cause-effect relationship. No one truly believes that changes in the stock market are the result of changes in women's hemlines. Conclusions concerning cause-effect relationships must be based on sound theoretical arguments and the judgment of the analyst.

EXERCISES

1. A high-school guidance counselor collected the following data regarding the grade point average (GPA) and the SAT mathematics test score for six seniors.

GPA	2.7	3.5	3.7	3.3	3.6	3.0
SAT	440	560	720	520	640	480

a. Develop a scatter diagram for these data with GPA on the horizontal axis.
b. Does there appear to be any relationship between the GPA and the SAT mathematics test score? Explain.
c. Compute and interpret the sample covariance for these data.
d. Compute the correlation coefficient for these data using (4.3). What does this value tell us about the relationship between the two variables?
e. Compute the correlation coefficient using (4.4). When using a calculator, why is this formula preferred over (4.3)?

2. Given are five observations taken for two variables.

x	4	6	11	3	16
y	50	50	40	60	30

a. Develop a scatter diagram with x on the horizontal axis.
b. What does the scatter diagram developed in part (a) indicate about the relationship between the two variables?
c. Compute and interpret the sample covariance for these data.
d. Compute and interpret the sample correlation coefficient for these data.

3. Tyler Realty collected the following data regarding the selling price of new homes and the size of the homes measured in terms of square footage of living space.

Square Footage	Selling Price
2500	$124,000
2400	$108,000
1800	$ 92,000
3000	$146,000
2300	$110,000

a. Develop a scatter diagram for these data with square footage on the horizontal axis.

b. Does there appear to be a linear relationship? Explain.

4. A sociologist collected the following data regarding the age of the wife and husband when they were married.

Wife's age	19	42	28	25	36
Husband's age	20	32	31	24	33

a. Develop a scatter diagram for these data with the wife's age on the horizontal axis.

b. Does there appear to be a linear relationship? Explain.

5. Given are five observations taken for two variables.

x	6	11	15	21	27
y	6	9	6	17	12

Comment on the relationship between these two variables.

4.2 REGRESSION ANALYSIS

In the chapter introduction we stated that one objective in bivariate data analysis is to assess the strength of the relationship between two variables. We learned that the correlation coefficient provides a measure of the strength of the linear relationship between the variables. However, even if we find that two variables are linearly related, the correlation coefficient does not help in using knowledge of one variable to predict the value of the other.

A second common objective in the analysis of bivariate data is to develop an equation whereby we can utilize knowledge of the value of one variable to predict the value of the other. Recall that in Example 4.2 we described a situation where a firm was interested in predicting the time required to translate technical and scientific documents from English into German. Clearly, if an equation could be developed relating the two variables (x = number of words in a document and y = translation time), it could be used to predict translation time given knowledge of the number of words in a document. Regression analysis provides a means for developing such an equation. In this section we show how simple linear regression can be used to develop a linear equation relating two variables. In Example 4.6 we present sample data that is used later to develop such a prediction equation.

EXAMPLE 4.6

Armand's Pizza Parlors operates a chain of Italian-food restaurants located in a five-state area. One of the most successful types of location for Armand's has been near a college campus. To investigate the relationship between student population (x) and annual sales (y) at these locations, Armand's collected data from a sample of ten of its restaurants. This data and the associated scatter diagram are shown below:

x_i	y_i
2	58
6	105
8	88
8	118
12	117
16	137
20	157
20	169
22	149
26	202

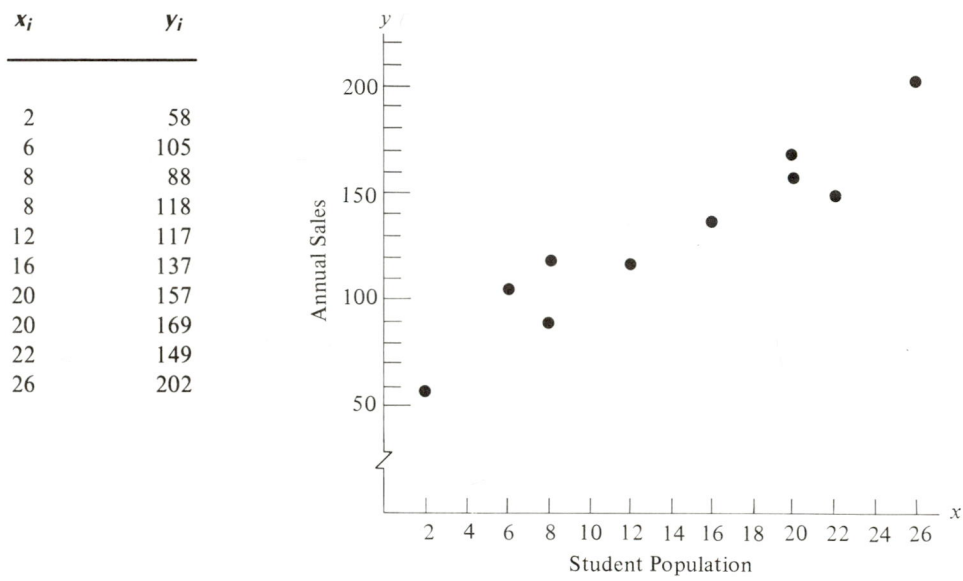

What preliminary conclusions can we draw from the given scatter diagram? It appears that lower sales volumes are associated with smaller student populations and higher sales volumes are associated with larger student populations. It also appears that the relationship between the two variables can be approximated by a straight line.

• • •

In simple linear regression analysis, a procedure called the *least squares method* is used to find the straight line that provides the best approximation for the relationship between the two variables. Recall that the slope-intercept form for the equation of a straight line is given by:

$$y = b_0 + b_1 x$$

In this equation b_0 is the y-intercept, the point at which the line intersects the y-axis when $x = 0$; b_1 is the slope, the amount of change in y per unit change in x. The least squares method provides the following formulas for the intercept b_0 and the slope b_1.

Least Squares Equations for Slope and Intercept

$$b_1 = \frac{\Sigma x_i y_i - (\Sigma x_i)(\Sigma y_i)/n}{\Sigma x_i^2 - (\Sigma x_i)^2/n} \qquad (4.6)$$

$$b_0 = \bar{y} - b_1 \bar{x} \qquad (4.7)$$

The straight line resulting from applying (4.6) and (4.7) is called the *estimated regression line;* to indicate that the value of y is to be predicted from a known value of x, we write $\hat{y}$ (pronounced y hat) on the left-hand side of the equation, as shown next.

Estimated Regression Line

$$\hat{y} = b_0 + b_1 x \qquad (4.8)$$

where

$$b_0 = y\text{-intercept of the estimated regression line}$$

$$b_1 = \text{slope of the estimated regression line}$$

In the continuation of Example 4.6 we show how the estimated regression line is computed for the Armand's Pizza problem.

EXAMPLE 4.6 (continued)

The calculations for computing b_0 and b_1 for the data set presented in Example 4.6 are as follows:

	x_i	y_i	$x_i y_i$	x_i^2
	2	58	116	4
	6	105	630	36
	8	88	704	64
	8	118	944	64
	12	117	1,404	144
	16	137	2,192	256
	20	157	3,140	400
	20	169	3,380	400
	22	149	3,278	484
	26	202	5,252	676
Totals	140	1,300	21,040	2,528

$$b_1 = \frac{\sum x_i y_i - (\sum x_i)(\sum y_i)/n}{\sum x_1^2 - (\sum x_i)^2/n}$$

$$= \frac{21{,}040 - (140)(1300)/10}{2528 - (140)^2/10} = \frac{2840}{568} = 5$$

$$b_0 = \bar{y} - b_1 \bar{x}$$

$$= \frac{1300}{10} - 5\left(\frac{140}{10}\right) = 130 - 5(14) = 60$$

Substituting these values for b_0 and b_1 into (4.8) yields the estimated regression line

$$\hat{y} = 60 + 5x$$

In Figure 4.5, we show the graph of this equation.

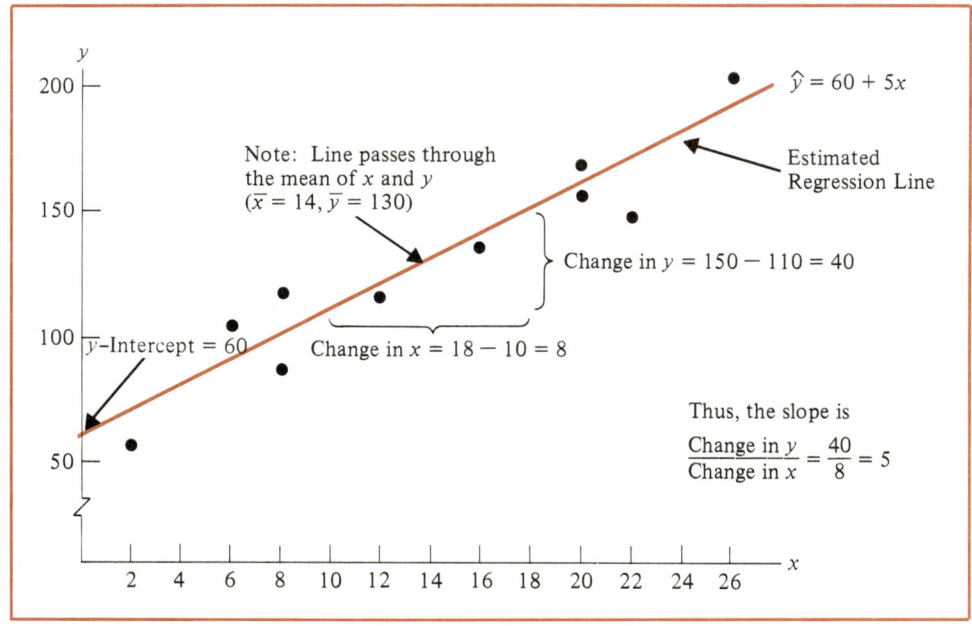

FIGURE 4.5
Illustration of Estimated Regression Line for Example 4.6

• • •

In Example 4.6 the slope of the estimated regression line ($b_1 = 5$) is positive, implying that as student population increases, annual sales increase. In fact, we can conclude (since sales are measured in $1000s and student population in 1000s) that an increase in the student population of 1000 is associated with an increase of $5000 in expected annual sales; that is, sales are expected to increase by $5.00 per student. We can now use this estimated regression line to predict annual sales given knowledge of the student population.

EXAMPLE 4.6 (continued)

To predict annual sales for a new restaurant Armand's is considering building near a campus with 16,000 students, we compute

$$\hat{y} = 60 + 5(16)$$

$$= 140$$

Hence, we predict annual sales of $140,000 for the new restaurant.

• • •

It is not recommended that you use an estimated regression line for predictions beyond the range of the data. For instance, the Armand's Pizza data was collected for campuses ranging in size from 2000 to 26,000 students. It might be misleading to use this equation to predict sales for a campus with 40,000 students, for instance. The same relationship might not hold for larger campuses.

Returning to the Armand's Pizza estimated regression line, we note that the mean student population is $\bar{x} = 14,000$. Substituting $\bar{x} = 14$ into the estimated regression line, we obtain

$$\hat{y} = 60 + 5(14) = 130$$

Hence the predicted value of y is its mean, $\bar{y} = 130$. Thus we see that the estimated regression line passes through the means of the data. That is, the estimated regression line passes through the point whose coordinates are $(\bar{x}, \bar{y})$. This provides a check on the calculations in computing b_0 and b_1. We can substitute $\bar{x}$ for x and $\bar{y}$ for y in Equation (4.8); if the left-hand side and right-hand side are not equal, we have made an error and should go back and check our calculations.

Rounding Inaccuracies

While it was not necessary to do any rounding in computing b_0 and b_1 for the Armand's Pizza problem, we recommned using as many significant digits as possible when solving for b_1 using (4.6). Once b_1 has been computed and used in Equation (4.7) to find b_0, the values of b_0 and b_1 can be rounded to whatever number of significant digits seems appropriate. However, rounding during intermediate calculations can cause a buildup of round-off error and result in significantly different results for b_0 and b_1 (especially b_0). As a rule of thumb, we recommend carrying at least four significant digits in intermediate calculations. Let us now consider another example of computing an estimated regression line.

EXAMPLE 4.7

An instructor collected the following data, which shows the number of hours a student spent studying for an exam (x) and the student's score on the exam (y). The data and a scatter diagram are shown.

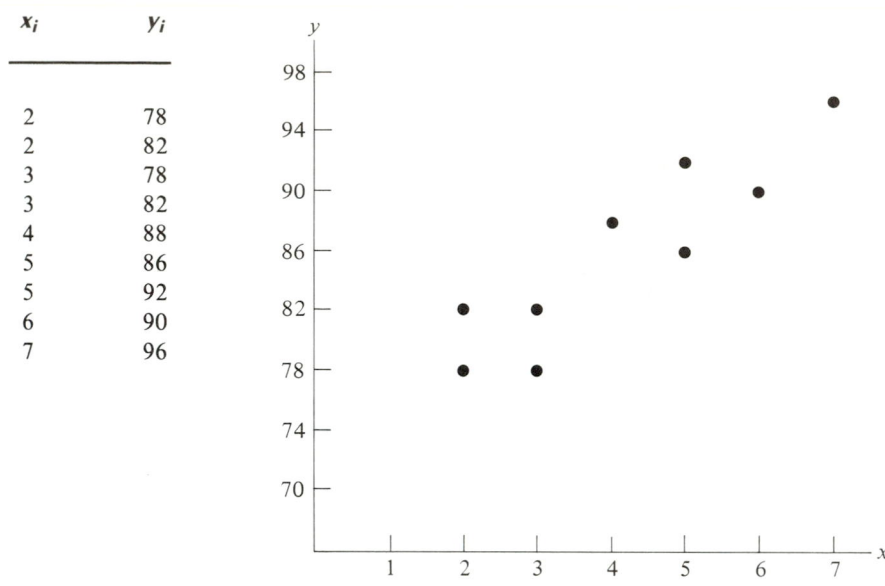

x_i	y_i
2	78
2	82
3	78
3	82
4	88
5	86
5	92
6	90
7	96

The scatter diagram suggests that the relationship between the two variables is linear. We show the calculations necessary to compute b_1 and b_0, the slope and intercept of the estimated regression line.

	x_i	y_i	x_iy_i	x_i^2
	2	78	156	4
	2	82	164	4
	3	78	234	9
	3	82	246	9
	4	88	352	16
	5	86	430	25
	5	92	460	25
	6	90	540	36
	7	96	672	49
Totals	37	772	3254	177

Using (4.6) and (4.7), we obtain

$$b_1 = \frac{3254 - (37)(772)/9}{177 - (37)^2/9} = \frac{80.2222}{24.8889} = 3.2232$$

$$\bar{x} = \frac{37}{9} = 4.1111$$

$$\bar{y} = \frac{772}{9} = 85.7778$$

$$b_0 = 85.7778 - (3.2232)(4.1111)$$

$$= 72.5269$$

· · ·

In these computations, we have followed our recommended practice of computing the values of b_0 and b_1 to at least four significant digits. Hence, we can now round b_0 and b_1 to whatever degree we feel is appropriate, such as $b_0 = 72.5$ and $b_1 = 3.2$. Using these rounded values, we can write our estimated regression line as $\hat{y} = 72.5 + 3.2x$.

EXAMPLE 4.7 (continued)

The estimated regression line $\hat{y} = 72.5 + 3.2x$ shows that an increase of 1 hour in time studying (x) is associated with an increase of 3.2 points in the exam score. If we believe that this line adequately describes the relationship between x and y, then the following predictions could be made.

Study Hours x	Predicted Test Score $\hat{y} = 72.5 + 3.2x$
5	88.5
6	91.7
7	94.9
8	98.1
9	101.3

Upon seeing results such as this, a student might conclude that in order to obtain a 100 on the exam, he or she would have to study between 8 and 9 hours. Another student who looks at these results might conclude that if he or she doesn't study at all ($x = 0$), then the exam score will be $\hat{y} = 72.5 + 3.2(0) = 72.5$. Conclusions such as this should not be made, however, since the data set used to develop the estimated regression line included values only between $x = 2$ and $x = 7$. In general, conclusions made about the relationship are only valid within the range of the x values in the data set.

· · ·

In Chapter 16 we study many issues involving regression analysis in more detail. At that time we will see that b_0 and b_1 are like all other sample statistics in that they are estimates of corresponding population parameters themselves.

Computer Analysis

In Chapter 3 we described how the Minitab statistical computing system can be used to generate descriptive statistics for a data set involving one variable. To provide an illustration of computer analysis involving two variables, we use the data presented in Example 4.6. Referring to Figure 4.6, we see that the data are first read into columns 1 and 2 of the worksheet; column 1 contains the size of the student population (x) and

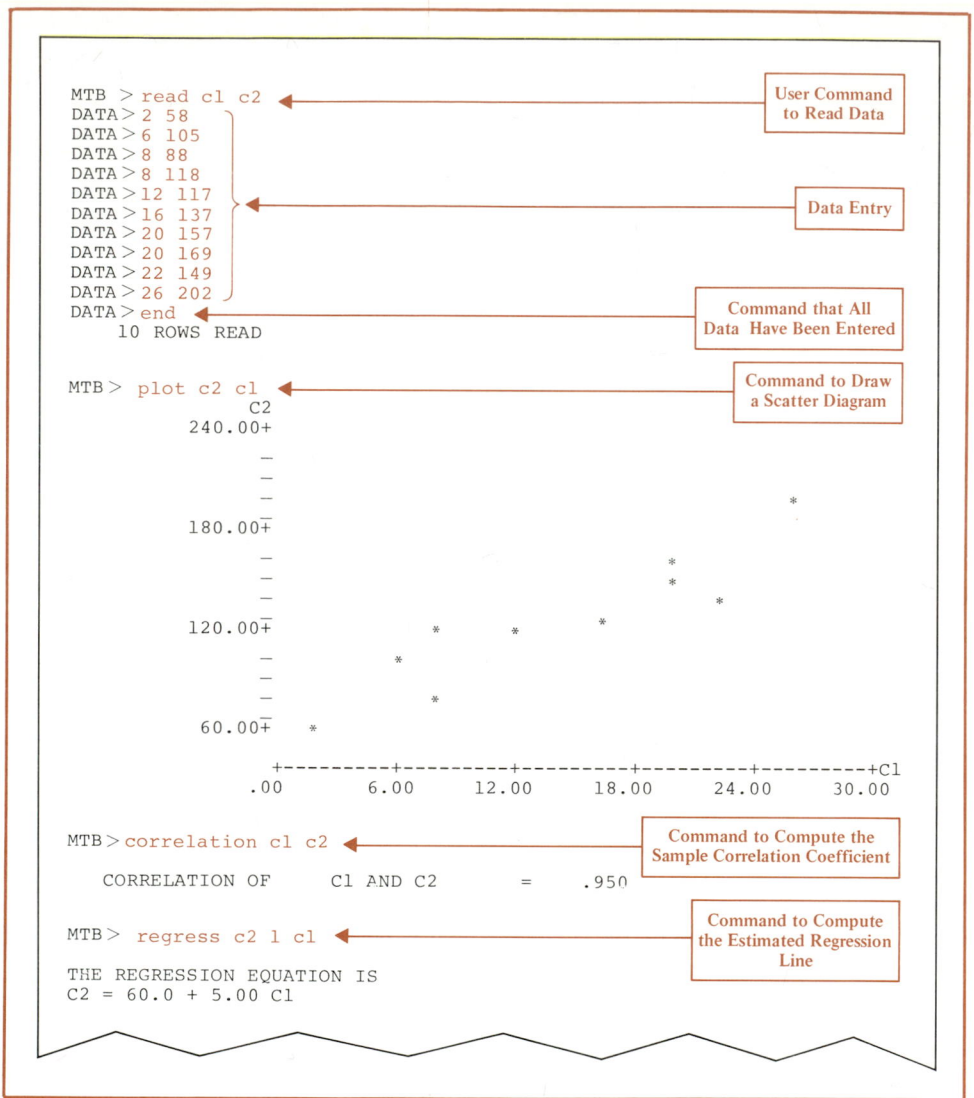

FIGURE 4.6
Computer Analysis Using Minitab

column 2 contains the annual sales (y). The command plot c2 c1 tells Minitab to construct the scatter diagram (with c2 on the vertical axis), the command correlation c1 c2 provides the sample correlation coefficient between c1 and c2, and the command regress c2 1 c1 is a command that tells Minitab to develop a regression equation with c2 as the dependent variable.* Considering the amount of hand calculations that we had to go through to obtain these results previously, we see that the use of a statistical computer package can result in significant savings of time and effort.

*The regression command shown in Figure 4.6 produces additional output, which is illustrated in Chapter 16.

EXERCISES

6. The following data were collected regarding the weights (pounds) of women swimmers and their heights (inches).

Height	68	64	62	65	66
Weight	132	108	102	115	128

a. Draw a scatter diagram for these data with weight on the horizontal axis.
b. What does the scatter diagram developed in part (a) indicate about the relationship between the two variables?
c. Try to approximate the relationship between height and weight by drawing a straight line through the data.
d. Write the equation of the estimated regression line by computing the values of b_0 and b_1 using (4.6) and (4.7).
e. If a swimmer weighs 125 pounds, what would you estimate her height to be?

7. The college grade point average (GPA) and the corresponding monthly starting salary for six recent graduates are shown.

GPA	Monthly Starting Salary ($)
2.6	1100
3.4	1400
3.6	1800
3.2	1300
3.5	1600
2.9	1200

a. Draw a scatter diagram for these data with GPA on the horizontal axis.
b. Try to approximate the relationship between GPA and monthly starting salary by drawing a straight line through the data.
c. Write the equation of the estimated regression line by computing the values of b_0 and b_1 using (4.6) and (4.7).
d. If a graduate had a 3.3 GPA, what would you estimate his or her monthly starting salary to be?
e. If a graduate had a 2.2 GPA, what would you estimate his or her monthly starting salary to be?

8. The following data were collected for two variables.

x	4	10	7	9	5	8	6	7
y	22	8	13	10	18	8	19	14

    a. Does there appear to be any relationship between the two variables? Explain.

    b. Develop the estimated regression line by computing the values of b_0 and b_1 using (4.6) and (4.7).

    c. Estimate the value of y for a value of $x = 8$.

9. A university medical center has developed a test designed to measure a patient's stress level. The test is designed so that the higher scores on the test correspond to higher levels of stress. As part of a research study, the blood pressure (low reading) of patients who took the test was recorded. The following results were obtained.

Stress Test Score	Blood Pressure
53	70
94	91
64	78
73	78
82	85
90	84

    a. Draw a scatter diagram for these data with stress test score on the horizontal axis. Does a linear relationship between the two variables appear to be appropriate?

    b. Write the equation of the estimated least squares line for these data.

    c. Use the line developed in part (b) to estimate an individual's blood pressure if the individual scored 85 on the stress test.

10. In Exercise 4 the following data were collected regarding the age of the wife and husband when they were married.

Wife's age	19	42	28	25	36
Husband's age	20	32	31	24	33

    a. Write the equation of the estimated least squares line that can be used to estimate the age of the husband given the age of the wife.

    b. Use the line developed in part (a) to estimate a husband's age if the wife's age is 30.

4.3 Crosstabular Analysis

Many applications involve two or more qualitative variables. As we will see in later chapters, it is possible to deal with qualitative variables using other methods. However, in this section we briefly introduce some descriptive procedures that are effective with two or more qualitative variables. In particular we look at crosstabulation procedures involving contingency tables.

EXAMPLE 4.8

Alber's Brewery of Clearwater, Arizona, manufactures and distributes three types of beers: a low-calorie light beer, a regular beer, and a dark beer. The market research group has raised questions concerning differences in preferences for the three beers among male and female beer drinkers. A sample of 150 beer drinkers has been selected. After taste-testing each beer, the individuals in the sample are asked to state their preference or first choice. A partial listing of the data is shown below:

Individual	Sex	Beer Preference
1	Female	Dark
2	Male	Light
3	Male	Regular
.	.	.
.	.	.
.	.	.
149	Female	Regular
150	Female	Light

Clearly there are two qualitative variables of interest in this study: sex and beer preference. We are working with nominal data and not much insight can be gained from the data in this unorganized form.

• • •

Organizing the data for two qualitative variables into a table often provides valuable insights. Suppose we utilize the format of Table 4.1. Every individual in the sample can be classified as belonging to one of the six cells in the table. For example, an individual may be a male preferring regular beer (cell 2), a female preferring light beer (cell 4), a female preferring dark beer (cell 6), and so on. Since we have included all

TABLE 4.1

**Contingency Table—Beer Preference
and Sex of Beer Drinker**

		Beer Preference		
		Light	Regular	Dark
Sex	Male	(Cell 1)	(Cell 2)	(Cell 3)
	Female	(Cell 4)	(Cell 5)	(Cell 6)

possible combinations of beer preference and sex—or, in other words, listed all possible contingencies—Table 4.1 is called a *contingency table*.

EXAMPLE 4.8 (continued)

A summary of the data in the Alber's Brewery study in the form of a contingency table is shown.

		Beer Preference			
		Light	Regular	Dark	Total
Sex	Male	20 (13.33)	40 (26.67)	20 (13.33)	80 (53.33)
	Female	30 (20.00)	30 (20.00)	10 (6.67)	70 (46.67)
	Total	50 (33.33)	70 (46.67)	30 (20.00)	150 (100.00)

Percentage   Number of individuals

Thus, of the 150 individuals in the sample, 20, or 13.33%, were men who favored light beer, 40, or 26.67%, were men who favored regular beer, and so on. The contingency table presentation clearly facilitates data interpretation. For example, we now see that a higher percentage of female drinkers prefer light beer and that male drinkers prefer dark beer—as compared to female drinkers—by about a 2-to-1 ratio.

Contingency table analysis is also referred to as *crosstabulation analysis*. Many of the computerized packages for statistical analysis provide generalized crosstabulation procedures that facilitate the analysis of data sets involving two or more qualitative variables.

Quantitative data can also be analyzed using crosstabulation procedures by first breaking the quantitative variable into a series of discrete classes. For example, if one of the variables in a particular study is income, one way to handle this variable using crosstabulation procedures is to first break income into a series of discrete classes (a frequency distribution), such as less than 10,000, 10,000–19,999.99, 20,000–29,999.99, and so on. Each class can then represent a row or column in a contingency table. By developing a frequency distribution, then, a quantitative variable can be treated as a qualitative variable for purposes of crosstabulation.

EXERCISES

11. A study of starting positions for business and engineering graduates, when classified by industry, resulted in the following data: 30 business graduates were employed in the oil industry; 15 business graduates were employed in the chemical industry; 15 business, in electrical; 40 business, in computer; 30

engineering, in oil; 30 engineering, in chemical; 20 engineering, in electrical; and 20 engineering, in computer. Summarize the data by developing a contingency table similar to the one developed in Example 4.8. What preliminary conclusions can be drawn from a careful inspection of this table?

12. A study of educational levels of voters and their political affiliations showed the following results: Of the 70 voters who did not complete high school, 40 are registered Democrats, 20 are Republicans, and 10 are Independents; of the 80 voters who had a high-school degree, 30 are Democrats, 35 are Republicans, and 15 are Independents; of the 100 voters having a college degree, 30 are Democrats, 45 are Republicans, and 25 are Independents. Develop a contingency table for these data similar to the one developed in Example 4.8. What preliminary conclusions can be drawn from this analysis?

13. A sport-preference poll shows the following data.

Sex	Favorite Sport
M	Basketball
F	Football
F	Baseball
M	Baseball
M	Football
M	Football
F	Basketball
F	Basketball
M	Football
M	Baseball
F	Basketball
F	Football
M	Baseball
M	Football
F	Basketball
F	Baseball
F	Football
M	Basketball
M	Football
F	Basketball

Summarize the data by developing a contingency table similar to the one developed in Example 4.8. What preliminary conclusions can be drawn from a careful inspection of this table?

14. A study of the grade distributions for three different professors teaching an introductory calculus course resulted in the following data: 18 of the 90 students enrolled received an A, 34 received a B, 23 received a C, 6 received a D, and 9 failed; 4 of the 26 students in Professor Johnson's section received an A, 12

received a B, 5 received a C, 2 received a D, and 3 failed; in Professor Pray's section, the distribution was 2 A's, 14 B's, 8 C's, 1 D, and 5 F's. There were 34 students enrolled in Professor Evans' section. Summarize the data by developing a contingency table. What preliminary conclusions can be drawn from this contingency table?

Summary

In this chapter we introduced some methods for summarizing data sets involving more than one variable. First, covariance was introduced as a measure of the linear association between two variables. However, since scaling problems make it difficult to interpret covariance as a measure of the strength of a relationship between two variables, the correlation coefficient was introduced as another measure of linear association that takes on values between -1 and $+1$. A correlation coefficient of $+1$ corresponds to a perfect positive linear association between two variables, x and y, and a value of -1 corresponds to a perfect negative or inverse linear association. Values of the correlation coefficient near zero indicate that x and y are not linearly related.

Regression analysis was then introduced as a statistical technique for developing a mathematical equation representing the type of relationship between two variables. Specifically, we showed how the least squares method could be used in simple linear regression to identify the coefficients b_0 and b_1 in the estimated regression line. Given the estimated regression line, known values of the independent variable can be used to predict values for the dependent variable.

Finally, the chapter concluded with a brief introduction to the use of contingency tables as a method of crosstabulation for qualitative variables. In later chapters we have much more to say about all the procedures introduced in this chapter. The focus then is on the use of these procedures for statistical inference from a sample to a population.

Statistics in Practice

RESEARCH IN EDUCATION

The field of education is widely studied and researched. Validity and reliability of testing methods, use of aptitude tests, effect of socioeconomic status on academic achievement, and the evaluation of alternative teaching methods are just a few of the topics studied in modern educational research.

A recent study conducted by Thomas K. Crowl investigated the relationship between an elementary-school teacher's need for social approval and his or her grading behavior. The research hypothesis was that teachers who possess higher needs for social approval tend to give higher grades to students in order to reduce the likelihood students will dislike or hold grudges against them. A total of 51 elementary-school teachers participated in the study.

The Crowne/Marlowe Social Desirability Scale (SDS) was used to measure a teacher's need for social approval. Along with completing the SDS instrument, each teacher was asked to grade 10 short answers for each of two typical school questions. The answers to a question about the difference between a discovery and an invention were provided by junior-high-school students. Answers to a question about the reason to celebrate Thanksgiving were provided by elementary-school students. The correlation coefficients between the teacher's SDS score and the grades given to the student answers are as follows:

Discovery/Invention Grade	Thanksgiving Grade	Combined Grade
.26	.34	.39

Although the correlations were not strong, they were all positive. As a result, the research study supported the conclusion that teachers with a high need for social approval tend to assign higher grades than teachers with a low need for social approval.

"Grading Behavior and Teachers' Need for Social Approval," Thomas K. Crowl, *Education* 104, no. 3 (1984)

Glossary

Correlation analysis—A statistical technique that can be used to determine the strength of the linear relationship between two variables.

Bivariate data—A data set that involves two variables.

Multivariate data—a data set that involves two or more variables.

Regression analysis—A statistical technique that can be used to develop a mathematical equation showing how two or more variables are related.

Dependent variable—The variable that is being predicted by the mathematical equation in regression analysis.

Independent variable—The variable being used to predict the value of the dependent variable in regression analysis.

Scatter diagram—A graph of a set of bivariate data in which one variable appears on the horizontal axis and the other appears on the vertical axis.

Sample covariance (s_{xy})—A measure of linear association between the two variables x and y.

Pearson Product Moment correlation coefficient—A measure of linear association that always takes on values between -1 and $+1$; a value of $+1$ indicates that x and y are perfectly related in a positive linear sense and a value of -1 indicates that x and y are perfectly related in a negative linear sense. A value close to zero indicates that x and y are not linearly related.

Least squares method—A mathematical procedure for computing the coefficients (b_0 and b_1) in an estimated regression line.

Contingency table—A table used to display data involving two or more qualitative variables.

Crosstabulation analysis—Analysis of data sets involving two or more qualitative variables.

Key Formulas

Sample Covariance

$$s_{xy} = \frac{\Sigma(x_i - \bar{x})(y_i - \bar{y})}{n - 1} \qquad (4.1)$$

Population Covariance

$$\sigma_{xy} = \frac{\Sigma(x_i - \mu_x)(y_i - \mu_y)}{N} \qquad (4.2)$$

Pearson Product Moment Correlation Coefficient—Sample Data

$$r = \frac{s_{xy}}{s_x s_y} \qquad (4.3)$$

Pearson Product Moment Correlation Coefficient—Sample Data Alternate Formula

$$r = \frac{\Sigma x_i y_i - (\Sigma x_i \, \Sigma y_i)/n}{\sqrt{\Sigma x_i^2 - (\Sigma x_i)^2/n} \, \sqrt{\Sigma y_i^2 - (\Sigma y_i)^2/n}} \qquad (4.4)$$

Pearson Product Moment Correlation Coefficient—Population Data

$$\rho = \frac{\sigma_{xy}}{\sigma_x \sigma_y} \qquad (4.5)$$

Least Squares Equations for Slope and Intercept

$$b_1 = \frac{\Sigma x_i y_i - (\Sigma x_i)(\Sigma y_i)/n}{\Sigma x_i^2 - (\Sigma x_i)^2/n} \qquad (4.6)$$

$$b_0 = \bar{y} - b_1 \bar{x} \qquad (4.7)$$

Estimated Regression Line

$$\hat{y} = b_0 + b_1 x \qquad (4.8)$$

Review Quiz

TRUE/FALSE

1. In regression analysis the variable being predicted is called the independent variable.
2. Covariance is a measure of linear association.
3. If the points on a scatter diagram are concentrated in quadrants I and III, this indicates a positive linear association between x and y.
4. If the covariance is positive, the correlation coefficient is negative.
5. The sample covariance must be greater than or equal to -1 and less than or equal to 1.
6. If all the points on a scatter diagram fall on a straight line, the value of the correlation coefficient must equal 1.
7. If $b_1 > 0$ for the estimated regression line, the independent and dependent variables are positively related.
8. If $b_1 > 0$ for an estimated regression line, the correlation coefficient will be negative.
9. Crosstabular analysis should never be used with qualitative data.
10. A bivariate data set is one involving two or more variables.

MULTIPLE CHOICE

In answering questions 11 and 12, use the following four observations taken for two variables.

x	2	5	8	9
y	20	14	8	6

11. The sample covariance is closest to
    a. 10
    b. -15
    c. 20
    d. 25
12. The correlation coefficient is closest to
    a. 1
    b. .5
    c. 0
    d. $-.7$
13. An estimated regression line developed by the method of least squares is $\hat{y} = 160 + 8.51x$. If $x = 17$, the predicted value of y is closest to
    a. 250
    b. 300
    c. 350
    d. 400

In answering questions 14 and 15, use the following data collected for two variables.

x	3	5	8
y	7	10	13

14. The value of b_1 in the estimated regression line is closest to
   a. 1.00
   b. 1.50
   c. 2.00
   d. 2.50
15. The value of b_0 in the estimated regression line is closest to
   a. 1.50
   b. 2.50
   c. 3.50
   d. 4.50
16. The number of cells in a contingency table with 6 rows and 5 columns is
   a. 6
   b. 20
   c. 11
   d. 30

Supplementary Exercises

15. Eight observations on two variables are given.

x	2	9	6	8	4	7	5	6
y	11	4	6	5	9	4	9	7

   a. Develop a scatter diagram for these data with x on the horizontal axis.
   b. What does the scatter diagram developed in part (a) indicate about the relationship between the two variables?
   c. Compute and interpret the sample covariance for these data.
   d. Compute and interpret the sample correlation coefficient for these data.
16. The following data are shown for two variables x and y.

x	4	10	7	9	5	8	6	7
y	22	8	13	10	18	8	19	14

   a. Compute and interpret the sample covariance for these data.
   b. Compute and interpret the sample correlation coefficient for these data.
17. Shown are some data that a sales manager has collected concerning annual sales and years of experience.

Salesperson	Years of Experience	Annual Sales ($1,000's)
1	1	80
2	3	97
3	4	92
4	4	102
5	6	103
6	8	111
7	10	119
8	10	123
9	11	117
10	13	136

    a. Develop a scatter diagram for these data with years of experience on the horizontal axis.

    b. What does the scatter diagram developed in (a) indicate about the relationship between the two variables?

    c. Compute and interpret the sample covariance for these data.

    d. Compute and interpret the sample correlation coefficient for these data.

18. In a manufacturing process the assembly line speed (feet per minute) was thought to affect the number of defective parts found during the inspection process. To test this theory, management devised a situation where the same batch of parts was inspected visually at a variety of line speeds. The following data were collected.

Line Speed	Number of Defective Parts Found
20	21
20	19
40	15
30	16
60	14
40	17

    a. Develop a scatter diagram for these data with line speed on the horizontal axis.

    b. Compute and interpret the sample correlation coefficient for these data.

19. Reconsider the data regarding the line speed and number of defective parts given in exercise 18.

    a. Develop the estimated regression line that relates line speed to the number of defective parts found.

    b. Use the line developed in (a) to estimate the number of defective parts for a line speed of 50 feet per minute.

20. The PJH&D Company is in the process of deciding whether or not to purchase a maintenance contract for its new word processing system. They feel that maintenance expense should be related to usage and have collected the following information on weekly usage (hours) and annual maintenance expense.

Weekly Usage (Hours)	Annual Maintenance Expense ($100s)
13	17.0
10	22.0
20	30.0
28	37.0
32	47.0
17	30.5
24	32.5
31	39.0
40	51.5
38	40.0

a. Develop the estimated regression equation that relates annual maintenance expense, in hundreds of dollars, to weekly usage.

b. PJH&D expects to operate the word processor 30 hours per week. Estimate the company's annual maintenance expense.

21. The management of a chain of fast-food restaurants would like to investigate the relationship between the daily sales volume of a company restaurant and the number of competitor restaurants within a 1-mile radius of the firm's restaurant. The following data have been collected.

Number of Competitors Within 1 Mile	Sales ($)
1	3600
1	3300
2	3100
3	2900
3	2700
4	2500
5	2300
5	2000

a. Develop the least squares estimated regression line that relates daily sales volume to the number of competitor restaurants within a 1 mile radius.

b. Use the estimated regression line developed in (a) to develop an estimate of the daily sales volume for a particular company restaurant that has four competitors within a 1-mile radius.

22. The regional transit authority for a major metropolitan area would like to determine if there is any relationship between the age of a bus and the annual maintenance cost. A sample of ten buses resulted in the following data.

Age of Bus (years)	Maintenance Cost ($)
1	350
2	370
2	480
2	520
2	590
3	550
4	750
4	800
5	790
5	950

a. Develop the least squares estimated regression line.

b. Use the estimated regression line developed in (a) to estimate the annual maintenance cost for a bus that is 4 years old.

23. A large amusement park surveyed park visitors at the end of the day in order to investigate what effect (if any) the distance traveled had on how satisfied visitors are with regard to the food service facilities at the park. Responses were coded as follows: satisfied (S), not satisfied (NS), less than 25 miles traveled (1), 25 to 100 miles traveled (2), and over 100 miles traveled (3). The following data were obtained.

Opinion	Distance
S	2
S	1
NS	1
NS	3
S	3
NS	1
S	1
S	1
S	3
NS	2
NS	1
S	1
NS	3
S	3
S	2
S	1
NS	2
S	1
NS	3
S	1

Construct a contingency table for these data and develop whatever conclusions appear to be appropriate.

24. Eastern Pharmaceutical Corporation is testing a new drug intended to help relieve the symptoms associated with hay fever. One hundred patients were given different levels of the drug (A, B, or C) and then observed for any possible side effects. The following results were obtained: 60 patients experienced no side effects, and of this group 25 had been given level A of the drug and 30 had been given level B; 20 of the patients that were given level B and 15 of the patients that were given level C also experienced some side effects. Construct a contingency table for these data and develop whatever conclusions appear to be appropriate.

Computer Exercise

An administrator at a high school in Vermont has collected data showing how graduating seniors perform on the mathematics section of the Scholastic Aptitude Test (SAT). In addition, the administrator has access to records that show each student's intelligence quotient (IQ) score. A question of interest is to what extent the performance on the SAT mathematics exam is related to the student's IQ score. The following data were collected for a sample of 50 students.

IQ Score	Math SAT Score	IQ Score	Math SAT Score
109	365	104	533
112	461	113	608
104	380	114	624
115	683	103	389
122	619	116	624
118	444	118	625
140	699	109	493
105	497	114	570
116	578	107	526
104	409	106	436
112	548	108	544
104	413	115	471
129	782	103	395
110	506	111	518
124	651	104	373
111	556	117	493
110	412	102	435
101	525	109	454
115	482	117	585
129	697	118	558
104	385	109	542
113	406	119	626
122	616	114	411
107	376	110	482
108	512	116	567

QUESTIONS

1. Use the descriptive statistical methods of Chapters 2 and 3 to summarize each of the variables in the given data set.
2. Does there appear to be a relationship between a student's IQ score and his or her performance on the SAT mathematics test? What is the statistical measure of the strength of the relationship?
3. Develop an equation that could be used to predict performance on the SAT mathematics test based on the student's IQ score. What is the predicted SAT mathematics score for a student with an IQ of 112?

CHAPTER 5

Introduction to Probability

What You Will Learn in This Chapter

- an understanding of the basic concepts of probability

- the three methods commonly used for assigning probabilities

- how to use the basic rules of probability to compute the probability of an event

- how to use Bayes' theorem

- important terms such as experiment, sample space, event, mutually exclusive event, independent event, and conditional probability

Contents

Statistics in the News

PROBABILITY: WHAT IT TELLS ABOUT YOUR FUTURE

Americans are the most widely researched people on earth. Every day, between 20,000 and 30,000 of us participate in some sort of survey. Based on these surveys, estimates are made of the proportion, or percentage of people who have certain experiences, participate in various activities, and so on. If you consider yourself an average American, research findings are available to estimate the probability of many events in your life. Some of these events and their probability estimates are shown below.

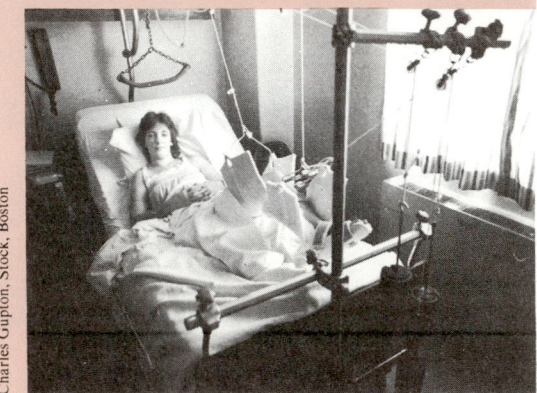

Charles Gupton, Stock, Boston

Statistics show that there is a .12 probability of being hospitalized sometime during a one-year period.

Event	Probability For Men	For Women
Will visit a doctor next year	.71	.79
Will be hospitalized next year	.12	.12
Active in sports/exercise	.63	.54
Drink alcoholic beverages	.78	.60
Married (20–24 age group)	.29	.44
If married, have an older spouse	.14	.73
If married, have a younger spouse	.73	.14
Will lose job next year	.16	.20
Will earn over $20,000 next year	.45	.10
Will die next year (15–24 age group)	.0017	.0005

Based on *How You Rate*, by Tom Biracree. New York: Dell Publishing Co., 1984.

Throughout life we must deal with situations involving uncertainty; for instance, some examples of the types of situations with which we must deal are the events referred to in "Statistics in the News." Some other types of situations are the following:

1. What is the chance that your car will be towed away if you park illegally in the visitors' parking lot?
2. What is the likelihood that you will receive an A on the first exam if you wait until the night before to begin studying?

3. How likely is it that you will complete your college program on time if you take a part-time job that requires you to work 20 hours per week?

4. What are the chances you will get a 7 on the first roll of the dice at a casino in Las Vegas?

Probabilities are most useful in effectively dealing with such uncertainties. In everyday terminology, *probability* can be thought of as a numerical measure of the chance, or likelihood, that a particular event will occur. The probabilities for several events involving many Americans are presented at the end of "Statistics in the News."

Probability values are always assigned on a scale from 0 to 1. A probability near 0 indicates that an event is very unlikely to occur; a probability near 1 indicates that an event is almost certain to occur. Other probabilities between 0 and 1 represent varying degrees of likelihood that an event will occur. Figure 5.1 depicts this view of probability as a numerical measure of the likelihood an event will occur.

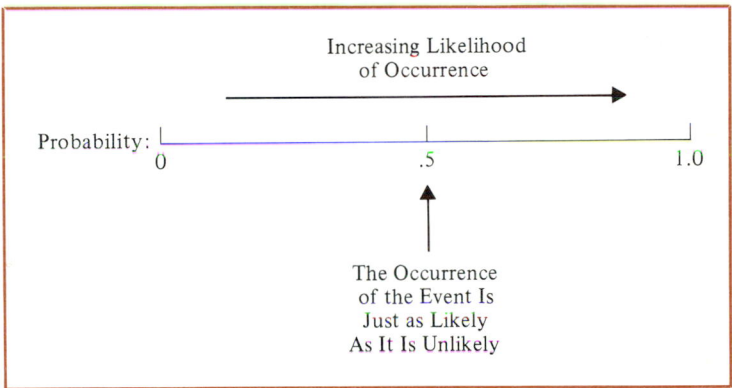

FIGURE 5.1
Probability as a Numerical Measure of the Likelihood of an Event's Occurrence

EXAMPLE 5.1

If we consider the event "rain tomorrow," we understand that when the television weather report indicates a near-zero probability of rain, there is almost no chance of rain. However, if a 90% probability of rain is reported, we know that it is very likely or almost certain that rain will occur. A 50% probability indicates that rain is just as likely to occur as not.

• • •

5.1 EXPERIMENTS AND THE SAMPLE SPACE

Using the terminology of probability, we define an *experiment* to be any process that generates well-defined outcomes. By this we mean that on any single repetition of the experiment, one and only one of the possible experimental outcomes will occur. Several examples of experiments and their associated outcomes are as follows:

Experiment	Experimental Outcomes
Toss a coin	Head, tail
Apply for a job	Hired, not hired
Roll a die	1, 2, 3, 4, 5, 6
Play a football game	Win, lose, tie

The first step in analyzing a particular experiment is to define carefully the experimental outcomes. When we have defined all possible experimental outcomes, we have specified the *sample space* for the experiment.

Sample Space

The sample space for an experiment is the set of all experimental outcomes.

Elements of the sample space are called *sample points*. Since the experimental outcomes are elements of the sample space, each experimental outcome corresponds to a sample point.

EXAMPLE 5.2

Consider the experiment of tossing a coin, as mentioned earlier. The experimental outcomes are defined by the upward face of the coin—a head or a tail. If we let S denote the sample space, we can describe the sample space and sample points for the coin-tossing experiment with the following notation:

$$S = \{\text{head, tail}\}$$

Listing of the sample points

• • •

EXAMPLE 5.3

Consider the experiment of rolling a die, with the experimental outcomes defined as the number of dots appearing on the upward face of the die. In this experiment, the numerical values 1, 2, 3, 4, 5, and 6 represent the possible experimental outcomes, or sample points, for the experiment. We can describe the sample space and sample points for this experiment as follows:

$$S = \{1, 2, 3, 4, 5, 6\}$$

• • •

EXAMPLE 5.4

To illustrate a slightly more complex example, consider the experiment of tossing two coins. Let the experimental outcomes be defined in terms of the pattern of heads and tails appearing on the upward faces of the two coins. How many experimental outcomes (sample points) are possible for this experiment? This can be thought of as a two-step experiment in which step 1 is the tossing of the first coin and step 2 is the tossing of the second coin. If we use H to denote a head and T to denote a tail, (H, H) indicates the sample point with a head on the first coin and a head on the second coin. Continuing this notation, we can describe the sample space (S) for the two-coin-tossing experiment as follows:

$$S = \{(H, H), (H, T), (T, H), (T, T)\}$$

Thus we see that there are four outcomes for this experiment.

• • •

Example 5.4 describes an experiment consisting of two steps, in which there are two possible outcomes (head or tail) on the first step and two possible outcomes (head or tail) on the second step. Let us introduce a rule that is helpful in determining the number of sample points for an experiment consisting of multiple steps.

A Counting Rule for Multiple-Step Experiments

If an experiment can be described as a sequence of k steps in which there are n_1 possible outcomes on the first step, n_2 possible outcomes on the second step, and so on, then the total number of experimental outcomes is given by $(n_1)(n_2) \cdots (n_k)$.

EXAMPLE 5.4 (continued)

Looking at the experiment of tossing two coins as a sequence of first tossing one coin ($n_1 = 2$) and then tossing the other coin ($n_2 = 2$), we can see from the counting rule that there must be $(2)(2) = 4$ distinct sample points or experimental outcomes.

• • •

EXAMPLE 5.5

A fast-food franchise is considering opening a new restaurant in Penn Yan, New York. Two possible locations (Main Street and Elm Street) and three possible seating capacities (seating for 100, seating for 150, seating for 200) are being considered. Let us consider the experiment of observing the choice made for location and seating capacity. We can view this process as a multiple-step experiment involving two possible outcomes on the first step (selecting a location) and three possible outcomes on the second step (selecting a seating capacity). Thus, according to the counting rule, there must be $(2)(3) = 6$ distinct sample points, or experimental outcomes.

Suppose we let M denote a decision to locate on Main Street and E denote a decision to locate on Elm Street and denote the seating capacities as 100, 150, and 200 respectively. Then $(M, 100)$ indicates the sample point corresponding to a Main Street location with a seating capacity of 100 customers, $(M, 150)$ indicates a sample point corresponding to a Main Street location with a seating capacity of 150 customers, and so on. Using this notation, we can denote the sample space for this experiment as follows:

$$S = \{(M, 100), (M, 150), (M, 200), (E, 100), (E, 150), (E, 200)\}$$

• • •

A graphical device that is helpful in visualizing an experiment and enumerating sample points in a multiple-step experiment is a *tree diagram*. Figure 5.2 shows a tree diagram for the two-coin-tossing experiment described in Example 5.4. The sequence of steps is depicted by moving from left to right through the tree. Step 1 corresponds to tossing the first coin, and there are two branches corresponding to the two possible outcomes. Step 2 corresponds to tossing the second coin, and for each possible outcome at step 1, there are two branches corresponding to the two possible outcomes at step 2. Finally, each of the points on the right-hand end of the tree corresponds to a sample point or experimental outcome. Each path through the tree from the leftmost node to one of the nodes at the right hand side of the tree corresponds to a unique sequence of outcomes for each step.

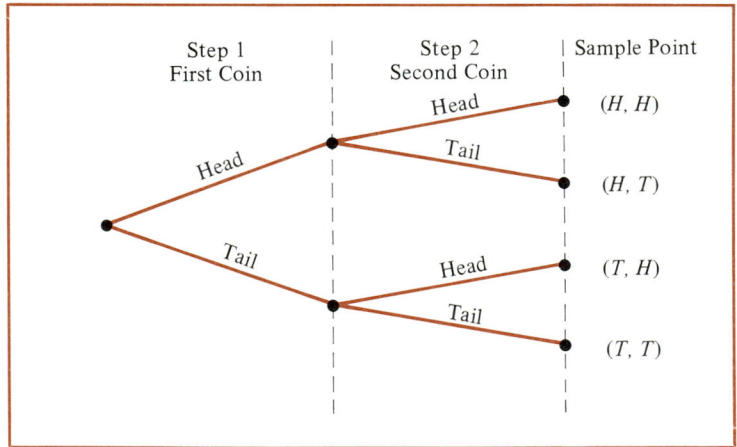

FIGURE 5.2
Tree Diagram for the Experiment of Tossing Two Coins

EXERCISES

1. In a long-term study designed to examine the relationship between heart disease and behavior type, the blood pressure (low, normal, high) was recorded for individuals who have already been identified as having either type A or type B behavior characteristics.
   a. How many experimental outcomes are possible?
   b. Develop a tree diagram for this experiment.

2. An airline has offered a special vacation package to Hawaii. The length of stay is either for 3 days or 7 days, and the type of accommodations can be either economy, regular, or deluxe.
   a. How many experimental outcomes are possible?
   b. Develop a tree diagram for this experiment.

3. A major resort in Florida is concerned about weather conditions in both the Northeast as well as in Florida. In characterizing the temperature in both areas, the following three categories are used: below-average, average, or above-average. Their interest stems from the fact that a combination of below-average temperatures in the Northeast with above-average temperatures in Florida means an increased volume of business.
   a. How many experimental outcomes are possible?
   b. Develop a tree diagram for this experiment.

4. As part of a special new car loan, a savings and loan association offers customers two down-payment options (10% or 20%), as well as two length-of-loan options (36 months or 48 months).
   a. How many experimental outcomes are possible?
   b. Develop a tree diagram for this experiment.

5. A research scientist is experimenting with a new drug that contains varying amounts of two chemicals. If the percentage of the first chemical used is either 1%, 2%, or 3% and the percentage of the second chemical used is either 1%, 2%, 3%, 4% or 5%, how many experimental outcomes are possible.

6. A true-false exam consists of 10 questions. How many answers, or experimental outcomes, are possible?

7. A department store has advertised a special sale for a particular television model at two of its regional warehouses. There are four sets in inventory at warehouse 1 and three sets in inventory at warehouse 2. How many experimental outcomes are possible if the experiment consists of two steps: observing the number of sets sold at warehouse 1 and the number of sets sold at warehouse 2? Develop a tree diagram for this experiment.

8. In the city of Milford applications for zoning changes go through a two-step process: a review by the planning commission and a final decision by the city council. At step 1 the planning commission will review the zoning change request and make a positive or negative recommendation concerning the change. At step 2 the city council will review the planning commission's recommendation and then vote to approve or to disapprove the zoning change. In some instances the city council vote has agreed with the planning commission's recommendation. However, in other instances the council vote has been the opposite of the planning commission's recommendation. An application for a zoning change has just been submitted by the developer of an apartment complex. Consider the application process as an experiment.
   a. How many sample points are there for this experiment? List the sample points.
   b. Construct a tree diagram for the experiment.

9. An investor has two stocks: stock A and stock B. Each stock may increase in value, decrease in value, or remain unchanged. Consider the experiment as the investment in the two stocks.
   a. How many experimental outcomes are possible?
   b. Show a tree diagram for the experiment.
   c. How many of the experimental outcomes result in an increase in value for at least one of the two stocks?
   d. How many of the experimental outcomes result in an increase in value for both of the stocks?

10. Consider the experiment of rolling a pair of dice. Each die has six possible results (the number of dots on its face).
    a. How many sample points are possible for this experiment?
    b. Show a tree diagram for the experiment.
    c. How many experimental outcomes provide a sum of 7 for the dots on the dice?

11. Many states design their automobile license plates such that space is available for up to six letters or numbers.
    a. If a state decides to use only numerical values for the license plates, how many different license plate numbers are possible? Assume that 000000 is an acceptable license plate number, although it will be used only for display purposes at the license bureau. (*Hint:* Use the counting rule.)
    b. If the state decides to use two letters followed by four numbers, how many different license plate numbers are possible? Assume that the letters I and O will not be used because of their similarity to numbers 1 and 0.
    c. Would larger states, such as New York or California tend to use more or fewer letters in license plates? Explain.

5.2 ASSIGNING PROBABILITIES TO EXPERIMENTAL OUTCOMES

We now have an understanding of the concept of an experiment and of the sample space as the set of all experimental outcomes. Now let us see how probabilities for the experimental outcomes (sample points) can be determined. Recall the discussion at the beginning of this chapter. The probability of an experimental outcome was said to be a numerical measure of the likelihood that the experimental outcome would occur. In assigning probabilities to experimental outcomes, there are various acceptable approaches; however, regardless of the approach taken, the following two *basic requirements of probability* must be satisfied.

1. The probability values assigned to each experimental outcome (sample point) must be between 0 and 1, inclusive. That is, if we let E_i indicate an experimental outcome and $P(E_i)$ indicate the probability of the experimental outcome, we must have

$$0 \le P(E_i) \le 1, \qquad \text{for all } i \qquad (5.1)$$

2. The sum of the probabilities for all the experimental outcomes must equal 1. For instance, if a sample space has k sample points (experimental outcomes), we must have

$$P(E_1) + P(E_2) + \cdots + P(E_k) = 1 \qquad (5.2)$$

Any method of assigning probability values to the experimental outcomes that satisfies these two requirements and yields a numerical measure of the likelihood of the outcome is acceptable. In practice, one of the following three methods are used:

1. Classical method
2. Relative frequency method
3. Subjective method

Classical Method

The *classical method* of assigning probabilities is based upon the assumption that the experimental outcomes are all equally likely. In general, if an experiment has n possible outcomes, following the classical approach we would assign a probability of $1/n$ to each outcome.

EXAMPLE 5.2 (continued)

For the experiment of tossing a single coin, the two experimental outcomes—head and tail—are equally likely. Therefore, since one of the two equally likely outcomes is head, the probability of observing a head is $1/2$, or .50. Similarly, the probability of observing tail is also $1/2$, or .50.

· · ·

EXAMPLE 5.3 (continued)

For the experiment of rolling a die it would also seem reasonable to conclude that the six possible outcomes are equally likely, and hence each outcome is assigned a probability of $1/6$. If $P(1)$ denotes the probability that one dot appears on the upward face of the die, then $P(1) = 1/6$. Similarly, $P(2) = 1/6$, $P(3) = 1/6$, $P(4) = 1/6$, $P(5) = 1/6$, and $P(6) = 1/6$.

· · ·

In these examples, the probability assignments shown satisfy the two basic requirements for assigning probabilities. That is, the probabilities assigned to each outcome are between 0 and 1, and the sum of the probabilities is 1. In fact, requirements (5.1) and (5.2) will always be satisfied when the classical method is used since each of the n sample points is assigned a probability of $1/n$.

EXAMPLE 5.4 (continued)

Recall that there are four experimental outcomes for the experiment involving the tossing of two coins. Since it seems reasonable to assume that the four possible experimental outcomes are equally likely, we assign a probability of $1/4$ to each sample point; thus the probability of sample point (H, H) is $\frac{1}{4}$, the probability of sample point (H, T) is $\frac{1}{4}$, the probability of sample point (T, H) is $\frac{1}{4}$, and the probability of sample point (T, T) is $\frac{1}{4}$.

• • •

The classical method was developed originally in the analysis of gambling problems, where the assumption of equally likely outcomes often is reasonable. In many situations, however, this assumption is not valid. Hence, alternative methods of assigning probabilities are required.

Relative Frequency Method

When using the *relative frequency method* of assigning probabilities, the probability of an experimental outcome is defined to be the proportion of the time the experimental outcome occurs when the experiment is repeated a "large" number of times.

EXAMPLE 5.6

As part of a study of the X-ray department for a local hospital, the number of patients waiting for service at 9:00 A.M. was recorded for 20 successive days. The following results were obtained.

Number Waiting	Number of Days Outcome Occurred
0	2
1	5
2	6
3	4
4	3
	Total 20

This data shows that on 2 of the 20 days, 0 patients were waiting for service, on 5 of the days, 1 patient was waiting for service, and so on. Using the relative frequency method, we would thus assign a probability of $\frac{2}{20} = .10$ to the experimental outcome of 0 patients waiting for service, $\frac{5}{20} = .25$ to the experimental outcome of 1 patient waiting, $\frac{6}{20} = .30$ to 2 patients waiting, $\frac{4}{20} = .20$ to 3 patients waiting, and $\frac{3}{20} = .15$ to 4 patients waiting. Note that the two basic requirements for assigning probabilities are satisfied with this probability assignment.

• • •

EXAMPLE 5.7

In the test market evaluation of a new product, 400 potential customers were contacted; 100 actually purchased the product, but 300 did not. In effect, then, we have repeated the experiment of contacting a customer 400 times and have found that the product was purchased 100 times. Thus using the relative frequency approach, we assign a probability of $100/400 = .25$ to the experimental outcome of purchasing the product. Similarly, $300/400 = .75$ is assigned to the experimental outcome of not purchasing the product.

• • •

Subjective Method

When the *subjective method* is used to assign probabilities to the experimental outcomes, we may use any information available, such as our experience, intuition, etc. After considering all available information, a probability value that expresses our degree of belief that the experimental outcome will occur is specified. Since subjective probability expresses a person's degree of belief, it is personal. Using the subjective method, different people can be expected to assign different probabilities to the same experimental outcome.

When using the subjective probability assignment method, care must be taken to ensure that requirements (5.1) and (5.2) are satisfied. That is, regardless of a person's degree of belief, the probability value assigned to each experimental outcome must be between 0 and 1, inclusive, and the sum of all the experimental outcome probabilities must equal 1.

EXAMPLE 5.8

Consider the next football game that the Pittsburgh Steelers will play. What is the probability that the Steelers will win? The experimental outcomes of win, lose, and tie are obviously not equally likely. Also, since the teams involved will not have played several times previously in the same year, there are no relative frequency data available relevant to the game. Thus if we want an estimate of the probability of the Steelers' winning, we must use the subjective method and state a value that expresses our degree of belief that they will win.

• • •

EXAMPLE 5.5 (continued)

Suppose that the subjective method had been used to assign the following probabilities to the six experimental outcomes concerning location and seating capacity for the fast-food franchise in Example 5.5.

Experimental Outcome	Probability
$(M, 100)$	.40
$(M, 150)$	.15
$(M, 200)$	.10
$(E, 100)$	.20
$(E, 150)$	.10
$(E, 200)$	.05
Total	1.00

Since the probability value assigned to each experimental outcome is between 0 and 1 and the sum of all the probabilities is 1, this assignment meets the two requirements for assigning probabilities to the experimental outcomes.

• • •

EXERCISES

12. The president of a college stated that the probability that next year's enrollment will be less than 5000 students is .35, the probability that enrollment will be 5000 but less than 6000 is .25, and the probability that enrollment will exceed 6000 is .50. Comment on the president's statement.

13. The final grades in a course in contemporary science resulted in the following grades.

Grade	A	B	C	D	F
Number	7	12	16	5	3

How can you estimate the probability of a randomly selected student obtaining a specific grade using these data?

14. An investor forecasts that the probabilities that a certain stock will either go down, remain the same, or go up are .20, .60, and .30, respectively. Does this seem reasonable? Explain.

15. Consider the experiment of selecting a card from a deck of 52 cards.
    a. How many sample points are possible?
    b. Which method (classical, relative frequency, or subjective) would you recommend for assigning probabilities to the sample points?
    c. What are the probability assignments?
    d. Show that your probability assignments satisfy the two basic requirements for assigning probabilities.

16. Faced with the question of determining the probability of obtaining either 0 heads, 1 head, or 2 heads when flipping a coin twice, an individual argued that since it seems reasonable to treat the outcomes as equally likely, the probability of each event is $\frac{1}{3}$. Do you agree? Explain.

17. Planes flying from New York City to Chicago are listed as either arriving early, on time, or late. Discuss how you could develop estimates of the probabilities for each of these events.

18. A company that manufactures toothpaste has five different package designs they want to study. Assuming that one design is just as likely to be selected by a consumer as any other design, what probability would you assign to each of the package designs? In an actual experiment, 100 consumers were asked to pick the design they preferred. The following data were obtained.

Design	1	2	3	4	5
Total	5	15	30	40	10

Do the data appear to confirm the belief that one design is just as likely to be selected as another? Explain.

19. A small-appliance store in Madeira has collected data on refrigerator sales for the last 50 weeks.

Number of Refrigerators Sold	Number of Weeks
0	6
1	12
2	15
3	10
4	5
5	2
	50

Suppose that we are interested in the experiment of observing the number of refrigerators sold in 1 week of store operations.
a. How many sample points (experimental outcomes) are there?
b. Which approach would you recommend for assigning probabilities to the sample points?
c. Assign probabilities to the sample points and verify that your assignments satisfy the two basic requirements.

20. Strom Construction has made a bid on two contracts. The owner has identified the possible outcomes and subjectively assigned probabilities as follows:

Experimental Outcome	Obtain Contract 1	Obtain Contract 2	Probability
1	Yes	Yes	.15
2	Yes	No	.15
3	No	Yes	.30
4	No	No	.25

a. Are these valid probability assignments? Why or why not?
b. What would have to be done to make the probability assignments valid?

5.3 EVENTS AND THEIR PROBABILITIES

In the introduction to this chapter we used the term *event* much as it would be used in everyday language. However, at this point we introduce the formal definition of an event as it relates to probability.

Event

An *event* is a collection of sample points.

EXAMPLE 5.3 (continued)

Recall that for the experiment of rolling a die, the sample space was $S = \{1, 2, 3, 4, 5, 6\}$. If we define E to be the event that an even number of dots appear on the upward face of the die, then we can describe event E as follows:

$$E = \{2, 4, 6\}$$

Similarly, if O is the event that an odd number of dots appear on the upward face of the die, then

$$O = \{1, 3, 5\}$$

We see that events E and O are simply different collections of the sample points.

• • •

EXAMPLE 5.9

A cab company has analyzed its operating records for the past 20 days. On 8 of these days, no vehicle breakdowns were observed; on 6 of the days one cab had a breakdown; on 3 days there were 2 breakdowns; on 2 days there were 3 breakdowns; and on 1 of the days 4 cabs had breakdowns. Let

$$S = \{0, 1, 2, 3, 4\}$$

denote the sample space for the experiment of observing the number of cab breakdowns on a day. The numerical values 0, 1, 2, 3, and 4 denote the number of breakdowns (the experimental outcomes). If A is defined as the event that 2 or more vehicle breakdowns are observed on a typical day, then

$$A = \{2, 3, 4\}$$

If B is defined as the event that less than 2 breakdowns are observed, then

$$B = \{0, 1\}$$

Similarly, if C is defined as the event that no breakdowns are observed, then

$$C = \{0\}$$

We see that the event C consists of just one sample point.

• • •

Given the probabilities of the sample points (experimental outcomes), we can use the following definition to compute the probability of any event.

Probability of an Event

The probability of any event is equal to the sum of the probabilities of the sample points in the event.

Using this definition, we calculate the probability of a particular event by adding the probabilities of the experimental outcomes that make up the event.

EXAMPLE 5.3 (continued)

Recall that for the experiment of rolling a die, we used the classical approach to assign probabilities of $P(1) = \frac{1}{6}, P(2) = \frac{1}{6}, \ldots, P(6) = \frac{1}{6}$. Thus to compute the probability of event $E = \{2, 4, 6\}$, we sum the probabilities of the sample points 2, 4, and 6:

$$P(E) = P(2) + P(4) + P(6)$$
$$= \frac{1}{6} + \frac{1}{6} + \frac{1}{6} = \frac{3}{6} = \frac{1}{2}$$

Similarly, for $O = \{1, 3, 5\}$,

$$P(O) = P(1) + P(3) + P(5)$$
$$= \frac{1}{6} + \frac{1}{6} + \frac{1}{6} = \frac{3}{6} = \frac{1}{2}$$

• • •

EXAMPLE 5.9 (continued)

Using the relative frequency method, we can use the data provided to estimate the probability of a specific number of breakdowns. The results are shown below.

Number of Breakdowns	Number of Occurrences	Probability
0	8	8/20 = .40
1	6	6/20 = .30
2	3	3/20 = .15
3	2	2/20 = .10
4	1	1/20 = .05
Totals	20	1.00

Thus the probability of event $A = \{2, 3, 4\}$ is given by

$$P(A) = P(2) + P(3) + P(4)$$
$$= .15 + .10 + .05 = .30$$

Similarly, the probability of event $B = \{0, 1\}$ is

$$P(B) = P(0) + P(1)$$
$$= .40 + .30 = .70$$

Finally, we see that for event $C = \{0\}$, $P(C) = P(0) = .40$.

• • •

Any time we can identify all the sample points of an experiment and assign the corresponding probabilities, we can use the definition of this section to compute the probability of an event of interest. However, in many experiments the number of sample points is large, and the identification of the sample points—as well as determining their associated probabilities—becomes extremely cumbersome, if not impossible. In the remaining sections of this chapter we present probability relationships and rules that can be used to compute the probability of an event without requiring knowledge of individual sample point probabilities. These probability relationships require a knowledge of the probabilities for some events in the experiment. Probabilities of other events are then computed directly from the known event probabilities using one or more of the probability relationships.

EXERCISES

21. Suppose that a manager of a large apartment complex provides the following subjective probability estimate about the number of vacancies that will exist next month.

Vacancies	Probability
0	.05
1	.15
2	.35
3	.25
4	.10
5	.10

List the sample points in each of the following events and provide the probability of each event.
   a. no vacancies
   b. at least four vacancies
   c. two or fewer vacancies

22. Consider the experiment of rolling a pair of dice. Suppose that we are interested in the sum of the face values showing on the dice.
  a. How many sample points are possible? (*Hint:* Use the counting rule.)
  b. List the sample points.
  c. What is the probability of obtaining a value of 7?
  d. What is the probability of obtaining a value of 9 or greater?
  e. Since there are six possible even values (2, 4, 6, 8, 10, and 12) and only five possible odd values (3, 5, 7, 9, and 11), the dice should show even values more often than odd values. Do you agree with this statement? Explain.
  f. What method did you use to assign the probabilities shown above?

23. A sample of 100 customers of Montana Gas and Electric resulted in the following frequency distribution of monthly charges.

Amount $	Number
0–49	13
50–99	22
100–149	34
150–199	26
200–249	5

  a. Let A be the event that monthly charges are $150 or more. Find $P(A)$.
  b. Let B be the event that monthly charges are less than $150. Find $P(B)$.

24. A survey of 50 students at Tarpon Springs College regarding the number of extracurricular activities resulted in the following data.

Number of activities	0	1	2	3	4	5
Frequency	8	20	12	6	3	1

  a. Let A be the event that a student participates in at least 1 activity. Find $P(A)$.
  b. Let B be the event that a student participates in 3 or more activities. Find $P(B)$.
  c. What is the probability a student participates in exactly 2 activities?

25. A marketing manager is attempting to assign probability values to the possible profits and losses resulting from a new product. Relying on subjective probabilities, the manager's probability estimates are as follows.

$$P(\text{profit over } \$10,000) = .25$$

$$P(\text{profit from } \$0 \text{ to } \$10,000) = .50$$

$$P(\text{loss}) = .15$$

What advice would you offer before the manager uses these estimates to perform further probability calculations?

26. A telephone survey was used to determine viewer response to a new television show. The following data were obtained.

Rating	Frequency
Poor	4
Below average	8
Average	11
Above average	14
Excellent	13

a. What is the probability that a randomly selected viewer rates the new show as average or better?

b. What is the probability that a randomly selected viewer rates the new show below average or worse?

27. A bank has observed that credit card account balances have been growing over the past year. A sample of 200 customer accounts resulted in the following data.

Amount Owed $	Frequency
0–99	62
100–199	46
200–299	24
300–399	30
400–499	26
500–599	12

a. Let A be the event that a customer's balance is less than $200. Find $P(A)$.

b. Let B be the event that a customer's balance is $300 or more. Find $P(B)$.

28. The manager of a furniture store sells from 0 to 4 china hutches each week. Based on past experience, the following probabilities are assigned to sales of 0, 1, 2, 3, or 4 hutches.

$$P(0) = .08$$

$$P(1) = .18$$

$$P(2) = .32$$
$$P(3) = .30$$
$$P(4) = \underline{.12}$$
$$1.00$$

a. Are these valid probability assignments? Why or why not?
b. Let A be the event that two or fewer are sold in one week. Find $P(A)$.
c. Let B be the event that four or more are sold in one week. Find $P(B)$.

5.4 INTRODUCTORY PROBABILITY RELATIONSHIPS

Complement of an Event

Given an event A, the *complement* of A is defined to be the event consisting of all sample points that are *not* in A. The complement of A is denoted by $\overline{A}$. Figure 5.3 provides a diagram known as a *Venn diagram*, which illustrates the concept of a complement. The rectangular area represents the sample space for the experiment and, as such, contains all possible sample points. The circle represents event A and contains only the sample points that belong to A. The shaded region of the diagram contains all sample points not in event A, which is the definition of the complement of A, $\overline{A}$.

In any probability application, event A and its complement $\overline{A}$ must satisfy

$$P(A) + P(\overline{A}) = 1 \tag{5.3}$$

Solving for $P(A)$, we obtain the following result:

Computing Probability Using the Complement

$$P(A) = 1 - P(\overline{A}) \tag{5.4}$$

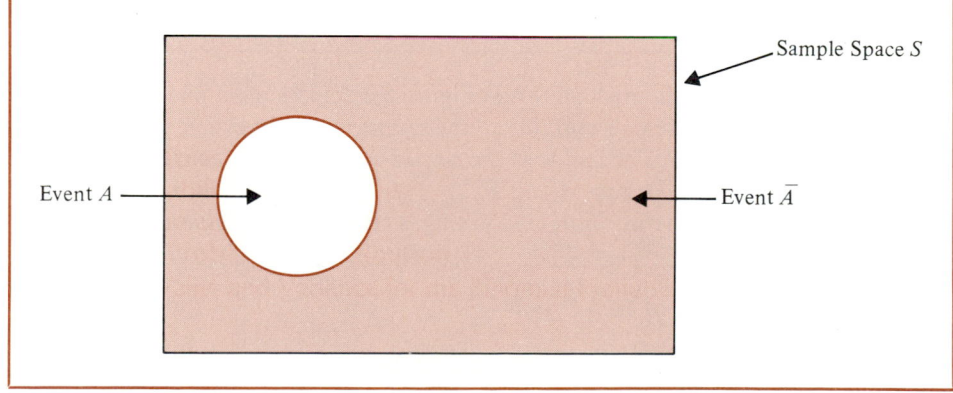

FIGURE 5.3
Complement of Event A

Equation (5.4) shows that the probability of an event A can easily be computed if the probability of its complement, $P(\overline{A})$, is known.

EXAMPLE 5.10

Based upon an analysis of student records, the placement director at a university states that 98% of the students who interview with a particular firm are not given a job offer. Letting A denote the event of a job offer and $\overline{A}$ denote the event of no job offer, the placement director is stating that $P(\overline{A}) = .98$. Using Equation (5.4), we see that

$$P(A) = 1 - P(\overline{A}) = 1 - .98 = .02$$

This shows that there is a .02 probability that a student who interviews with the firm will receive a job offer.

• • •

EXAMPLE 5.11

A purchasing agent states that there is a .90 probability that a supplier will send a shipment that is free of defective parts. Using the complement, we can conclude that there is a $1 - .90 = .10$ probability that the shipment will contain at least one defective part.

• • •

Union and Intersection of Events

Given two or more events, we often are interested in more complex events obtained by combining two or more of the original events. To begin with let us consider the concept referred to as the *union* of two events.

Union of Two Events

Given two events A and B, the *union of A and B* is the event containing all sample points belonging to *A or B or both.* The union is denoted by $A \cup B$.

The Venn diagram shown in Figure 5.4 depicts the union of events A and B. Note that the shaded region contains all the sample points in event A as well as all the sample points in event B. The fact that the circles overlap indicates that there are some sample points contained in both A and B.

Another way of combining events is referred to as the *intersection*.

Intersection of Two Events

Given two events A and B, the *intersection of A and B* is the event containing the sample points belonging to *both A and B.* The intersection is denoted by $A \cap B$.

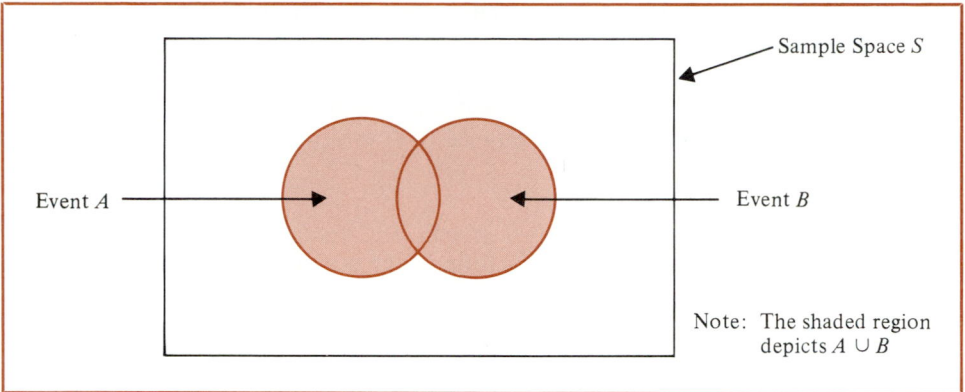

FIGURE 5.4
Union of Events *A* and *B*

The Venn diagram depicting the intersection of the events *A* and *B* is shown in Figure 5.5. The area where the two circles overlap is the intersection; it contains the sample points that are in both *A* and *B*.

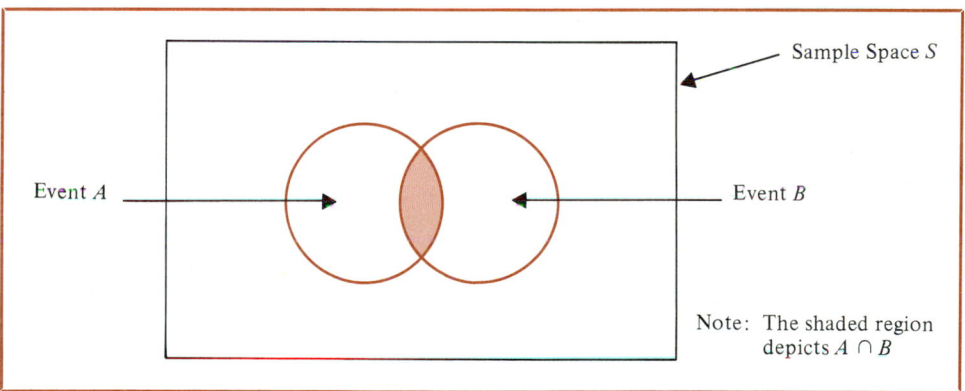

FIGURE 5.5
Intersection of Events *A* and *B*

Addition Rule

The *addition rule* is a helpful probability relationship when we have two events and are interested in knowing the probability that at least one of the events occurs. That is, with events *A* and *B*, we are interested in knowing the probability that event *A* or event *B* or both occur. In other words, the addition rule is used to compute the probability of the union of two events, $A \cup B$.

Addition Rule

$$P(A \cup B) = P(A) + P(B) - P(A \cap B) \qquad (5.5)$$

To obtain an intuitive understanding of the addition rule, note that the first two terms in the addition rule, $P(A) + P(B)$, account for all the sample points in A and all the sample points in B. However, since the sample points in the intersection, $A \cap B$, are in both A and B, when we compute $P(A) + P(B)$ we are in effect counting each of the sample points in $A \cap B$ twice. We correct for this by subtracting $P(A \cap B)$.

EXAMPLE 5.12

As an example of the application of the addition rule, consider the following grades obtained in an introductory psychology course. Of 200 students taking the course, 160 passed the midterm exam and 140 passed the final exam; 124 students passed both exams. Letting

$$M = \text{event of passing the midterm exam}$$

$$F = \text{event of passing the final exam}$$

the given relative frequency information leads to the following probabilities:

$$P(M) = \frac{160}{200} = .80$$

$$P(F) = \frac{140}{200} = .70$$

$$P(M \cap F) = \frac{124}{200} = .62$$

After reviewing the grades, the professor of the course decides to give a passing grade to any student who passed at least one of the two exams. That is, a passing grade will be given to any student who passes the midterm, to any student who passes the final, and to any student who passes both exams. What is the probability of receiving a passing grade in this course?

While your first reaction may be to try to count how many of the 200 students passed at least one exam, that information is not available; even if it was, the counting process would be tedious. However, note that the question concerns the union of the events M and F. That is, we want to know the probability a student passes the midterm (M), passes the final (F), or both. Thus we want to know $P(M \cup F)$. Here is where the basic relationships of probability can be helpful. Using the addition rule (5.5) for the events M and F, we have

$$P(M \cup F) = P(M) + P(F) - P(M \cap F)$$

Knowing the three probabilities on the right-hand side of the above expression, we can write

$$P(M \cup F) = .80 + .70 - .62 = .88$$

This tells us there is a .88 probability of passing the course because there is a .88 probability of passing at least one of the exams.

• • •

EXAMPLE 5.13

Consider a study involving the television-viewing habits of married couples. It was found that 30% of the husbands and 20% of the wives were regular viewers of a particular Friday evening program. For 12% of the couples in the study, both husband and wife were regular viewers of the program. What is the probability that at least one of the family members is a regular viewer of the program?

Letting

$$H = \text{husband is a regular viewer}$$

$$W = \text{wife is a regular viewer}$$

we have $P(H) = .30$, $P(W) = .20$, and $P(H \cap W) = .12$. Using the addition rule, we have

$$P(H \cup W) = P(H) + P(W) - P(H \cap W) = .30 + .20 - .12 = .38$$

This shows there is a .38 probability that at least one of the spouses is a regular viewer of the program.

• • •

Let us now see how the addition rule is applied to *mutually exclusive events*. First, we define mutually exclusive events.

Mutually Exclusive Events

Two or more events are said to be *mutually exclusive* if the events do not have any sample points in common.

That is, events A and B are mutually exclusive if when one event occurs the other cannot occur. Thus a requirement for A and B to be mutually exclusive is that their intersection must contain no sample points. A Venn diagram for the mutually exclusive events A and B

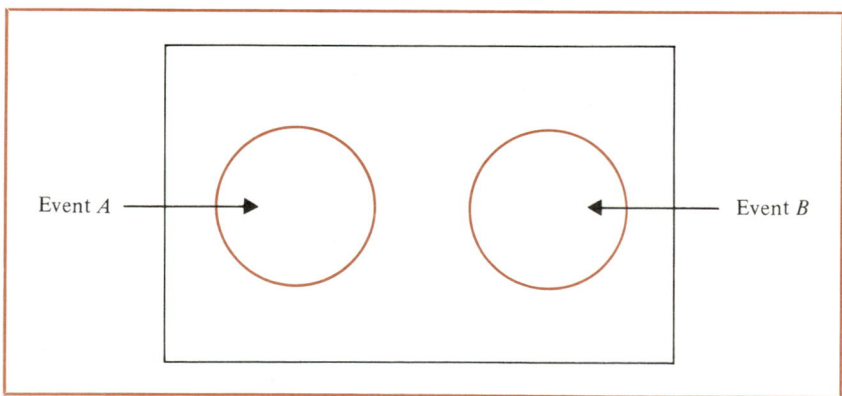

FIGURE 5.6
Mutually Exclusive Events

is shown in Figure 5.6. In the case of mutually exclusive events, $P(A \cap B) = 0$, and hence the addition rule can be shortened to:

Addition Rule for Mutually Exclusive Events

$$P(A \cup B) = P(A) + P(B) \tag{5.6}$$

EXAMPLE 5.14

If A denotes the event that you get a grade of A for a term paper and B denotes the event that you get a grade of B, then A and B are mutually exclusive events. Thus if $P(A) = .20$ and $P(B) = .50$, then

$$P(A \cup B) = P(A) + P(B)$$
$$= .20 + .50 = .70$$

• • •

We note that experimental outcomes are actually mutually exclusive events. Each experimental outcome corresponds to a single sample point and no two experimental outcomes correspond to the same sample point. Thus no two experimental outcomes may have any sample points in common.

EXERCISES

29. Suppose that we have a sample space $S = \{E_1, E_2, E_3, E_4, E_5, E_6, E_7\}$, where E_1, $E_2, \ldots E_7$ denotes the sample points, and the following probability assignments for the sample points.

$$P(E_1) = .05$$

$$P(E_2) = .20$$

$$P(E_3) = .20$$

$$P(E_4) = .25$$

$$P(E_5) = .15$$

$$P(E_6) = .10$$

$$P(E_7) = \underline{.05}$$
$$\text{Total} \quad 1.00$$

Let

$$A = \{E_1, E_4, E_6\}$$

$$B = \{E_2, E_4, E_7\}$$

$$C = \{E_2, E_3, E_5, E_7\}$$

a. Find $P(A)$, $P(B)$, and $P(C)$.
b. Find $A \cup B$ and $P(A \cup B)$.
c. Find $A \cap B$ and $P(A \cap B)$.
d. Are events A and C mutually exclusive?
e. Find $\bar{B}$ and $P(\bar{B})$.

30. An automotive store sells from zero to four car batteries each week. Based on past experience, the following probabilities are assigned to sales of zero, one, two, three, or four batteries.

$$P(0) = .19$$

$$P(1) = .24$$

$$P(2) = .35$$

$$P(3) = .14$$

$$\begin{array}{ll} P(4) = & .08 \\ \text{Total} & 1.00 \end{array}$$

a. Are these valid probability assignments? Why or why not?
b. Let A be the event that two or fewer are sold in one week. Find $P(A)$.
c. Let B be the event that four or more are sold in one week. Find $P(B)$.
d. Are A and B mutually exclusive? Find $P(A \cap B)$ and $P(A \cup B)$.

31. Let

$$A = \text{the event that a person runs 5 miles or more per week}$$

$$B = \text{the event that a person dies of heart disease}$$

$$C = \text{the event that a person dies of cancer}$$

Further, suppose that $P(A) = .01$, $P(B) = .25$, and $P(C) = .20$.
a. Are events A and B mutually exclusive? Can you find $P(A \cap B)$?
b. Are events B and C mutually exclusive? Find the probability that a person dies of heart disease or cancer.
c. Find the probability that a person dies from causes other than cancer.

32. In a study conducted to evaluate the effect of an allergy relief medicine, 250 patients with symptoms that included itchy eyes and a skin rash were given the new drug. The results of the study are as follows: 90 of the patients treated experienced eye relief, 135 had their skin rash clear up, and 45 experienced both relief from itchy eyes and the skin rash. What is the probability that a patient that takes the drug will experience relief for at least one of the two symptoms?

33. In a study of 100 students that had been awarded university scholarships, it was found that 40 had part-time jobs, 25 had made the dean's list the previous semester, and 15 had both a part-time job and had made the dean's list. What was the probability that a student had a part-time job or was on the dean's list?

34. During winter in Cincinnati, Mr. Krebs experiences difficulty in starting his two cars. The probability that the first car starts is .80, and the probability that the second car starts is .40. There is a probability of .30 that both cars start.
   a. Define the events involved and use probability notation to show the probability information given above.
   b. What is the probability that at least one car starts?
   c. What is the probability that Mr. Krebs cannot start either of the two cars?

35. Consider an experiment where eight possible outcomes exist. We will denote the experimental outcomes as $E_1, E_2, \ldots, E_8$. Suppose the following events are defined.

$$A = (E_1, E_2, E_3, E_5)$$

$$B = (E_2, E_4, E_5, E_8)$$

Note that experimental outcomes E_6 and E_7 are in neither event A nor B. List the sample points making up the following events:
   a. $A \cup B$
   b. $A \cap B$
   c. $\overline{A}$
   d. Are A and B mutually exclusive events? Explain.

36. Let A be an event that a person's primary method of transportation to and from work is an automobile and B be an event that a person's primary method of transportation to and from work is a bus. Suppose that in a large city we find $P(A) = .45$ and $P(B) = .35$.
   a. Are events A and B mutually exclusive? What is the probability that a person uses an automobile or a bus in going to and from work?
   b. Find the probability that a person's primary method of transportation is something other than a bus.

5.5 CONDITIONAL PROBABILITY, INDEPENDENCE, AND THE MULTIPLICATION RULE

Conditional Probability

In many probability situations it is important to be able to determine the probability of one event *given* that another related event is known to have occurred, which is called the *conditional probability* of an event.

EXAMPLE 5.15

A major metropolitan police force in the Eastern United States consists of 1200 officers—960 men and 240 women. Over the past 2 years, 324 officers on the police force have been awarded promotions. The breakdown of promotions for male and female officers is shown in Table 5.1.

   After reviewing the data in Table 5.1, a committee of female officers charged discrimination on the basis that 288 male officers had received promotions, whereas only 36 female officers had received promotions. The police administration countered with the argument that the relatively low number of promotions for female officers was not due to discrimination but due to the fact that there are fewer female officers on the police force.

   After reflecting on this situation, we see that the real issue involves not the number of promotions but the probabilities of promotion given the officer is male and given the officer is female.

•   •   •

   In dealing with conditional probabilities, our interest is in computing probabilities that can be used to answer questions such as those raised in Example 5.15. Suppose that we have an event A with probability denoted by $P(A)$. If we should learn that a related event, B, has occurred, we would want to take advantage of this additional information in computing the probability for event A.

TABLE 5.1

**Promotional Status of Police
Officers Over the Past 2 Years**

	Men	Women	Total
Promoted	288	36	324
Not Promoted	672	204	876
Totals	960	240	1200

The probability of event *A given* that event *B* is known to have occurred is written $P(A|B)$. The vertical line, |, between *A* and *B* is used to denote the fact that we are considering the probability of event *A given* the condition that event *B* has occurred. The notation $P(A|B)$ is read "the probability of *A given B*."

EXAMPLE 5.15 (continued)

Returning to the example concerned with the police department discrimination case we let

$$M = \text{event that a randomly selected officer is a man}$$

$$W = \text{event that a randomly selected officer is a woman}$$

$$A = \text{event that a randomly selected officer is promoted}$$

The conditional probabilities of relevance in the discrimination charge are $P(A|M)$, the probability that an officer is promoted given the officer is a man, and $P(A|W)$, the probability that an officer is promoted given the officer is a woman. Referring to Table 5.1, we can compute these probabilities using the relative frequency approach.

$$P(A|M) = \frac{288}{960} = .30$$

$$P(A|W) = \frac{36}{240} = .15$$

From these calculations we see that the probability of promotion is twice as great for male officers. These calculations do not necessarily prove that the female officers have been discriminated against, but they do support the female officers' argument.

• • •

In Example 5.15 we were able to estimate the conditional probabilities directly from a table showing frequency of occurrence for the event. In many cases this information is not available and we must compute conditional probabilities from other probability information available concerning the event. In such cases the following definition of conditional probability is useful.

Conditional Probability

$$P(A|B) = \frac{P(A \cap B)}{P(B)} \tag{5.7}$$

or

$$P(B|A) = \frac{P(A \cap B)}{P(A)} \tag{5.8}$$

EXAMPLE 5.15 (continued)

Dividing the data values in Table 5.1 by the total of 1200 officers permits us to summarize the available information in the following probability values:

$$P(M \cap A) = \frac{288}{1200} = .24 = \text{probability that an officer is a man } \textit{and} \text{ is promoted}$$

$$P(M \cap \overline{A}) = \frac{672}{1200} = .56 = \text{probability that an officer is a man } \textit{and} \text{ is not promoted}$$

$$P(W \cap A) = \frac{36}{1200} = .03 = \text{probability that an officer is a woman } \textit{and} \text{ is promoted}$$

$$P(W \cap \overline{A}) = \frac{204}{1200} = .17 = \text{probability that an officer is a woman } \textit{and} \text{ is not promoted}$$

Since each of these values gives the probability of the intersection of two events, the probabilities are given the name of *joint probabilities*. Table 5.2, which provides a summary of the probability information for the police officer promotion situation, is referred to as a *joint probability table*.

The values in the margins of the joint probability table provide the probabilities of each event separately. That is, $P(M) = .80$, $P(W) = .20$, $P(A) = .27$, and $P(\overline{A}) = .73$. Thus we see that 80% of the force is male, 20% of the force is female, 27% of all officers received promotions, and 73% were not promoted. These probabilities are referred to as *marginal probabilities* because of their location in the margins of the joint probability table.

TABLE 5.2

Joint Probability Table for Promotion of Police Officers

Joint probabilities appear in the body of the table.

	Men (M)	Women (W)	Total
Promoted (A)	.24	.03	.27
Not Promoted ($\overline{A}$)	.56	.17	.73
Totals	.80	.20	1.00

Marginal probabilities appear in the margins of the table.

Using these joint and marginal probabilities, we can apply the definition of conditional probability to find the probability of promotion given a male officer and the probability of promotion given a female officer.

$$P(A|M) = \frac{P(A \cap M)}{P(M)} = \frac{.24}{.80} = .30$$

$$P(A|W) = \frac{P(A \cap W)}{P(W)} = \frac{.03}{.20} = .15$$

•　•　•

EXAMPLE 5.16

A research study concerning the relationship between smoking and heart disease in men over 50 years old led to the finding that 10% of the men smoked and had experienced heart disease. Furthermore, it was known that 30% of the men in the study were smokers. Let

$$H = \text{has experienced heart disease}$$

$$S = \text{smoker}$$

Using Equation (5.7) we can compute the conditional probability of a man over 50 experiencing heart disease given that he smokes.

$$P(H|S) = \frac{P(H \cap S)}{P(S)} = \frac{.10}{.30} = .33$$

•　•　•

Independent Events

In Example 5.15 involving promotional practices for male and female police officers, we saw that $P(A) = .27$, $P(A|M) = .30$, and $P(A|W) = .15$. As you will recall, the events were defined as follows: A = promotion, M = male officer, and W = female officer. This data shows that the probability of a promotion (event A) is affected or influenced by whether the officer is male or female. In particular, since $P(A|M) \neq P(A)$, we say events A and M are *dependent* events. That is, the probability of event A (promotion) is altered or affected by knowing whether or not M (the officer is male) occurs. Similarly, with $P(A|W) \neq P(A)$, we say events A and W are *dependent* events. On the other hand, if the probability of event A was not changed by the occurrence of event M, that is, $P(A|M) = P(A)$, we say events A and M are *independent* events. This leads us to the following definition of the independence of two events.

Independent Events

Two events A and B are independent if

$$P(A|B) = P(A) \qquad (5.9)$$

or

$$P(B|A) = P(B) \qquad (5.10)$$

Otherwise, the events are dependent.

EXAMPLE 5.17

Suppose that $P(A) = .30$, $P(B) = .25$, and $P(A|B) = .20$. Are events A and B independent? Since $P(A|B) \neq P(A)$, the events are dependent. Note that $P(B)$ was not even needed to answer the question.

$$\bullet \quad \bullet \quad \bullet$$

Multiplication Rule

Recall that the addition rule of probability is used to compute the probability of a union of two events. We now show how the *multiplication rule* can be used to find the probability of an intersection of two events (ie., the joint probability of the two events). The multiplication rule is based upon the definition of conditional probability. Using (5.7) and (5.8) and solving for $P(A \cap B)$, we obtain the multiplication rule:

Multiplication Rule

$$P(A \cap B) = P(B) \, P(A|B) \qquad (5.11)$$

or

$$P(A \cap B) = P(A) \, P(B|A) \qquad (5.12)$$

EXAMPLE 5.18

A newspaper circulation department knows that 84% of its customers subscribe to the daily edition of the paper. Letting D denote the event that a customer subscribes to the daily edition, we have $P(D) = .84$. In addition, it is known that the probability that a customer already holding a daily subscription also subscribes to the Sunday edition (event S) is .75; that is, $P(S|D) = .75$. What is the probability that a customer subscribes to both the daily and Sunday editions of the newspaper? Using the multiplication rule, we compute the desired value, $P(D \cap S)$, as follows:

$$P(D \cap S) = P(S|D)P(D) = .75(.84) = .63$$

This tells us that 63% of the newspaper's customers take both the daily and Sunday editions.

• • •

Before concluding this section, let us consider the special case of the multiplication rule when the events involved are independent. Recall earlier in this section we said events A and B were independent whenever $P(A|B) = P(A)$ or $P(B|A) = P(B)$. Hence, using (5.11) and (5.12) for the special case of independent events, the multiplication rule becomes the following.

Multiplication Rule for Independent Events

$$P(A \cap B) = P(A)P(B) \qquad (5.13)$$

Thus to compute the probability of the intersection of two independent events, we simple multiply the corresponding probabilities. The multiplication rule for independent events provides another way to determine if A and B are independent. That is, if $P(A \cap B) = P(A)P(B)$, then A and B are independent; if $P(A \cap B) \neq P(A)P(B)$, then A and B are dependent.

EXAMPLE 5.19

A service station manager knows from past experience that 80% of the customers use a credit card when purchasing gasoline. What is the probability that the next two customers purchasing gasoline will both use credit cards? If we let

A = event that the first customer uses a credit card

B = event that the second customer uses a credit card

then the event of interest is $A \cap B$. It seems reasonable to assume that A and B are independent events. Thus $P(A \cap B) = P(A)P(B) = (.80)(.80) = .64$.

• • •

EXERCISES

37. A Daytona Beach nightclub has the following data on the age and marital status of 140 customers.

		Marital Status	
		Single	Married
Age	Under 30	77	14
	30 or Over	28	21

a. Develop a joint probability table using the given data.

b. Use the marginal probabilities to comment on the ages of customers attending the club.

c. Use the marginal probabilities to comment on the marital status of customers attending the club.

d. What is the probability of finding a customer who is single and under the age of 30?

e. If a customer is under 30, what is the probability that he or she is single?

f. Is marital status independent of age? Explain, using probabilities.

38. A survey of automobile ownership was conducted for 200 families in Houston. The results of the study showing ownership of automobiles of United States and foreign manufacture are summarized in the following table.

		Do You Own a U.S. Car?		
		Yes	No	Totals
Do you own a foreign car?	Yes	30	10	40
	No	150	10	160
	Totals	180	20	200

a. Show the joint probability table for the given data.

b. Use the marginal probabilities to compare U.S. and foreign car ownership.

c. What is the probability that a family will own both a U.S. car and a foreign car?

d. What is the probability that a family owns a car, U.S. or foreign?

e. If a family owns a U.S. car, what is the probability that it also owns a foreign car?

f. If a family owns a foreign car, what is the probability that it also owns a U.S. car?

g. Are U.S. and foreign car ownership independent events? Explain.

39. The probability that Ms. Smith will get an offer on the first job she applies for is .5, and the probability that she will get an offer on the second job she applies for is .6. The probability that she will get an offer on both jobs is .15.
    a. Define the events involved, and use probability notation to show the probability information given above.
    b. What is the probability that Ms. Smith gets an offer on the second job given that she receives an offer for the first job?
    c. What is the probability that Ms. Smith gets an offer on at least one of the jobs she applies for?
    d. What is the probability that Ms. Smith does not get an offer on either of the two jobs she applies for?
    e. Are the job offers independent? Explain.

40. Shown are data from a sample of 80 families in a midwestern city. The data shows the record of college attendance by fathers and their oldest sons.

		Son	
		Attended College	Did Not Attend College
Father	Attended College	18	7
	Did Not Attend College	22	33

    a. Show the joint probability table.
    b. Use the marginal probabilities to comment on the comparison between fathers and sons in terms of attending college.
    c. What is the probability that a son attends college given that his father attended college?
    d. What is the probability that a son attends college given that his father did not attend college?
    e. Is attending college by the son independent of whether or not his father attended college? Explain, using probability values.

41. The Texas Oil Company provides a limited partnership arrangement whereby small investors can pool resources in order to invest in large scale oil exploration programs. In the exploratory drilling phase, locations for new wells are selected based on the geologic structure of the proposed drilling sites. Experience shows that there is a .40 probability of a type A structure present at the site given a productive well. It is also known that 50% of all wells are drilled in locations with type A structure. Finally, 30% of all wells drilled are productive.
    a. What is the probabilty of a well being drilled in a type A structure *and* being productive?
    b. If the drilling process begins in a location with a type A structure, what is the probability of having a productive well at the location?
    c. Is finding a productive well independent of the type A geologic structure? Explain.

42. In a study involving a manufacturing process, 10% of all parts tested were defective, and 30% of all parts were produced on machine A. Given that a part was produced on machine A, there is a .15 probability that it is defective.

    a. What is the probability that a part tested is both defective and produced by machine A?

    b. If a part is found to be defective, what is the probability that it came from machine A?

    c. Is finding a defective part independent of its being produced on machine A? Explain.

    d. What is the probability of the part being either defective or produced by machine A?

    e. Are the events "a defective part" and "produced by machine A" mutually exclusive events? Explain.

43. Assume that we have two events, A and B, which are mutually exclusive. Assume further that it is known that $P(A) = .30$ and $P(B) = .40$.

    a. What is $P(A \cap B)$?

    b. What is $P(A \mid B)$?

    c. A student in statistics argues that the concepts of mutually exclusive events and independent events are really the same and that if events are mutually exclusive they must be independent. Do you agree with this statement? Use the probability information in this problem to justify your answer.

    d. What general conclusion would you make about mutually exclusive and independent events given the results of this problem?

44. A hospital has placed two rush orders for a particular drug from two different suppliers A and B. If neither order arrives in 4 days, a research project must be stopped until at least one of the orders arrives. The probability that supplier A can deliver the material in 4 days is .55. The probability that supplier B can deliver the material in 4 days is .35.

    a. What is the probability that both suppliers deliver the material in 4 days? Since two separate suppliers are involved, we are willing to assume independence.

    b. What is the probability that at least one supplier delivers the material in 4 days?

    c. What is the probability the research project is shut down in 4 days because of a shortage in raw material (i.e., both orders are late)?

5.6 BAYES' THEOREM

In the discussion of conditional probability, we indicated that it was possible to revise or update probabilities given new information. Often, we begin our probability analysis with initial, or *prior, probability* estimates for specific events of interest. Then, from sources such as a sample, a special report, a product test, and so on, we obtain additional information affecting the probability of the events. Given this new information, we want to revise or update the prior probability values. The updated, or revised, probabilities for the events are referred to as *posterior* probabilities. *Bayes' theorem* provides a means for computing posterior probabilities. The steps of the probability revision process are shown in Figure 5.7.

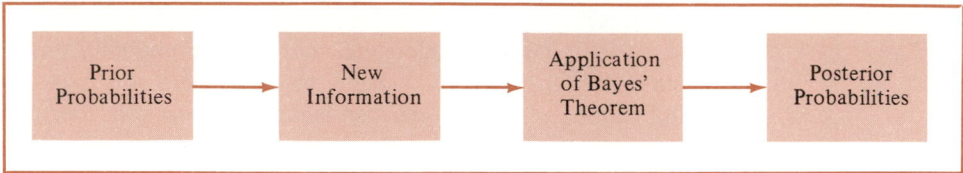

FIGURE 5.7
Probability Revision Using Bayes' Theorem

EXAMPLE 5.20

A manufacturing firm receives 65% of its parts from one supplier and 35% from a second supplier. The quality of the purchased parts varies with the supplier. Table 5.3 shows the percentages of good and bad parts received from the two suppliers. Let A_1 denote the event a part comes from supplier 1 and A_2 the event that a part comes from supplier 2. If we let G denote the event that a part is good and B denote the event that a part is defective, or bad, the information in Table 5.3 leads to the following conditional probability values:

$$P(G \mid A_1) = .98 \qquad P(B \mid A_1) = .02$$
$$P(G \mid A_2) = .95 \qquad P(B \mid A_2) = .05$$

Furthermore, given the percentages of parts received from each supplier, if a part is selected at random, the probability it came from supplier 1 is $P(A_1) = .65$ and the probability it came from supplier 2 is $P(A_2) = .35$.

Suppose that the manufacturing firm has just had a machine breakdown due to a faulty part and wants to determine the probability that the part came from supplier 1 and the probability that the part came from supplier 2. Thus it is desired to compute $P(A_1 \mid B)$ and $P(A_2 \mid B)$. Since $P(A_1 \cap B)$ and $P(A_2 \cap B)$ are not known, we cannot use the conditional probability law for these calculations. This is a problem in which Bayes' theorem is needed; we return to make the calculations shortly.

$$\bullet \quad \bullet \quad \bullet$$

TABLE 5.3

**Percentages of Good and Bad
Parts Received from Suppliers**

	Percentage Good Parts	Percentage Bad Parts
Supplier 1	98%	2%
Supplier 2	95%	5%

Suppose we have two mutually exclusive events A_1 and A_2 that represent the only possible outcomes in a situation. Furthermore, suppose that we know the values of the prior probabilities $P(A_1)$ and $P(A_2)$; moreover, it is also known that a related event B has occurred. The following formulas, known as Bayes' theorem, can be used to compute the posterior probabilities $P(A_1 \mid B)$ and $P(A_2 \mid B)$.

Bayes' Theorem

$$P(A_1 \mid B) = \frac{P(B \mid A_1)P(A_1)}{P(B \mid A_1)P(A_1) + P(B \mid A_2)P(A_2)} \tag{5.14}$$

and

$$P(A_2 \mid B) = \frac{P(B \mid A_2)P(A_2)}{P(B \mid A_1)P(A_1) + P(B \mid A_2)P(A_2)} \tag{5.15}$$

EXAMPLE 5.20 (continued)

Recall that we are interested in finding the posterior probabilities that the bad part came from supplier 1, $P(A_1 \mid B)$, and from supplier 2, $P(A_2 \mid B)$. With the prior probabilities $P(A_1) = .65$ and $P(A_2) = .35$ and the conditional probabilities $P(B \mid A_1) = .02$ and $P(B \mid A_2) = .05$, we can utilize Bayes' theorem to compute the probabilities in question. Using (5.14) we find

$$P(A_1 \mid B) = \frac{P(B \mid A_1)P(A_1)}{P(B \mid A_1)P(A_1) + P(B \mid A_2)P(A_2)}$$

$$= \frac{(.02)(.65)}{(.02)(.65) + (.05)(.35)} = \frac{.0130}{.0130 + .0175}$$

$$= \frac{.0130}{.0305} = .426$$

Using (5.15) we find $P(A_2 \mid B)$ as follows:

$$P(A_2 \mid B) = \frac{(.05)(.35)}{(.02)(.65) + (.05)(.35)}$$

$$= \frac{.0175}{.0130 + .0175} = \frac{.0175}{.0305} = .574$$

In this application we initially had a probability of .65 that a part selected at random was from supplier 1. However, given information that the part is bad, the probability that the part is from supplier 1 drops to .426. Thus, if the part is bad, there is a better than 50–50 chance that the part came from supplier 2; that is, $P(A_2 \mid B) = .574$.

• • •

TABLE 5.4

Summary of Bayes' Theorem Calculations for the Two-Supplier Problem

Column 1 A_i	Column 2 $P(A_i)$	Column 3 $P(B\|A_i)$	Column 4 $P(A_i \cap B)$	Column 5 $P(A_i\|B)$
A_1	.65	.02	.0130	$\dfrac{.0130}{.0305} = .426$
A_2	.35	.05	.0175	$\dfrac{.0175}{.0305} = .574$
	1.00		$P(B) = .0305$	1.000

The Tabular Approach

A tabular approach helpful in organizing and conducting the Bayes' theorem calculations is shown in Table 5.4 for the data presented in Example 5.20. The computations shown in that table are conducted as follows:

Step 1 Prepare the following three columns.

> Column 1: The list of all mutually exclusive events that can occur in the problem.
> Column 2: The prior probabilities for the events. Note that since the mutually exclusive events include all the events that can occur, the probabilities in column 2 must sum to 1.
> Column 3: The conditional probabilities of the new information (event B) *given* each of the mutually exclusive events in column 2.

Step 2 In column 4 compute the joint probabilities for each mutually exclusive event and the new information, event B. These joint probabilities are found by multiplying the values in column 2 by the corresponding values in column 3. That is, $P(A_i \cap B) = P(A_i)P(B \mid A_i)$.

Step 3 Add the joint probability column (column 4) to find the probability of the new information (event B), $P(B)$. We see that in Example 5.20, there is a .0130 probability of a bad part and supplier 1 and there is a .0175 probability of a bad part and supplier 2. Since these are the only two ways a bad part can be obtained, the sum .0130 + .0175 = .0305 shows there is an overall probability of .0305 of finding a bad part from the combined shipments of both suppliers.

Step 4 In column 5, compute the posterior probabilities using the basic relationship of conditional probability

$$P(A_i \mid B) = \frac{P(A_i \cap B)}{P(B)}$$

The joint probabilities $P(A_i \cap B)$ are found in column 4, whereas the probability $P(B)$ appears as the sum of column 4.

As a final note, we can generalize Bayes' theorem to the case where there are n mutually exclusive events $A_1, A_2, \ldots, A_n$ and where one of the n events must occur when the experiment is conducted. In such a case, Bayes' theorem for the computation of any posterior probability $P(A_i \mid B)$ appears as follows:

$$P(A_i \mid B) = \frac{P(B \mid A_i)\, P(A_i)}{P(B \mid A_1)\, P(A_1) + P(B \mid A_2)\, P(A_2) + \cdots + P(B \mid A_n)\, P(A_n)} \quad (5.16)$$

With the prior probabilities $P(A_1), P(A_2), \ldots, P(A_n)$ and the appropriate conditional probabilities $P(B \mid A_1), P(B \mid A_2), \ldots, P(B \mid A_n)$, (5.16) can be used to compute the posterior probability of the events $A_1, A_2, \ldots, A_n$.

EXERCISES

45. The prior probabilities for events A_1, A_2, and A_3 are $P(A_1) = .20$, $P(A_2) = .50$, and $P(A_3) = .30$. The conditional probabilities of event B given A_1, A_2, and A_3 are $P(B \mid A_1) = .50$, $P(B \mid A_2) = .40$, and $P(B \mid A_3) = .30$.
   a. Compute $P(B \cap A_1)$, $P(B \cap A_2)$, and $P(B \cap A_3)$.
   b. Apply Bayes' theorem (5.16), to compute the posterior probability $P(A_2 \mid B)$.
   c. Use the tabular approach to applying Bayes' theorem to compute $P(A_1 \mid B)$, $P(A_2 \mid B)$, and $P(A_3 \mid B)$.

46. A consulting firm has submitted a bid for a large research project. The firm's management initially felt there was a 50–50 chance of getting the bid. However, the agency to which the bid was submitted has subsequently requested additional information on the bid. Past experience indicates that on 75% of the successful bids and 40% of the unsuccessful bids the agency requested additional information.
   a. What is your prior probability the bid will be successful (i.e., prior to receiving the request for additional information)?
   b. What is the conditional probability of a request for additional information given that the bid will ultimately be successful?
   c. Compute a posterior probability that the bid will be successful given that a request for additional information has been received.

47. A local bank is reviewing its credit card policy with a view toward recalling some of its credit cards. In the past approximately 5% of cardholders have defaulted, and the bank has been unable to collect the outstanding balance. Thus management has established a prior probability of .05 that any particular cardholder will default. The bank has further found that the probability of missing one or more monthly payments for those customers who do not default is .20. Of course the probability of missing one or more payments for those who default is 1.
   a. Given that a customer has missed a monthly payment, compute the posterior probability that the customer will default.

b. The bank would like to recall its card if the probability that a customer will default is greater than .20. Should the bank recall its card if the customer misses a monthly payment? Why or why not?

48. In a major eastern city, 60% of the automobile drivers are 30 years of age or older, and 40% of the drivers are under 30 years of age. Of all drivers 30 years of age or older, 4% will have a traffic violation in a 12-month period. Of all drivers under 30 years of age, 10% will have a traffic violation in a 12-month period. Assume that a driver has just been charged with a traffic violation; what is the probability that the driver is under 30 years of age?

49. A certain college football team plays 55% of their games at home and 45% of their games away. Given that the team has a home game, there is a .80 probability that it will win. Given that the team has an away game, there is a .65 probability that it will win. If the team wins on a particular Saturday, what is the probability that the game was played at home?

Summary

In this chapter we introduced probability. We described how probability can be interpreted as a numerical measure of the likelihood that an event will occur. In addition, we showed that the probability of an event could be computed either by summing the probabilities of the experimental outcomes (sample points) comprising the event or by using the relationships established by the rules of probability. For cases where additional information is available, we demonstrated how Bayes' theorem could be used to obtain revised, or posterior, probabilities.

Statistics in Practice

IMPROVING CUSTOMER SERVICE

Carstab Corporation, a subsidiary of Morton Thiokol, Inc., provides a variety of specialty chemical products for its customers. In one instance, a customer made small but repeated orders for an expensive catalyst product used in its chemical processing. Because of the nature of its operation, the customer placed unique specifications on the product. Some, but not all, of the lots produced by Carstab would meet the customer's exact specifications.

The customer agreed to test each lot as it was received to determine whether or not the catalyst would perform the desired function. Carstab agreed to ship lots to the customer with the understanding that the customer would perform the test and return the lots that did not pass the customer's specification test. The problem encountered was that only 60% of the lots sent to the customer passed the customer's test. This meant that although the product was still good and usable to other customers, approximately 40% of the shipments sent to this customer were returned.

Carstab explored the possibility of duplicating the customer's test and shipping only lots that passed the test. However, the test was unique to that one customer, and it was infeasible to purchase the expensive testing equipment needed to perform the customer's test. Therefore, Carstab designed a new test, one that was believed to indicate whether or not the lot would eventually pass the customer's test. The question was whether the new Carstab test would increase the probability that a lot shipped to the customer would pass the customer's test.

A sample of lots were tested under both the customer's procedure and the new proposed procedure. Of the lots tested, 55% passed the company's test and 50% passed both the customer's and the company's tests. In probability notation, we have

$$A = \text{the event the lot passes the customer's test}$$

$$B = \text{the event the lot passes the company's test}$$

where

$$P(B) = .55 \quad \text{and} \quad P(A \cap B) = .50$$

The probability information sought was the conditional probability $P(A \mid B)$, which was given by

$$P(A \mid B) = \frac{P(A \cap B)}{P(B)} = \frac{.50}{.55} = .909$$

Prior to the company's test, the probability that a lot would pass the customer's test was .60. The new results showed that when a lot passed the company's test, it had a .909 probability of passing the customer's test. This was good supporting evidence for the use of the company's test prior to shipment as a way of improving service to the customer.

Glossary

Probability—A numerical measure of the likelihood that an event will occur.

Experiment—Any process that generates well-defined outcomes.

Sample space—The set of all possible sample points (experimental outcomes).

Sample points—The individual outcomes of an experiment.

Tree diagram—A graphical device helpful in defining sample points of an experiment involving multiple steps.

Basic requirements of probability—Two requirements that restrict the manner in which probability assignments can be made:
1. For each experimental outcome E_i, we must have $0 \le P(E_i) \le 1$.
2. If there are k experimental outcomes, then $P(E_i) + P(E_2) + \cdots + P(E_i) = 1$.

Classical method—A method of assigning probabilities which assumes that the experimental outcomes are equally likely.

Relative frequency method—A method of assigning probabilities based upon experimentation or historical data.

Subjective method—A method of assigning probabilities based upon judgment.

Event—A set consisting of a collection of sample points or experimental outcomes.

Complement of event A—The event containing all sample points that are not in A.

Venn diagram—A graphical device for symbolically representing the sample space and operations involving events.

Union of events A **and** B—The event containing all sample points that are in A, in B, or in both.

Intersection of A **and** B—The event containing all sample points that are in both A and B.

Mutually exclusive events—Events that have no sample points in common; that is, $A \cap B$ is empty and $P(A \cap B) = 0$.

Addition rule—A probability law used to compute the probability of a union, $P(A \cup B)$. It is $P(A \cup B) = P(A) + P(B) - P(A \cap B)$ in general. For mutually exclusive events, since $P(A \cap B) = 0$, it reduces to $P(A \cup B) = P(A) + P(B)$.

Conditional probability—The probability of an event given that another event has occurred. The conditional probability of A given B is $P(A \mid B) = P(A \cap B)/P(B)$.

Independent events—Two events A and B where $P(A \mid B) = P(A)$ or $P(B \mid A) = P(B)$; that is, the events have no influence on each other.

Multiplication rule—A probability rule used to compute the probability of an intersection, $P(A \cap B)$. It is $P(A \cap B) = P(A)P(B \mid A)$ or $P(A \cap B) = P(B)P(A \mid B)$. For independent events it reduces to $P(A \cap B) = P(A)P(B)$.

Prior probabilities—Probabilities for a set of mutually exclusive events prior to being updated by Bayes' theorem.

Posterior probabilities—The revised probabilities for events resulting from application of Bayes' theorem.

Bayes' theorem—A formula for revising prior probabilities concerning mutually exclusive events that include all possible outcomes. The revised probabilities are called posterior probabilities.

Key Formulas

Computing Probability Using the Complement

$$P(A) = 1 - P(\overline{A}). \tag{5.4}$$

Addition Rule

$$P(A \cup B) = P(A) + P(B) - P(A \cap B). \tag{5.5}$$

Addition Rule for Mutually Exclusive Events

$$P(A \cup B) = P(A) + P(B) \tag{5.6}$$

Definition of Conditional Probability

$$P(A \mid B) = \frac{P(A \cap B)}{P(B)} \qquad (5.7)$$

or

$$P(B \mid A) = \frac{P(A \cap B)}{P(A)} \qquad (5.8)$$

Multiplication Rule

$$P(A \cap B) = P(B) P(A \mid B) \qquad (5.11)$$

or

$$P(A \cap B) = P(A) P(B \mid A) \qquad (5.12)$$

Multiplication Rule for Independent Events

$$P(A \cap B) = P(A) P(B) \qquad (5.13)$$

Bayes' Theorem

$$P(A_i \mid B) = \frac{P(B \mid A_i) P(A_i)}{P(B \mid A_1) P(A_1) + P(B \mid A_2) P(A_2) + \cdots + P(B \mid A_n) P(A_n)} \qquad (5.16)$$

Review Quiz

TRUE/FALSE

1. Probabilities can never be greater than 1 or less than 0.
2. For each experimental outcome, there is exactly one sample point.
3. The sum of the probabilities for all experimental outcomes may be any number between 0 and 1, inclusive.
4. The subjective method of assigning probabilities is one of the two methods permitting probabilities less than 0.
5. The sum of the probabilities of the sample points in an event must equal 1.
6. If we know the probability of an event, then the probability of the complement of the event can also be computed.
7. To use the addition rule to compute the probability of the union of two events, we must know the probability of the intersection of the events.
8. If two events are independent, they must be mutually exclusive.
9. If two events are independent, we need only know each event's probability in order to compute the probability of the intersection of the events.
10. Posterior probabilities must be known before Bayes' theorem can be applied.

MULTIPLE CHOICE

The following event probabilities for a statistical experiment are utilized in questions 11–13.

$$P(A) = .60 \qquad P(B) = .40$$
$$P(A \cap B) = .25$$

11. $P(A \cup B)$ is closest to
    a. .65
    b. .72
    c. .79
    d. .82

12. $P(A \mid B)$ is closest to
    a. .60
    b. .67
    c. .74
    d. .81

13. $P(\overline{A})$ is closest to
    a. .50
    b. .60
    c. .70
    d. .80

In questions 14–16, assume events A and B are independent, $P(A \mid B) = .70$, and $P(A \cap B) = .21$.

14. $P(A)$ is closest to
    a. .30
    b. .50
    c. .70
    d. .90

15. $P(B)$ is closest to
    a. .20
    b. .30
    c. .50
    d. .70

16. $P(A \cup B)$ is closest to
    a. .50
    b. .60
    c. .70
    d. .80

17. If A and B are independent events with $P(A) = .3$ and $P(B) = .5$, then $P(A \mid B) =$
    a. 0
    b. .15
    c. .20
    d. .30

18. If J and K are mutually exclusive events with $P(J) = .4$ and $P(K) = .5$, then $P(J \cap K) =$
    a. 0
    b. .2
    c. .7
    d. .9

19. If J and K are mutually exclusive events with $P(J) = .4$ and $P(K) = .5$, then $P(J \cup K) =$
    a. 0
    b. .2
    c. .7
    d. .9

Supplementary Exercises

50. A school district has decided to begin two new programs; one program involves reading improvement and the other program enriched mathematics. After a 1-year period, each of the programs will be classified as either successful or unsuccessful. Consider the decision to begin the two programs as an experiment.
    a. How many sample points exist for this experiment?
    b. Show a tree diagram and list the sample points.
    c. Let R = the event that the reading program is successful and M = the event that the mathematics program is successful. List the sample points in R and M.
    d. List the sample points in the union of the events R and M.
    e. List the sample points in the intersection of the events R and M.
    f. Are events R and M mutually exclusive? Explain.

51. Consider an experiment where eight experimental outcomes exist. We denote the experimental outcomes as $E_1, E_2, \ldots, E_8$. Suppose that the following events are identified.

$$A = \{E_1, E_2, E_3\}$$
$$B = \{E_2, E_4\}$$
$$C = \{E_1, E_7, E_8\}$$
$$D = \{E_5, E_6, E_7, E_8\}$$

Determine the sample points making up the following events.
    a. $A \cup B$
    b. $C \cup D$
    c. $A \cap B$
    d. $C \cap D$
    e. $B \cap C$
    f. $\overline{A}$
    g. $\overline{D}$
    h. $A \cup \overline{D}$

i. $A \cap \overline{D}$

j. Are A and B mutually exclusive?

k. Are B and C mutually exclusive?

52. Referring to Exercise 51 and assuming that the classical method is an appropriate way of establishing probabilities, find the following probabilities.

a. $P(A), P(B), P(C),$ and $P(D)$

b. $P(A \cap B)$

c. $P(A \cup B)$

d. $P(A \mid B)$

e. $P(B \mid A)$

f. $P(B \cap C)$

g. $P(B \mid C)$

h. Are B and C independent events?

53. A history professor has been asked by two students to review the grades that were assigned on their term papers. The review of each paper results in one of three outcomes: a change of grade, a request to meet with the student to discuss the paper in more detail, or no change in grade. Consider the review of the two papers as an experiment.

a. How many sample points are possible?

b. Show a tree diagram and list the sample points.

c. Let $A =$ the event that the review of the first student's paper results in a change of grade and $B =$ the event that the review of the second student's paper results in a grade change. List the sample points in the event A and then list the points in the event B.

d. List the sample points in the intersection of events A and B.

e. List the sample points in the union of the two events A and B.

54. In a particular resort area on the west coast of Florida the probability of the sun's shining on a given day is .80 and the probability of rain is .10. In addition, the probability that the resort experiences both sunshine and rain during the same day is .05. Assume that the experiment involves the weather possibilities during a 1-day period.

a. Are sun and rain mutually exclusive events? Explain.

b. Are sun and rain independent events?

c. We know that it rained on a given day. What is the probability that the resort also had sunshine during that day?

55. Suppose that $P(A) = .30, P(B) = .25,$ and $P(A \cap B) = .20.$

a. Find $P(A \cup B), P(A \mid B),$ and $P(B \mid A).$

b. Are events A and B independent? Why or why not?

56. Suppose that $P(A) = .40, P(A \mid B) = .60,$ and $P(B \mid A) = .30.$

a. Find $P(A \cap B)$ and $P(B).$

b. Are events A and B independent? Why or why not?

57. Suppose that $P(A) = .60, P(B) = .30,$ and events A and B are mutually exclusive.

a. Find $P(A \cup B)$ and $P(A \cap B).$

b. Are events A and B independent?

c. Can you make a general statement about whether or not mutually exclusive events can be independent?

58. A survey of 800 people found the following facts about the ability to recall a television commercial for a particular product and actual purchase of the product.

	Could Recall Television Commercial	Could Not Recall Television Commercial	Totals
Purchased the Product	160	80	240
Had Not Purchased the Product	240	320	560
Totals	400	400	800

Let T be the event of the person recalling the television commercial and B the event of buying or purchasing the product.
a. Find $P(T)$, $P(B)$, and $P(T \cap B)$.
b. Are T and B mutually exclusive events? Use probability values to explain.
c. What is the probability that a person who could recall seeing the television commercial has actually purchased the product?
d. Are T and B independent events? Use probability values to explain.
e. Comment on the value of the commercial in terms of its relationship to purchasing the product.

59. A large consumer goods company has been running a television advertisement for one of its soap products. A survey was conducted. On the basis of this survey, probabilities were assigned to the following events.

B = individual purchased the product

S = individual recalls seeing the advertisement

$B \cap S$ = individual purchased the product and recalls seeing the advertisement

The probabilities assigned were $P(B) = .20$, $P(S) = .40$, and $P(B \cap S) = .12$. The following problems relate to this situation.
a. What is the probability of an individual's purchasing the product given that the individual saw the advertisement? Does seeing the advertisement increase the probability the individual will purchase the product? As a decision maker, would you recommend continuing the advertisement (assuming that the cost is reasonable)?
b. Assume that those individuals who do not purchase the company's soap product buy from its competitors. What would be your estimate of the company's market share? Would you expect that continuing the advertisement will increase the company's market share? Why or why not?
c. The company has also tested another advertisement and assigned it values of $P(S) = .30$ and $P(B \cap S) = .10$. What is $P(B \mid S)$ for this other advertisement? Which advertisement seems to have had the bigger effect on customer purchases?

60. Western Airlines has done an analysis of a price promotion it is offering to frequent air travelers in order to increase the number of people on their New York to San Francisco route. Some 20% of the people in a large sample of individuals identified as frequent travelers from New York to San Francisco were aware of the Western promotion and elected to fly with Western on their next trip. It was further found that 80% were aware of the promotion and that prior to the promotion, 25% of these travelers flew with Western.

   a. What is the probability that a person will fly with Western on their next trip given that he or she is aware of the price promotion?
   b. For a randomly selected traveler, are the events "fly with Western" and "aware of the price promotion" independent? Why or why not?
   c. On the basis of these results, does the promotion appear to be successful in terms of increasing business? Why or why not?

61. Cooper Realty is a small real estate company specializing primarily in residential listings. They have recently become interested in the possibility of determining the likelihood of one of their listings being sold within a certain number of days. An analysis of company sales of 800 homes for the previous years produced the following data.

| Initial Asking Price | Days Listed Until Sold | | | |
	Under 30	31–90	Over 90	Totals
Under $25,000	50	40	10	100
$25,000–50,000	20	150	80	250
$50,000–75,000	20	280	100	400
Over $75,000	10	30	10	50
Totals	100	500	200	800

Total Homes Sold

   a. If A is defined as the event that a home is listed for over 90 days before being sold, estimate the probability of A.
   b. If B is defined as the event that the initial asking price is under $25,000, estimate the probability of B.
   c. What is the probability of $A \cap B$?
   d. Assuming that a contract has just been signed to list a home that has an initial asking price of less than $25,000, what is the probability the home will take Cooper Realty more than 90 days to sell?
   e. Are events A and B independent?

62. In the evaluation of a sales training program, a firm found that of 50 salespersons making a bonus last year, 20 had attended a special sales training program. The firm has 200 salespersons. Let B = the event that a salesperson makes a bonus and S = the event a salesperson attends the sales training program.
   a. Find $P(B)$, $P(S \mid B)$, and $P(S \cap B)$.

b. Assume that 40% of the salespersons have attended the training program. What is the probability that a salesperson makes a bonus given that the salesperson attended the sales training program, $P(B\,|\,S)$?

c. If the firm evaluates the training program in terms of the effect it has on the probability of a salesperson's making a bonus, what is your evaluation of the training program? Comment on whether B and S are dependent or independent events.

63. A company has studied the number of lost-time accidents occurring at its Brownsville, Texas plant. Historical records show that 6% of the employees had lost-time accidents last year. Management believes that a special safety program will reduce the accidents to 5% during the current year. In addition, it is estimated that 15% of those employees having lost-time accidents last year will have a lost-time accident during the current year.

a. What percentage of the employees will have lost-time accidents in both years?

b. What percentage of the employees will have at least one lost-time accident over the 2-year period?

64. In a study of television viewing habits among married couples, a researcher found that for a popular Saturday night program 25% of the husbands viewed the program regularly and 30% of the wives viewed the program regularly. The study found that for couples where the husband watches the program regularly 80% of the wives also watch regularly.

a. What is the probability that both the husband and wife watch the program regularly?

b. What is the probability that at least one—husband or wife—watches the program regularly?

c. What percentage of married couples do not have at least one regular viewer of the program?

65. A statistics professor has noted from past experience that students who do the homework for the course have a .90 probability of passing the course. On the other hand, students who do not do the homework for the course have a .25 probability of passing the course. The professor estimates that 75% of the students in the course do the homework. Given a student who passes the course, what is the probability that she or he completed the homework?

66. A salesperson for Business Communication Systems, Inc., sells automatic envelope-addressing equipment to medium- and small-size businesses. The probability of making a sale to a new customer is .10. During the initial contact with a customer, sometimes the salesperson will be asked to call back later. Of the 30 most recent sales, 12 were made to customers who initially told the salesperson to call back later. Of 100 customers who did not make a purchase, 17 had initially asked the salesperson to call back later. If a customer asks the salesperson to call back later, should the salesperson do so? What is the probability of making a sale to a customer who has asked the salesperson to call back later?

67. Migliori Industries, Inc., manufactures a gas-saving device for use on natural gas forced-air residential furnaces. The company is currently trying to determine the probability that sales of this product will exceed 25,000 units during next year's winter sales period. The company believes that sales of the product depend to a large extent on the winter conditions. Management's best estimate is that the

probability that sales will exceed 25,000 units if the winter is severe is .8. This probability drops to .5 if the winter conditions are moderate. If the weather forecast is .7 for a severe winter and .3 for moderate conditions, what is Migliori's best estimate that sales will exceed 25,000 units?

68. The Dallas IRS auditing staff is concerned with identifying potential fraudulent tax returns. From past experience they believe that the probability of finding a fraudulent return given that the return contains deductions for contributions exceeding the IRS standard is .20. Given that the deductions for contributions do not exceed the IRS standard, the probability of a fraudulent return decreases to .02. If 8% of all returns exceed the IRS standard for deductions due to contributions, what is the best estimate of the percentage of fraudulent returns?

69. An oil company has purchased an option on land in Alaska. Preliminary geologic studies have assigned the following prior probabilities.

$$P(\text{high quality oil}) = .50,$$

$$P(\text{medium quality oil}) = .20,$$

$$P(\text{no oil}) = .30.$$

a. What is the probability of finding oil?
b. After 200 feet of drilling on the first well, a soil test is taken. The probabilities of finding this particular type of soil are as follows.

$$P(\text{soil}\,|\,\text{high quality oil}) = .20,$$

$$P(\text{soil}\,|\,\text{medium quality oil}) = .80,$$

$$P(\text{soil}\,|\,\text{no oil}) = .20.$$

How should the firm interpret the soil test? What are the revised probabilities, and what is the new probability of finding oil?

CHAPTER 6

Random Variables and Discrete Probability Distributions

What You Will Learn in This Chapter

- what random variables are and how they are used

- what is meant by the probability distribution of a random variable

- how to compute and interpret the expected value and variance of a random variable

- when it is appropriate to use the binomial probability distribution

- how to use the binomial probability function and binomial tables to obtain probabilities

Contents

Statistics in the News

PETE ROSE VERSUS JOE DIMAGGIO: BASEBALL GREATS

Many baseball enthusiasts will agree that the most difficult record for modern baseball players to challenge is Joe DiMaggio's 56-game hitting streak in 1941. In baseball circles, the prevailing view is that Joe's performance set an unbreakable major league standard. However, for 6 weeks of the 1978 season (June 14 to August 1) Pete Rose of the Cincinnati Reds was unstoppable as he knocked down consecutive-game hitting records. In game 38 of the hitting streak, Rose passed Tommy Holmes' 1945 modern National League record. In game 44 he became the National League's all-time consecutive-game hitting record-holder when he matched Willie Keeler's 1897 streak. Joe DiMaggio's 56-game record was next.

However, on August 1, Pete Rose's hitting streak of 44 consecutive games came to an end when he went 0 for 4 in a game with the Atlanta Braves in Atlanta. During the streak Pete hit .376 with 56 singles, 14 doubles, and 13 walks. He provided plenty of excitement when on six occasions he saved the streak in his last at-bat. On four occasions during the streak, Rose's only hit was a bunt. Reds manager Sparky Anderson exclaimed, "Watching Pete break the National League record is the biggest thrill I've had as a manager, but Joe's 56-game record is an impossibility."

When Pete set the 44-game record, Las Vegas odds makers were still strong in their belief that he could not continue the hitting streak to DiMaggio's 56-game record. Probability specialists say that a special probability distribution, known as the binomial distribution, can be used to estimate the probability, or odds, of continuing a hitting streak for 12 more games. Using Pete's .376 batting average as the probability of a hit on each at-bat and assuming Pete would come to bat four times in a game, the probability of Pete having at least one hit in the four at-bats can be computed to be .8484. The probability of hitting in 12 successive games can be computed from this to be only .1391. The probability specialists say there was a .1391 probability of Pete reaching Joe's record

National Baseball Library, Cooperstown, NY

Joe Dimaggio, the Yankee Clipper, hit in 56 consecutive games in 1941.

© Cincinnati Reds 1985

Pete Rose, baseball's all-time total hit leader, challenged Dimaggio's record in 1978 with a 44 consecutive game hitting streak.

and a .8609 probability that he would not. Thus the odds were still better than 6 to 1 that Pete could not match DiMaggio's performance.

While Pete Rose's 44-game hitting streak provided much interest and excitement, Joe DiMaggio's 56-game record stands as baseball's "most unbreakable" record.

Based on *"Doing Much," Sports Illustrated* (August 7, 1978).

In this chapter we continue the study of probability by introducing the concept of a random variable and its probability distribution. The emphasis here is on discrete random variables. We first define what is meant by a random variable and its probability distribution and then show how to compute the expected value and variance. The binomial probability distribution is the subject of the rest of the chapter. It is a discrete probability distribution that has been successfully applied in a variety of practical situations. One such situation, Pete Rose's hitting streak, is described in "Statistics in the News," where the binomial probability distribution was used to compute the probability of Pete getting a hit in 12 consecutive games.

6.1 RANDOM VARIABLES

In Chapter 5 we studied the role of an experiment and the associated experimental outcomes in statistics. Random variables provide a means of assigning numerical values to experimental outcomes. These numerical values are used in computing means, variances, and other measures used to describe populations of interest. The definition of a random variable is as follows.

> **Random Variable**
>
> A *random variable* is a numerical description of the outcome of an experiment.

For any experiment a random variable can be defined such that each possible experimental outcome generates one and only one numerical value for the random variable. The particular numerical value that the random variable takes on depends upon the outcome of the experiment. That is, the value of the random variable is not known until the experimental outcome is observed.

EXAMPLE 6.1

Consider the experiment that consists of tossing a coin twice. The sample space for this experiment is

$$S = \{(H, H), (H, T), (T, H), (T, T)\}$$

Suppose we let x = number of heads occurring on the two coin tosses. Then x is a random variable (it provides a numerical description of the experimental outcome) that can assume the values 0, 1, and 2.

• • •

EXAMPLE 6.2

To receive state certification as medical lab technicians, candidates must pass a series of three examinations. If we define the random variable x as the number of examinations any one candidate passes, then x can assume the values 0, 1, 2, and 3.

• • •

EXAMPLE 6.3

The construction of a new library has just gotten underway at Lakeland Community College. If we define a random variable x as the percentage of the project that is completed after 6 months, the possible values of x range from 0 to 100. In other words, $0 \leq x \leq 100$.

• • •

A random variable is classified as either discrete or continuous depending upon the numerical values it can assume. A random variable that may assume only a finite or infinite sequence (e.g., 1, 2, 3, . . .) of values is referred to as a *discrete random variable*. The number of units sold, the number of defects observed, and the number of customers that enter a bank during one day of operation are examples of discrete random variables. Examples 6.1 and 6.2 involve discrete random variables. The distinguishing feature of a discrete random variable is the separation between successive values it may assume.

Random variables such as weight, time, and temperature, which may take on all values in a certain interval or collection of intervals, are referred to as *continuous random variables*. For instance, the random variable in Example 6.3 (percentage of project completed after 6 months) is a continuous random variable because it may take on any value in the interval from 0 to 100 (e.g., 56.33 or 64.227).

EXERCISES

1. Three students have interviews scheduled for summer employment at the Brookwood Institute. In each case the result of the interview will either be that a position is offered or not offered.
   a. List the experimental outcomes.
   b. Define a random variable that represents the number of offers made. Is this a discrete or continuous random variable?
   c. Show what value the random variable will assume for each of the experimental outcomes.

2. A new treatment for lower back pain has been developed and is being tested on two patients. The results of the experiment are a rating for each patient of either no improvement, moderate improvement, or strong improvement.
   a. List the experimental outcomes.
   b. Define a random variable that represents the number of patients that show at least some improvement. Show what value the random variable will assume for each of the experimental outcomes.

3. In order to perform a certain type of blood analysis, lab technicians have to perform two procedures. The first procedure requires either 1 or 2 separate steps, and the second procedure requires either 1, 2, or 3 steps.
   a. List the experimental outcomes associated with performing an analysis.
   b. If the random variable of interest is the total number of steps required to do the complete analysis, show what value the random variable will assume for each of the experimental outcomes.

4. Listed is a series of experiments and associated random variables. In each case, identify the values that the random variable can take on and state whether the random variable is discrete or continuous.

Experiment	Random Variable (x)
a. Take a 20-question examination	Number of questions answered correctly
b. Observe cars arriving at a tollbooth for 1 hour	Number of cars arriving at tollbooth
c. Audit 50 tax returns	Number of returns containing errors
d. Observe an employee's work	Number of nonproductive hours in an 8-hour workday
e. Weigh a shipment of goods	Number of pounds

6.2 DISCRETE PROBABILITY DISTRIBUTIONS

For any discrete random variable x, the *probability function,* denoted by $f(x)$, gives the probabilities associated with the values the random variable may assume. A *probability distribution* is a table, graph, or mathematical formula that shows all possible values of the random variable, x, and the associated probability function, $f(x)$.

EXAMPLE 6.4

As part of a study of 300 households in a village on the coast of Maine, a sociologist collected data showing the number of children in each household. The following data were obtained: 54 of the households had no children, 117 had one child, 72 had two children, 42 had three children, 12 had four children, and 3 had five children each.

Suppose we consider the experiment of randomly selecting one of these households to participate in a follow-up study. If we let x denote the number of children in the household selected, possible values of x are 0, 1, 2, 3, 4, and 5. Thus $f(0)$ provides the probability that a randomly selected household has no children, $f(1)$ provides the probability that a randomly selected household has one child, and so on. Since 54 of the 300 households have no children, we assign the value $54/300 = .18$ to $f(0)$. Similarly, since 117 of the 300 households have one child, we assign the value $117/300 = .39$ to $f(1)$. Continuing in this fashion for the other values the random variable x may assume, we obtain Table 6.1. This table, showing the values the random variable may assume and the associated probabilities $f(x)$, is the probability distribution for the random variable x.

● ● ●

We can also present the probability distribution of x graphically. In Figure 6.1 the values of the random variable x from Example 6.4 are shown on the horizontal axis and the probability that x assumes these values is shown on the vertical axis. For many discrete random variables, the probability distribution can also be given as a formula that yields $f(x)$ for every possible value of x.

TABLE 6.1

Probability Distribution for the Number of Children per Household

x	f(x)
0	.18
1	.39
2	.24
3	.14
4	.04
5	.01
Total	1.00

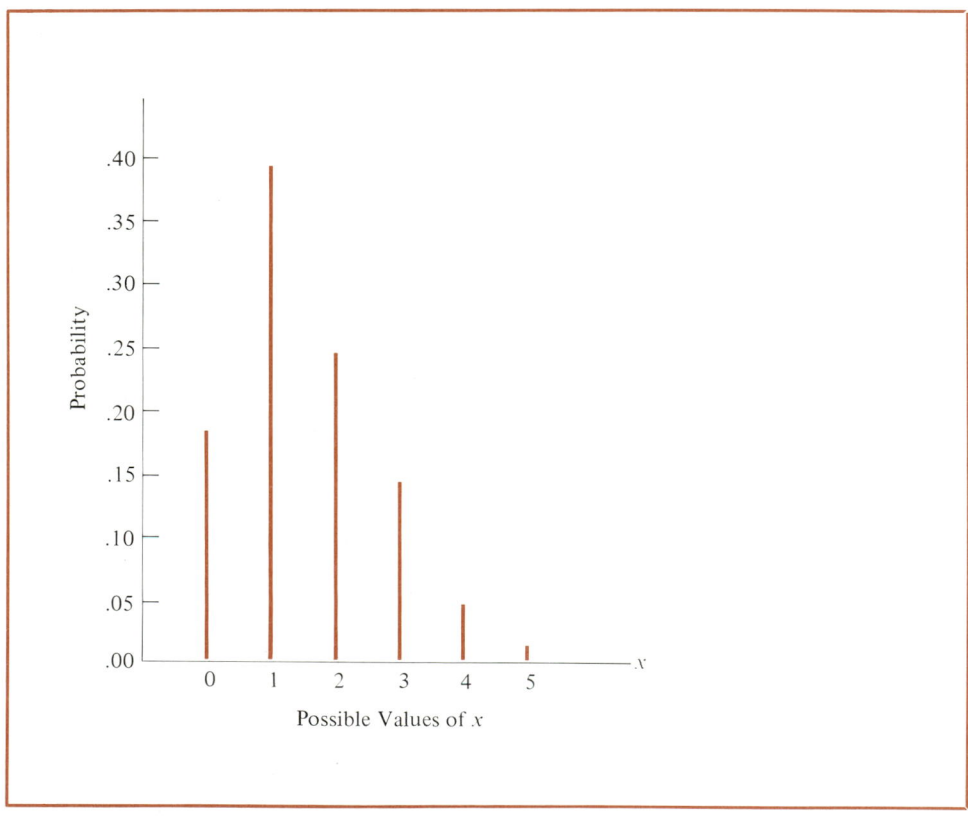

FIGURE 6.1
Graphical presentation of the Probability Distribution for Number of Children Per Household

EXAMPLE 6.5

Consider the random variable x and its probability distribution, as shown by the following table:

x	1	2	3	4
$f(x)$	$\frac{1}{4}$	$\frac{1}{4}$	$\frac{1}{4}$	$\frac{1}{4}$

This probability distribution can also be given by the formula

$$f(x) = 1/4 \qquad \text{for } x = 1, 2, 3, 4$$

• • •

A more complex example of a discrete random variable with its probability distribution given by a formula is the binomial probability distribution; it is introduced in Section 6.4.

In the development of the probability function for a discrete random variable, the following two conditions must always be satisfied:

Required Conditions for a Discrete Probability Function

$$f(x) \geq 0 \qquad\qquad (6.1)$$

$$\Sigma f(x) = 1 \qquad\qquad (6.2)$$

The symbol Σ in (6.2) is used to indicate that the summation is over all the values x may assume.

EXAMPLE 6.4 (continued)

Table 6.1 shows that the probabilities for the random variable x are all greater than or equal to 0. In addition, we note that

$$\Sigma f(x) = f(0) + f(1) + f(2) + f(3) + f(4) + f(5)$$
$$= .18 + .39 + .24 + .14 + .04 + .01 = 1.00$$

Since (6.1) and (6.2) are satisfied, the probability function developed by the sociologist is a valid discrete probability function.

• • •

EXERCISES

5. The following data were collected by counting the number of operating rooms in use at Tampa General Hospital over a 20-day period: On 3 of the days only 1

operating room was used, on 5 of the days 2 were used, on 8 of the days 3 were used, and on 4 days all 4 of the hospital's operating rooms were used.

   a. Use the relative frequency approach to construct a probability distribution for the number of operating rooms in use on any given day.

   b. Draw a graph of the probability distribution.

   c. Show that your probability distribution satisfies the required conditions for a valid discrete probability distribution.

6. A stockbroker has given the following probability estimates for the price of Mills Corporation stock at the end of next week: $P(\$22) = .10$, $P(\$23) = .40$, $P(\$24) = .30$, $P(\$25) = .20$.

   a. Identify an appropriate probability function and specify a probability distribution for the price of the stock at the end of next week.

   b. Draw a graph of the probability distribution.

   c. Show that the probability distribution satisfies (6.1) and (6.2).

7. QA Properties is considering making an offer to purchase an apartment building. Management has subjectively assessed a probability distribution for x, the purchase price:

x	$f(x)$
$148,000	.20
$150,000	.40
$152,000	.40

   a. Determine if this is a proper probability distribution. (Check (6.1) and (6.2).)

   b. What is the probability that the apartment house can be purchased for $150,000 or less?

8. The cleaning and changeover operation for a production system requires from 1 to 4 hours, depending upon the specific product that will begin production. Let x be a random variable indicating the time in hours required to make the changeover. The following probability function can be used to compute the probability associated with any changeover time x:

$$f(x) = \frac{x}{10} \qquad \text{for } x = 1, 2, 3, \text{ or } 4$$

   a. Show that the probability function meets the required conditions of (6.1) and (6.2).

   b. What is the probability that the changeover will take 2 hours?

   c. What is the probability that the changeover will take more than 2 hours?

   d. Graph the probability distribution for the changeover times.

9. The director of admissions at Lakeville Community College has subjectively assessed a probability distribution for x, the number of entering students.

x	$f(x)$
1000	.15
1100	.20
1200	.30
1300	.25
1400	.10

a. Is this a valid probability distribution?
b. What is the probability there will be 1200 or fewer entering students?

10. A psychologist has determined that the number of hours required to obtain the trust of a new patient is either 1, 2, or 3. Let x be a random variable indicating the time in hours required to gain the patient's trust. The following probability function has been proposed.

$$f(x) = \frac{x}{6} \quad \text{for } x = 1, 2, \text{ or } 3$$

a. Is this a valid probability function? Explain.
b. What is the probability that it takes exactly 2 hours to gain the patient's trust?
c. What is the probability that it takes at least 2 hours to gain the patient's trust?

11. Shown below is a partial probability distribution for the MRA Company's projected profits (in $1000s) for the first year of operation (the negative value shows a loss).

x	$f(x)$
−100	.10
0	.20
50	.30
100	.25
150	.10
200	

a. What is the value of $f(200)$? What is your interpretation of this value?
b. What is the probability that MRA will be profitable?
c. What is the probability that MRA will make at least $100,000?

6.3 EXPECTED VALUE AND VARIANCE

Expected Value

The mean or *expected value* of a random variable gives us the average, or central, value for the random variable. The mathematical expression for the expected value of a discrete random variable x is as follows.

Expected Value of a Discrete Random Variable

$$E(x) = \mu = \Sigma\, xf(x) \tag{6.3}$$

Both the notations $E(x)$ and μ are used to refer to the expected value of a random variable.

Equation (6.3) shows that in order to compute the expected value of a discrete random variable, we must multiply each value of the random variable by the corresponding value of its probability function, and then add the resulting products.

EXAMPLE 6.4 (continued)

In Table 6.2 we show the calculation of the expected value of the random variable x, which is the number of children in a randomly selected household. We see that 1.50 is the expected value of the number of children per household. Since it is impossible for

TABLE 6.2

Expected Value of Random Variable for Example 6.4

x	$f(x)$	$xf(x)$
0	.18	0(.18) = .00
1	.39	1(.39) = .39
2	.24	2(.24) = .48
3	.14	3(.14) = .42
4	.04	4(.04) = .16
5	.01	5(.01) = .05
		1.50

$$E(x) = \mu = \Sigma\, xf(x)$$

any household to have 1.5 children, we see that the expected value of a random variable does not have to be one of the values the random variable can assume. The expected value is thought of as an average value and not necessarily some value we expect the random variable to assume.

• • •

Variance

While the expected value gives us the average value for the random variable, we often need a measure of the dispersion, or variability, of the random variable. Just as we used variance in Chapter 3 to summarize the dispersion in a data set, we now use the variance measure to summarize the variability in the values of a random variable. The mathematical expression for the variance of a discrete random variable is as follows.

Variance of a Discrete Random Variable

$$\text{Var}(x) = \sigma^2 = \Sigma(x - \mu)^2 f(x) \qquad (6.4)$$

As (6.4) shows, an essential part of the variance formula is the deviation, $x - \mu$, which measures how far a particular value of the random variable is from the expected value or mean, μ. In computing the variance of a random variable, the deviations are squared and then weighted by the corresponding value of the probability function. The sum of these weighted squared deviations for all values of the random variable is referred to as the *variance*.

EXAMPLE 6.4 (continued)

The calculation of the variance for the probability distribution of the number of children per household is summarized in Table 6.3. We see that the variance for the number of children per household is 1.25. The *standard deviation, σ,* is defined as the positive square root of the variance. Thus the standard deviation of the number of children per household is

$$\sigma = \sqrt{1.25} = 1.118$$

The standard deviation is measured in the same units as the random variable ($\sigma = 1.118$ children per household); for this reason σ is often preferred in describing the

---

TABLE 6.3

Calculation of Variance for Example 6.4

x	$x - \mu$	$(x - \mu)^2$	$f(x)$	$(x - \mu)^2 f(x)$
0	$0 - 1.50 = -1.50$	2.25	.18	$2.25(.18) = .4050$
1	$1 - 1.50 = - .50$	.25	.39	$.25(.39) = .0975$
2	$2 - 1.50 = .50$	.25	.24	$.25(.24) = .0600$
3	$3 - 1.50 = 1.50$	2.25	.14	$2.25(.14) = .3150$
4	$4 - 1.50 = 2.50$	6.25	.04	$6.25(.04) = .2500$
5	$5 - 1.50 = 3.50$	12.25	.01	$12.25(.01) = .1225$
				$\overline{1.2500}$

$$\sigma^2 = \Sigma(x - \mu)^2 f(x)$$

variability of a random variable. The variance (σ^2) is measured in squared units and is thus more difficult to interpret.

• • •

An alternate formula for the variance of a random variable that is usually preferred for making computations is shown next.

Computational Formula for Variance of a Discrete Random Variable

$$\text{Var}(x) = \sigma^2 = \Sigma\, x^2 f(x) - \mu^2 \qquad (6.5)$$

This formula for the variance of a random variable is preferred because it is less likely to result in significant rounding errors as compared to (6.4).

EXAMPLE 6.4 (continued)

The calculations necessary to use (6.5) to compute the variance of the number of children per household are summarized in the accompanying table.

x	x^2	$f(x)$	$x^2 f(x)$
0	0	.18	.00
1	1	.39	.39
2	4	.24	.96
3	9	.14	1.26
4	16	.04	.64
5	25	.01	.25
15	55		3.50

$\mu = 1.50$

$$\text{Var}(x) = \sigma^2 = \Sigma\, x^2 f(x) - \mu^2 = 3.50 - (1.50)^2 = 1.25$$

As we should expect, this is the same answer we obtained previously using (6.4).

EXERCISES

12. A volunteer ambulance service handles from 0 to 5 service calls on any given day. The following probability distribution for the number of service calls is assumed:

Number of Service Calls	Probability
0	.10
1	.15
2	.30
3	.20
4	.15
5	.10

a. What is the expected number of service calls?

b. What is the variance in the number of service calls? What is the standard deviation?

13. Glazer's Winton Woods apartment building has 20 two-bedroom apartments. The number of apartment air-conditioner units that must be replaced during the summer season has the probability distribution shown.

Air Conditioners Replaced	Probability
0	.30
1	.35
2	.20
3	.10
4	.05

a. What is the expected number of air-conditioner units that will be replaced during a summer season?

b. What is the variance in the number of air-conditioner replacements?

c. What is the standard deviation?

14. A roulette wheel at a Las Vegas casino has 18 red numbers, 18 black numbers, and 2 green numbers. Assume that a $5 bet is placed on the black numbers. If a black number comes up, the player wins the bet; otherwise the player loses the $5.

a. Let x be a random variable indicating the player's net winnings on one bet. Show the probability distribution for x.

b. What is the expected amount won on a bet? What is your interpretation of this value?

c. What is the variance in the amount won on a bet? What is the standard deviation?

d. If a player places 100 bets of $5 each, what are the expected winnings? Comment on why casinos like a high volume of betting.

15. The probability distribution for collision insurance claims paid by the Newton Automobile Insurance Company is as follows.

Claims ($)	Probability
(No claims) 0	.90
200	.04
500	.03
1000	.01
2000	.01
3000	.01

   a. Use the expected collision claim amount to determine the collision insurance premium that would allow the company to break even on the collision portion of the policy.

   b. The insurance company charges an annual rate of $130 for the collision coverage. What is the expected value of the collision policy for the policyholder? Why does the policyholder purchase a collision policy with this expected value?

16. The number of dots up on the roll of a die has the following probability function.

$$f(x) = 1/6 \quad \text{for } x = 1, 2, 3, 4, 5, 6$$

   a. Show that this probability function possesses the properties necessary for probability distributions.

   b. Draw a graph of the probability distribution.

   c. What is the expected value? What is the interpretation of this value?

   d. What are the variance and the standard deviation for the number of dots up on the roll of a die?

17. The demand for a product of Carolina Industries varies greatly from month to month. Based on the past 2 years of data, the following probability distribution shows the company's monthly demand.

Unit demand	300	400	500	600
Probability	.20	.30	.35	.15

   a. If the company places monthly orders based on the expected value of the monthly demand, what should Carolina's monthly order quantity be for this product?

b. Assume that each unit demanded generates $70 in revenue and that each unit ordered costs $50. How much will the company gain or lose in a month if it places an order based on your answer to part (a) and where the actual demand for the item is 300 units?

18. What are the variance and the standard deviation for the number of units demanded in Exercise 17?

19. The J.R. Ryland Computer Company is considering a plant expansion that will enable the company to begin production of a new computer product. The company's president must determine whether to make the expansion a medium- or large-scale project. An uncertainty involves the demand for the new product, which for planning purposes may be low demand, medium demand, or high demand. The probability estimates for the demands are .20, .50 and .30, respectively. Letting x indicate the annual profit in $1000s, the firm's planners have developed profit forecasts for the medium- and large-scale expansion projects.

Demand	Medium-Scale Expansion Profits		Large-Scale Expansion Profits	
	x	$f(x)$	x	$f(x)$
Low	50	.20	0	.20
Medium	150	.50	100	.50
High	200	.30	300	.30

a. Compute the expected value for the profit associated with the two expansion alternatives. Which decision is preferred for the objective of maximizing the expected profit?

b. Compute the variance for the profit associated with the two expansion alternatives. Which decision is preferred for the objective of minimizing the risk or uncertainty?

6.4 THE BINOMIAL PROBABILITY DISTRIBUTION

The *binomial probability distribution* is a discrete probability distribution that has many applications. It is associated with a multiple-step experiment that we call the binomial experiment.

Binomial Experiment

For a probability experiment to be classified as a *binomial experiment,* it must have the following four properties.

Properties of a Binomial Experiment

1. The experiment consists of a sequence of *n* identical trials.
2. Two outcomes are possible on each trial. We refer to one as a *success* and the other as a *failure*.
3. The probability of a success, denoted by p, does not change from trial to trial. Consequently, the probability of failure, denoted by $1 - p$, does not change from trial to trial. Also, since there are only two possible outcomes on each trial, the probability of a success plus the probability of a failure must equal 1.
4. The trials are independent.

Figure 6.2 depicts the outcome of a binomial experiment involving eight trials.

In a binomial experiment, our interest is in the *number of successes occurring in the n trials*. If we let x denote the number of successes occurring in the n trials, we see that x can assume the values of $0, 1, 2, 3, \ldots, n$. Since the number of values is finite, x is a *discrete* random variable. The probability distribution associated with this random variable is called the *binomial probability distribution*.

EXAMPLE 6.6

Consider the experiment of tossing a coin five times and on each toss observing whether the coin lands with a head or a tail on its upward face. Suppose we are interested in counting the number of heads appearing during the five tosses. Does this experiment have the properties of a binomial experiment? What is the random variable of interest? Note that:

1. The experiment consists of five identical trials, where each trial involves the tossing of one coin.
2. There are two outcomes possible for each trial. The possible outcomes are a head and a tail. We can designate head as success and tail as failure.
3. The probability of a head and the probability of a tail are the same for each trial, with $p = .5$ and $1 - p = .5$.

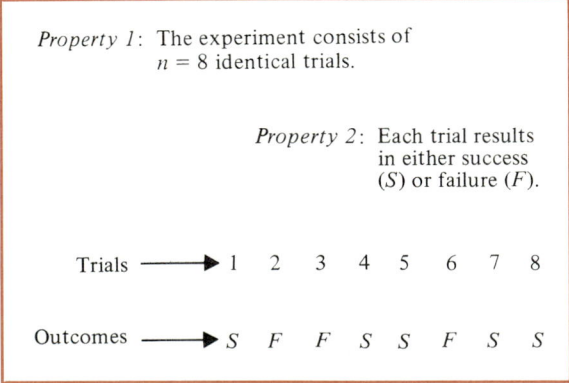

FIGURE 6.2
Diagram of Eight-Trial Binomial Experiment

4. The trials or tosses are independent, since the outcome on any one trial is not affected by what happens on other trials or tosses.

Thus the properties of a binomial experiment are satisfied. The random variable of interest is x = the number of heads appearing in the five trials. In this case, x can assume the values of 0, 1, 2, 3, 4, or 5.

• • •

Property 2 is not as restrictive as it may first appear. For instance, consider rolling a die and observing whether or not a 5 comes up. Defining a success to be "5 comes up" and a failure to be "5 does not come up" we have defined the experiment in such a fashion that there are exactly two outcomes possible on each trial, 5 or not 5. This is true even though there are actually six numbers that may appear on the upward face of a die.

EXAMPLE 6.7

An insurance salesperson pays a visit to 10 randomly selected families. An outcome associated with a visit is classified as a success if the family purchases an insurance policy and a failure if the family does not. From past experience, the salesperson knows the probability that a randomly selected family purchases an insurance policy is .10. Show that the process of the salesperson contacting the ten families and recording the number of families that purchase an insurance policy is a binomial experiment.

Checking the properties of a binomial experiment, we observe the following:

1. The experiment consists of 10 identical experiments, where each experiment, or trial, involves contacting one family.
2. There are two outcomes possible on each trial: the family purchases a policy or the family does not purchase a policy.
3. The probabilities of a purchase and a nonpurchase are assumed to be the same for each family, with $p = .10$ and $1 - p = .90$.
4. The trials are independent since the families are randomly selected.

Since the four assumptions are satisfied, this is a binomial experiment. The random variable of interest is the number of sales obtained in contacting the 10 families. In this case, x can assume the values of 0, 1, 2, 3, 4, 5, 6, 7, 8, 9 and 10.

• • •

Property 3 of the binomial experiment is often called the *stationarity assumption* and is sometimes confused with Property 4, independence of trials. To see how they differ, consider again the case of the salesperson in Example 6.7 calling on families to sell insurance policies. If, as the day wore on the salesperson got tired and lost enthusiasm, then the probability of success (selling a policy) might drop to .05 by the tenth call. In such a case Property 3 (stationarity) would not be satisfied, and we would not have a binomial experiment. This would be true even if Property 4 held—that is, the purchase decisions of each family were made independently.

In applications involving binomial experiments, a special mathematical formula, called the *binomial probability function,* can be used to compute the probability of x

successes in the *n* trials. We develop the binomial probability function by considering a situation that can be analyzed using the methods from Chapter 5. We then show that if the properties of a binomial experiment are satisfied, the binomial probability function can be used to compute the desired probabilities.

EXAMPLE 6.8

A moving target at a policy academy target range can be hit 80% of the time by a particular individual classified as an expert shot. Suppose this person takes three shots at the target. What is the probability the expert makes exactly two hits?

Using a tree diagram, we can see from Figure 6.3 that the experiment of taking three shots at the target has eight possible outcomes. Using S to denote success (hitting the target) and F to denote failure (missing the target), we are interested in outcomes having two successes in the three trials, or shots.

Next, let us verify that the experiment of taking three shots at the target has the

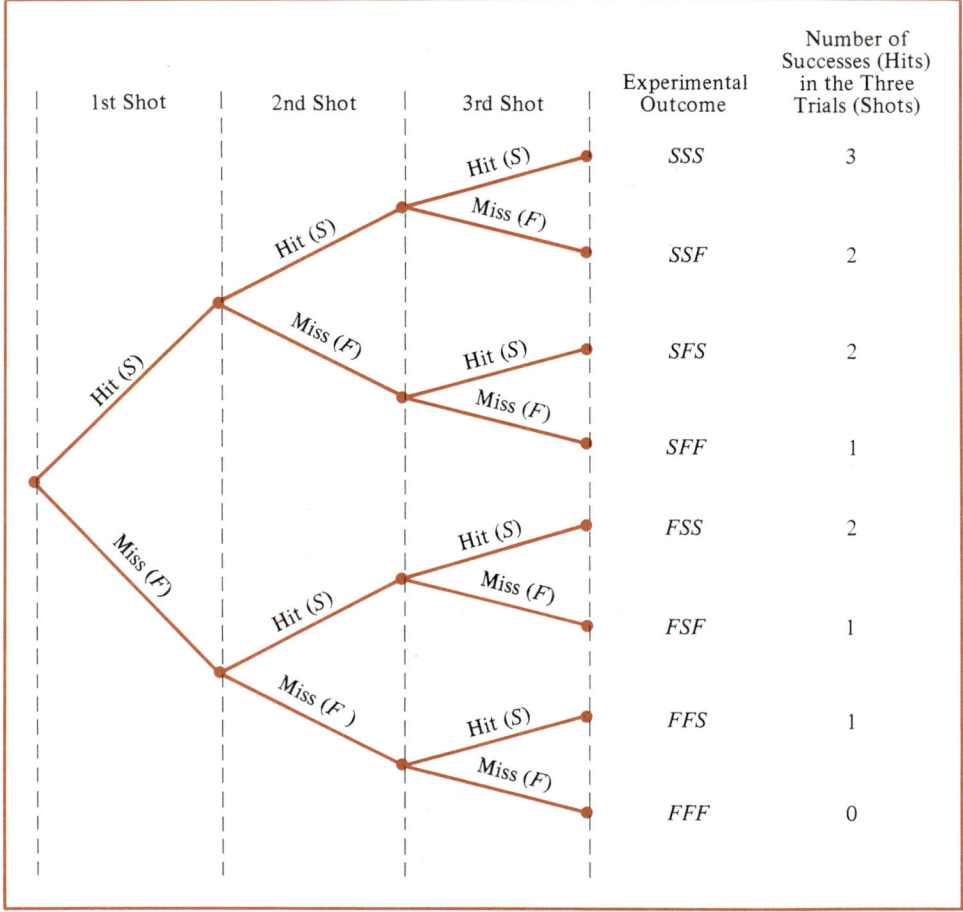

FIGURE 6.3
Tree Diagram of Experiment Involving Taking Three Shots at a Target

properties of a binomial experiment. Check to see if you agree with the following conclusions:

1. The experiment consists of three identical trials or shots at the target.
2. The two outcomes per trial are a hit (S) or a miss (F).
3. The probability of a hit and the probability of a miss are the same for each trial, with $p = .80$ and $1 - p = 1 - .80 = .20$.
4. The trials, or shots, are independent.

$$\bullet \quad \bullet \quad \bullet$$

The number of outcomes of a binomial experiment that result in exactly x successes in n trials can be computed from the following formula.*

$$\begin{matrix} \text{Number of Experimental Outcomes Providing} \\ \text{Exactly } x \text{ Successes in } n \text{ Trials} \end{matrix} = \frac{n!}{x!(n-x)!} \qquad (6.6)$$

where

$$n! = n(n-1)(n-2) \cdots (2)(1) \qquad (6.7)$$

with 0! defined to be 1.

The term $n!$ is called *n factorial*. For example, $5! = (5)(4)(3)(2)(1) = 120$.

EXAMPLE 6.8 (continued)

Now let us return to the experiment of taking three shots at a target. Equation (6.6) can be used to determine the number of experimental outcomes involving two hits; that is, the number of ways of obtaining $x = 2$ successes in the $n = 3$ trials. From (6.6) we have

$$\frac{n!}{x!(n-x)!} = \frac{3!}{2!(3-2)!} = \frac{(3)(2)(1)}{[(2)(1)](1)} = \frac{6}{2} = 3$$

Formula (6.6) shows that three of the outcomes yield two successes. From Figure 6.3 we see these three outcomes are denoted by *SSF, SFS,* and *FSS.*

Using (6.6) to determine how many experimental outcomes have three successes (hits) in the three trials, we obtain:

$$\frac{n!}{x!(n-x)!} = \frac{3!}{3!(3-3)!} = \frac{3!}{3!0!} = \frac{(3)(2)(1)}{[(3)(2)(1)](1)} = \frac{6}{6} = 1$$

From Figure 6.3 we see that the one experimental outcome with three successes is identified by *SSS.*

$$\bullet \quad \bullet \quad \bullet$$

*This formula is commonly used to determine the number of combinations of n objects selected x at a time. For the binomial experiment, this combinatorial formula provides the number of experimental outcomes having x successes in n trials.

We know that (6.6) can be used to determine the number of experimental outcomes that result in x successes. But, if we are to determine the probability of x successes in n trials, we must also know the probability associated with each experimental outcome. Since the trials of a binomial experiment are independent, we can simply multiply the probabilities associated with each trial outcome to find the probability of a particular sequence of outcomes.

EXAMPLE 6.8 (continued)

With three shots at a target, the probability of hitting on the first and second shots but missing on the third is given by

$$pp(1 - p)$$

With a .80 probability of hitting the target on any one shot, the probability of hitting the target with the first two shots and missing the target with the third shot is given by

$$(.80)(.80)(.20) = (.80)^2(.20) = .128$$

There are two other sequences of outcomes resulting in two successes and one failure. The probabilities for all three sequences involving 2 successes are as shown.

Trial Outcomes				
1st Shot	2nd Shot	3rd Shot	Success-Failure Notation	Probability of Outcome
Hit	Hit	Miss	SSF	$pp(1 - p) = p^2(1 - p) = (.80)^2(.20) = .128$
Hit	Miss	Hit	SFS	$p(1 - p)p = p^2(1 - p) = (.80)^2(.20) = .128$
Miss	Hit	Hit	FSS	$(1 - p)pp = p^2(1 - p) = (.80)^2(.20) = .128$

Observe that all three outcomes with two successes have exactly the same probability. This observation holds in general. In any binomial experiment, each sequence of trial outcomes yielding x successes in n trials has the *same probability* of occurrence. The probability of each sequence of trials yielding x successes in n trials is as follows:

$$\text{Probability of a Particular Sequence of Trial Outcomes} = p^x(1 - p)^{(n-x)} \qquad (6.8)$$
$$\text{with } x \text{ Successes in } n \text{ Trials}$$

For the target practice situation of Example 6.8, this formula shows that any outcome with two successes has a probability of $p^2(1 - p)^{(3-2)} = p^2(1 - p)^1 = (.80)^2(.20)^1 = .128$, as shown.

Since (6.6) shows the number of outcomes in a binomial experiment with x successes and since (6.8) gives the probability for each sequence involving x successes, we combine (6.6) and (6.8) to obtain the following *binomial probability function.*

Binomial Probability Function

$$f(x) = \frac{n!}{x!(n-x)!} \, p^x (1-p)^{(n-x)} \qquad (6.9)$$

where

$$f(x) = \text{the probability of } x \text{ successes in } n \text{ trials}$$

$$n = \text{the number of trials}$$

$$p = \text{the probability of a success on any one trial}$$

$$(1-p) = \text{the probability of a failure on any one trial}$$

EXAMPLE 6.8 (continued)

In the experiment of taking three shots at the target, what is the probability for each of the following: hits on all three shots, hits on exactly two shots, a hit on exactly one shot, and misses on all three shots?

The binomial probability function can be used to answer these questions.

$$f(3) = \frac{3!}{3!(3-3)!} \, (.80)^3 (.20)^0 = \frac{6}{6}(.512)(1) = .512$$

$$f(2) = \frac{3!}{2!(3-2)!} \, (.80)^2 (.20)^1 = \frac{6}{2}(.64)(.20) = .384$$

$$f(1) = \frac{3!}{1!(3-1)!} \, (.80)^1 (.20)^2 = \frac{6}{2}(.80)(.04) = .096$$

$$f(0) = \frac{3!}{0!(3-0)!} \, (.80)^0 (.20)^3 = \frac{6}{6}(1)(.008) = .008$$

Summarizing these calculations in a tabular form provides a tabular presentation of the probability distribution for the number of hits in three shots at the target.

x	f(x)
3	.512
2	.384
1	.096
0	.008
	1.000

A graphical representation of this probability distribution is also shown.

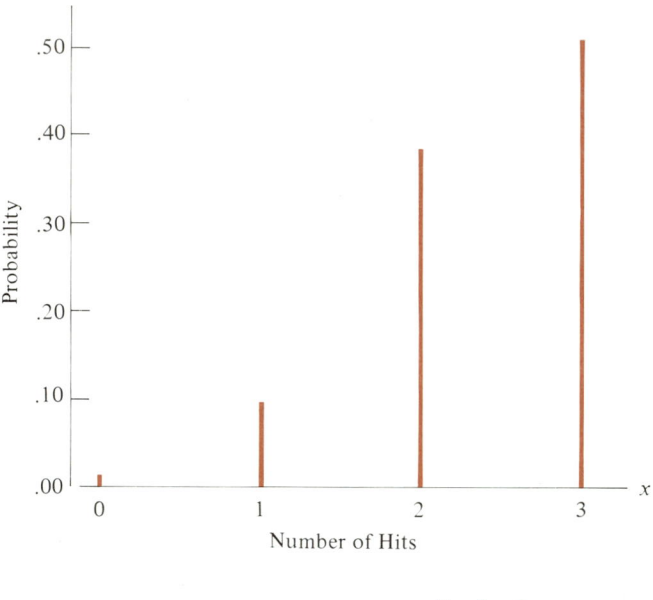

The binomial probability function can be applied to *any* binomial experiment. If we are satisfied that a situation has the properties of a binomial experiment and if we know the values of n, p, and $(1 - p)$, (6.9) can be used to compute the probability of x successes in the n trials.

EXAMPLE 6.9

A surgical operation is successful 90% of the time. The operation is performed on five patients.

a.   What is the probability all five operations are successful?
b.   What is the probability at least four operations are successful?

a. Assuming the properties of the binomial experiment apply, we have $n = 5$ trials with $p = .90$ and $1 - p = 1 - .90 = .10$:

$$f(5) = \frac{5!}{5!(5 - 5)!} (.90)^5(.10)^{(5-5)} = 1(.590)(1) = .5900$$

b. We must consider two binomial probabilities, since the probability of *at least* 4 successes consists of the probability of 4 successes plus the probability of 5 successes. Computing the probability of 4 successes, we have:

$$f(4) = \frac{5!}{4!(5 - 4)!} (.90)^4(.10)^{(5-4)} = 5(.0656) = .3281$$

Therefore, the probability of at least four successful operations is $f(4) + f(5) = .3281 + .5900 = .9181$.

$$\bullet \quad \bullet \quad \bullet$$

EXAMPLE 6.10

Eight customers enter a clothing store during a 1-hour period. From past experience, it is known that approximately 30% of the people entering the store make a purchase. Answer the following questions.

a. What is the probability exactly three of the eight customers make a purchase?
b. What is the probability at least one customer makes a purchase?

a. Assuming the properties of the binomial experiment apply, we have $n = 8$ trials with $p = .30$ and $1 - p = 1 - .30 = .70$:

$$f(3) = \frac{8!}{3!(8-3)!}(.30)^3(.70)^{(8-3)}$$

$$= \frac{40,320}{720}(.027)(.16807) = .2541$$

b. The probability of *at least* one customer purchase is the sum of the probabilities of 1, 2, 3, 4, 5, 6, 7 and 8 customer purchases. While we could compute each of these probabilities separately and add them together, it is helpful to note that the probability of at least one person making a purchase is 1 minus the probability of no customer purchases. Computing the probability of no successes in the eight trials, we have:

$$f(0) = \frac{8!}{0!(8-0)!}(.30)^0(.70)^{(8-0)}$$

$$= \frac{8!}{1(8!)}(1)(.0576) = .0576$$

Therefore, the probability of at least one success must be:

$$P(\text{at least one success}) = 1 - f(0)$$

$$= 1 - .0576 = .9424$$

$$\bullet \quad \bullet \quad \bullet$$

Using Tables of Binomial Probabilities

Tables have been developed that give the probability of x successes in n trials for a binomial experiment. These tables are generally easy to use and quicker than Equation (6.9), especially when the number of trials involved is large. A table of binomial probabilities is provided as Table 5 of Appendix B. A portion of this table is given in Table 6.3. In order to use this table it is necessary to specify the values of n, p, and x for the binomial experiment of interest. In the example at the top of Table 6.3, we see that the probability of $x = 3$ successes in a binomial experiment with $n = 10$ and $p = .40$ is .2150. You might want to use (6.9) to verify that this is the answer you would obtain using the binomial probability function directly.

EXAMPLE 6.10 (continued)

Assume that 10 customers enter the clothing store during a 1-hour period and that the probability of a customer purchase is .30. Use the table of binomial probabilities to answer the following questions.

a. What is the probability exactly 3 of the 10 customers make a purchase?
b. What is the probability at least 1 customer makes a purchase?
c. What is the probability 3 or fewer customers make a purchase?

Using Table 6.4 to find the probability values, the answers are as follows;

a. $f(3) = .2668$
b. $P(\text{at least 1 success}) = 1 - f(0) = 1 - .0282 = .9718$
c. $P(3 \text{ or fewer successes}) = f(3) + f(2) + f(1) + f(0) = .2668 + .2335 + .1211 + .0282 = .6496$

Using all the probabilities corresponding to $n = 10$ trials with $p = .30$, we can construct a graphical representation of the binomial probability distribution for the number of customer purchases:

• • •

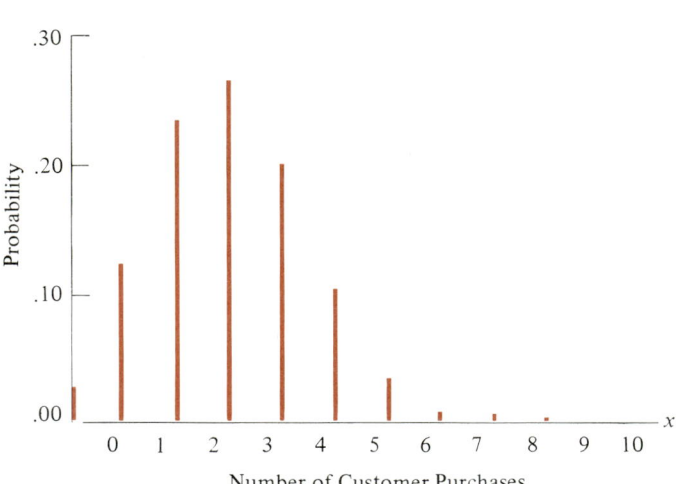

Number of Customer Purchases

While this example demonstrates the relative ease of using the tables of binomial probabilities, it is impossible to have tables that show all possible values of n and p that might be encountered in a binomial experiment. In cases where the appropriate probabilities are not available, one can interpolate to arrive at an approximation. For instance, suppose we had $p = .275$, $n = 10$, and $x = 3$. Using Table 6.3 for $p = .25$, $n = 10$, and $x = 3$, we find that $f(3) = .2503$. Similarly for $p = .30$, $n = 10$, and $x = 3$ we

TABLE 6.4

Selected values of the Binomial Probability Table
Example: $n = 10$, $x = 3$, $p = .40$; $P(x = 3$ successes $= .2150)$

						p					
n	x	.05	.10	.15	.20	.25	.30	.35	.40	.45	.50
9	0	.6302	.3874	.2316	.1342	.0751	.0404	.0207	.0101	.0046	.0020
	1	.2985	.3874	.3679	.3020	.2253	.1556	.1004	.0605	.0339	.0176
	2	.0629	.1722	.2597	.3020	.3003	.2668	.2162	.1612	.1110	.0703
	3	.0077	.0446	.1069	.1762	.2336	.2668	.2716	.2508	.2119	.1641
	4	.0006	.0074	.0283	.0661	.1168	.1715	.2194	.2508	.2600	.2461
	5	.0000	.0008	.0050	.0165	.0389	.0735	.1181	.1672	.2128	.2461
	6	.0000	.0001	.0006	.0028	.0087	.0210	.0424	.0743	.1160	.1641
	7	.0000	.0000	.0000	.0003	.0012	.0039	.0098	.0212	.0407	.0703
	8	.0000	.0000	.0000	.0000	.0001	.0004	.0013	.0035	.0083	.0176
	9	.0000	.0000	.0000	.0000	.0000	.0000	.0001	.0003	.0008	.0020
10	0	.5987	.3487	.1969	.1074	.0563	.0282	.0135	.0060	.0025	.0010
	1	.3151	.3874	.3474	.2684	.1877	.1211	.0725	.0403	.0207	.0098
	2	.0746	.1937	.2759	.3020	.2816	.2335	.1757	.1209	.0763	.0439
	3	.0105	.0574	.1298	.2013	.2503	.2668	.2522	.2150	.1665	.1172
	4	.0010	.0112	.0401	.0881	.1460	.2001	.2377	.2508	.2384	.2051
	5	.0001	.0015	.0085	.0264	.0584	.1029	.1536	.2007	.2340	.2461
	6	.0000	.0001	.0012	.0055	.0162	.0368	.0689	.1115	.1596	.2051
	7	.0000	.0000	.0001	.0008	.0031	.0090	.0212	.0425	.0746	.1172
	8	.0000	.0000	.0000	.0001	.0004	.0014	.0043	.0106	.0229	.0439
	9	.0000	.0000	.0000	.0000	.0000	.0001	.0005	.0016	.0042	.0098
	10	.0000	.0000	.0000	.0000	.0000	.0000	.0000	.0001	.0003	.0010
11	0	.5688	.3138	.1673	.0859	.0422	.0198	.0088	.0036	.0014	.0005
	1	.3293	.3835	.3248	.2362	.1549	.0932	.0518	.0266	.0125	.0054
	2	.0867	.2131	.2866	.2953	.2581	.1998	.1395	.0887	.0531	.0269
	3	.0137	.0710	.1517	.2215	.2581	.2568	.2254	.1774	.1259	.0806
	4	.0014	.0158	.0536	.1107	.1721	.2201	.2428	.2365	.2060	.1611
	5	.0001	.0025	.0132	.0388	.0803	.1321	.1830	.2207	.2360	.2256
	6	.0000	.0003	.0023	.0097	.0268	.0566	.0985	.1471	.1931	.2256
	7	.0000	.0000	.0003	.0017	.0064	.0173	.0379	.0701	.1128	.1611
	8	.0000	.0000	.0000	.0002	.0011	.0037	.0102	.0234	.0462	.0806
	9	.0000	.0000	.0000	.0000	.0001	.0005	.0018	.0052	.0126	.0269
	10	.0000	.0000	.0000	.0000	.0000	.0000	.0002	.0007	.0021	.0054
	11	.0000	.0000	.0000	.0000	.0000	.0000	.0000	.0000	.0002	.0005

find that $f(3) = .2668$. Interpolating halfway between these values, we obtain $f(3) = (.2503 + .2668)/2 = .2586$ for the case of $p = .275$. Alternatively, with today's calculators it is not too difficult to calculate the desired probability using (6.9), especially if the number of trials is not too large. In the exercises, you should practice using (6.9) to compute the binomial probabilities unless the problem specifically requests that you use the binomial probability table.

EXERCISES

20. Suppose that a newly married couple is planning to have three children and that the couple is interested in knowing the probabilities of having no girls, one girl, two girls, and three girls. Assume that the probability of having a girl is .50 on any one birth.
    a. What are the trials of the experiment in this application? How many trials are there?
    b. How many outcomes are possible on each trial and what are they?
    c. What are the probabilities associated with the outcomes for each trial? Are these probabilities the same for each trial?
    d. What additional assumption must be made about the trials in order for this to be a binomial experiment?
    e. What is the random variable of interest in this problem? Is it a discrete or a continuous random variable and what values can it assume?

21. A die is to be rolled five times. Assume that we are interested in the number of times the upward face of the die is a 6. Define success and failure for the trials of this experiment and describe the conditions existing in this example that make it a binomial experiment.

22. A study found that for 60% of the couples who have been married 10 years or less, both spouses work. A sample of 20 couples who have been married 10 years or less will be selected from marital records available at a local courthouse. We will be interested in the number of couples in the sample in which both spouses work. Describe the conditions necessary for this sampling process to be viewed as a binomial experiment.

23. The New York State Bar Examination is the basis for admitting law school graduates into the law profession. Historically, 30% of the individuals taking the examination pass on their first attempt. Suppose a group of 15 individuals will be taking the examination for the first time and that we are interested in the number of individuals in this group who will pass the exam. Describe the conditions necessary for this situation to be a binomial experiment.

24. When a new machine is functioning properly, only 3% of the items produced are defective. Assume that we will randomly select two parts produced on the machine and that we are interested in the number of defective parts found.
    a. Describe the conditions under which this situation would be a binomial experiment.
    b. Draw a tree diagram similar to Figure 6.3 showing this as a two-trial experiment.

    c. How many experimental outcomes result in exactly one defect being found?

    d. Compute the probabilities associated with finding no defects, exactly 1 defect, and 2 defects.

25. Forty-five percent of the residents in a township who are of voting age are not registered to vote.

    a. In a sample of 10 people, what is the probability 5 are not registered to vote?

    b. In a sample of 10 people, what is the probability 2 or fewer are not registered to vote?

26. A baseball player with a batting average of .300 comes to bat four times in a game. What is the probability the player obtains exactly 1 hit? Exactly 2 hits? No hits?

27. National Oil Company conducts exploratory oil drilling operations in the southwestern United States. In order to fund the operation, investors form partnerships, which provide the financial support necessary to drill a fixed number of oil wells. Each well drilled is classified as a producer well or a dry well. Past experience shows that this type of exploratory operation provides producer wells for 15% of all wells drilled. A newly formed partnership has provided the financial support for drilling at 12 exploratory locations.

    a. What is the probability that all 12 wells will be producer wells?

    b. What is the probability that all 12 wells will be dry wells?

    c. What is the probability that exactly 1 well will be a producer well?

    d. In order to make the partnership venture profitable, at least 3 of the exploratory wells must be producer wells. What is the probability that the venture will be profitable?

28. Military radar and missile detection systems are designed to warn a country against enemy attacks. A reliability question deals with the ability of the detection system to identify an attack and issue the warning. Assume that a particular detection system has a .90 probability of detecting a missile attack. Answer the following questions using the binomial probability distribution.

    a. What is the probability that a single detection system will detect an attack?

    b. If two detection systems are installed in the same area and operate independently, what is the probability that at least one of the systems will detect the attack?

    c. If three systems are installed, what is the probability that at least one of the systems will detect the attack?

    d. Would you recommend that multiple detection systems be operated? Explain.

29. Assume that the binomial distribution applies for the case of a college basketball player shooting free throws. Late in a basketball game, a team will sometimes foul intentionally in the hope that the player shooting the free throw will miss and the team committing the foul will get the ball. Assume that the best player on the opposing team has a .82 probability of making a free throw and that the worst player has a .56 probability of making a free throw.

    a.   What are the probabilities that the best player makes 0, 1, and 2 points if fouled and given two free throws?

    b.   What are the probabilities that the worst player makes 0, 1, and 2 points if fouled and given two free throws?

    c.   Does it make sense for a coach to have a preset plan about which player to foul intentionally late in a basketball game? Explain.

30.   A firm estimates the probability of having employee disciplinary problems on any day to be .10.

    a.   What is the probability that the company experiences 5 days without a disciplinary problem?

    b.   What is the probability of exactly 2 days with disciplinary problems in a 10-day period?

    c.   What is the probability of at least 2 days with disciplinary problems in a 20-day period?

6.5 THE EXPECTED VALUE AND VARIANCE FOR THE BINOMIAL PROBABILITY DISTRIBUTION

In Section 6.3 we provided formulas for computing the expected value and variance of a discrete random variable. In the special case where the random variable has a binomial probability distribution with a known number of trials (n) and known probabilities, the general formulas for the expected value and variance (6.3) and (6.4) can be simplified; the results are as follows.

Expected Value and Variance for the Binomial Probability Distribution

$$E(x) = \mu = np \tag{6.10}$$

$$\text{Var}(x) = \sigma^2 = np(1 - p) \tag{6.11}$$

EXAMPLE 6.11

A statistics class has 25 students. From past experience it is known that 20% of the students withdraw from the course before the end of the term. What is the expected number of withdrawals, and what is the variance in the number of withdrawals?

    Assuming each student makes a decision independently, the process can be viewed as a binomial experiment with $n = 25$, $p = .20$, and $(1 - p) = 1 - .20 = .80$. The random variable of interest is the number of withdrawals during the term.

    Using (6.10) and (6.11), the expected value and variance of the number of withdrawals are as follows:

$$\mu = np = (25)(.20) = 5$$

$$\sigma^2 = np(1 - p) = (25)(.20)(.80) = 4$$

The corresponding standard deviation is $\sigma = \sqrt{4} = 2$.

• • •

EXAMPLE 6.12

A shipment of 500 parts is received from a supplier. From past experience, it is known that the probability that any particular part is defective is .03. What is the expected number of defective parts in the shipment? What are the variance and standard deviation in the number of defective parts?

Viewing this as a binomial experiment with $n = 500$, $p = .03$, and $1 - p = 1 - .03 = .97$, we have:

$$\mu = (500)(.03) = 15$$

$$\sigma^2 = (500)(.03)(.97) = 14.5500$$

$$\sigma = \sqrt{14.5500} = 3.8144$$

• • •

EXERCISES

31. Suppose a salesperson makes a sale on 20% of customer contacts. A normal work week will enable the salesperson to contact 25 customers. What is the expected number of sales for the week? What is the variance for the number of sales for the week? What is the standard deviation for the number of sales for the week?

32. Eighty-five percent of the next-day express mailings handled by the U.S. Postal Service are actually received by the addressee one day after the mailing. What is the expected value and variance for the number of 1-day deliveries in a group of 250 express mailings?

33. Betting on the color red in the game of roulette has a 18/38 chance of winning. What is the expected value and variance for the number of wins in a series of 100 bets on red?

Summary

The concepts of a random variable and its probability distribution were introduced. Although there are two types of random variables, discrete and continuous, the emphasis in this chapter was on discrete random variables. We saw that probability distributions for discrete random variables can be given by tables or graphs that show the values the random variable may assume, together with the associated probabilities. In some cases the probability distribution is defined by a special mathematical formula, which, for each value of the random variable x, provides the associated probability $f(x)$. The binomial probability distribution is one widely used distribution defined by such a special mathematical formula.

The binomial probability distribution can be used to determine the probability of x successes in n trials whenever the experiment has the following properties:

1. The experiment consists of a sequence of n identical trials.
2. Two outcomes are possible on each trial, one called success and the other failure.

3. The probability of a success, p, does not change from trial to trial. Consequently, the probability of failure, $1 - p$, does not change from trial to trial.

4. The trials are independent.

When the above conditions hold, a binomial probability function, or a table of binomial probabilities, can be used to determine the probability of x successes in n trials. Formulas were also presented for the mean and variance of the binomial probability distribution.

Statistics in Practice

ACCEPTANCE SAMPLING

In an attempt to obtain high-quality products for its customers, many manufacturing firms are using statistical sampling as a quality control procedure to reduce the number of defective items in their manufacturing processes. Sampling both incoming raw materials and outgoing finished products provides opportunities to reduce the number of defective products passed on to the customers.

In cases where large quantities of raw materials and/or finished goods are handled, it is time-consuming and expensive to inspect or test every item. This is where the notion of acceptance sampling comes into practice. In acceptance sampling, a sample of items is selected from a shipment or a production batch; each item sampled is then inspected. If the number of defective items found in the sample is too large the shipment or batch is rejected and either returned to the supplier, scrapped, or held for future testing. If the number of defective items found in the sample is reasonably small, the entire shipment or batch is accepted and passed on to the customer or to the next phase of the production process.

In the electronics industry it is common to have component parts shipped from suppliers in large lots. Inspection of a sample of n components can be viewed as n trials of a binomial experiment. The outcome for each component tested (trial) will be that the component is either good or defective. Assuming that the entire shipment has a defective proportion denoted by p, the binomial probability distribution can be used to find the probability of $0, 1, 2, \ldots, n$ defective items appearing in the sample of n components.

Based on binomial probabilities, a quality control analyst can determine an acceptance number (maximum acceptable number of defective items), which will balance the costs of accepting a lot with a large number of defective items with the cost of rejecting a lot with a small number of defective items. The acceptance sampling decision rule used by inspectors will be to accept the lot if the number of defective items in the sample does not exceed the acceptance number and reject the lot if more defective items are found in the sample.

Considering the competitive pressures to produce high-quality products, acceptance sampling offers an economical way to minimize the number of defective products being passed on to the consumer. The binomial probability distribution plays an important role in establishing decision rules for the acceptance sampling procedure.

Glossary

Random variable—A numerical description of the outcome of an experiment

Discrete random variable—A random variable that can assume only a finite or infinite sequence of values.

Continuous random variable—A random variable that may assume all values in an interval or collection of intervals.

Probability function—A function, denoted $f(x)$, that gives the probability that the discrete random variable x assumes a particular value.

Discrete probability distribution—A table, graph, or equation showing the values of a discrete random variable and the associated probabilities.

Expected value—A measure of the average, or central, value of a random variable.

Variance—A measure of the dispersion, or variability, of a random variable.

Standard deviation—The positive square root of the variance.

Binomial experiment—A probability experiment possessing the four properties stated in Section 6.4.

Binomial probability distribution—A probability distribution showing the probability of x successes in n trials of a binomial experiment.

Binomial probability function—The function used to compute probabilities in a binomial experiment.

Key Formulas

Expected Value of a Discrete Random Variable

$$E(x) = \mu = \Sigma\, xf(x) \tag{6.3}$$

Variance of a Discrete Random Variable

$$\mathrm{Var}(x) = \sigma^2 = \Sigma\, (x - \mu)^2 f(x) \tag{6.4}$$

Computational Formula for Variance of a Discrete Random Variable

$$\mathrm{Var}(x) = \sigma^2 = \Sigma\, x^2 f(x) - \mu^2 \tag{6.5}$$

Number of Experimental Outcomes Providing Exactly x Successes in n Trials

$$\frac{n!}{x!(n - x)!} \tag{6.6}$$

Binomial Probability Function

$$f(x) = \frac{n!}{x!(n - x)!}\, p^x (1 - p)^{(n-x)} \tag{6.9}$$

Expected Value for the Binomial Probability Distribution

$$E(x) = \mu = np \tag{6.10}$$

Variance for the Binomial Probability Distribution

$$\mathrm{Var}(x) = \sigma^2 = np(1 - p) \tag{6.11}$$

Review Quiz

TRUE/FALSE

1. A random variable may assume only numerical values.
2. A random variable that may assume any value between 5 and 6 is a discrete random variable.
3. The sum of the probabilities for all values that a discrete random variable may assume cannot be greater than 1.
4. The expected value for a random variable must be a value the random variable can assume.
5. The variance of a random variable is the sum of the squared deviations from the mean.
6. In a binomial experiment the probability of success is 1 minus the probability of failure.
7. The binomial random variable is a discrete random variable.
8. In a 4-trial binomial experiment, there are 12 experimental outcomes.
9. In a 5-trial binomial experiment, there are 6 possible values for the random variable.
10. In a binomial experiment involving 4 trials, there are 3 experimental outcomes yielding 3 successes.
11. In a binomial experiment involving 3 trials with a success probability of .3, the probability of 1 success is .441.
12. In a binomial experiment involving 100 trials with a success probability of .22, the expected number of successes is less than or equal to 20.

MULTIPLE CHOICE

For questions 13–17, consider the random variable x, which gives the number of successes in six identical trials, each of which has a probability of success of .3.

13. The random variable x is
    a. discrete
    b. continuous
    c. normal
    d. uniform
14. The probability of no successes is
    a. .0000
    b. .1176
    c. .3025
    d. .1780

15. The probability of one success is
    a. .0000
    b. .1176
    c. .3025
    d. .1780

16. the probability of at least four successes is
    a. .0704
    b. .0595
    c. .0109
    d. not able to be computed from the information given

17. The expected value of x is closest to
    a. 1.50
    b. 1.75
    c. 2.00
    d. 3.00

18. A random variable assumes the values 1, 2, and 3 with probabilities .10, .60, .30, respectively. The expected value of this random variable is
    a. 1.80
    b. 2.00
    c. 2.20
    d. 3.00

19. The variance of the random variable in question 18 is
    a. 2.00
    b. 2.20
    c. 8.22
    d. 9.16

20. Consider a binomial random variable with $p = .4$ and $n = 18$. The standard deviation is
    a. 7.20
    b. 4.32
    c. 2.08
    d. none of the above

Supplementary Exercises

34. An automobile agency located in Beverly Hills specializes in the rental of luxury automobiles. Assume that the probability distribution of daily demand at their agency is as follows.

x	$f(x)$
0	.15
1	.30
2	.40
3	.10
4	.05
	1.00

a. Compute the expected value of daily demand.

b. If the daily rental cost for an automobile is $75 per day, what is the expected value of daily automobile rental?

35. At a large university, the number of student problems handled by the dean for student affairs varies from semester to semester. Assume that the number of student problems (x) handled by the dean has the following probability distribution.

x	f(x)
0	.10
1	.15
2	.30
3	.25
4	.10
5	.10
Total	1.00

What are the mean and variance of the number of student problems handled by the dean each semester?

36. Which of the following are and which are not probability distributions? Explain.

x	f(x)	y	f(y)	z	f(z)
0	.20	0	.25	−1	.20
1	.30	2	.05	0	.50
2	.25	4	.10	1	−.10
3	.35	6	.60	2	.40

37. The number of weekly lost-time injuries at a particular plant (x) has the following probability distribution.

x	f(x)
0	.05
1	.20
2	.40
3	.20
4	.15

a. Compute the expected value.

b. Compute the variance.

38. Assume that the plant in Exercise 37 initiated a safety training program and that the number of lost-time injuries during the 20 weeks following the training program was as follows.

Number of Injuries	Number of Weeks
0	2
1	8
2	6
3	3
4	1
	Total 20

a. Construct a probability distribution for weekly lost-time injuries based on these data.

b. Compute the expected value and the variance and use both to evaluate the effectiveness of the safety training program.

39. The Hub Real Estate Investment stock is currently selling for $16 per share. An investor plans to buy shares and hold the stock for 1 year. Let x be the random variable indicating the price of the stock after 1 year. The probability distribution for x is shown.

Price of Stock (x)	$f(x)$
16	.35
17	.25
18	.25
19	.10
20	.05

a. Show that the above probability distribution possesses the properties of all probability distributions.

b. What is the expected price of the stock after 1 year?

c. What is the expected gain per share of the stock over the 1-year period? What percent return on the investment is reflected by this expected value?

d. What is the variance in the price of the stock over the 1-year period?

e. Another stock with a similar expected return has a variance of 3. Which stock appears to be the better investment in terms of minimizing risk or uncertainty associated with the investment? Explain.

40. The budgeting process for a midwestern college resulted in expense forecasts for the coming year (in 1,000,000s) of $9, $10, $11, $12, and $13. Since the actual expenses were unknown, the following respective probabilities were assigned: .3, .2, .25, .05, and .2.

a. Show the probability distribution for the expense forecast.

b. What is the expected value of the expenses for the coming year?

c.   What is the variance in the expenses for the coming year?

d.   If income projections for the year are estimated at $12 million, comment on the financial position of the college.

41. Exercise 8 provided a probability function for x, the hours required to change over a production system, as follows.

$$f(x) = \frac{x}{10} \qquad \text{for } x = 1, 2, 3, \text{ or } 4.$$

a.   What is the expected value of the changeover time?

b.   What is the variance of the changeover time?

42. The police department of a major midwestern city makes arrests on 40% of its reported robberies. Assume that we are interested in the number of arrests that will be made in the next 20 reported robberies. Describe whether or not you feel the properties of a binomial experiment are satisfied.

43. It is estimated that 45% of all marriages end in a divorce.

a.   Consider a family with two children. Assume that both children will eventually enter into a marriage relationship. What is the probability at least one of the children will experience a divorce?

b.   Repeat part (a) for a family with three children.

c.   Repeat part (a) for a family with four children.

d.   Repeat part (a) for a family with five children.

e.   Do you feel any of the properties of the binomial experiment may not hold in this situation? What ones and why?

44. A large church has 56% female and 44% male members. Assume 6 individuals are randomly selected for a worship committee.

a.   What is the probability that the committee will consist of three women and three men?

b.   What is the probability that the committee will consist of more women than men?

c.   What is the probability that the committee will consist of more men than women?

45. A new clothes-washing compound is found to satisfactorily remove excessive dirt and stains on 88% of the items washed. Assume that 10 items are to be washed with the new compound.

a.   What is the probability of satisfactory results on all 10 items?

b.   What is the probability at least 2 items are found with unsatisfactory results?

46. In an audit of a company's billings, an auditor randomly selects five bills. If 3% of all bills contain an error, what is the probability that the auditor will find the following?

a.   exactly one bill in error

b.   at least one bill in error

47. A salesperson contacts 8 potential customers per day. From past experience we know that the probability of a potential customer making a purchase is .10.

a.   What is the probability the salesperson makes *exactly* two sales in a day?

b.   What is the probability the salesperson makes *at least* two sales in a day?

c.   What percentage of the days will the salesperson not make a sale?

d.   What is the expected number of sales per day? Over a 5-day week, how many sales are expected?

48. In the Butler Football League, 5% of the games end in a tie. With 208 games played during the regular season, what are the mean and variance for the number of games that end in a tie?

49. A particular television show has captured 25% of the viewing audience.

    a. In a sample of 20 television-viewing households, what is the probability at least 3 households watch the show?

    b. In a sample of 20 television-viewing households, what is the probability at least 5 households watch the show?

    c. In a sample of 200 television viewing households, what are the mean and variance for the number of households that watch the show?

CHAPTER 7

The Normal Probability Distribution

What You Will Learn in This Chapter

- the properties of the normal probability distribution

- how to use the standard normal probability distribution to compute probabilities

- how to use the normal probability distribution to approximate binomial probabilities

Contents

Statistics in the News

IQ SCORES: ARE YOU NORMAL?

Intelligence test scores, referred to as intelligence quotient, or IQ scores, are based on characteristics such as verbal skill, abstract reasoning power, numerical ability, and spatial visualization. An IQ score of 100 is considered average. If plotted on a graph with IQ scores on the horizontal axis, the distribution of intelligence test scores approximates a bell-shaped normal probability curve. This distribution shows that the greatest concentration of scores are near 100 and that the frequency of scores decreases gradually and symmetrically as the extremes of intelligence are approached.

Knowing your IQ score gives you an indication of how you rate or compare to other individuals in terms of intelligence. An IQ score above 115 is considered superior. Studies of intellectually gifted children have generally set the lower limit at an IQ score of 140. Approximately 1% of the population has IQ scores of 140 or more. The average IQ score for a high-school graduate is 110. The average IQ score of a college graduate is 120, and the average IQ score of a person with a Ph.D. is 130.

IQ test and achievement test scores help teachers determine if students are working up to their abilities.

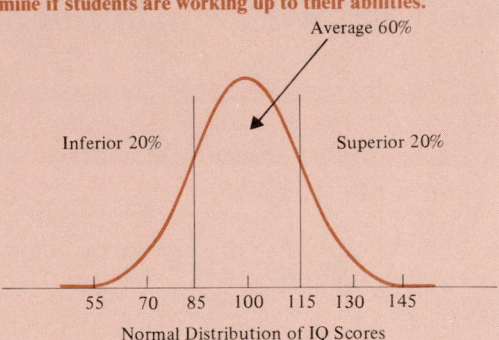

Normal Distribution of IQ Scores

Based on "Your Intelligence Quotient," Tom Biracree, in *How You Rate.* New York: Dell Publishing Co., Inc., 1984.

In this chapter we introduce the *normal probability distribution.* The normal random variable is continuous; it can assume an infinite number of values. Since it can assume an infinite number of values, we cannot list each value the random variable and then identify the associated probability. Indeed, with a continuous random variable, the probability of any single value is zero. Interest focuses on computing the probability the random variable assumes any value in a given interval.

The graph of the normal probability distribution takes the form of a bell-shaped curve mentioned in "Statistics in the News." Because of its many practical applications, the normal probability distribution is the one most widely used in statistical analysis.

7.1 THE NORMAL CURVE

The normal probability distribution has been applied in a wide variety of practical applications in which the random variables involved are heights and weights of people, IQ scores, scientific measurements, amounts of rainfall, and so on. In order to use this

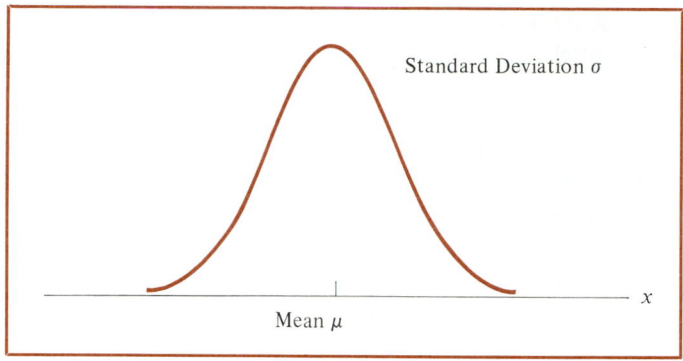

FIGURE 7.1

Bell-Shaped Curve for the Normal Probability Distribution

probability distribution, the random variable must be continuous. However, as we shall see, a continuous normal random variable is often used as an approximation in situations involving discrete random variables. The form, or shape, of the normal probability distribution is illustrated by the bell-shaped curve shown in Figure 7.1. The mathematical equation that describes the bell-shaped curve of the normal probability distribution is given by

$$f(x) = \frac{1}{\sqrt{2\pi}\,\sigma}\, e^{-(x-\mu)^2/2\sigma^2}$$

where μ is the mean, σ is the standard deviation, $\pi = 3.14159$ and $e = 2.71828$. The value of $f(x)$ for any choice of x gives the height of the curve (see Figure 7.1).

We make some observations about the characteristics of the normal probability distribution:

1. There is an entire family of normal probability distributions with each specific normal distribution being differentiated by its mean μ and its standard deviation σ.

2. The highest point on the normal curve occurs at the mean, which is also the median and mode of the distribution.

3. The mean of the distribution can be any numerical value: negative, zero, or positive. Three normal curves with the same standard deviation but three different means (-10, 0, and 20) are shown.

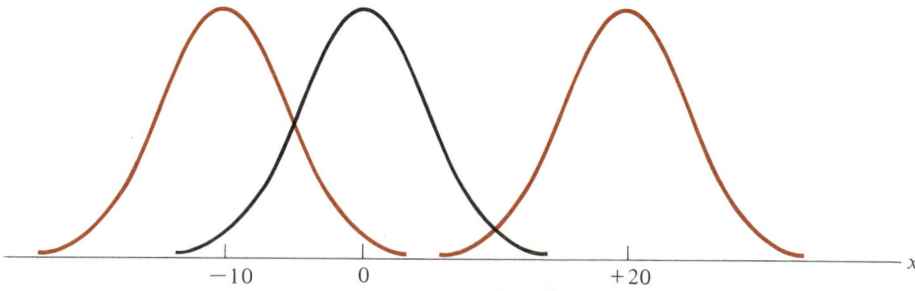

4. The normal probability distribution is symmetric, with the tails of the curve extending indefinitely in both directions and theoretically never touching the horizontal axis.

5. The standard deviation determines the width of the curve. Larger values of the standard deviation result in wider, flatter curves, showing more dispersion in the data. Two normal distributions with the same mean but different values for the standard deviation are shown.

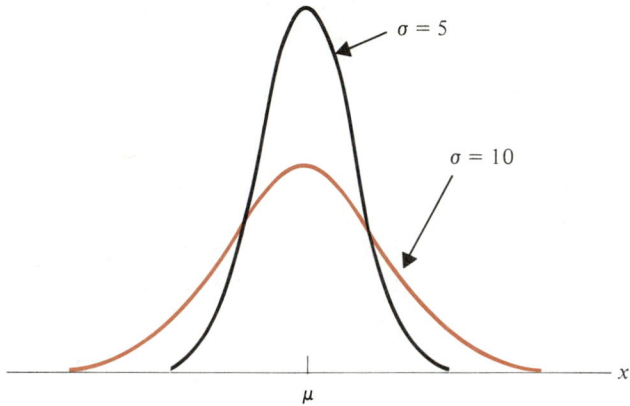

6. The total area under the curve for the normal probability distribution is 1. (This is true for all continuous probability distributions.)

7. Regardless of the value of the mean, μ, and the standard deviation, σ, probabilities for the normal random variable are given by areas under the curve. Probabilities for some commonly used intervals are:

   a. 68.26% of the time, a normal random variable assumes a value within plus or minus 1 standard deviation of its mean.

   b. 95.44% of the time, a normal random variable assumes a value within plus or minus 2 standard deviations of its mean.

   c. 99.72% of the time, a normal random variable assumes a value within plus or minus 3 standard deviations of its mean. Figure 7.2 shows properties (a), (b), and (c) graphically.

EXERCISES

1. Using Figure 7.2 as a guide, sketch a normal curve for a random variable x that has a mean $\mu = 100$ and a standard deviation $\sigma = 10$. Label the horizontal axis with values of 70, 80, 90, 100, 110, 120 and 130.

2. The length of time required to complete a college examination is normally distributed with a mean of $\mu = 50$ minutes and a standard deviation of $\sigma = 5$ minutes.

   a. Sketch a normal curve for the length of the examination. Label the horizontal axis with values of 35, 40, 45, 50, 55, 60 and 65 minutes. Figure 7.2 shows that the normal curve almost touches the horizontal line at three standard deviations below and at three standard deviations above the mean (in this case at 35 and 65).

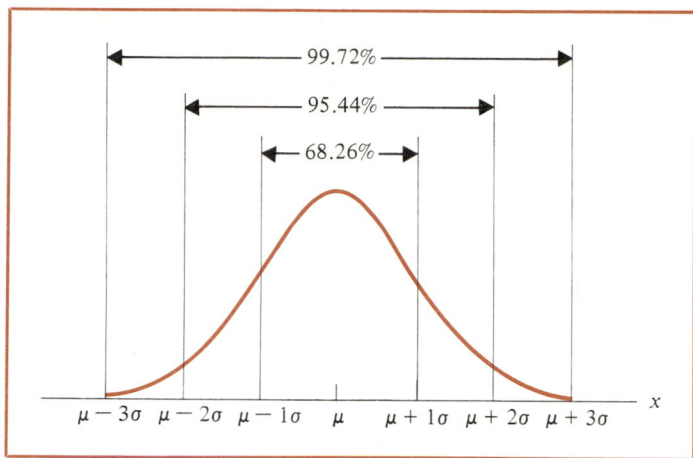

FIGURE 7.2
Areas Under the Curve for Any Normal Probability Distribution

    b.   What is the probability that a student will take between 45 and 55 minutes to complete the exam?

    c.   What is the probability that a student will take between 40 and 60 minutes to complete the exam?

3. Assume that scores on a verbal skills test are normally distributed with a mean of 500 and a standard deviation of 100.

    a.   Sketch a normal curve for the test scores. Refer to Figure 7.2 to see that the normal curve almost touches the horizontal axis at three standard deviations below and three standard deviations above the mean.

    b.   What is the probability that a randomly selected student who takes the test scores between 400 and 600?

    c.   What is the probability that a randomly selected student who takes the test scores between 300 and 700?

    d.   What is the probability that a randomly selected student scores 500 or better on the exam?

    e.   What is the probability that a randomly selected student scores between 500 and 700 on the exam?

    f.   What is the probability that a randomly selected student who takes the test scores 700 or higher? (*Hint:* Use your answers to (d) and (e) to help answer this question.)

4. The mean annual precipitation in the state of Ohio is 32 inches with a standard deviation of 4 inches.

    a.   Sketch a normal curve for the annual precipitation in the state of Ohio. Label the horizontal axis at the mean as well as at 1, 2, and 3 standard deviations above and below the mean.

    b.   What is the probability that the precipitation in any one year will be between 24 and 40 inches?

    c.   What is the probability that the precipitation in any one year will be between 32 and 40 inches?

    d. What is the probability that the precipitation in any one year will be 32 inches or more?

    e. What is the probability that the precipitation in any one year will be 36 inches or more?

5. Automobile painting times at the Jay Nickerson Auto Body Painting Shop are believed to be normally distributed. It is estimated that 68.26% of the automobiles are painted in from 2.5 to 3.5 hours. In addition, it is estimated that 95.44% of the automobiles are painted in from 2 to 4 hours.

    a. What is the mean painting time for the automobiles?

    b. What is the standard deviation of painting times for the automobiles?

    c. What is the probability that an automobile will be painted in 3 hours or less?

    d. What is the probability that an automobile will require between 2 and 3 hours for painting?

    e. What is the probability that an automobile will be painted in 2 hours or less?

    f. What is the probability that an automobile will require 2 hours or more to be painted?

7.2 THE STANDARD NORMAL PROBABILITY DISTRIBUTION

A random variable that has a normal distribution with a mean of 0 and a standard deviation of 1 is said to have a *standard normal probability distribution*. The letter z is commonly used to designate this particular normal random variable. The graph of the standard normal probability distribution is shown in Figure 7.3. This too is a normal probability distribution; hence it has the same general appearance as other normal distributions but with the special properties of $\mu = 0$ and $\sigma = 1$.

With a continuous random variable, probability calculations are always concerned with finding the probability that the random variable assumes any value in an interval between two specific points a and b. The probability that a continuous random

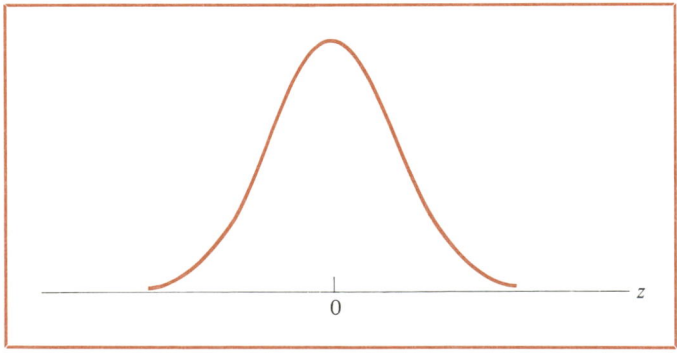

FIGURE 7.3
The Standard Normal Probability Distribution

TABLE 7.1

Areas, or Probabilities, for the Standard Normal Distribution

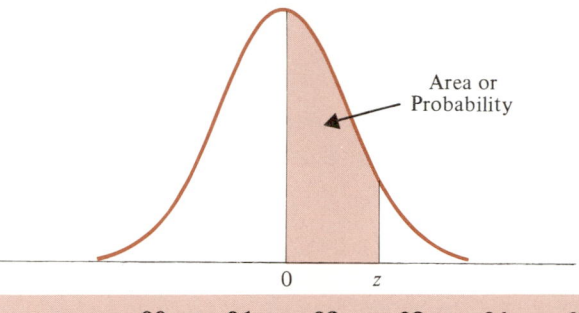

Area or Probability

z	.00	.01	.02	.03	.04	.05	.06	.07	.08	.09
.0	.0000	.0040	.0080	.0120	.0160	.0199	.0239	.0279	.0319	.0359
.1	.0398	.0438	.0478	.0517	.0557	.0596	.0636	.0675	.0714	.0753
.2	.0793	.0832	.0871	.0910	.0948	.0987	.1026	.1064	.1103	.1141
.3	.1179	.1217	.1255	.1293	.1331	.1368	.1406	.1443	.1480	.1517
.4	.1554	.1591	.1628	.1664	.1700	.1736	.1772	.1808	.1844	.1879
.5	.1915	.1950	.1985	.2019	.2054	.2088	.2123	.2157	.2190	.2224
.6	.2257	.2291	.2324	.2357	.2389	.2422	.2454	.2486	.2518	.2549
.7	.2580	.2612	.2642	.2673	.2704	.2734	.2764	.2794	.2823	.2852
.8	.2881	.2910	.2939	.2967	.2995	.3023	.3051	.3078	.3106	.3133
.9	.3159	.3186	.3212	.3238	.3264	.3289	.3315	.3340	.3365	.3389
1.0	.3413	.3438	.3461	.3485	.3508	.3531	.3554	.3577	.3599	.3621
1.1	.3643	.3665	.3686	.3708	.3729	.3749	.3770	.3790	.3810	.3830
1.2	.3849	.3869	.3888	.3907	.3925	.3944	.3962	.3980	.3997	.4015
1.3	.4032	.4049	.4066	.4082	.4099	.4115	.4131	.4147	.4162	.4177
1.4	.4192	.4207	.4222	.4236	.4251	.4265	.4279	.4292	.4306	.4319
1.5	.4332	.4345	.4357	.4370	.4382	.4394	.4406	.4418	.4429	.4441
1.6	.4452	.4463	.4474	.4484	.4495	.4505	.4515	.4525	.4535	.4545
1.7	.4554	.4564	.4573	.4582	.4591	.4599	.4608	.4616	.4625	.4633
1.8	.4641	.4649	.4656	.4664	.4671	.4678	.4686	.4693	.4699	.4706
1.9	.4713	.4719	.4726	.4732	.4738	.4744	.4750	.4756	.4761	.4767
2.0	.4772	.4778	.4783	.4788	.4793	.4798	.4803	.4808	.4812	.4817
2.1	.4821	.4826	.4830	.4834	.4838	.4842	.4846	.4850	.4854	.4857
2.2	.4861	.4864	.4868	.4871	.4875	.4878	.4881	.4884	.4887	.4890
2.3	.4893	.4896	.4898	.4901	.4904	.4906	.4909	.4911	.4913	.4916
2.4	.4918	.4920	.4922	.4925	.4927	.4929	.4931	.4932	.4934	.4936
2.5	.4938	.4940	.4941	.4943	.4945	.4946	.4948	.4949	.4951	.4952
2.6	.4953	.4955	.4956	.4957	.4959	.4960	.4961	.4962	.4963	.4964
2.7	.4965	.4966	.4967	.4968	.4969	.4970	.4971	.4972	.4973	.4974
2.8	.4974	.4975	.4976	.4977	.4977	.4978	.4979	.4979	.4980	.4981
2.9	.4981	.4982	.4982	.4983	.4984	.4984	.4985	.4985	.4986	.4986
3.0	.4986	.4987	.4987	.4988	.4988	.4989	.4989	.4989	.4990	.4990

variable assumes a value between the two points a and b is the area under the graph of the probability distribution between a and b. Areas under the normal curve have been computed and are available in tables that can be used in computing the probability values for the standard normal probability distribution. Table 7.1 is such a table. This table is also available as Table 1 of Appendix B as well as inside the back cover of the text.

Let us show how the table of areas for the standard normal probability distribution (Table 7.1) can be used to find areas or probabilities by considering some examples. Later we will see how this same table can be used to compute probabilities for any normal distribution.

EXAMPLE 7.1

What is the probability that the z value for the standard normal random variable will be between 0.00 and 1.00? That is, what is $P(0.00 \leq z \leq 1.00)$? The shaded region in the following graph shows this area or probability.

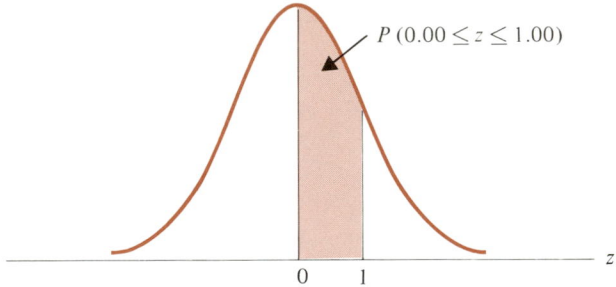

The entries in Table 7.1 give the area under the standard normal curve between the mean, $z = 0$, and a specified positive value of z. In this case we are interested in the area between $z = 0$ and $z = 1.00$. Thus we must find the entry in the table corresponding to $z = 1.00$. To do this, we first find 1.0 in the left-hand column of the table and then find .00 in the top row of the table. Then by looking in the body of the table we find that the 1.0 row of the table and the .00 column of the table intersect at the value of .3413. We have found the desired probability; $P(0.00 \leq z \leq 1.00) = .3413$. A portion of Table 7.1 showing these steps is shown below.

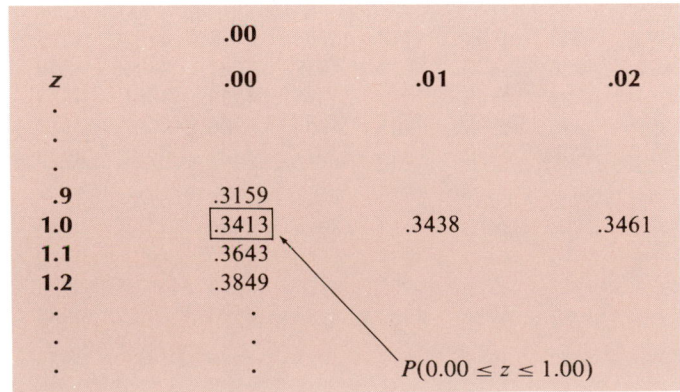

As another example, use Table 7.1 to show that the area between $z = 0.00$ and $z = 1.25$ is .3944. This area or probability value is found by using the $z = 1.2$ row and the .05 column of the table.

• • •

EXAMPLE 7.2

What is the probability of obtaining a z value between $z = -1.00$ and $z = 1.00$? That is, what is $P(-1.00 \leq z \leq 1.00)$?

First note that we have already used Table 7.1 to show that the probability of a z value between $z = 0.00$ and $z = 1.00$ is .3413. Recall now that the normal probability distribution is *symmetric*. That is, the shape of the curve to the left of the mean is the mirror image of the shape of the curve to the right of the mean. Thus the probability of a z value between $z = 0.00$ and $z = -1.00$ is the *same* as the probability of a z value between $z = 0.00$ and $z = +1.00$. Hence the probability of a z value between $z = -1.00$ and $z = +1.00$ is

$$P(-1.00 \leq z \leq 0.00) + P(0.00 \leq z \leq 1.00) = .3413 + .3413 = .6826.$$

This area is shown graphically as follows.

• • •

In a manner similar to Example 7.2 we can use the values in Table 7.1 to show that the probability of a z value between -2.00 and $+2.00$ is $.4772 + .4772 = .9544$ and that the probability of a z value between -3.00 and $+3.00$ is $.4986 + .4986 = .9972$. Since we know that the total probability or total area under the curve for any continuous random variable must be 1.0000, the probability .9972 tells us that the value of z will almost always fall between -3.00 and $+3.00$.

EXAMPLE 7.3

What is the probability of obtaining a z value of at least 1.58? That is, what is $P(z \geq 1.58)$?

First, we use the $z = 1.5$ row and the .08 column of Table 7.1 to find that $P(0.00 \leq z \leq 1.58) = .4429$. Now, since the normal probability distribution is symmetric and the total area under the curve equals 1, we know that 50% of the area must be above the mean (i.e., $z = 0$) and 50% of the area must be below the mean. Since .4429 is the area between the mean and $z = 1.58$, the area or probability corresponding to $z \geq 1.58$ must be $.5000 - .4429 = .0571$. This probability is shown below.

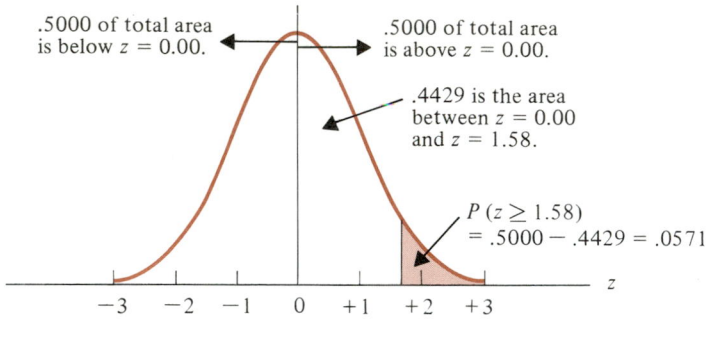

. . .

EXAMPLE 7.4

What is the probability the random variable z assumes a value of $-.50$ or larger? That is, what is $P(z \geq -.50)$.

To make this computation, we note that the probability we are seeking can be written as the sum of two probabilities: $P(z \geq -.50) = P(-.50 \leq z \leq 0.00) + P(z \geq 0.00)$. We have previously seen that $P(z \geq 0.00) = .50$. Also, we know that since the normal distribution is symmetric, $P(-.50 \leq z \leq 0.00) = P(0.00 \leq z \leq .50)$. Referring to Table 7.1 we find that $P(0.00 \leq z \leq .50) = .1915$. Therefore, $P(-.50 \leq z \leq 0.00) = .1915$. Thus $P(z \geq -.50) = P(-.50 \leq z \leq 0.00) + P(z \geq 0.00) = .1915 + .5000 = .6915$. The graph shows this area.

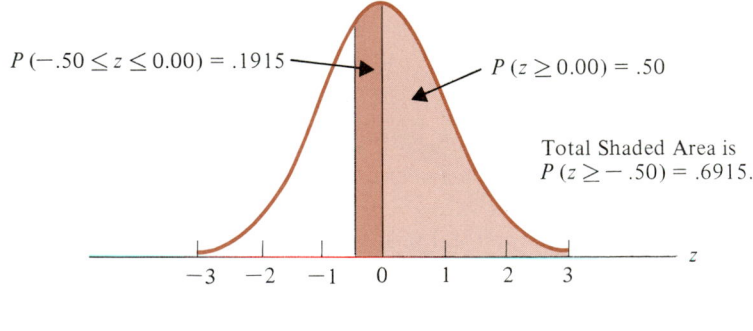

. . .

EXAMPLE 7.5

What is the probability of obtaining a z value between 1.00 and 1.58. That is, what is $P(1.00 \leq z \leq 1.58)$?

From Examples 7.1 and 7.3 we know that there is a .3413 probability of a z value between $z = 0.00$ and $z = 1.00$ and that there is a .4429 probability of a z value between $z = 0.00$ and $z = 1.58$. Thus there must be a $.4429 - .3413 = .1016$ probability of a z

value between $z = 1.00$ and $z = 1.58$. Thus $P(1.00 \leq z \leq 1.58) = .1016$. This situation is shown graphically in the following figure.

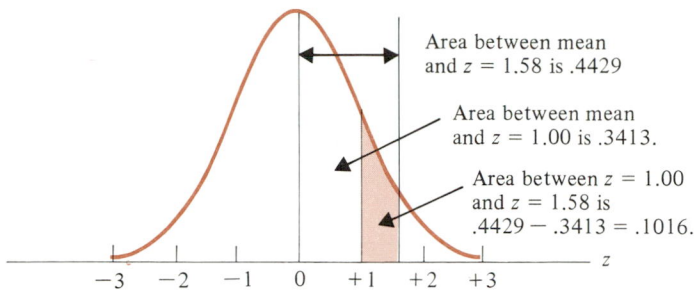

$\bullet \quad \bullet \quad \bullet$

EXAMPLE 7.6

Find a z value such that the probability of obtaining a larger z value is only $.10$. This situation is shown graphically as follows:

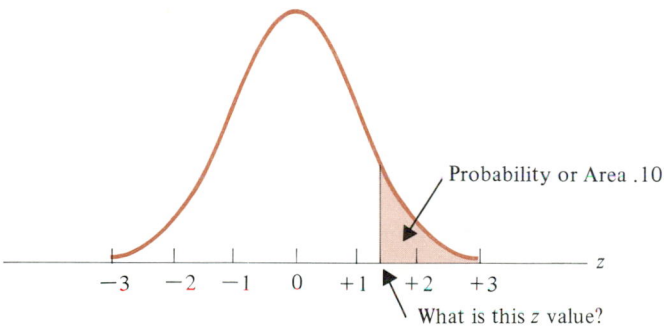

This problem is somewhat different from the examples we have considered thus far. The difference is that previously we specified the z value of interest and then found the corresponding probability, or area. In this example we are given the probability, or area, information and asked to find the corresponding z value. This can be found by using the table of areas for the standard normal probability distribution (Table 7.1) a little differently.

Recall that the body of Table 7.1 gives the area under the curve between the mean and a particular z value. In the above example we are given the information that the area in the upper tail of the curve is $.10$. Thus we must determine how much of the area is between the mean and the z value of interest. Since we know $.5000$ of the area is above the mean, $.5000 - .1000 = .4000$ must be the area under the curve *between* the mean and the desired z value. Scanning the body of the table, we find $.3997$ as the probability value closest to .4000. The section of the table providing this result is shown below:

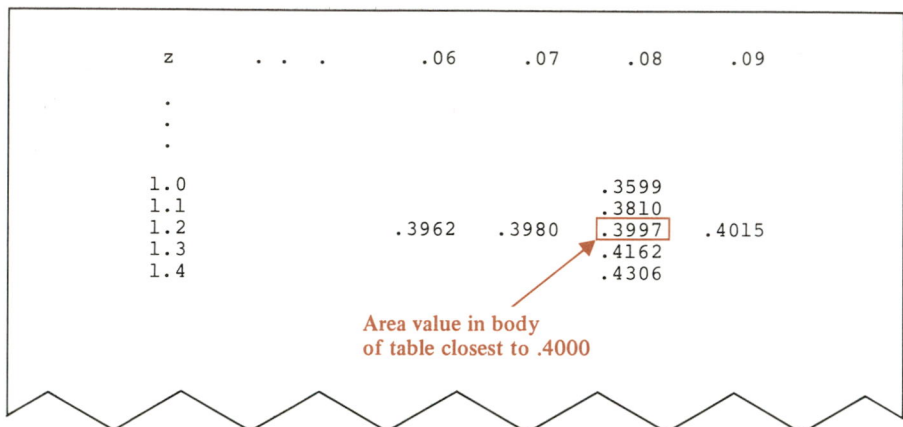

z	. . .	.06	.07	.08	.09
.					
.					
.					
1.0				.3599	
1.1				.3810	
1.2		.3962	.3980	.3997	.4015
1.3				.4162	
1.4				.4306	

Area value in body
of table closest to .4000

Reading the z value from the left column and the top row of the table, we find that the corresponding z value is 1.28. Thus there will be an area of approximately .4000 (actually .3997) between the mean and $z = 1.28$*. In terms of the question originally asked, there is approximately a .1000 probability of a z value larger than 1.28.

• • •

The examples illustrate that the table of areas for the standard normal probability distribution can be used to find probabilities associated with values of the standard normal random variable z. Two types of questions can be asked. The first type of question specifies a value, or values, for z and asks us to use the table to determine the corresponding areas, or probabilities. The second type of question provides an area, or probability, and asks us to use the table to determine the corresponding z value. Thus we need to remain flexible in terms of using the standard normal probability table to answer the desired probability question. In most cases, sketching a graph of the standard normal probability distribution and shading the appropriate area helps to visualize the situation and aid in determining the correct answer.

EXERCISES

6. Given that z is the standard normal random variable, sketch the standard normal curve. Label the horizontal axis at values of $-3, -2, -1, 0, 1, 2,$ and 3. Then use the table of probabilities for the standard normal distribution to compute the following probabilities.
    a. $P(0 \leq z \leq 1)$
    b. $P(0 \leq z \leq 1.5)$
    c. $P(0 \leq z \leq 2)$
    d. $P(0 \leq z \leq 2.5)$

*We could use interpolation in the body of the table to get a better approximation of the z value that cuts off an area of .4000. Doing so to provide one more decimal place of accuracy would yield a z value of 1.282. However, in most practical situations sufficient accuracy is obtained by simply using the table value closest to the desired probability.

7. Given that z is the standard normal random variable, compute the following probabilities.
   a. $P(-1 \leq z \leq 0)$
   b. $P(-1.5 \leq z \leq 0)$
   c. $P(-2 \leq z \leq 0)$
   d. $P(-2.5 \leq z \leq 0)$
   e. $P(-3 \leq z \leq 0)$

8. Given that z is the standard normal random variable, compute the following probabilities.
   a. $P(0 \leq z \leq .83)$
   b. $P(-1.57 \leq z \leq 0)$
   c. $P(z \geq .44)$
   d. $P(z \geq -.23)$
   e. $P(z \leq 1.20)$
   f. $P(z \leq -.71)$

9. Given that z is the standard normal random variable, compute the following probabilities.
   a. $P(-1.98 \leq z \leq .49)$
   b. $P(.52 \leq z \leq 1.22)$
   c. $P(-1.75 \leq z \leq -1.04)$

10. Given that z is the standard normal random variable, find z for each situation.
    a. The area between 0 and z is .4750.
    b. The area between 0 and z is .2291.
    c. The area to the right of z is .1314.
    d. The area to the left of z is .6700.

11. Given that z is the standard normal random variable, find z for each situation.
    a. The area to the left of z is .2119.
    b. The area between $-z$ and z is .9030.
    c. The area between $-z$ and z is .2052.
    d. The area to the left of z is .9948.
    e. The area to the right of z is .6915.

12. Given that z is the standard normal random variable, find z for each situation.
    a. The area to the right of z is .01.
    b. The area to the right of z is .025.
    c. The area to the right of z is .05.
    d. The area to the right of z is .10.

7.3 COMPUTING PROBABILITIES FOR ANY NORMAL DISTRIBUTION

The reason that we have been discussing the standard normal distribution so extensively is that probabilities for all normal distributions are computed using the standard normal distribution. That is when we have a normal distribution with any mean μ and any standard deviation σ, we answer probability questions about the distribution by first converting to the standard normal distribution. Then we can use Table 7.1 and the appropriate z values to find the desired probabilities. The formula used to convert any normal random variable x with mean μ and standard deviation σ to the standard normal distribution is as follows.

Converting to a Standard Normal Random Variable

$$z = \frac{x - \mu}{\sigma} \qquad\qquad (7.1)$$

A value of x equal to its mean μ results in $z = (\mu - \mu)/\sigma = 0$. Thus we see that a value of x equal to its mean μ corresponds to a value of z at its mean 0. Now suppose that x is one standard deviation above its mean; that is, $x = \mu + \sigma$. Applying Equation (7.1) we see that the corresponding z value is $z = [(\mu + \sigma) - \mu]/\sigma = \sigma/\sigma = 1$. Thus a value of x that is one standard deviation above its mean yields $z = 1$. In other words, we can interpret the z value as *the number of standard deviations that the normal random variable, x, is from its mean μ.*

To see how this conversion enables us to compute probabilities for any normal distribution, suppose we have a normal distribution with $\mu = 10$ and $\sigma = 2$. What is the probability that the random variable, x, is between 10 and 14? Using (7.1) we see that at $x = 10$, $z = (x - \mu)/\sigma = (10 - 10)/2 = 0$ and that at $x = 14$, $z = (14 - 10)/2 = 4/2 = 2$. Thus the answer to our question about the probability of x being between 10 and 14 is given by the equivalent probability that z is between 0 and 2 for the standard normal distribution. In other words, the probability that we are seeking is the probability that the random variable x is between its mean and two standard deviations above the mean. Using $z = 2.00$ and Table 7.1, we see that the probability is .4772. Hence the probability that x is between 10 and 14 is .4772.

EXAMPLE 7.7

IQ scores for a group of sixth graders are normally distributed with a mean of 100 and a standard deviation of 12. What is the probability of randomly selecting a student with an IQ score between 90 and 110?

Letting x be the normally distributed IQ score, we must compute $P(90 \leq x \leq 110)$ based on the information $\mu = 100$ and $\sigma = 12$. Converting the x values to the corresponding z values we have:

$$\text{For } x = 90, \qquad z = \frac{x - \mu}{\sigma} = \frac{90 - 100}{12} = -.83$$

$$\text{For } x = 110, \qquad z = \frac{x - \mu}{\sigma} = \frac{110 - 100}{12} = +.83$$

Using Table 7.1 for $z = +.83$, we find that the probability of z being between zero and $+.83$ is .2967. Also, since $P(-.83 \leq z \leq 0.00) = .2967$, we have $P(-.83 \leq z \leq +.83) = .2967 + .2967 = .5934$. In terms of the IQ scores, we now know the probability of randomly selecting a student with an IQ score between 90 and 110 is

.5934. The graphical representation of this probability with the corresponding z values is shown.

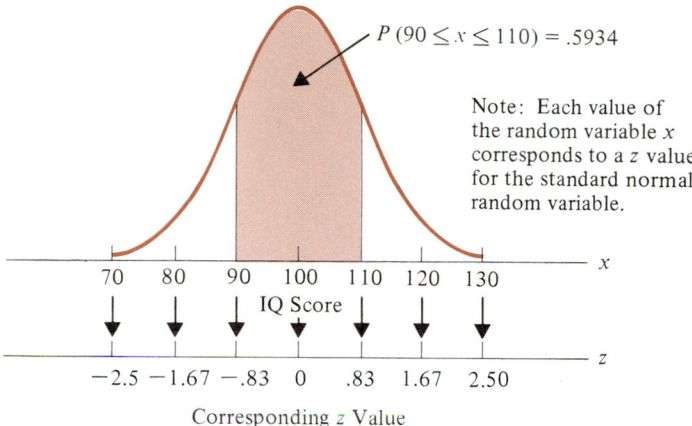

$P(90 \leq x \leq 110) = .5934$

Note: Each value of the random variable x corresponds to a z value for the standard normal random variable.

Corresponding z Value

Using the IQ scores with mean $\mu = 100$ and standard deviation $\sigma = 12$, what is the probability of randomly selecting a student with an IQ score of 120 or more?

For $x = 120$, we have

$$z = \frac{x - \mu}{\sigma} = \frac{120 - 100}{12} = 1.67$$

Using Table 7.1, we find an area of .4525 between $z = 0$ and $z = 1.67$. Thus .4525 is the probability that a student's IQ score is between the mean $\mu = 100$ and the IQ score of 120. This is not the answer to the question seeking the probability of randomly selecting a student with an IQ score of 120 or more. However, since .5000 of the area under a normal curve is above the mean, we see the probability of selecting a student with an IQ score of 120 or more must be $.5000 - .4525 = .0475$. Less than 5% of the students will have an IQ score of 120 or more.

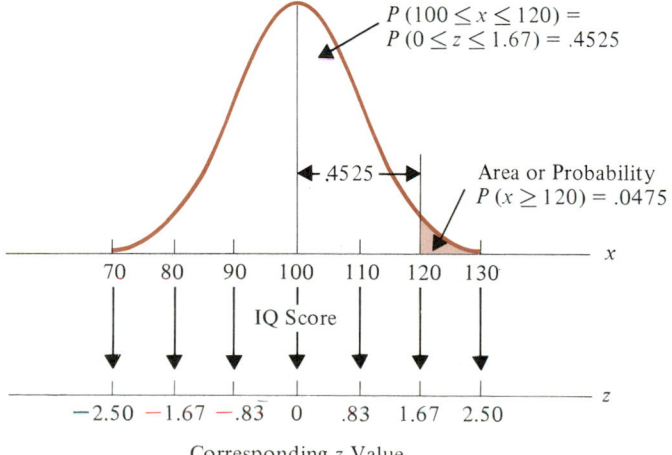

$P(100 \leq x \leq 120) =$
$P(0 \leq z \leq 1.67) = .4525$

.4525

Area or Probability
$P(x \geq 120) = .0475$

Corresponding z Value

EXAMPLE 7.8

The Grear Tire Company has just developed a new steel-belted radial tire that will be sold through a national chain of discount stores. From road tests with the tires, it is found that tire mileage is normally distributed with a mean tire mileage of $\mu = 36,500$ miles and a standard deviation of $\sigma = 5000$ miles. What is the probability that a tire will last at least 30,000 miles?

The z value corresponding to $x = 30,000$ miles is

$$z = \frac{x - \mu}{\sigma} = \frac{30,000 - 36,500}{5000} = \frac{6500}{5000} = -1.30$$

Using Table 7.1, we find that $P(-1.30 \leq z \leq 0.00) = .4032$. Thus the probability a tire will provide at least 30,000 miles of usage is $P(x \geq 30,000) = .4032 + .5000 = .9032$. That is, better than 90% of the tires can be expected to wear for at least 30,000 miles. This situation is shown graphically. Again, values for the corresponding standard normal random variable are also shown.

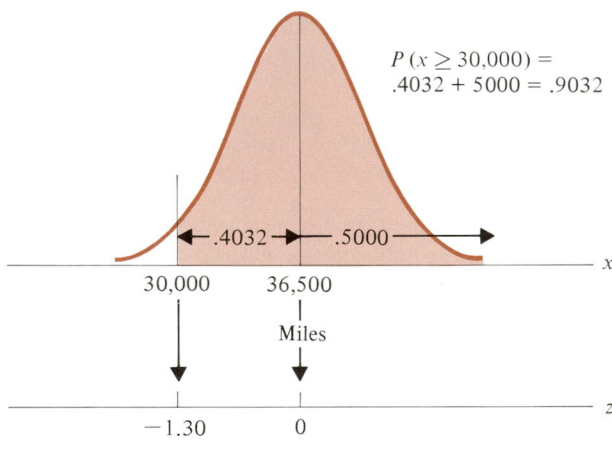

• • •

EXAMPLE 7.9

Test scores for a college midterm examination are normally distributed with a mean of $\mu = 72$ and a standard deviation of $\sigma = 13$. Suppose the professor wishes to assign the grade of A to the 15% of the students obtaining the highest scores on the exam. What is the cutoff score for the A grade?

This is a situation in which the probability is known and the question concerns finding a particular value for the random variable, the exam score. With 15% of the area in the upper tail of the normal distribution, we know that the area between the mean score of $\mu = 72$ and the exam score required to obtain the grade of A must be $.5000 - .1500 = .3500$. Using the *body* of Table 7.1, we find the probability value closest to .3500 is .3508. Using the left-hand column and the top row of the table, we find that the z value corresponding to .3508 is $z = 1.04$. This tells us that the midterm exam score required to obtain the grade of A must be at least 1.04 standard deviations

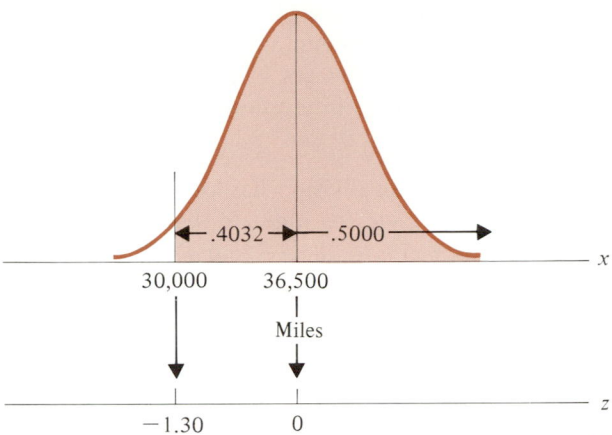

above the mean. Computing the corresponding value of x, we find

$$x = \mu + 1.04\sigma = 72 + 1.04(13) = 85.52$$

as the minimum score a student must obtain in order to receive the grade of A for the exam.

• • •

As the examples of this section show, the key to answering probability questions about the normal distribution is converting values of the normal random variable x to the corresponding z values and interpreting the table of probabilities for the standard normal probability distribution.

EXERCISES

13. The demand for a new product is assumed to be normally distributed with $\mu = 200$ and $\sigma = 40$. Letting x be the number of units demanded, find the following.
   a. $P(180 \leq x \leq 220)$
   b. $P(x \geq 250)$
   c. $P(x \leq 100)$
   d. $P(225 \leq x \leq 250)$

14. The Webster National Bank is reviewing its service charge and interest-paying policies on checking accounts. The bank has found that the average daily balance on personal checking acounts is $550.00, with a standard deviation of $150.00. In addition, the average daily balances have been found to be normally distributed.
   a. What percentage of personal checking account customers carry average daily balances in excess of $800.00?
   b. What percentage of the bank's customers carry average daily balances below $200.00?
   c. What percentage of the bank's customers carry average daily balances between $300.00 and $700.00?

d.   The bank is considering paying interest to customers carrying average daily balances in excess of a certain amount. If the bank does not want to pay interest to more than 5% of its customers, what is the minimum average daily balance it should be willing to pay interest on?

15.   General Hospital's patient account division has compiled data on the age of accounts receivables. The data collected indicate that the age of the accounts follows a normal distribution, with $\mu = 28$ days and $\sigma = 8$ days.
   a.   What percentage of the accounts are between 20 and 40 days old ($P(20 \leq x \leq 40)$?
   b.   The hospital administrator is interested in sending reminder letters to the oldest 15% of accounts. How many days old should an account be before a reminder letter is sent?
   c.   The hospital administrator would like to give a 5% discount to those accounts that pay their balance by the 21st day. What percentage of the accounts will receive the discount?

16.   The time required to complete a final examination in a particular college course is normally distributed, with a mean of 80 minutes and a standard deviation of 10 minutes. Answer the following questions:
   a.   What is the probability of completing the exam in 1 hour or less?
   b.   What is the probability a student will complete the exam in more than 60 minutes but less than 75 minutes?
   c.   Assume that the class has 60 students and that the examination period is 90 minutes in length. How many students do you expect will be unable to complete the exam in the allotted time?

17.   The useful life of a computer terminal at a university computer center is known to be normally distributed, with a mean of 3.25 years and a standard deviation of .5 years.
   a.   Historically 22% of the terminals have had a useful life less than the manufacturer's advertised life. What is the manufacturer's advertised life for the computer terminals?
   b.   What is the probability that a computer terminal will have a useful life of at least 3 but less than 4 years?

18.   From past experience, the management of a well known fast-food restaurant estimates that the number of weekly customers at a particular location is normally distributed, with a mean of 5000 and a standard deviation of 800 customers.
   a.   What is the probability that on a given week the number of customers will be between 4760 and 5800?
   b.   What is the probability of more than 6500 customers?
   c.   For 90% of the weeks the number of customers should exceed what amount?

7.4 NORMAL APPROXIMATION OF BINOMIAL PROBABILITIES

In Chapter 6 we introduced the binomial probability distribution as a means for determining the probability of x successes in n trials of a binomial experiment. In cases where the number of trials, n, is large, binomial tables are not usually available and the computations associated with the binomial probability function are not practical. For instance with $n = 120$ and $p = .34$, the probability of 40 successes is given by

$$P(x = 40) = \frac{n!}{x!(n-x)!} p^x (1-p)^{n-x} = \frac{120!}{40!\, 80!} (.34)^{40} (.66)^{80}$$

Evaluating such an expression is laborious and can lead to rounding errors.

In situations such as this it is often possible to use the normal probability distribution to obtain good approximations of binomial probabilities. In this section we show how the normal probability distribution can be used for this purpose. The normal approximation provides acceptable accuracy whenever the number of trials, n, and the probability of success on each trial, p, have values such that both $np \geq 5$ and $n(1-p) \geq 5$.

EXAMPLE 7.10

A particular company has a history of making errors on 10% of its invoices. In a sample of 100 invoices, what is the probability that exactly 12 invoices have an error?

Note that this is a binomial experiment with $n = 100$ trials, $p = .10$ and $x = 12$. Rather than using the binomial probability function directly, we want to show how the normal probability distribution can be used to approximate the desired probability. Checking the requirements for the normal approximation we find that $np = 100 (.10) = 10$ and $n(1-p) = 100(.90) = 90$ are both at least 5. Thus, as previously stated, the normal approximation should provide good results.

Recall from Chapter 6 that the mean of a binomial random variable is $\mu = np$ and standard deviation is $\sigma = \sqrt{np(1-p)}$. For this example, we have $\mu = np = 100(.10) = 10$ and $\sigma = \sqrt{100(.10)\ .90} = 3$. A normal distribution with this mean and standard deviation is shown in Figure 7.4. This is the normal distribution that is used to approximate the probabilities in this situation.

Also recall that with a continuous probability distribution, probabilities are computed as areas under the curve. As a result, the probability of any particular value for a continuous random variable is *zero*. Thus to approximate the binomial probability of 12 successes, we compute the area under the corresponding normal curve between 11.5 and 12.5. The .5 that we add to and subtract from 12 to enable the use of a continuous distribution to approximate discrete probabilities is called the *continuity correction*. Thus the interval $11.5 \leq x \leq 12.5$ for the normal random variable is used to approximate the probability that $x = 12$ for the discrete binomial distribution. Thus the binomial probability of $f(12)$ is approximated by the normal probability, $P(11.5 \leq x \leq 12.5)$.

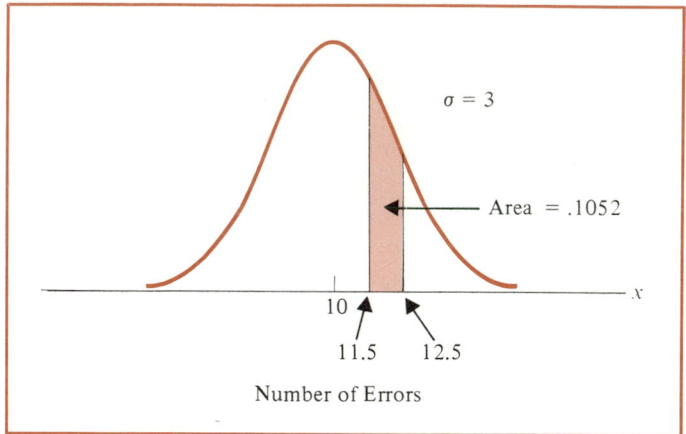

FIGURE 7.4
Normal Approximation of a Binomial Probability with *n* = 100 and *p* = .10:
The Probability of 12 Errors in 100 Trials Is Approximately .1052

Using the normal distribution shown in Figure 7.4, we use the following *z* values.

$$\text{At } x = 12.5, \qquad z = \frac{x - \mu}{\sigma} = \frac{12.5 - 10}{3} = .83$$

$$\text{At } x = 11.5, \qquad z = \frac{x - \mu}{\sigma} = \frac{11.5 - 10}{3} = .50$$

Using Table 7.1, we find the area under the curve between 0 and .83 is .2967. Similarly, the area under the normal curve between 0 and .50 is .1915. Therefore, the area between $z = .50$ and $z = .83$ is $.2967 - .1915 = .1052$. Thus the normal approximation of exactly 12 errors in the 100 invoices is .1052. As it turns out, the actual binomial probability is .0988. Thus the error in our approximation is $.1052 - .0988 = .0064$. This is a pretty good approximation.

•   •   •

EXAMPLE 7.11

It is believed that 45% of a large population of registered voters favor a particular candidate for the state senate. A public opinion poll uses randomly selected samples of voters and asks each person polled to indicate his or her preference for the candidates. What is the probability that a weekly poll based on the responses of 200 registered voters will show at least 50% of the voters favoring the candidate? That is, what is the probability at least 100 of the 200 voters will favor the candidate?

First note that this is a binomial probability experiment with $n = 200$ voters and $p = .45$. The binomial probability question asks for the probability of at least 100 successes. With $np = 200(.45) = 90$ and $n(1-p) = 200(.55) = 110$, we see that the normal probability distribution can be used to approximate the desired binomial probability. The mean for the distribution is $\mu = np = 200(.45) = 90$ and the standard deviation is $\sigma = \sqrt{np(1 - p)} = \sqrt{200(.45)(.55)} = 7.04$. Using the continuity correction

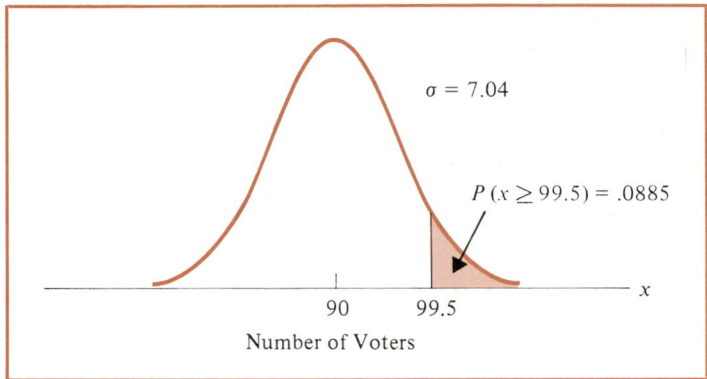

FIGURE 7.5
Normal Approximation to a Binomial Probability Distribution with $n = 200$ and $p = .45$:
The Approximation to the Binomial Probability of at Least 100 Voters is .0885

we see that the interval 99.5 to 100.5 under the normal curve with $\mu = 90$ and $\sigma = 7.04$ provides the approximation of the binomial probability of 100 voters favoring the candidate. Since we are asking for the probability of 100 or more voters, we must compute the corresponding normal distribution probability of $x \geq 99.5$. The graph of the normal probability distribution approximation is shown in Figure 7.5.

Converting to the appropriate z value, we find

$$\text{At } x = 99.5, \qquad z = \frac{x - \mu}{\sigma} = \frac{99.5 - 90}{7.04} = 1.35$$

Using Table 7.1, the area between $z = 0$ and $z = 1.35$ is found to be .4115. Thus we know there must be a $.5000 - .4115 = .0885$ probability of a value of 99.5 or more. We conclude that there is approximately a .0885 probability that the sample of 200 voters will show at least 100 voters favoring the candidate.

$\bullet \quad \bullet \quad \bullet$

EXERCISES

19. In order to obtain cost savings, a company is considering offering an early retirement incentive for its older management personnel. The consulting firm that designed the early retirement program has found that approximately 22% of the employees qualifying for the program will select early retirement during the first year of eligibilty. Assume that the company offers the early retirement program to 50 of its management personnel.

   a. What is the expected number of employees who will elect early retirement in the first year?

   b. What is the probability at least 8 but not more than 12 employees will elect early retirement in the first year?

   c. What is the probability that 15 or more employees will select the early retirement option in the first year?

   d. For the program to be judged successful, the company believes that it should entice at least 10 management employees to elect early retirement in the first year. What is the probability that the program is successful?

20. Thirty percent of the students at a particular university attended Catholic high schools. A random sample of 50 of this university's students has been taken. Use the normal approximation to the binomial probability distribution to answer the following questions:
    a. What is the probability that exactly 10 of the students selected attended Catholic high schools?
    b. What is the probability that 20 or more of the students attended Catholic high schools?
    c. What is the probability that the number of students from Catholic high schools is between 10 and 20 inclusively?

21. A Myrtle Beach Resort Hotel has 120 rooms. In the spring months, hotel room occupancy is approximately 75%. Use the normal approximation to the binomial distribution to answer the following questions:
    a. What is the probability that at least half the rooms are occupied on a given day?
    b. What is the probability that 100 or more rooms are occupied on a given day?
    c. What is the probability that 70 or fewer rooms are occupied on a given day?

22. It is known that 30% of all customers of a major national charge card pay their bills in full before any interest charges are incurred. Answer the following questions for a group of 150 credit card holders:
    a. What is the probability that between 40 and 60 customers pay their account charges before any interest charges are incurred? That is, find $P(40 \leq x \leq 60)$.
    b. What is the probability that 30 or fewer customers pay their account charges before any interest charges are incurred?

Summary

This chapter was devoted to the study of the normal probability distribution. Because of its wide range of applicability, the normal probability distribution is considered by many to be the most important distribution in probability and statistics. Each normal probability distribution belongs to a family of similar bell-shaped distributions with the specific normal distribution depending upon the value of the mean μ and standard deviation σ.

Tables of probabilities are available for a normal probabilty distribution with a mean of 0 and a standard deviation of 1. This special normal probability distribution is referred to as the standard normal probability distribution. Common notation is to use z to denote the standard normal random variable. The relationship between z and any other normal random variable x with mean μ and standard deviation σ is given by $z =$

$(x - \mu)/\sigma$. The interpretation of a z value is that it indicates the number of standard deviations the normal random variable x is from its mean μ.

Probability questions about a random variable x with a normal distribution can be answered by first converting the random variable x to its corresponding z value and then using the table of areas for the standard normal probability distribution to determine the appropriate probabilities. We saw that two types of probability questions can be asked: An interval of values of the random variable is given and the question is to determine the probability the random variable assumes a value in the interval; alternatively, a probability value is given and we want to determine the values of the random variable yielding the given probabilities. Both types of questions can be answered by using the table of areas for the standard normal probability distribution. In making the probability calculations, it is recommended that a sketch of the appropriate normal curve be made as an aid to visualizing the probability information desired.

Finally, we noted that binomial probabilities can be difficult to compute whenever the number of trials, n, is large. However, if both $np \geq 5$ and $np(1 - p) \geq 5$, the normal probability distribution with $\mu = np$ and $\sigma = \sqrt{np(1 - p)}$ provides a good approximation to the binomial probabilities. A continuity correction must be utilized to account for the fact that the discrete binomial probability is being approximated by the continuous normal probability distribution.

Statistics in Practice

CREDIT CARDS FOR BANKING

The business of the Burroughs Corporation is information management. The Office Products Group (OPG) is one of the major operating groups of the corporation. One of the products manufactured by OPG is a plastic credit card that is used in automatic bank teller machines. Some of the banks using the Burroughs credit cards were having problems with the credit cards being rejected by the automatic teller equipment. A study was initiated in an attempt to determine the cause of the credit card rejection problem.

In the manufacturing process, the plastic credit cards were cut to final dimensions from larger plastic sheets using a die-cutting machine. Four different dies could be used, but the dies were all designed to produce cards that met the same product specifications. Approximately 250 cards from each of the four dies were sampled and the height and length of each card were measured.

The distribution of the height and length dimensions for the cards produced by each die followed a normal probability distribution. The probabilities that the height and length of the cards produced by these dies would not meet product specifications were calculated based on the normal distribution. For example, the minimum acceptable length of a credit card is 3.365 inches. For die 1, the sample results showed a mean of 3.367 inches and a standard deviation of .001 inches. Assuming a normal probability distribution with a mean of 3.367 inches and a standard deviation of .001 inches, the probability of a length as small as 3.365 inches or smaller was computed to be .0228. Thus approximately 2.3% of the cards manufactured using die 1 were expected to have a length less than the minimum length specification.

The final results of the study showed that the probabilities that the cards would not meet the length specifications were relatively low for all four dies. However, probabilities that the cards would not meet the height specifications were unacceptably large for all four dies. Further investigation showed that the entire die-cutting operation had to be upgraded to meet the height specifications. As a result, new process equipment was installed. Further tests and probability calculations based on the normal probability distribution showed that the new equipment would be able to produce credit cards that met all specifications.

Glossary

Normal probability distribution—A continuous probability distribution. It is described by a bell-shaped curve, which depends upon a mean μ and a standard deviation σ.

Standard normal probability distribution—A normal probability distribution with a mean of 0 and a standard deviation of 1.

z value—The value of the standard normal random variable. When the value of a normal random variable, x, has been converted to a z value, the z value represents the number of standard deviations x is from its mean, μ.

Continuity correction—A value, .5, that is added to and/or subtracted from a value of x when the continuous normal probability distribution is used to approximate the discrete binomial probability distribution.

Key Formulas

Converting to a Standard Normal Random Variable

$$z = \frac{x - \mu}{\sigma} \tag{7.1}$$

Review Quiz

TRUE/FALSE

1. The normal probability distribution involves a discrete random variable.
2. As the standard deviation increases, the height of the normal curve increases.
3. The highest point on the normal probability distribution occurs at the mean.
4. The area under the curve to the left of the mean for a normal probability distribution is .50.

5. The probability the normal random variable assumes a value within one standard deviation of the mean is .50.

6. A standard normal probability distribution has a mean of zero and a variance of 1.

7. In order to compute probabilities for a normal random variable, one must first convert to a standard normal probability distribution.

8. A continuity correction is needed whenever normal probabilities are computed.

9. The normal probability distribution may be used to approximate the binomial, provided that $np \geq 5$ and $n(1 - p) \geq 5$.

10. The probability of a normal random variable assuming a value within two standard deviations of its mean is approximately .95.

MULTIPLE CHOICE

For questions 11–14, consider the normally distributed random variable x, which has a mean of 17 and a standard deviation of 3.

11. The random variable x is
   a. discrete
   b. continuous
   c. not enough information is given

12. The probability that x is less than or equal to 14 is
   a. .1587
   b. .1765
   c. .3414
   d. .4986

13. If a value of x is randomly selected, the probability that it will be between 20 and 22 is
   a. .1112
   b. .3413
   c. .4525
   d. .7938

14. The probability that two randomly selected values of x will both have a value less than 22 is
   a. .2048
   b. .4525
   c. .9073
   d. .9525

15. Assume that a normal probability distribution is being used to approximate binomial probabilities for the case of $n = 500$ and $p = .10$. The value of σ that should be used is closest to
   a. 3
   b. 6
   c. 9
   d. 40

16. Scores on a reading skills test are normally distributed with a mean of $\mu = 500$ and a standard deviation of 50. An agency hires only people whose scores are in

the top 5% of individuals taking the test. This company should consider hiring anyone who achieves at least a score of

a. 500
b. 538
c. 550
d. 583

Supplementary Exercises

23. Given that z is a standard normal random variable, compute the following probabilities.
    a. $P(-.72 \leq z \leq 0)$
    b. $P(-.35 \leq z \leq .35)$
    c. $P(.22 \leq z \leq .87)$
    d. $P(z \leq -1.02)$

24. Given that z is a standard normal random variable, compute the following probabilities.
    a. $P(z \geq -.88)$
    b. $P(z \geq 1.38)$
    c. $P(-.54 \leq z \leq 2.33)$
    d. $P(-1.96 \leq z \leq 1.96)$

25. Given that z is a standard normal random variable, find z if it is known that
    a. the area between $-z$ and z is .90
    b. the area to the right of z is .20
    c. the area between -1.66 and z is .25
    d. the area to the left of z is .40
    e. the area between z and 1.80 is .20

26. Medical research has concluded that individuals experience a common cold roughly two times each year. Assume that the time between colds is normally distributed with a mean of 160 days and a standard deviation of 40 days.
    a. What is the probability of going 200 or more days between colds?
    b. What is the probability of getting a cold within 80 days of a previous cold?
    c. What percentage of the population will go 120 to 240 days between colds?

27. The lifetime of a washing machine can be approximated by a normal probability distribution with a mean of 4 years and a standard deviation of .75 years.
    a. Sketch a normal curve for the lifetime of a washing machine. Label the horizontal axis with various values for the lifetime of the washing machine.
    b. What is the probability that the lifetime of the washing machine is between 2.5 and 5.5 years?
    c. What is the probability that the lifetime of the washing machine is 5.5 years or more?
    d. What is the probability that the washing machine operates 21 months (1.75 years) or less?

28. A soup company markets eight varieties of homemade soup throughout the Eastern states. The standard-size soup can holds a maximum of 11 ounces, while the label on each can advertises contents of 10¾ ounces. The extra ¼ ounce is to allow for the possibility of the automatic filling machine placing more soup than

the company actually wants in a can. Past experience shows that the number of ounces placed in a can is approximately normally distributed, with a mean of 10¾ and a standard deviation of .1 ounces. What is the probability that the machine will attempt to place more than 11 ounces in a can, causing an overflow to occur?

29. The sales of High-Brite Toothpaste are believed to be approximately normally distributed, with a mean of 10,000 tubes per week and a standard deviation of 1500 tubes per week.
    a. What is the probability that more than 12,000 tubes will be sold in any given week?
    b. In order to have a .95 probability that the company will have sufficient stock to cover the weekly demand, how many tubes should be produced?

30. Points scored by the winning team in NCAA college football games are approximately normally distributed, with a mean of 24 and a standard deviation of 6.
    a. What is the probability that a winning team in a football game scores between 20 and 30 points; that is, $P(20 \leq x \leq 30)$?
    b. How many points does a winning team have to score to be in the highest 20% of scores for college football games?

31. Ward Doering Auto Sales is considering offering a special service contract that will cover the total cost of any service work required on leased vehicles. From past experience the company manager estimates that yearly service costs are approximately normally distributed, with a mean of $150 and a standard deviation of $25.
    a. If the company offers the service contract to customers for a yearly charge of $200, what is the probability that any one customer's service costs will exceed the contract price of $200?
    b. What is Ward's expected profit per service contract?

32. The attendance at football games at a certain stadium is normally distributed, with a mean of 45,000 and a standard deviation of 3,000.
    a. What percentage of the time should attendance be between 44,000 and 48,000?
    b. What is the probability of the attendance exceeding 50,000?
    c. Eighty percent of the time the attendance should be at least how many?

33. Assume that the test scores from a college admissions test are normally distributed, with a mean of 450 and a standard deviation of 100.
    a. What percentage of the people taking the test score between 400 and 500?
    b. Suppose that someone receives a score of 630. What percentage of the people taking the test score better? What percentage score worse?
    c. If a particular university will not admit anyone scoring below 480, what percentage of the persons taking the test would be acceptable to the university?

34. The lifetime of a color television picture tube is normally distributed, with a mean of 7.8 years and a standard deviation of 2 years.
    a. What is the probability that a picture tube will last more than 10 years?
    b. If the firm guarantees the picture tube for 2 years, what percentage of the television sets sold will have to be replaced because of picture tube failure?
    c. If the firm is willing to replace the picture tubes in a maximum of 1% of the television sets sold, what guarantee period can be offered for the television picture tubes?

35. A machine fills containers with a particular product. The standard deviation of filling weights is known from past data to be .6 ounces. If only 2% of the containers hold less than 18 ounces, what is the mean filling weight for the machine? That is, what must μ equal? Assume the filling weights have a normal distribution.

36. Suppose that 54% of a large population of registered voters favor the Democratic candidate for state senator. A public opinion poll uses randomly selected samples of voters and asks each person in the sample his or her preference; the Democratic candidate or the Republican candidate. The weekly poll is based on the response of 100 voters.
    a. What is the expected number of voters who will favor the Democratic candidate?
    b. What is the standard deviation for the number of voters who will favor the Democratic candidate?
    c. What is the probability that the poll will show that *less than* 50% of the voters favor the Democratic candidate when in fact 54% of the population of registered voters favor the candidate? That is, what is the probability that 49 or fewer individuals in the sample express support for the Democratic candidate?

37. It is estimated that in criminal trials, the jury will reach the correct decision (guilty or not guilty) 90% of the time. Consider a group of 100 cases that are brought to trial before a jury.
    a. What is the expected number of cases where the jury will reach the correct decision?
    b. What is the probability the jury will judge 95 or more cases correctly?
    c. What is the probability an incorrect decision is reached in 12 or more cases?
    d. Answer the question in (c) if the jury system reaches the correct decision 95% of the time.

38. Consider a multiple-choice examination with 50 questions. Each question has four possible answers. Assume that a student who has done the homework and attended lectures has a .75 probability of answering any question correctly.
    a. A student must answer 43 or more questions correctly in order to obtain a grade of A. What percentage of the students who have done their homework and attended lectures will obtain a grade of A on this multiple-choice examination?
    b. A student who answers 35 questions to 39 questions correctly will receive a grade of C. What percentage of students who have done their homework and attended lectures will obtain a grade of C on this multiple-choice examination?
    c. A student must answer 30 or more questions correctly in order to pass the examination. What percentage of the students who have done their homework and attended lectures will pass the examination?
    d. Assume that a student has not attended class and has not done the homework for the course. Furthermore, assume that the student will simply guess at the answer to each question. What is the probability that this student answers 30 or more questions correctly and passes the examination?

CHAPTER 8

Sampling and Sampling Distributions

What You Will Learn in This Chapter

- the importance of sampling

- how to select a simple random sample

- how results from samples can be used to provide estimates of population parameters

- what a sampling distribution is

- what the central limit theorem is and the important role it plays in statistics

- other sampling techniques, such as stratified sampling, cluster sampling, systematic sampling, convenience sampling, and judgment sampling

Contents

Statistics in the News

U.S. SENATORS PROPOSE COLLEGE STUDENTS BE HELD TO ACADEMIC STANDARDS

A recent study conducted by the General Accounting Office (GAO) revealed that many students who are receiving aid are not meeting the minimum academic standards. Specifically, the GAO report, based on a random sample of 19 colleges and universities, showed that 10% of the students receiving aid had an F average. In addition, the report found that 20% of the students were doing less than C work.

Based on the results of the sample, a number of senators stated a concern for the apparent waste of providing federal aid dollars for students who are not meeting minimum academic standards. A bill sponsored by Claiborne Pell (D–Rhode Island) and Don Nickles (R–Oklahoma) has been introduced that will place on probation any student whose first-year grade-point average is less than a C. Failure to achieve a C average during the next grading period would result in the student's aid being cut off.

Federal aid programs currently cost tax-

College students waiting in line for financial aid.

payers $7 billion annually. Senators Pell and Nickles contend that money must be made available to assist young people in obtaining a higher education. However, the senators want to be sure that the financial support is being used by individuals who are willing and able to do the academic work.

Based on "College Aid," *Cincinnati Enquirer* (September 9, 1982).

As stated in Chapter 1, the primary purpose of statistics is to provide information about a *population* based upon information contained in a *sample* of the population. In "Statistics in the News," students from a sample of 19 colleges and universities were used to draw conclusions about the population of all students receiving financial aid from the federal government.

In this chapter we introduce simple random sampling and the process of using a sample mean to provide an estimate of a population mean. In addition we introduce the important concept of a sampling distribution. It is knowledge of the sampling distribution that enables us to make statements concerning the potential errors in using sample results to draw conclusions about a population. In the final section of the chapter, we present a variety of other sampling procedures.

8.1 SIMPLE RANDOM SAMPLING

The primary objective of sampling is to select a sample that is *representative* of the population being studied. There are several ways of selecting samples from a population; one of the most common of these is *simple random sampling*. A simple random sample of size n from a population of size N is defined as follows:

Simple Random Sample

A simple random sample of size n from a population of size N is a sample selected such that each possible sample of size n has the same probability of being selected.

EXAMPLE 8.1

Consider a population consisting of four college friends: Bob, Cathy, Krista, and Mark. Assume we would like to select a simple random sample of size 2 from this population in order to estimate the grade average for the population of friends. How should we select two friends to include in the simple random sample?

To begin, we can list all possible samples of size 2 from the population of four friends. The possible samples are:

Bob and Cathy
Bob and Krista
Bob and Mark
Cathy and Krista
Cathy and Mark
Krista and Mark

The definition of simple random sampling implies that in selecting one of the above six samples, we must ensure that each sample has the same $\frac{1}{6}$ probability of being selected. One way this can be accomplished is by writing the names corresponding to each of the six samples on six separate, but identical, pieces of paper. By shuffling the pieces of paper and then randomly selecting one, we can identify a simple random sample of two friends. This process is depicted in Figure 8.1. In this illustration, Cathy and Mark are selected as the 2 friends in the random sample.

● ● ●

EXAMPLE 8.2

A local elementary school uses five buses to transport students to and from school each day. The principal of the school would like to take a simple random sample of two of the five buses in order to learn about the number of students riding the buses each day. If the five buses in the population are labeled A, B, C, D, and E, we can use the following procedure to select a simple random sample of two buses.

With a population of size $N = 5$, there are 10 different samples of size $n = 2$. These 10 samples consist of buses AB, AC, AD, AE, BC, BD, BE, CD, CE, and DE. If we select a sample in such a way that each of these 10 samples has the same $\frac{1}{10}$ probability of being selected, the sample selected would be a simple random sample. As in Example 8.1, we could do this by writing the letters corresponding to each of the possible samples on 10 separate, but identical, pieces of paper, thoroughly shuffling the

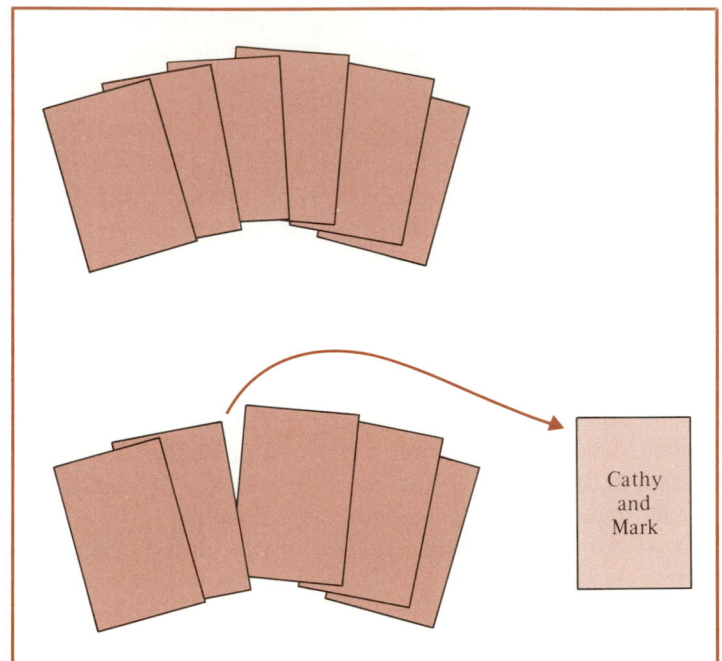

FIGURE 8.1
Selecting a Simple Random Sample of Two Friends

10 pieces of paper, and then randomly selecting one. Suppose that we do this and select the piece of paper with *BC* written on it. The buses labeled *B* and *C* would then be the two buses in the simple random sample.

• • •

In both Examples 8.1 and 8.2 we used the same procedure to select a simple random sample. The entire sample was selected in one random drawing. Another approach to choosing a simple random sample is based upon selecting the items for the sample *one at a time*. At each selection we make sure that each of the items remaining in the population has the same probability of being selected.

EXAMPLE 8.2 (continued)

The one-at-a-time approach to selecting a simple random sample of two school buses can be accomplished as follows:

STEP 1 Each of the letters *A*, *B*, *C*, *D*, and *E*—corresponding to each of the five buses—is written on a separate piece of paper. The five pieces of paper are shuffled thoroughly and placed face down.

STEP 2 One piece of paper is randomly selected. Each piece has a probability of $\frac{1}{5}$ of being selected. Suppose that the piece of paper with a *B* on it is selected; bus *B* is then one of the two buses in the simple random sample.

STEP 3 Using the *remaining* four pieces of paper, we randomly select a second piece of paper. Each remaining piece of paper has a probability of $\frac{1}{4}$ of being selected. Suppose this time the piece of paper with a C on it is selected; bus C would then be the second bus in the simple random sample.

The probability of selecting bus B as the first item in the sample was $\frac{1}{5}$. Given that B was selected, the probability of selecting bus C as the second item in the sample was $\frac{1}{4}$. Thus the probability of selecting B as the first item and C as the second item is given by $(\frac{1}{5})(\frac{1}{4}) = \frac{1}{20} = .05$. Of course, the same sample is obtained if C is the first item selected and B is the second; the probability of this happening is again .05. Thus using this one-at-a-time procedure, the probability of obtaining a sample consisting of buses B and C is $.05 + .05 = .10$. This is the same as the $\frac{1}{10}$ probability of selecting this sample when both items are selected simultaneously, as we did earlier when 2 letters were placed on each of 10 sheets of paper.

• • •

In using the one-at-a-time procedure in Example 8.2 we did not replace the first piece of paper after it was selected from the population; this is called sampling *without replacement*. If we had replaced the first piece of paper selected prior to selecting the second, we would have been *sampling with replacement*. If this method of sampling had been used, the same piece of paper—and thus the same bus—could have been selected both times. Thus, samples consisting of AA, BB, CC, DD, and EE would have to be added to our list of possible sample results. *Sampling without replacement is the predominant sampling procedure used in practice and is assumed throughout the text.*

EXERCISES

1. A family has four children: Robert, Lynn, Jamie, and Lindsay. Suppose that a study on family values requires the sampling of three of the four children in the family.
   a. How many samples of size 3 are possible? List the different possible samples.
   b. Using simple random sampling, what is the probability each possible sample will be selected?

2. In Example 8.2, a local elementary school uses five buses to transport students to and from school. The five buses were identified by the letters A, B, C, D, and E.
   a. How many samples of size 3 are possible? List the different possible samples.
   b. Using simple random sampling, what is the probability each possible sample will be selected?

3. Consider the following six Midwestern states as a population: Iowa, Illinois, Wisconsin, Michigan, Indiana, and Ohio. Assume that a sample of four states will be selected from this population in order to study employment trends in the Midwest.

a. How many samples of size 4 are possible? List the samples.

b. How many of the samples contain the state of Illinois?

c. Using simple random sampling, the possible samples have the same probability of being selected. If this is the case, what is the probability of selecting a sample that contains the state of Illinois?

d. Repeat part (b) for the state of Indiana.

e. Using the results from parts (b) and (d), what can you say about the probability that a given state will be included in the simple random sample?

4. In Example 8.2. we demonstrated a three-step procedure that can be used to select a simple random sample of size 2 by selecting items for the sample *one at a time*. We will see later that this is the preferred procedure for identifying a simple random sample. What is the primary advantage of the one-at-a-time method of simple random sampling as compared to the method of listing all possible samples?

8.2 SAMPLING DISTRIBUTION OF $\bar{x}$

Once we obtain a simple random sample from a population, we want to summarize the data in order to make estimates or draw conclusions about the population. In general, we use characteristics of the sample, referred to as *sample statistics,* to estimate characteristics of the population, referred to as *population parameters*.

One of the most common statistical procedures is the use of a sample mean $\bar{x}$ to make inferences about an unknown population mean μ. In this case, the sample mean $\bar{x}$ is referred to as the sample statistic. The population mean μ is referred to as the population parameter. This statistical process is shown in Figure 8.2.

It is important to realize that if we were to repeat the sampling process shown in

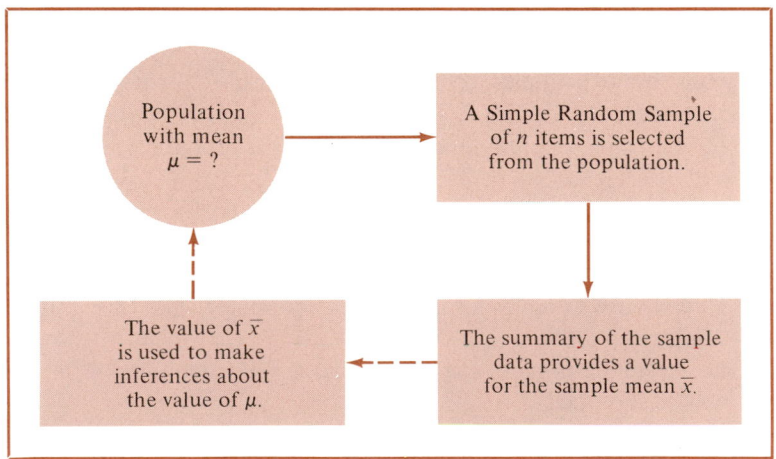

FIGURE 8.2
The Statistical Process of Using a Sample Mean
to Make Inferences About a Population Mean

Figure 8.2, we can anticipate obtaining a different value for the sample mean $\bar{x}$. As we showed in Section 8.1, several different simple random samples of size n are possible. Since each sample consists of different items from the population, we can expect different samples to provide different values for the sample statistic $\bar{x}$.

Since each sample has the same probability of being selected, we can associate a known probability with every possible sample and every possible value of $\bar{x}$. As a result we can identify the probability distribution for the sample mean $\bar{x}$. This probability distribution is called the *sampling distribution of* $\bar{x}$. The name of the probability distribution comes from the fact that the different possible values of $\bar{x}$ are due to the variety of different *samples* that can be selected from the population. Because of the importance of the sampling distribution of $\bar{x}$, we restate its definition.

Sampling Distribution of $\bar{x}$

The *sampling distribution of $\bar{x}$* is the probability distribution for all possible values of the sample mean $\bar{x}$.

EXAMPLE 8.2 (continued)

To illustrate the concept of a sampling distribution, let us reconsider the population of five school buses described in Example 8.2. The number of students riding on each of the buses in this population is shown in Table 8.1. Using the formulas for a population mean and a population standard deviation that were presented in Chapter 3, we can use the data in Table 8.1 to compute the mean and standard deviation for the population as follows:

$$\mu = \frac{\Sigma x_i}{N} = \frac{120}{5} = 24$$

$$\sigma = \sqrt{\frac{\Sigma(x_i - \mu)^2}{N}} = \sqrt{\frac{90}{5}} = \sqrt{18} = 4.24$$

TABLE 8.1

Number of Students per Bus for the Population of Five Buses

Bus	Number of Students
A	24
B	30
C	21
D	18
E	27

TABLE 8.2

Computation of the Population Mean and Standard Deviation for Example 8.2

Bus	Number of Students (x_i)	$(x_i - \mu)$	$(x_i - \mu)^2$
A	24	$(24 - 24) = 0$	0
B	30	$(30 - 24) = 6$	36
C	21	$(21 - 24) = -3$	9
D	18	$(18 - 24) = -6$	36
E	27	$(27 - 24) = 3$	9
Totals	120		90

$$\mu = \frac{120}{5} = 24 \qquad\qquad \sigma = \sqrt{\frac{90}{5}} = \sqrt{18} = 4.24$$

TABLE 8.3

Different Possible Simple Random Samples for Example 8.2

Buses Selected in Sample	Probability of Sample	Sample Mean ($\bar{x}$)
A and B	$\frac{1}{10}$	$\frac{24 + 30}{2} = 27.0$
A and C	$\frac{1}{10}$	$\frac{24 + 21}{2} = 22.5$
A and D	$\frac{1}{10}$	$\frac{24 + 18}{2} = 21.0$
A and E	$\frac{1}{10}$	$\frac{24 + 27}{2} = 25.5$
B and C	$\frac{1}{10}$	$\frac{30 + 21}{2} = 25.5$
B and D	$\frac{1}{10}$	$\frac{30 + 18}{2} = 24.0$
B and E	$\frac{1}{10}$	$\frac{30 + 27}{2} = 28.5$
C and D	$\frac{1}{10}$	$\frac{21 + 18}{2} = 19.5$
C and E	$\frac{1}{10}$	$\frac{21 + 27}{2} = 24.0$
D and E	$\frac{1}{10}$	$\frac{18 + 27}{2} = 22.5$

Details of the computations of the values of μ and σ are shown in Table 8.2.

In this situation the population size is small; thus it is easy to compute the mean and standard deviation for the population. However, to illustrate how a sample mean can be used to estimate a population mean, let us assume for the moment that μ is unknown and that we will have to use a simple random sample of two buses to estimate μ. Recall that there are 10 different samples of size 2 that could be selected. Table 8.3 lists these 10 possible samples and their corresponding sample means.

The column labeled "Sample Mean $(\bar{x})$" in Table 8.3 shows that the value of the sample mean depends upon the sample selected. Since we are using simple random sampling, we know that each possible sample—and, therefore, its corresponding sample mean—has the same $1/10$ probability of being selected. Figure 8.3 is a graph showing each possible value of $\bar{x}$ and the corresponding probability of occurrence. For instance, note that a value of $\bar{x} = 24.0$ has a $2/10$ probability of occurring because two samples, *BD* and *CE*, provide this value. Thus we see that Figure 8.3 is simply a probability distribution that shows all possible values of a random variable and their probabilities of occurrence. The random variable in this case is the sample mean, $\bar{x}$. The probability distribution of $\bar{x}$, as shown in Figure 8.3, is called the sampling distribution of $\bar{x}$.

$$\bullet \quad \bullet \quad \bullet$$

EXAMPLE 8.3

At the top of page 262 we show a population of five families with the data indicating the family size.

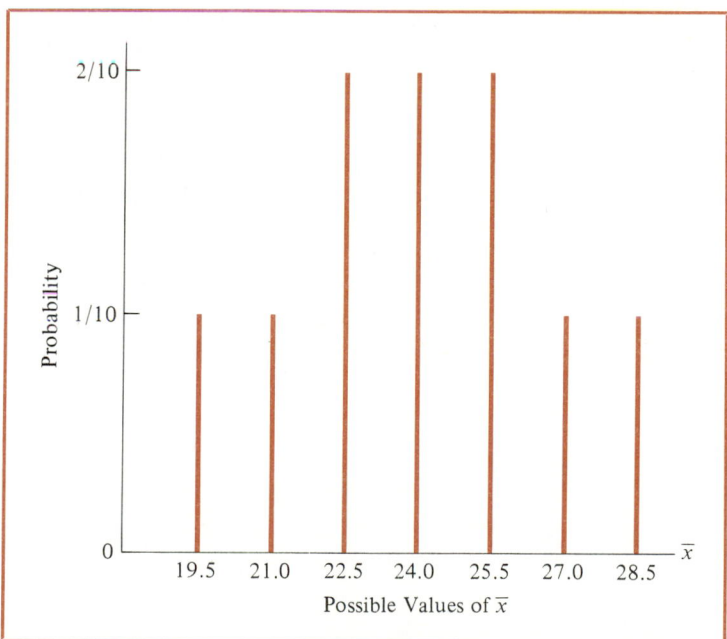

FIGURE 8.3
The Sampling Distribution of $\bar{x}$ for Example 8.2 Based on Samples of Size 2

Family	Family Size
A	2
B	4
C	3
D	5
E	3

If a simple random sample of size 3 is used to estimate the mean family size for the population, show the sampling distribution of $\bar{x}$.

First, we list all possible samples of size 3. There are 10 such samples consisting of families *ABC*, *ABD*, *ABE*, *ACD*, *ACE*, *ADE*, *BCD*, *BCE*, *BDE* and *CDE*. The sample mean for each possible sample is shown below.

Sample	Data Values	Sample Mean ($\bar{x}$)
ABC	2, 4, 3	3.00
ABD	2, 4, 5	3.67
ABE	2, 4, 3	3.00
ACD	2, 3, 5	3.33
ACE	2, 3, 3	2.67
ADE	2, 5, 3	3.33
BCD	4, 3, 5	4.00
BCE	4, 3, 3	3.33
BDE	4, 5, 3	4.00
CDE	3, 5, 3	3.67

With a probability of $\frac{1}{10}$ for each sample, the probabilities for each possible value of $\bar{x}$ are plotted below. This graph shows the sampling distribution of $\bar{x}$.

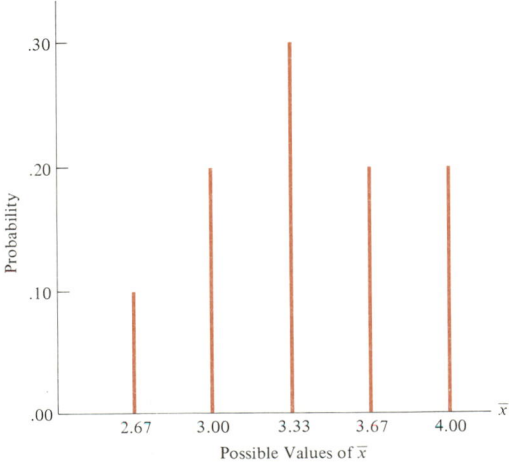

Possible Values of $\bar{x}$

Before continuing the discussion of the sampling distribution of $\bar{x}$, we note that in practice only one sample is actually taken; hence, only one value of $\bar{x}$ is computed and used to make inferences about the population mean μ. The purpose of this section has been to show that there are many possible samples that could be selected. Understanding this provides a much better perspective on the properties and the importance of the sampling distribution of $\bar{x}$. Knowledge of this sampling distribution provides the background to understand the material in Chapter 9; there the focus is on estimating the population mean μ using the information contained in just one sample.

Expected Value of $\bar{x}$

As we have seen, different samples may result in different values for the sample mean. We are often interested in the mean of all possible $\bar{x}$ values that can be generated by the various simple random samples. Let $E(\bar{x})$ denote the expected value of $\bar{x}$, or simply the mean of all possible $\bar{x}$ values. When using simple random sampling from a population with mean μ, the expected value of the sample mean is *equal to μ*.

Expected Value of $\bar{x}$

$$E(\bar{x}) = \mu \tag{8.1}$$

This result states that the expected value of $\bar{x}$—or, stated another way, the mean for all possible $\bar{x}$ values—is the same as the mean of the population from which the samples are taken.

EXAMPLE 8.2 (continued)

Recall that in example 8.2 we computed the population mean for the number of students on school buses to be $\mu = 24$. Equation (8.1) implies that the mean of the various $\bar{x}$ values must also be 24. Using the 10 values of $\bar{x}$ shown in Table 8.3, we can compute $E(\bar{x})$ as follows:

$$E(\bar{x}) = \frac{27.0 + 22.5 + 21.0 + 25.5 + 25.5 + 24.0 + 28.5 + 19.5 + 24.0 + 22.5}{10}$$

$$= \frac{240.0}{10} = 24$$

Thus we see that (8.1) holds with $E(\bar{x}) = \mu$.

• • •

Standard Deviation of $\bar{x}$

We use the standard deviation of $\bar{x}$ as the measure of dispersion in the possible $\bar{x}$ values. Let us explore what sampling theory says about the standard deviation of $\bar{x}$. We use the following notation:

$$\sigma_{\bar{x}} = \text{standard deviation of the } \bar{x} \text{ values}$$

$$\sigma = \text{standard deviation of the population being sampled}$$

$$n = \text{sample size}$$

$$N = \text{population size}$$

The formula for the standard deviation of $\bar{x}$ is as follows:

Standard Deviation of $\bar{x}$

$$\sigma_{\bar{x}} = \sqrt{\frac{N - n}{N - 1}} \frac{\sigma}{\sqrt{n}} \qquad (8.2)$$

Later we will see that when only one sample is selected, the value of $\sigma_{\bar{x}}$ is helpful in determining how far the sample mean may be from the population mean. Because of the role that $\sigma_{\bar{x}}$ plays in computing possible estimation errors, $\sigma_{\bar{x}}$ is referred to as the *standard error of the mean.* Thus the standard error of the mean is another name for the standard deviation of $\bar{x}$.

EXAMPLE 8.2 (continued)

Recall that in Example 8.2 the standard deviation for the population of schoolbuses was computed to be $\sigma = 4.24$ students (see Table 8.2). With a population of size $N = 5$ and a sample of size $n = 2$, (8.2) shows that the standard deviation of $\bar{x}$ must be

$$\sigma_{\bar{x}} = \sqrt{\frac{N - n}{N - 1}} \frac{\sigma}{\sqrt{n}} = \sqrt{\frac{5 - 2}{5 - 1}} \frac{4.24}{\sqrt{2}} = 2.60$$

• • •

Table 8.4 shows the computation of the standard deviation of $\bar{x}$ values using the 10 values of $\bar{x}$ generated by the 10 possible samples of two buses. As the computations show, we obtain the same value of $\sigma_{\bar{x}}$ as we found using (8.2). However, note that (8.2) provides the value for $\sigma_{\bar{x}}$ without having to generate all possible $\bar{x}$ values.

Consider for a moment the factor $\sqrt{(N - n)/(N - 1)}$ that appears in the formula for $\sigma_{\bar{x}}$. This factor is commonly referred to as the *finite population correction factor.* In many practical sampling situations we find that the population being sampled, although finite, is "large," whereas the sample size is relatively "small." In such cases the value of $\sqrt{(N - n)/(N - 1)}$ is close to 1. When this occurs, $\sigma_{\bar{x}} = \sigma/\sqrt{n}$ becomes a very good approximation to the standard deviation of $\bar{x}$. We give the following as a general guideline or rule of thumb for computing the standard deviation of $\bar{x}$.

TABLE 8.4

Computation of the Standard Deviation of $\bar{x}$ using All Possible Sample Means for the School Bus Example

Buses Selected in Sample	Sample Mean ($\bar{x}$)	($\bar{x} - 24$)	($\bar{x} - 24$)^2
A and B	27.0	$(27.0 - 24) = \;\;\;3.0$	9.00
A and C	22.5	$(22.5 - 24) = -1.5$	2.25
A and D	21.0	$(21.0 - 24) = -3.0$	9.00
A and E	25.5	$(25.5 - 24) = \;\;\;1.5$	2.25
B and C	25.5	$(25.5 - 24) = \;\;\;1.5$	2.25
B and D	24.0	$(24.0 - 24) = \;\;\;0.0$	0.00
B and E	28.5	$(28.5 - 24) = \;\;\;4.5$	20.25
C and D	19.5	$(19.5 - 24) = -4.5$	20.25
C and E	24.0	$(24.0 - 24) = \;\;\;0.0$	0.00
D and E	22.5	$(22.5 - 24) = -1.5$	2.25
Totals	240.0		67.50

$$E(\bar{x}) = \frac{240}{10} = 24 \qquad \sigma_{\bar{x}} = \sqrt{\frac{\Sigma(\bar{x} - 24)^2}{10}} = \sqrt{\frac{67.50}{10}} = 2.60$$

Guideline for Computing the Standard Deviation of $\bar{x}$:

Whenever the sample size is less than or equal to 5% of the population size (that is, $n/N \leq .05$), use

$$\sigma_{\bar{x}} = \frac{\sigma}{\sqrt{n}} \tag{8.3}$$

In the following chapters, we generally assume the population is large relative to the sample size; that is, $n/N \leq .05$. Thus we use (8.3) to compute the standard deviation of $\bar{x}$. If this assumption is not satisfied in a particular application, we use (8.2) to compute $\sigma_{\bar{x}}$.*

EXAMPLE 8.4.

Assume that a simple random sample of size 49 is to be taken from a large population with mean $\mu = 100$ and standard deviation $\sigma = 21$. We know that repeating the sampling process will generate different sample means, $\bar{x}$, due to the different samples

*This assumption is always satisfied when the population is infinite. If sampling from a finite population is done *with replacement*, (8.3) is used regardless of the sample size because the population size and composition are not changed as the sample is taken.

selected. What are the mean and standard deviation of the values of the sample means?

Using (8.1), we see the mean of the $\bar{x}$ values is

$$E(\bar{x}) = \mu = 100$$

Since the population is large relative to the sample size, (8.3) provides the standard deviation of the $\bar{x}$ values, as follows.

$$\sigma_{\bar{x}} = \frac{\sigma}{\sqrt{n}} = \frac{21}{\sqrt{49}} = 3$$

• • •

EXERCISES

Note to student: In the exercises that follow, the size of the population is not always stated. When the population size is *not* provided, you may make the assumption that the population is "large" relative to the sample size and that $n/N \leq .05$. In such cases, (8.3) may be used to compute the standard deviation of $\bar{x}$.

5. Four college students are taking the following number of credit hours during the current term:

Student	Number of Credit Hours
Albert	15
Becky	17
Cindy	19
David	17

Treating this as a population of size 4, answer the following questions.
a. How many simple random samples of size 2 are possible? List the possible samples.
b. Compute the sample mean for each of the possible simple random samples of size 2.
c. Show a graphical representation of the probability distribution of all possible values of the sample mean. This is a graphical representation of the sampling distribution of $\bar{x}$.

6. A local pet shop has six puppies for sale. The weights of the puppies in pounds are 8, 7½, 6, 10, 8, and 8½. Treating the puppies as a population of size 6, answer the following questions.
a. How many simple random samples of size 2 are possible? List the possible samples.

b. Compute the sample mean weight for each possible sample.
c. Show a graphical representation of the sampling distribution of $\bar{x}$.

7. Using the population of buses as shown in Table 8.1, show the sampling distribution of $\bar{x}$ if a simple random sample of three buses is used to estimate the mean number of students riding a bus.

8. The following data show the number of automobiles owned in a population of five households.

Household	Number of Automobiles
1	2
2	1
3	0
4	2
5	3

a. If a simple random sample of two households is used to estimate the mean number of automobiles per household, show the sampling distribution of $\bar{x}$.
b. Repeat part (a) if the simple random sample selected contains three households.

9. Assume we have a population with mean $\mu = 32$ and standard deviation $\sigma = 5$. Furthermore, assume that the population has 500 items and that a simple random sample of 25 items used to obtain information about this population. Let $\bar{x}$ denote the sample mean that will be used to estimate the value of the population mean.
a. What is the expected value of $\bar{x}$?
b. What is the standard deviation of $\bar{x}$?

10. The four students mentioned in exercise 5 were taking the following number of credit hours during the current term: Albert, 15; Becky, 17; Cindy, 19; and David, 17.
a. Treating the four students as the population, use the computational procedure of Table 8.2 to compute the population mean μ and standard deviation σ.
b. There are six possible simple random samples of size 2 that can be selected from this population. Identify each possible sample and compute its corresponding sample mean.
c. Use the computational procedure of Table 8.4 to compute the mean and standard deviation of the $\bar{x}$ values.
d. Use your results from part (a) and Equations (8.1) and (8.2) to compute the mean and standard deviation of the $\bar{x}$ values. Compare your answers to parts (c) and (d).

11. Five sales representatives sell mobile telephone units to private and commercial customers. Assume that the number of units sold for each sales representative is as follows:

Salesperson	Units Sold
Adams (A)	14
Baker (B)	20
Collins (C)	12
Davis (D)	8
Edwards (E)	16

a. Treating the five sales representatives as the population, use the computational procedure of Table 8.2 to compute the population mean μ and the population standard deviation σ.
b. There are 10 possible simple random samples of size 2 that can be selected from this population. Identify each possible sample and compute its corresponding sample mean, $\bar{x}$.
c. Show the sampling distribution of $\bar{x}$.
d. Use the computational procedure of Table 8.4 to compute the mean and standard deviation of the $\bar{x}$ values.
e. Use Equations (8.1) and (8.2) to determine the expected value and standard deviation of $\bar{x}$. Compare your results with your answer to part (d).
f. If you wish to compute $E(\bar{x})$ and $\sigma_{\bar{x}}$, do you prefer the approach used in part (d) or in part (e)? Explain.

12. The sizes of the 10 offices on the twelfth floor of the new Crosley Tower Bank Building are as follows.

Office	Size (square feet)	Office	Size (square feet)
1	150	6	300
2	175	7	140
3	180	8	150
4	180	9	150
5	225	10	200

a. Compute the population mean μ and population standard deviation σ for the population of 10 offices.
b. There are many different possible simple random samples of size 3 that can be selected from this population. What are the values of the mean and standard deviation for the $\bar{x}$ values? That is, compute $E(\bar{x})$ and $\sigma_{\bar{x}}$.
c. Compute $E(\bar{x})$ and $\sigma_{\bar{x}}$ if the sample size is increased to four offices.

13. A statistics class has 80 students. The mean score on the midterm exam was $\mu = 72$ and the standard deviation was $\sigma = 12$. Assume that a simple random sample of 20 students will be selected and the sample mean exam score $\bar{x}$ will be computed. What is the expected value and standard deviation of $\bar{x}$?

14. Weights for males between the ages of 20 and 30 have a mean $\mu = 170$ pounds with a standard deviation of $\sigma = 28$. If a simple random sample of 40 males in this age group is to be selected and the sample mean weight $\bar{x}$ computed, what are the values of $E(\bar{x})$ and $\sigma_{\bar{x}}$?

15. Consider a population of 1000 items. Assume the population standard deviation is $\sigma = 25$. Use (8.2) to compute the standard error of the mean $\sigma_{\bar{x}}$ for sample sizes of 50, 100, 150, and 200. What can you say about the size of the standard error of the mean as the sample size is increased?

16. In a study of the growth rate of a certain plant, a botanist is planning to use a simple random sample of 25 plants for data collection purposes. After analyzing the data on plant growth rate, the botanist believes that the standard error of the mean is too large. What size simple random sample should the botanist use in order to reduce the standard error to one-half its current value?

17. Simple random samples of size 30 are to be selected from a population of 2000 items. The population standard deviation is $\sigma = 12$.
    a. Use Equation (8.2) to compute the standard error of the mean.
    b. Use Equation (8.3) to compute the standard error of the mean.
    c. Compare your answers to parts (a) and (b) and comment on why it is acceptable to use equation (8.3) whenever $n/N \le .05$.
    d. What is the value of the finite population correction factor $\sqrt{(N-n)/(N-1)}$ for this problem?

18. A simple random sample of size 50 is to be selected from a population with $\sigma = 10$. Find the value of the standard error of the mean in each case.
    a. the population size is infinite
    b. the population size is $N = 50,000$
    c. the population size is $N = 5,000$
    d. the population size is $N = 500$
    e. In which of the above cases is it desirable to use the finite population correction factor and Equation (8.2)?

8.3 THE CENTRAL LIMIT THEOREM

At this point we know that simple random samples taken from a population will provide different values for the sample statistic $\bar{x}$ due to the fact that different samples consist of different items from the population. The probability distribution showing all possible values of $\bar{x}$ is referred to as the sampling distribution of $\bar{x}$. The final step in identifying the characteristics of the sampling distribution of $\bar{x}$ is to determine the *form* of the probability distribution of $\bar{x}$. We will consider two cases: one where the population distribution is unknown, and one where the population distribution is known to be a normal probability distribution.

For the situation where the population distribution is unknown, we rely on one of the most important theorems in statistics—the *central limit theorem*. A statement of the central limit theorem as it applies to the sampling distribution of $\bar{x}$ is as follows.*

Central Limit Theorem

In selecting simple random samples of size *n* from a population with mean μ and standard deviation σ, the sampling distribution of $\bar{x}$ approaches a normal probability distribution with mean μ and standard deviation $\sigma/\sqrt{n}$ as the sample size becomes large.

Figure 8.4 shows how the central limit theorem works for three different populations; in each case the population clearly is not normal. However, note what begins to happen to the sampling distribution of $\bar{x}$ as the sample size is increased. When the samples are of size 2 we see that the sampling distribution of $\bar{x}$ begins to take on an appearance different from the population distribution. For samples of size 5 we see all three sampling distributions beginning to take on a bell-shaped appearance. Finally, the samples of size 30 show all three sampling distributions to be approximately normal. General statistical practice is to assume that regardless of the population distribution, the sampling distribution of $\bar{x}$ can be approximated by a normal probability distribution whenever the sample size is 30 or more. In effect the sample size of 30 is the rule of thumb that allows us to assume that the large sample conditions of the central limit theorem have been satisfied. This observation about the sampling distribution of $\bar{x}$ is so important that we restate it:

The sampling distribution of $\bar{x}$ can be approximated by a normal probability distribution whenever the sample size is large. The large sample size condition can be assumed for simple random samples of size 30 or more.

The central limit theorem is the key to identifying the form of the sampling distribution of $\bar{x}$ whenever the population distribution is unknown. However, we may encounter some sampling situations where the population is assumed or believed to have a normal probability distribution. When this condition occurs, it is not necessary to rely on the central limit theorem because the following result identifies the form of the sampling distribution of $\bar{x}$.

---

*The theoretical proof of the central limit theorem requires independent observations or items in the sample. This condition exists for infinite populations or for finite populations where sampling is done with replacement. Although the central limit theorem does not directly address sampling without replacement from finite populations, general statistical practice has been to apply the findings of the central limit theorem in this situation provided that the population size is large.

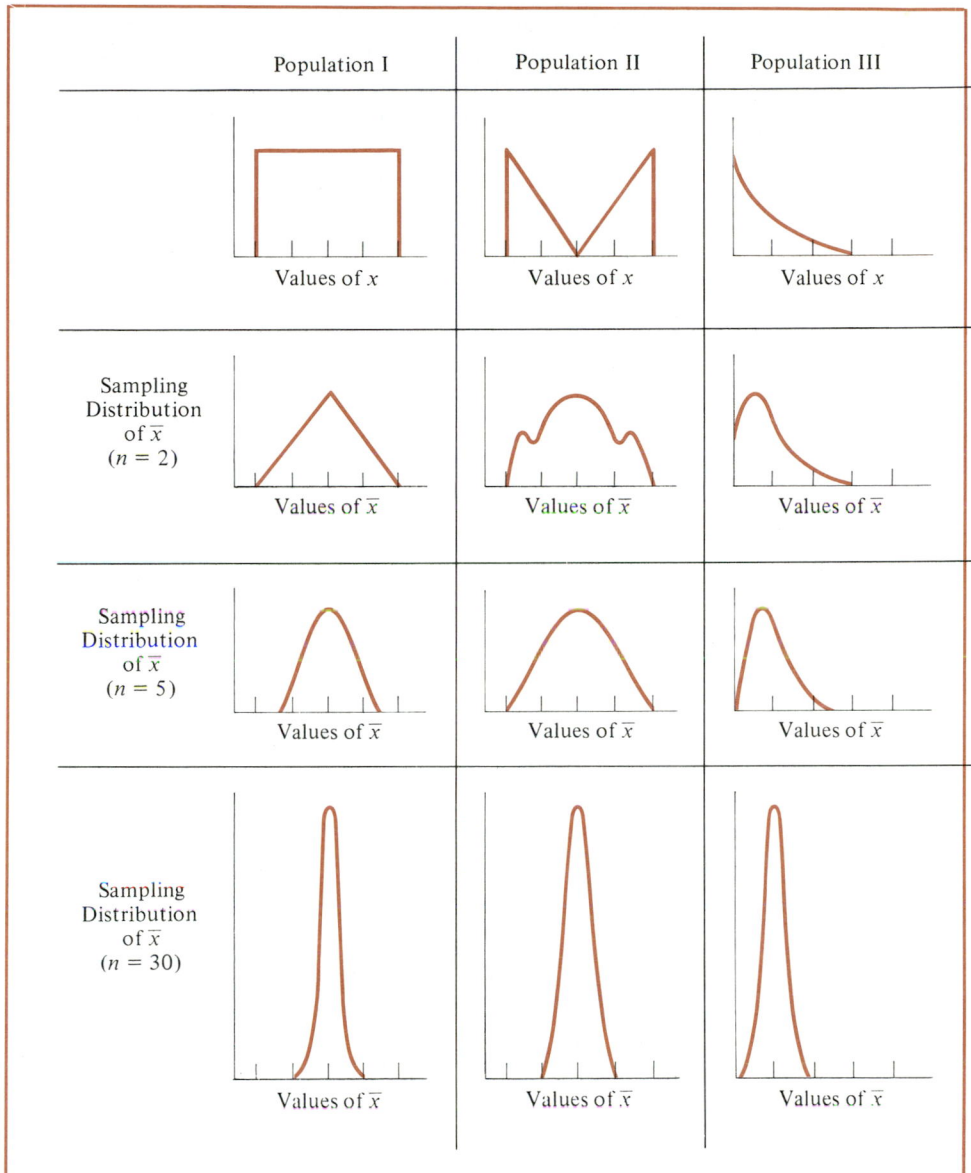

FIGURE 8.4
Illustration of the Central Limit Theorem for Three Populations

Whenever the population being sampled has a normal probability distribution, the sampling distribution of $\bar{x}$ is a normal probability distribution for any sample size.

In summary, whenever we are using a large simple random sample (*rule of thumb: n ≥ 30*), the central limit theorem enables us to conclude that the sampling distribution of $\bar{x}$ can be approximated by a normal probability distribution. In cases where the simple random sample is small ($n < 30$), the sampling distribution of $\bar{x}$ can be considered to be a normal probability only if we believe that the assumption of a normal probability distribution for the population is appropriate.

EXAMPLE 8.4 (continued)

In Example 8.4, we discussed a situation in which simple random samples of size 49 were to be taken from a population with mean $\mu = 100$ and standard deviation $\sigma = 21$. Show the sampling distribution of $\bar{x}$.

Recall that $E(\bar{x}) = 100$ and $\sigma_{\bar{x}} = 21/\sqrt{49} = 3$. The central limit theorem also tells us that for large samples ($n \geq 30$), the sampling distribution of $\bar{x}$ can be approximated by a normal distribution. In such cases the sampling distribution of $\bar{x}$ is as shown.

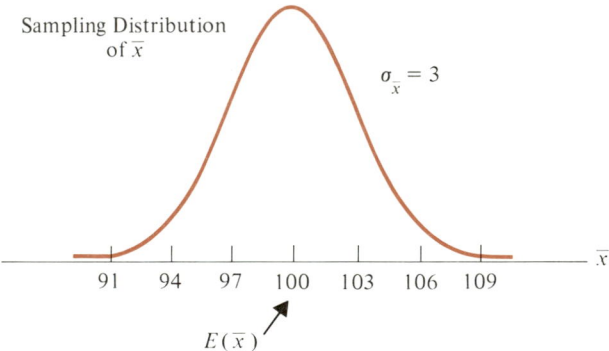

EXAMPLE 8.5

The heights of sixth-grade students in a particular school district are *normally distributed* with a mean of $\mu = 58$ inches and a standard deviation of $\sigma = 3.2$ inches. Show the sampling distribution of $\bar{x}$ if simple random samples of 16 students are to be used.

We know that

$$E(\bar{x}) = \mu = 58$$

and

$$\sigma_{\bar{x}} = \frac{\sigma}{\sqrt{n}} = \frac{3.2}{\sqrt{16}} = .80$$

Since $n < 30$, we cannot use the central limit theorem to conclude the sampling distribution of $\bar{x}$ is approximately normal. However, since the population of heights is described as being normally distributed, the sampling distribution of $\bar{x}$ will be normal for any sample size. Thus we have the distribution shown.

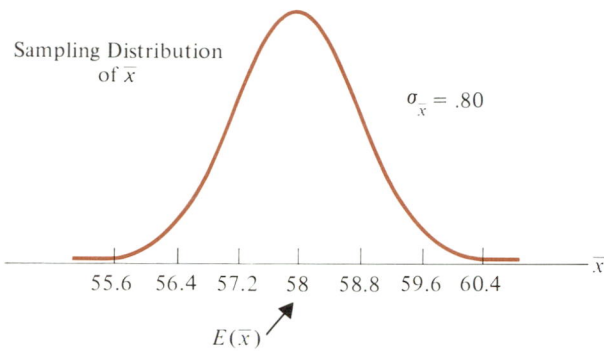

EXERCISES

19. The mean and standard deviation for the number of calories in a 12-ounce can of light beer are as follows: $\mu = 105$ and $\sigma = 3$. A random sample of 30 cans will be selected and a laboratory test conducted to determine the exact number of calories present in each of the 30 cans. The sample mean $\bar{x}$ will be computed.

a. What is the expected value of $\bar{x}$?

b. What is the standard deviation of $\bar{x}$?

c. What probability distribution can be used to approximate the sampling distribution of $\bar{x}$?

d. Sketch a graph of the sampling distribution of $\bar{x}$.

20. The length of time of long-distance telephone calls has a mean $\mu = 18$ minutes and a standard deviation $\sigma = 4$ minutes. Sketch the sampling distribution of $\bar{x}$ if a simple random sample of 50 telephone calls will be used to compute a sample mean length of long-distance telephone calls.

21. A population has a mean $\mu = 400$ and a standard deviation $\sigma = 50$. The probability distribution of the population is unknown.

a. A research study will use simple random samples of either 10, 20, 30, or 40 items to collect data about the population. In which of these sample-size alternatives will we be able to use a normal probability distribution to describe the sampling distribution of $\bar{x}$? Explain.

b. Sketch the sampling distribution of $\bar{x}$ for the instances where the normal probability distribution is appropriate.

22. The body length of a certain insect is believed to be *normally distributed* with a mean length of $\mu = 16.5$ mm and a standard deviation of $\sigma = .8$ mm. Describe the sampling distribution of the sample mean body length if 20 insects are to be used in the study. Is it necessary to use the central limit theorem to determine the shape of the sampling distribution? Explain.

23. Assume that the number of points scored in basketball games played by a particular college team is normally distributed with $\mu = 68$ and $\sigma = 5$. Show the sampling distribution of $\bar{x}$ for a sample of 10 games played by this team.

8.4 THE PRACTICAL VALUE OF THE SAMPLING DISTRIBUTION OF $\bar{x}$

With a knowledge of the sampling distribution of $\bar{x}$, let us demonstrate some of the practical uses of this distribution. As the following examples show, we can use this sampling distribution to compute the probability of selecting a sample that will provide a value of $\bar{x}$ in a prespecified range.

EXAMPLE 8.5 (continued)

In Example 8.5 we described the sampling distribution of $\bar{x}$ for simple random samples consisting of the heights of 16 sixth-grade students. Specifically, we showed that the sampling distribution of $\bar{x}$ is a normal distribution with $E(\bar{x}) = 58$ and $\sigma_{\bar{x}} = .80$. Suppose now we want to compute the probability of obtaining a simple random sample that results in a sample mean $\bar{x}$ between 57 and 59 inches.

Since the sampling distribution of $\bar{x}$ is a normal distribution, we can use the area under the standard normal curve to answer this question. First, we need to compute a z value where

$$z = \frac{\bar{x} - \mu}{\sigma_{\bar{x}}} \tag{8.4}$$

This z value is identical to the z value we used for the normal distribution in Chapter 7, with the exception of the notation $\bar{x}$ and $\sigma_{\bar{x}}$. This notation has been used to indicate that the normal random variable here is $\bar{x}$ and its standard deviation is denoted by $\sigma_{\bar{x}}$.

The probability of selecting a simple random sample with $\bar{x}$ between 57 and 59 is given by the shaded area under the normal curve shown.

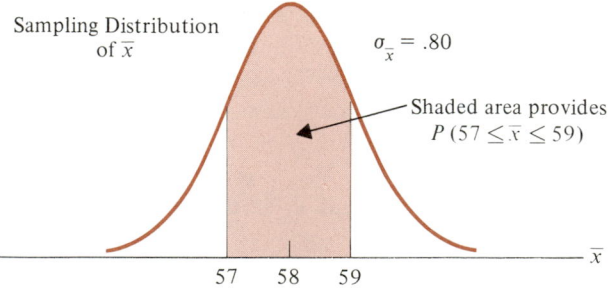

$$\text{At } \bar{x} = 59, \qquad z = \frac{59 - 58}{.80} = 1.25$$

Looking up $z = 1.25$ in the table of areas for the standard normal distribution shows an area of .3944 between 58 and 59.

$$\text{At } \bar{x} = 57, \qquad z = \frac{57 - 58}{.80} = -1.25$$

This z value shows that the area between 57 and 58 must also be .3944. Thus the total probability of selecting a sample with a mean $\bar{x}$ between 57 and 59 inches must be $.3944 + .3944 = .7888$.

Now, let us see how to compute the probability of selecting a simple random sample of 16 students and finding the probability of obtaining a sample mean $\bar{x}$ of 60 inches or more.

The graph of the sampling distribution of $\bar{x}$ for this situation is as follows:

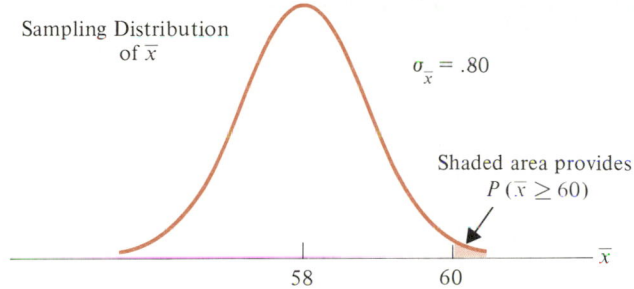

$$\text{At } \bar{x} = 60, \qquad z = \frac{\bar{x} - \mu}{\sigma_{\bar{x}}} = \frac{60 - 58}{.80} = 2.50$$

Using the table of areas for the standard normal distribution and $z = 2.50$, we find that the area between 0.00 and 2.50 is .4938. Thus the probability of $z \geq 2.50$ is a very low $.5000 - .4938 = .0062$. Thus the probability of obtaining a sample 16 students from the population with $\bar{x} \geq 60$ is .0062.

$$\bullet \quad \bullet \quad \bullet$$

EXAMPLE 8.6

Suppose that we have a population of high-school students with a mean IQ score of $\mu = 100$ and a standard deviation of $\sigma = 10$. If simple random samples of 36 students are to be taken from this population, compute an interval around 100 that will include 95% of all possible sample means that could be obtained.

The table of areas for the standard normal distribution shows that 95% of the values in any normal distribution fall between $z = -1.96$ and $z = +1.96$. In other words, 95% of all possible $\bar{x}$ values must be within $-1.96\sigma_{\bar{x}}$ and $+1.96\sigma_{\bar{x}}$ of 100. Thus the range containing 95% of all possible $\bar{x}$ values must be

$$100 - 1.96\sigma_{\bar{x}} = 100 - 1.96(1.67) = 96.7$$

$$100 + 1.96\sigma_{\bar{x}} = 100 + 1.96(1.67) = 103.3$$

That is, there is a .95 probability of selecting a simple random sample having a sample mean IQ score between 96.7 and 103.3. Only 5% of all possible sample means are outside this interval. This situation is shown below.

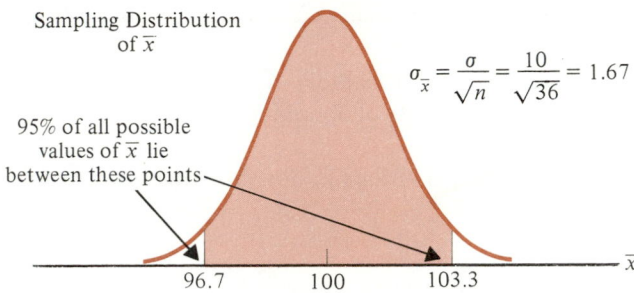

Sampling Distribution of $\bar{x}$

$$\sigma_{\bar{x}} = \frac{\sigma}{\sqrt{n}} = \frac{10}{\sqrt{36}} = 1.67$$

95% of all possible values of $\bar{x}$ lie between these points

96.7   100   103.3 $\bar{x}$

EXERCISES

24. An automatic machine used to fill cans of soup has the following characteristics: $\mu = 15.9$ ounces, $\sigma = .5$ ounces.
    a. Show the sampling distribution of $\bar{x}$, where $\bar{x}$ is the sample mean for 40 cans selected randomly by a quality control inspector.
    b. What is the probability of finding a sample of 40 cans with a mean $\bar{x}$ greater than 16 ounces?

25. A library checks out an average of 320 books per day, with a standard deviation of 75 books. Consider a sample of 30 days of operation, with $\bar{x}$ being the sample mean number of books checked out per day.
    a. Show the sampling distribution of $\bar{x}$.
    b. What is the standard error?
    c. What is the probability that the sample mean for the 30 days will be between 300 and 340 books?
    d. What is the probability the sample mean will show more than 325 books checked out per day?

26. A study of the time from computer program submission until program return (i.e., *turnaround* time) was conducted at a university computer center. Assume that under standard operating conditions the population mean is 120 minutes, with a population standard deviation of 40 minutes.
    a. Future studies of turnaround time are to be based on simple random samples of 30 programs. Show the sampling distribution of the sample mean turnaround time.
    b. What is the probability that the sample mean for 30 programs will be less than 100 minutes? Over 125 minutes?

27. An electrical component is designed to provide a mean service life of 3000 hours, with a standard deviation of 800 hours. A customer purchases a batch of 50 components, which can be considered a simple random sample of the population of components. What is the probability that the mean life for the group of 50 components will be at least 2750 hours? At least 3200 hours?

28. The grade-point average for all juniors at Strausser College has a standard deviation of .50.
    a. A random sample of 20 students is to be used to estimate the population mean grade-point average. What assumption is necessary in order to compute the probability of obtaining a sample mean within plus or minus .2 of the population mean?
    b. Provided that this assumption can be made, what is the probability of $\bar{x}$ being within plus or minus .2 of the population mean?
    c. If this assumption cannot be made, what would you recommend doing?

29. A simple random sample of 64 will be used to estimate the mean time required to perform a particular task in a mechanical aptitude test. If the standard deviation in times is $\sigma = 4$ minutes, use the sampling distribution of $\bar{x}$ to comment on the probability of each error shown. The error is defined as the difference between the observed sample mean and the actual population mean μ.
    a. error of 1 minute or less
    b. error of .5 minutes or less
    c. error greater than .25 minutes

30. Three firms have inventories that vary in size. Firm A has an inventory of 2000 items, firm B has an inventory of 5000 items, and firm C has an inventory of 10,000 items. The standard deviation for the cost of the items in inventory is $\sigma = 144$. A statistical consultant recommends that each firm take a sample of 50 items from their respective inventories in order to provide statistically valid estimates of the mean cost per item in inventory. Management of the small firm states that since it has the smallest inventory, it should be able to obtain the data from a smaller sample size than required by the larger firms. However, the consultant states that in order to obtain the same standard error and thus the same precision in the sample results, all firms should take the same sample size regardless of inventory size.
    a. Using the finite population correction factor, compute the standard error for each of the three firms, given a sample of size 50.
    b. For each firm, what is the probability that the sample mean $\bar{x}$ will be within ± 25 of the population mean, μ?
    c. Do you agree with the consultant's statement? Explain.

31. A survey reports its results by stating that the standard error of the mean was 20. The population standard deviation was 500.
    a. How large was the sample used in this survey?
    b. What is the probability that the estimate would be within ± 25 of the population mean?

32. In a study of annual salaries of managers, the population of managers of interest has annual salaries with $\mu = \$31,800$ and $\sigma = \$4000$.

 a. If a sample of 30 managers is selected, what is the probability the sample mean $\overline{x}$ will be within $\pm\$500$ of the population mean of \$31,800? That is, what is the probability the sample mean will be between \$31,300 and \$32,300?

 b. How large a sample should be selected in order for the probability of a sample mean $\overline{x}$ being within $\pm\$500$ of μ to be .95?

8.5 USING RANDOM NUMBERS TO SELECT A SIMPLE RANDOM SAMPLE

In Section 8.1 we discussed how a one-at-a-time approach to sampling could be used to produce a simple random sample. Recall that when using this method, we select the

TABLE 8.5

Random Numbers

63271	59986	71744	51102	15141	80714	58683	93108	13554	79945
88547	09896	95436	79115	08303	01041	20030	63754	08459	28364
55957	57243	83865	09911	19761	66535	40102	26646	60147	15702
46276	87453	44790	67122	45573	84358	21625	16999	13385	22782
55363	07449	34835	15290	76616	67191	12777	21861	68689	03263
69393	92785	49902	58447	42048	30378	87618	26933	40640	16281
13186	29431	88190	04588	38733	81290	89541	70290	40113	08243
17726	28652	56836	78351	47327	18518	92222	55201	27340	10493
36520	64465	05550	30157	82242	29520	69753	72602	23756	54935
81628	36100	39254	56835	37636	02421	98063	89641	64953	99337
84649	48968	75215	75498	49539	74240	03466	49292	36401	45525
63291	11618	12613	75055	43915	26488	41116	64531	56827	30825
70502	53225	03655	05915	37140	57051	48393	91322	25653	06543
06426	24771	59935	49801	11082	66762	94477	02494	88215	27191
20711	55609	29430	70165	45406	78484	31639	52009	18873	96927
41990	70538	77191	25860	55204	73417	83920	69468	74972	38712
72452	36618	76298	26678	89334	33938	95567	29380	75906	91807
37042	40318	57099	10528	09925	89773	41335	96244	29002	46453
53766	52875	15987	46962	67342	77592	57651	95508	80033	69828
90585	58955	53122	16025	84299	53310	67380	84249	25348	04332
32001	96293	37203	64516	51530	37069	40261	61374	05815	06714
62606	64324	46354	72157	67248	20135	49804	09226	64419	29457
10078	28073	85389	50324	14500	15562	64165	06125	71353	77669
91561	46145	24177	15294	10061	98124	75732	00815	83452	97355
13091	98112	53959	79607	52244	63303	10413	63839	74762	50289

Additional random numbers are provided in Table 6 of Appendix B.

elements for the sample one at a time, making sure that at each selection every one of the items remaining in the population has the same probability of being selected. For large populations, this approach is a practical way of selecting a simple random sample. In Example 8.7, we describe how random numbers can be used to select simple random samples from populations in a one-at-a-time fashion.

EXAMPLE 8.7

Suppose a university has received 7000 applications for admission. The director of admissions would like to use a sample of 50 applications in order to obtain information on Scholastic Aptitude Test (SAT) scores of incoming students. How could a simple random sample of 50 applications be selected?

Let us begin by numbering the 7000 applications in the population from 1 to 7000. Tables of random numbers are available from a variety of handbooks that contain page after page of random numbers.* We have included one such page of random numbers as Table 6 in Appendix B. A portion of this page is also shown in Table 8.5. The digit appearing in any position in the random number table is a random selection of the digits 0, 1, . . . , 9 with each digit having an equal chance of occurring. By selecting four-digit random numbers that range in value from 0001 to 7000, we have a random number corresponding to each of the numbered applications.

In order to use the random numbers to identify items for the sample, we enter the table at any *arbitrary point* and then select four-digit random numbers by moving systematically down a column or across a row of the table. For example, suppose that we arbitrarily start with the third column of random numbers in Table 8.5. Since we only need four-digit numbers, we ignore the first digit in the column. Starting at the top and moving downward, the four-digit random numbers and corresponding application numbers are as shown in the table.

	Random Number	Application Number to Include in Sample
	1744	1744
	5436	5436
	3865	3865
	4790	4790
	4835	4835
Numbers too high. Cannot use since applications 9902 and 8190 do not exist.	9902	—
	8190	—
	6836	6836
	.	.
	.	.
	.	.
	Continue until 50 different applications are selected.	

*For example, *A Million Random Digits with 100,000 Normal Deviates,* by the Rand Corporation (New York: The Free Press, 1955).

Since the numbers selected from the table in Example 8.7 are random, this procedure guarantees that each item in the population has the same probability of being included in the sample and that the sample selected will be a simple random sample.

In using random numbers for simple random sampling, a random number previously used to identify an item for the sample may reappear in the random number table. In selecting the simple random sample *without replacement,* previously used random numbers are ignored because the corresponding element is already in the sample.

A Note on Sampling from Infinite Populations

To this point we have restricted our attention to selecting a simple random sample from a finite population. Although most sampling situations involve finite populations, there are other situations in which the population is either infinite or so large that for practical purposes it must be treated as infinite. In sampling from an infinite population we must give a new definition for a simple random sample. In addition, since the items cannot be numbered, we must use a different process for selecting items for the sample.

Let us consider a situation which can be viewed as requiring a simple random sample from an infinite population. Suppose we want to estimate the average time between placing an order and receiving food for customers arriving at a fast food restaurant during the 11:30 A.M. to 1:30 P.M. lunch period. If we consider the population as being all possible customer visits, we see that it would be next to impossible to specify a finite limit on the number of possible visits. In fact, if we view the population as being all customer visits that could *conceivably* occur during the lunch period, we can consider the population as being infinite. Our task is now to select a simple random sample of *n* customers from this population. With this situation in mind we now state the definition of a simple random sample from an infinite population:

Simple Random Sample (Infinite Population)

A simple random sample from an infinite population is a sample selected such that the following conditions are satisfied:

1. Each item selected comes from the same population.
2. Each item is selected independently.

For our problem of selecting a simple random sample of customer visits at a fast-food restaurant, we find that the first condition is satisfied by any customer visit occurring during the 11:30 A.M. to 1:30 P.M. lunch period while the restaurant is operating with its regular staff under normal operating conditions. The second condition is satisfied by ensuring that the selection of a particular customer does not influence the selection of any other customer.

A well known fast-food restaurant has implemented a simple random sampling procedure for just such a situation. The sampling procedure is based on the fact that some customers present discount coupons, which provide special prices on sandwiches, drinks, french fries, and so on. Whenever a customer presents a discount coupon, the *next* customer is selected for the sample. Since the customers present discount coupons in a random and independent fashion, the firm is satisfied that the sampling plan satisfies the two conditions for a simple random sample from an infinite population.

In other sampling situations, such as the sampling of parts from a production line, the sampling of plants for biological study, the sampling of water for pollution control, and so on, the population can be considered to be infinite in size. In these cases, extra care must be taken to ensure that the sample is representative of the population. Thus selection patterns such as sampling one production machine, sampling only the plants near the window, and sampling the water supply only at 8:00 A.M. must be avoided. If such precautions can be taken, it is usually reasonable to assume that the properties of a simple random sample have been satisfied, and thus the methods introduced earlier in this chapter can be applied.

EXERCISES

33. A student government organization is interested in estimating the proportion of students who favor a mandatory pass-fail grading policy for elective courses. A list of names and addresses of the 645 students enrolled during the current quarter is available from the registrar's office. Using every digit in row 10 of Table 8.5 and moving across the row from left to right, identify the first 10 students who would be selected using simple random sampling. Note that when every digit in row 10 is used, the three-digit random numbers begin with 816, 283, and 610.

34. Consider a population of five salespersons selling mobile telephone units to a variety of customers. The individuals in the population are identified by the letters A, B, C, D, and E. Using the 15th row of random numbers in Table 8.5, use the random digits 1, 2, 3, 4, and 5 to correspond to the five salespersons and select a simple random sample of size 2 from the population.
    a. What sample is selected?
    b. What sample would have been selected if the random numbers in row 20 had been used?

35. Assume that we wanted to select five letters randomly from the alphabet, A–Z. Beginning with the first digits in Table 8.5 and moving down the column, the random number sequence would be 63, 88, 55, 46, and so on. Use this procedure to select five letters randomly from the alphabet. What are the letters chosen when the sampling is done without replacement?

36. Schuster's Interior Design, Inc., specializes in a variety of home-decorating services for its clients. During the past year, the firm has provided major decorating consultation for 875 homes. Schuster's management is interested in obtaining information about customer satisfaction 6 to 12 months after the project is completed. To obtain this information, the firm decided to sample 30 of the 875 clients and interview the group to learn about client satisfaction and ways that Schuster might improve its service. Begin in column 10 of Table 8.5 with the three-digit random number 945. Moving down the column, identify the first 10

clients that would be included in the sample. Assume that the 875 clients are numbered sequentially in the order in which the decorating projects were conducted.

37. Haskell Public Opinion Poll, Inc., conducts telephone surveys concerning a variety of political and general public-interest issues. The households included in the survey are identified by taking a simple random sample from telephone directories in selected metropolitan areas. The telephone directory for a major Midwest area contains 853 pages with 400 lines per page.

   a. Describe a two-stage random selection procedure that could be used to identify a simple random sample of 200 households. The selection process should involve first selecting a page at random and then selecting a line on the sampled page. Use the random numbers in Table 8.5 to illustrate this process. Select your own arbitrary starting point in the table.

   b. What would you do if the line selected in part (a) was clearly inappropriate for the study; that is, the line provided the phone number of a business, restaurant, etc.?

8.6 OTHER SAMPLING METHODS

We have described the simple random sampling procedure and discussed the properties of the sampling distribution of $\bar{x}$ when simple random sampling is used. It is important to realize that simple random sampling is not the only sampling method available. Sampling methods such as stratified random sampling, cluster sampling, and systematic sampling offer alternatives that in some situations have advantages over simple random sampling. In this section we briefly describe some of these alternative sampling methods.

Stratified Random Sampling

In *stratified random sampling,* the population is first divided into groups of elements called *strata,* such that each item in the population belongs to one and only one stratum. The basis for forming the various strata, such as department, location, age, industry type, and so on, is up to the discretion of the designer of the sample. Best results, however, are obtained whenever the elements within each stratum are as much alike as possible. Figure 8.5 shows a diagram with the population divided into H strata.

After the strata are formed, a simple random sample is taken from every stratum. Formulas are available for combining the results for the individual samples into one estimate of the population parameter of interest. The value of stratified random sampling depends upon how homogeneous the elements are within the strata. If units within strata are alike (homogeneity), the strata will have low variances. Thus relatively small sample sizes can be used to obtain good estimates of the strata characteristics. If homogeneous strata exist, the stratified random sampling procedure will provide results similar to simple random sampling but will do so with a smaller total sample size.

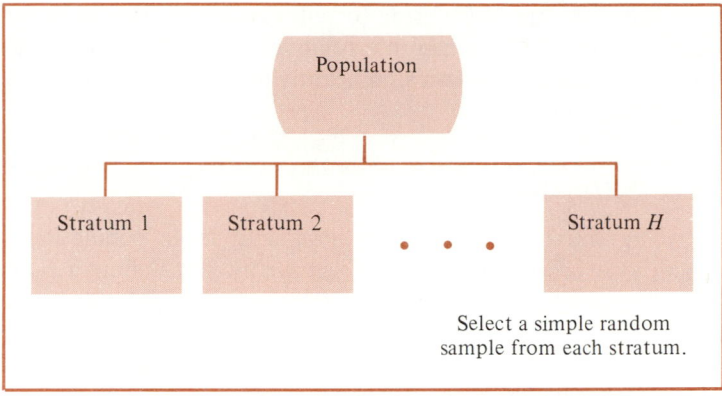

FIGURE 8.5
Diagram for Stratified Simple Random Sampling

Cluster Sampling

In *cluster sampling,* the population is first divided into separate groups of elements called *clusters.* Each element of the population belongs to one and only one cluster (see Figure 8.6) A simple random sample of the clusters is then taken. All elements within each sampled cluster form the sample. Cluster sampling tends to provide best results whenever the elements within the clusters are heterogeneous (not alike). In the ideal case, each cluster is a representative small-scale version of the entire population. The value of cluster sampling depends upon how representative each cluster is of the entire population. If each cluster is alike in this regard, sampling a small number of clusters will provide good estimates of the population parameters.

One of the primary applications of cluster sampling is area sampling, where clusters are city blocks or other well defined areas. Cluster sampling generally requires a larger total sample size than either simple random sampling or stratified random sampling. However, it can result in cost savings because of the fact that when an

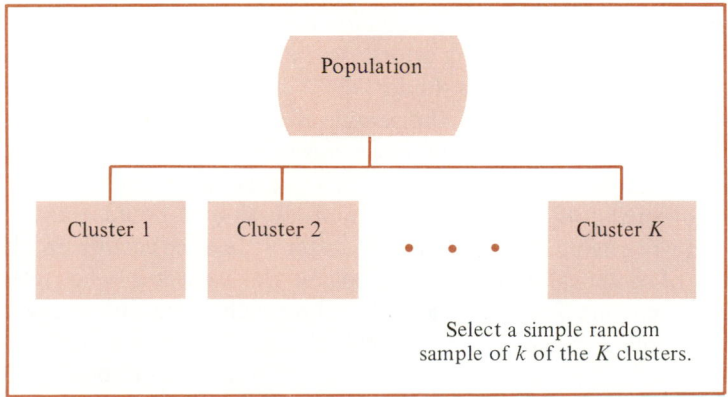

FIGURE 8.6
Diagram for Cluster Sampling

interviewer is sent to a sampled cluster (or city-block location) many interviews or sample observations can be obtained in a relatively short time. As a result, a larger sample size may be obtainable with a significantly lower cost per element sampled and thus possibly a lower total cost.

Systematic Sampling

In some sampling situations, especially those with large populations, it is often time-consuming to select a simple random sample by first finding a random number and then counting or searching through the list of population items until the corresponding element is found. An alternative to simple random sampling is a *systematic sampling procedure*. For example, if a sample size of 50 is desired from a population containing 5000 elements, we might sample one element for every $5000/50 = 100$ elements in the population. A systematic sample for this case would involve selecting randomly 1 of the first 100 elements from the population list. Other sample elements are identified by starting with the first sampled element and then selecting every 100th element that follows in the population list. In effect the sample of 50 is identified by moving systematically through the population and identifying every 100th element after the first randomly selected element. The sample of 50 usually will be easier to identify in this manner than would be the case if simple random sampling were used. Since the first element selected is a random choice, the assumption is usually made that a systematic sample has the properties of a simple random sample. This assumption is especially applicable when the list of the population elements is believed to be a random ordering of the elements.

Convenience Sampling

All the sampling methods discussed thus far fall under the heading of *probability sampling* techniques. By this we mean that elements selected from the population have a known probability of being included in the sample. The advantage of probability sampling is that the sampling distribution of the appropriate sample statistic generally can be identified. Formulas such as the ones for simple random sampling presented earlier in this chapter can be used to determine the properties of the sampling distribution. Then the sampling distribution can be used to make probability statements about possible sampling errors associated with the sample results.

Convenience sampling falls under the heading of *nonprobability sampling* techniques. As the name implies, the sample is identified primarily by convenience. Items are included in the sample without prespecified or known probabilities of being selected. For example, a professor conducting research at a university may ask student volunteers to constitute a sample for the study simply because they are readily available and will often participate as subjects for little or no cost. In another example, a shipment of oranges may be sampled by an inspector who selects oranges haphazardly from among several crates. Labeling each orange and using a probability method of sampling would be impractical. Samples such as wildlife captures and volunteer panels for consumer research are also convenience samples.

Convenience samples have the advantage of relatively easy sample selection and data collection; however, it is impossible to evaluate the "goodness" of the sample in terms of its representativeness of the population. A convenience sample may provide good results, or it may not. However, there is no statistically justified procedure that will allow a probability analysis and inference about the quality of the sample results. Nevertheless, at times you will see statistical methods designed for probability samples applied to a convenience sample. The researcher argues that the convenience sample may well provide a sample which may be treated as if it were a random sample. However, this argument cannot be supported, and we should be very cautious in interpreting convenience samples that are used to make inferences about populations.

Judgment Sampling

One additional nonprobability sampling technique is *judgment sampling*. In this situation the person most knowledgeable on the subject of the study selects individuals or other elements of the population that he or she feels are most representative of the population. Often this is a relatively easy way of selecting a sample. For example, a reporter may sample two or three senators based on the judgment that these senators reflect the general opinion of all senators. However, the quality of the sample results is dependent on the judgment of the person selecting the sample. Again great caution must be used in drawing conclusions based on judgment samples used to make inferences about populations.

Summary

In this chapter we have introduced the important concepts of simple random sampling and the sampling distribution of $\overline{x}$. We introduced the process of using a sample mean $\overline{x}$ (sample statistic) to provide information about a population mean μ (population parameter). Each simple random sample potentially provides a different value for the sample mean $\overline{x}$. The probability distribution for the population of $\overline{x}$ values is called the sampling distribution of $\overline{x}$.

In considering the characteristics of the sampling distribution of $\overline{x}$, we stated that $E(\overline{x}) = \mu$ and that $\sigma_{\overline{x}} = \sigma/\sqrt{n}$, provided $n/N \le .05$. In addition, the central limit theorem provided the basis for using a normal probability distribution to approximate the sampling distribution of $\overline{x}$. The rule of thumb of $n \ge 30$ provided the large-sample conditions necessary to use the normal probability distribution to approximate the sampling distribution of $\overline{x}$. We also noted that whenever the population being sampled has a normal probability distribution, the sampling distribution of $\overline{x}$ would be normal for any sample size. Knowledge of the sampling distribution of $\overline{x}$ was used to make probability statements about the values of $\overline{x}$ that may be obtained from a simple random sample.

The use of random numbers to select simple random samples was demonstrated. Finally, we concluded the chapter by discussing alternative sampling methods, including stratified random sampling, cluster sampling, systematic sampling, convenience sampling and judgment sampling.

Statistics in Practice

PUBLIC OPINION SURVEYS

Each year there is a large outpouring of research information about you, your attitudes, your family, your friends and your fellow Americans. People living in the United States are the most studied and researched people on earth. Every day thousands of us participate in surveys and polls designed to generate data for assessing public opinion. The samples of individuals reached in such public-opinion surveys are used to make inferences about the views of the larger population.

United Media Enterprises, a Scripps-Howard company that syndicates news features, commissioned a public-opinion sample survey designed to determine how Americans spend their spare time. A total of 1000 telephone interviews were used to collect the desired information. During the interviews, individuals were read a list of leisure activities and asked how often they engaged in each activity.

The survey found television-viewing to be the top leisure-time activity in America with 72% of the persons interviewed stating they watched television every day or almost every day. Newspaper reading was the second highest activity, with 70% listing it as a daily activity. Other popular leisure activities included listening to music, talking with friends or relatives, hobbies, gardening, and exercise.

Although public-opinion surveys provide both interesting and useful information, not all public opinion surveys are valid. In some cases, the survey findings may even provide misleading information. The experts in public-opinion polls warn that polls not based on scientifically selected samples should be disregarded. Television phone-in polls and newspaper clip-the-coupon-and-mail-it-in polls can yield distorted pictures of the public mood. In addition, experts suggest checking the source of the poll as the key to its validity. Professional polls by organizations such as Gallup and Harris are reliable, whereas private polls sponsored by political candidates seeking voter support may be suspect. Finally, check the sample size used in the survey. If 100 or less people were interviewed, the poll is subject to large errors. A typical Gallup survey samples about 1500 individuals.

Public-opinion surveys are a way of life for Americans. However, caution should be taken in interpreting public-opinion polls that have used questionable sampling methods.

Glossary

Simple random sample (finite population)—A sample selected such that each possible sample of size *n* has the same probability of being selected.

Simple random sample (infinite population)—A sample selected such that each item comes from the same population and each item is selected independently.

Sampling without replacement—Once an item from the population has been included in the sample, it is removed from further consideration and cannot be selected a second time.

Sampling with replacement—As each item is selected for the sample, it is returned to the population. It is possible that a previously selected item may be selected again and, therefore, appear in the sample more than once.

Parameter—A population characteristic such as a population mean μ.

Statistic—A sample characteristic such as a sample mean $\bar{x}$.

Sampling distribution—A probability distribution showing all possibles values of a sample statistic, such as a sample mean $\bar{x}$.

Standard error of the mean—The standard deviation of $\bar{x}$, denoted by $\sigma_{\bar{x}}$.

Finite population correction factor—The multiplier term $\sqrt{(N - n)/(N - 1)}$ that is used in the formula for $\sigma_{\bar{x}}$; whenever $n/N \leq .05$, the finite population factor is close to 1 and hence $\sigma_{\bar{x}} = \sigma/\sqrt{n}$.

Central limit theorem—A theorem that enables us to use the normal probability distribution to approximate the sampling distribution of $\bar{x}$ whenever the sample size is large. *Rule of thumb:* The central limit theorem applies whenever $n \geq 30$, where n is the sample size.

Probability sampling—Any sampling procedure wherein each element in the population has a known probability of being included in the sample. Simple random sampling, stratified random sampling, cluster sampling, and systematic sampling can be classified as probability sampling techniques.

Nonprobability sample—A sample selected such that the probability of each element being included in the sample is unknown. Convenience and judgment samples are nonprobability samples.

Key Formulas

Expected Value of $\bar{x}$

$$E(\bar{x}) = \mu \tag{8.1}$$

Standard Deviation of $\bar{x}$

$$\sigma_{\bar{x}} = \sqrt{\frac{N - n}{N - 1}} \frac{\sigma}{\sqrt{n}} \tag{8.2}$$

If the sample size is less than or equal to 5% of the population size, use

$$\sigma_{\bar{x}} = \frac{\sigma}{\sqrt{n}} \tag{8.3}$$

Review Quiz

TRUE/FALSE

1. A simple random sample of size n from a population of size N is a sample selected such that each possible sample of size n has the same probability of being selected.
2. A primary objective of sampling is to choose a sample that is representative of the population being studied.
3. The sample mean and sample standard deviation are not sample statistics.
4. The probability distribution of the sample mean is called the sampling distribution of $\bar{x}$.
5. The standard error of the mean cannot be computed unless the sample size is known.
6. In practice, sampling with replacement is more commonly employed than sampling without replacement.
7. The term $\sqrt{(N - n)/(n - 1)}$ in the formula for the standard deviation of $\bar{x}$ is called the continuity correction factor.
8. When sampling with replacement, we always use the formula $\sigma_{\bar{x}} = \sqrt{(N - n)/(N - 1)}\, \sigma/\sqrt{n}$ to compute the standard deviation of $\bar{x}$.
9. The central limit theorem ensures that the sampling distribution of $\bar{x}$ is a normal probability distribution regardless of the sample size.
10. When using cluster sampling, we need not be concerned with selecting a sample that is representative of the population.

MULTIPLE CHOICE

11. Consider a population with $\mu = 50$ and $\sigma = 10$. A simple random sample of size $n = 25$ will be used to provide a sample mean $\bar{x}$ to estimate μ. What is the mean value for the sampling distribution of $\bar{x}$.
    a. 2
    b. 10
    c. 50
    d. none of the above
12. Once an item from the population has been included in the sample, it is removed from further consideration and cannot be selected for the sample a second time. This is an example of which of the following?
    a. cluster sampling
    b. probability sampling
    c. sampling without replacement
    d. sampling with replacement
13. In sampling from a large population with $\sigma = 20$, the standard error of the mean is found to be 2. What was the size of the simple random sample used in this situation?
    a. 100
    b. 10

    c. 40

    d. none of the above

14. What condition is required before the central limit theorem justifies approximating the sampling distribution of $\bar{x}$ with a normal probability distribution?

    a. $n/N \geq .05$

    b. $n \geq 30$

    c. $n < 30$

    d. $N \geq 30$

15. Assume a sample of 49 items is taken from a population with $\mu = 16$ and $\sigma = 7$. What is the probability the sample mean, $\bar{x}$, will be within ± 1 of the population mean $\mu = 16$?

    a. .3413

    b. .6826

    c. .1114

    d. cannot be determined from the above information

16. Which of the following is not an example of a probability sample?

    a. simple random sampling

    b. stratified sampling

    c. cluster sampling

    d. convenience sampling

17. As the sample size increases, variability among the sample means

    a. increases

    b. decreases

    c. remains the same

    d. not enough information given

18. Random samples of size 17 are taken from a population that has 200 elements, a mean of 36, and a standard deviation of 8. The distribution of the population is unknown. The mean and standard deviation of the sample mean are

    a. 8.7 and 1.94

    b. 36 and 1.94

    c. 36 and 1.86

    d. 36 and 8

19. Which of the following best describes the form of the sampling distribution of the sample mean for the situation in question 18?

    a. approximately normal because the sample size is small relative to the population size

    b. approximately normal because of the central limit theorem

    c. exactly normal because the population is normally distributed

    d. nothing can be said with the information given

20. Random samples of size 32 are taken from an infinite population whose mean and standard deviation are 40 and 5, respectively. The distribution of the population is unknown. The mean and standard deviation of the sample mean are

    a. 40 and .78

    b. 40 and .88

    c. 24 and .78

    d. 24 and .88

Supplementary Exercises

38. Assume that a simple random sample of size 2 is to be taken from the following list of airlines: American, United, TWA, Delta, Eastern, Piedmont, and US Air.
    a. How many simple random samples of size 2 are possible? List the possible samples.
    b. What is the probability of each sample being selected?
    c. How many samples include United Airlines?
    d. What is the probability that United Airlines appears in the simple random sample selected?

39. An apartment complex consists of six buildings. The number of apartments rented in each of the six buildings are as follows: 5, 4, 6, 3, 5, and 4, respectively. Assume that a simple random sample of two buildings will be used to estimate the mean number of apartment rentals per building.
    a. Compute the mean μ and standard deviation σ for the population of six apartment buildings.
    b. There are 15 possible simple random samples of size 2 that can be selected from this population. Identify each possible sample and compute its corresponding sample mean $\overline{x}$.
    c. Show the sampling distribution of $\overline{x}$.
    d. Use the computational procedure of Table 8.4 to compute the mean and standard deviation of all possible $\overline{x}$ values.
    e. Use Equations (8.1) and (8.2) and the results of part (a) to determine the expected value and standard deviation of $\overline{x}$. Compare your results with your answer to part (d).

40. The time it takes a fire department to respond to a request for emergency aid has a mean of $\mu = 14$ minutes with a standard deviation of $\sigma = 4$ minutes. Suppose we randomly sample 50 emergency requests over the past two-month period. Records of aid-request times and arrival times will be used to compute a sample mean response time for the 50 requests.
    a. Show the sampling distribution of $\overline{x}$.
    b. What role does the central limit theorem play in identifying this sampling distribution?
    c. What is the probability that the sample mean will be 15 minutes or less?
    d. What is the probability that the sample mean will be within ± .5 minutes of the mean time for the population?

41. The speed of automobiles on a section of I-75 in northern Florida has a mean of $\mu = 57$ miles per hour with a standard deviation of $\sigma = 6$ miles per hour. Answer the following questions if the population can be assumed to have a *normal distribution* and if a sample of 16 automobiles will be selected to compute a sample mean automobile speed.
    a. What is the expected value of $\overline{x}$?
    b. What is the value of the standard error of the mean?
    c. Show the sampling distribution of $\overline{x}$.
    d. What is the probability that the value of the sample mean will be 55 miles per hour or more?
    e. What is the probability that the value of the sample mean will be between 56 and 58 miles per hour?

42. Consider a population of size $N = 500$ with a mean $\mu = 200$ and a standard deviation $\sigma = 40$. Assume that a simple random sample of size $n = 100$ will be selected from this population.

   a. Should the finite population correction factor be used in computing the standard error of the mean?

   b. What is the value of the standard error of the mean for this problem?

   c. What is the probability of selecting a simple random sample that provides a value of $\bar{x}$ that is within ± 5 of the population mean μ?

43. In a population of 4000 employees, a simple random sample of 40 employees is selected in order to estimate the mean age for the population.

   a. Would you use the finite population correction factor in calculating the standard error of the mean? Explain.

   b. If the population standard deviation is $\sigma = 8.2$ years, compute the standard error of the mean, first with and then without the finite population correction factor. What is the rationale for ignoring the finite population correction factor whenever $n/N \leq .05$?

44. In the preceding problem, what is the probability that the sample mean age of the employees will be within ± 2 years of the population mean age?

45. A production process is checked periodically by a quality control inspector. The inspector selects simple random samples of 30 finished products and computes the sample mean product weights, denoted by $\bar{x}$. If test results over a long period of time show that 5% of the $\bar{x}$ values are over 2.1 pounds and 5% are under 1.9 pounds, what are the mean and standard deviation for the population produced with this process?

46. Assume that we wish to identify a simple random sample of 12 of the 372 doctors located in a particular city. The doctors' names are available from a local medical organization. Use the eighth column of five-digit random numbers in Table 8.5 to identify the 12 doctors for the sample. Ignore the first two random digits in each five-digit grouping of the random numbers. This process begins with random number 108 and proceeds down the column of random numbers.

47. Assume that we have a listing of 5500 employees in a large company. Beginning with the four-digit random number 1102 in column 4 of the random numbers of Table 8.5, continue down the column to identify the first 10 employees to be included in the sample.

48. Comment on why each of the samples shown below do not constitute a simple random sample. What kind of samples are they?

   a. To obtain consumer reaction to a new product, a firm contacts women's groups at several local churches and offers to pay the organization for each person participating in the study.

   b. A psychology professor uses a freshman class in Psychology 101 as a sample of subjects for a research project.

   c. After a television debate for presidential candidates, viewers are encouraged to phone the television station to indicate the candidate of their preference.

CHAPTER 9

Inferences About a Population Mean

What You Will Learn in This Chapter

- how to construct and interpret an interval estimate of a population mean

- what is meant by the term confidence level

- the concept of sampling error

- how to determine the sample size when estimating a population mean

- what a hypothesis is and how it is formulated

- how to use sample results to test hypotheses about a population mean

- how to interpret the type I and type II errors in hypothesis testing

- what a *p* value is and how it is used in hypothesis testing

- the *t* distribution

- how to use the *t* distribution to make interval estimates and test hypotheses about a population mean

Contents

Statistics in the News

THE COST OF A COLLEGE EDUCATION CONTINUES TO RISE

How much does it cost per year to attend college in the United States? The College Board, which is an association of colleges and universities that coordinates joint efforts in guidance, admissions, and placement, provides an answer to this question by conducting an annual survey of the nation's two-year and four-year colleges.

Results of the most recent survey show an estimated mean cost for a student attending a public institution to be $4881 per year. The estimated mean cost for a student attending a private institution is $9022 per year. These cost figures include room, board, tuition, books, transportation, and related expenses for an undergraduate student. The new cost estimates are up from the figures of $2365 per year (public) and $3860 per year (private) of 10 years ago. The most expensive university to attend this fall will be Massachusetts Institute of Technology, where the estimated mean annual cost is $16,130. Bennington, Harvard, Princeton, Barnard, Yale, Brandeis, Tufts, and Brown make up the next most expensive group of schools.

If there is an element of good news in this year's cost estimates, it is that the rate of increase in the cost of a college education appears to be slowing. With 10% and 11% annual increases in the cost since 1980, the projected annual increase for the coming year is only 6%. From a cost point of view, the most economical way to attend college is to live at home and attend a nearby two-year institution; the estimated mean annual cost including allowances for living and eating at home is $3423 per year.

College students during a graduation ceremony.

Although the federal government is considering cuts in student-aid programs, the College Board estimates that $16 billion in various forms of financial aid will be available to students during the coming year. If a student can demonstrate eligibility for financial aid, the actual amount paid for a college education can be considerably lowered.

Based on "College Costs Up, But Rate Slows," *New York Times* (August 14, 1984).

In this chapter we continue the discussion of how a sample mean $\overline{x}$ can be used to make inferences about a population mean μ. First we consider the statistical process known as *estimation*, where the value of $\overline{x}$ is used to estimate the value of μ. This was the process used by the College Board in "Statistics in the News"; the College Board's sample results provided an estimate of the mean annual cost for the population of students attending public and private colleges.

Next we discuss the statistical process known as *hypothesis testing*. In this process, we hypothesize that the population mean has a specific numerical value; then we use the value of the sample mean $\bar{x}$ to *test the hypothesis* that the population mean is equal to the stated value.

In the first two sections of the chapter we consider inferences about a population mean based upon a simple random sample which consists of at least 30 items. We refer to this situation as the *large-sample* case. In the last section of the chapter we discuss inferences about a population mean for the *small-sample* case, where the sample consists of less than 30 items. As you will see, the methodology for the small-sample case requires the use of a new probability distribution known as the *t* distribution.

9.1 ESTIMATION OF A POPULATION MEAN

In Chapter 8, we presented the central limit theorem. This theorem enabled us to conclude that the sampling distribution of $\bar{x}$ can be approximated by a *normal probability distribution* with mean μ and standard deviation $\sigma/\sqrt{n}$ whenever the sample size n is large. The generally accepted rule for a large sample is that n must be at least 30. In this section, we show how a sample mean can be used to estimate a population mean whenever the large-sample condition exists.

In the estimation process, the observed value of the sample mean $\bar{x}$ is used as the estimate of the unknown value of the population mean μ. The value of $\bar{x}$ is referred to as the *point estimate* of μ. However, recall from the discussion of the sampling distribution of $\bar{x}$ in Chapter 8 that the sample mean can take on many different values, depending upon the specific simple random sample selected. Thus we cannot expect the point estimate, $\bar{x}$, to provide the exact value of μ. Since the point estimate does not contain information about how close it is to the population mean, we often need to develop a statistical statement about the precision of the point estimate. The *interval estimate* has been developed to provide this information.

EXAMPLE 9.1

In a large school district, a simple random sample of 100 sixth-grade students was used to study the mathematical skills of all sixth-grade students in the district. As part of this study, each of the 100 students in the sample completed a two-hour mathematics achievement test. The sample mean test score for this group was $\bar{x} = 72.5$. Extensive use of this test throughout the country has shown that although the mean test scores differ from district to district, the standard deviation of test scores has been approximately 12; thus in the current study it was assumed that the population standard deviation of test scores is $\sigma = 12$. An objective of the study is to provide a point and an interval estimate of the mean test score for the population of all sixth-grade students in the school district.

• • •

In Example 9.1, the value of the sample mean $\bar{x} = 72.5$ provides the point estimate of the mean test score μ for all sixth-grade students in the district. The central limit theorem enables us to conclude that the value of $\bar{x} = 72.5$ is from the sampling

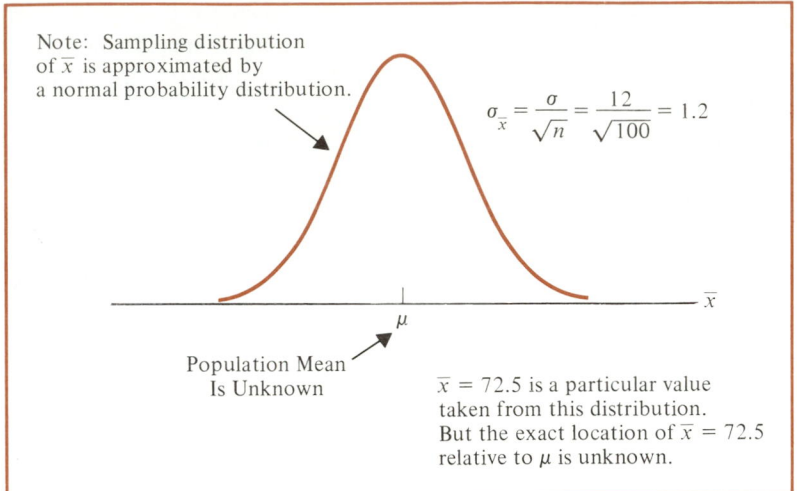

FIGURE 9.1
The Sampling Distribution of $\bar{x}$ for Example 9.1

distribution shown in Figure 9.1. We now want to show how an interval estimate of μ can be used to help determine how close the point estimate of $\bar{x} = 72.5$ is to the population mean μ.

To develop the general approach for computing an interval estimate of μ, first recall that for any normal distribution, 95% of the values are contained in the interval from 1.96 standard deviations below the mean to 1.96 standard deviations above the mean. Thus when using a normal distribution with mean μ and standard deviation $\sigma/\sqrt{n}$ to approximate the sampling distribution of $\bar{x}$, we can conclude that 95% of the samples yield a value of $\bar{x}$ in the interval from $\mu - 1.96(\sigma/\sqrt{n})$ to $\mu + 1.96(\sigma/\sqrt{n})$. This interval is shown in Figure 9.2.

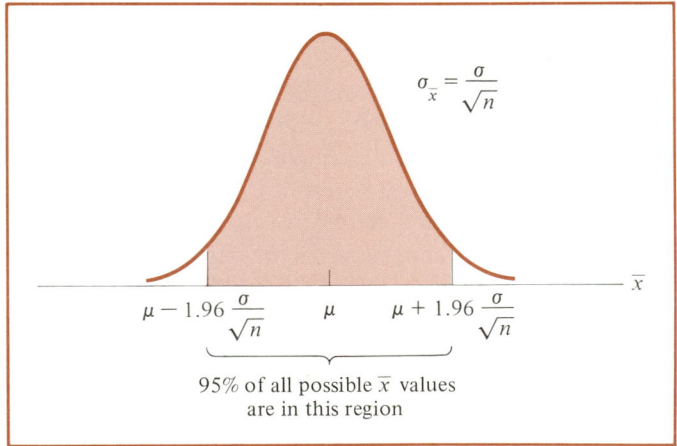

FIGURE 9.2
The Location of 95% of the Possible Sample Means

Looking at Figure 9.2, we see that if the value of the sample mean $\bar{x}$ is in the region corresponding to the shaded area of the distribution, it is *within* $\pm 1.96(\sigma/\sqrt{n})$ *of the value of* μ; thus for any sample mean that is in this region, we can add $1.96(\sigma/\sqrt{n})$ to $\bar{x}$ and subtract $1.96(\sigma/\sqrt{n})$ from $\bar{x}$ to form an interval that contains the population mean μ. Thus the interval estimate for the population mean μ is formed as follows:

$$\bar{x} \pm 1.96\,\frac{\sigma}{\sqrt{n}} \tag{9.1}$$

Since 95% of all possible values of $\bar{x}$ fall in the region corresponding to the shaded area of the distribution, we are 95% confident that the interval formed in this way contains the value of the population mean μ; in other words, 95% of the time that we use this approach to estimate μ, the resulting interval actually contains μ. This interval is called a *confidence interval,* and the value .95 is referred to as the *confidence coefficient* for the interval. Because 95% confidence intervals are commonly used, we restate this confidence interval as follows:

95% Confidence Interval Estimate of a Population Mean

$$\bar{x} \pm 1.96\,\frac{\sigma}{\sqrt{n}} \tag{9.2}$$

EXAMPLE 9.1 (continued)

Using the sample data for the mathematics achievement test scores, we have $\bar{x} = 72.5$, $\sigma = 12$, and $n = 100$. Using (9.2), the interval estimte of μ is computed as follows:

$$72.5 \pm 1.96\,\frac{12}{\sqrt{100}} = 72.5 \underbrace{\pm 2.35}$$

The measure of precision,
or closeness, of $\bar{x} = 72.5$
to the population mean μ.

Subtracting 2.35 and adding 2.35 to the value of the sample mean $\bar{x} = 72.5$ provides the interval 70.15 to 74.85 as the 95% confidence interval estimate of μ. In other words, we are 95% confident that the interval 70.15 to 74.85 contains the value of the population mean μ.

• • •

Other confidence levels are possible for interval estimation of a population mean. The value 1.96 in (9.2) was selected because it is the z value that corresponds to 95% of the values in a standard normal probability distribution. Using a table of areas for the standard normal distribution (Table 1 in Appendix B), other confidence levels can be used by finding the z value that provides the corresponding percentage of values. Such a z value would be substituted for 1.96 in (9.2). Some of the common confidence levels and their corresponding z values are as follows:

z Values for Selected Confidence Levels

90% confidence: $z = 1.645$
95% confidence: $z = 1.96$
98% confidence: $z = 2.33$
99% confidence: $z = 2.58$

From these z values, we note that as the confidence level increases, the z value also increases. This means that the confidence intervals become wider as the confidence level increases. This property of interval estimation points out that if we want to increase the confidence that the interval estimate contains the true value of μ, we need a wider interval. Example 9.2 will show this property of interval estimation.

Using the expression $(1 - \alpha)$ to denote the confidence coefficient, the following formula provides the general procedure for interval estimation of a population mean:

Interval Estimation of a Population Mean

$$\bar{x} \pm z_{\alpha/2} \frac{\sigma}{\sqrt{n}} \tag{9.3}$$

where $(1 - \alpha)$ is the confidence coefficient and $z_{\alpha/2}$ is the z value providing an area of $\alpha/2$ in the upper tail of the standard normal probability distribution.

For example, a 95% confidence interval has a confidence coefficient of .95. With $(1 - \alpha) = .95$, we have $\alpha = .05$. This means there must be an area of $\alpha/2 = .025$ in the upper tail of the standard normal probability distribution. From Table 1 of Appendix B, we find that the $z_{\alpha/2}$ value for a 95% confidence interval is $z_{.025} = 1.96$. Note that this is the z value we used in (9.2) when we computed a 95% confidence interval estimate of a population mean.

EXAMPLE 9.2

A large insurance company selected a simple random sample of 64 policyholders in order to estimate the mean age of individuals insured by the company. From past studies, it is assumed that the population standard deviation for age is $\sigma = 7.2$ years. If the sample mean age for the 64 policyholders is $\bar{x} = 39.5$ years, what are the 95% and 99% confidence interval estimates of the mean age for the population of policyholders?

For a 95% confidence interval, $\alpha = .05$ and $z_{\alpha/2} = z_{.025} = 1.96$. Thus, using (9.3), we obtain

$$\bar{x} \pm 1.96 \, \frac{\sigma}{\sqrt{n}} = 39.5 \pm 1.96 \, \frac{7.2}{\sqrt{64}}$$

$$= 39.5 \pm 1.76$$

or a confidence interval of 37.74 to 41.26 years.

For a 99% confidence interval, $\alpha = .01$ and $z_{\alpha/2} = z_{.005} = 2.58$. Thus (9.3) provides

$$\bar{x} \pm 2.58 \, \frac{\sigma}{\sqrt{n}} = 39.5 \pm 2.58 \, \frac{7.2}{\sqrt{64}}$$

$$= 39.5 \pm 2.32$$

or a confidence interval of 37.18 to 41.82 years.

As we mentioned earlier, these two confidence interval estimates of μ show that if we want to be *more confident* that the interval estimate contains the population mean, we have to use a *wider* interval estimate.

• • •

Interval Estimation When σ Is Unknown

The general expression for calculating the interval estimate of a population mean is given in (9.3). A difficulty in using (9.3) is that in many sampling situations, the value of the population standard deviation σ is *unknown*. In these instances we simply use the value of the sample standard deviation, s, as the point estimate of the population standard deviation σ. The formula for s as provided in Chapter 3 is as follows:

Sample Standard Deviation

$$s = \sqrt{\frac{\Sigma(x_i - \bar{x})^2}{n - 1}} \qquad (9.4)$$

After computing the value of s from the sample data, the interval estimate of the population mean can be computed as follows:

Interval Estimation of a Population Mean—σ Unknown

$$\bar{x} \pm z_{\alpha/2} \, \frac{s}{\sqrt{n}} \qquad (9.5)$$

where $(1 - \alpha)$ is the confidence coefficient, $z_{\alpha/2}$ is the z value providing an area of $\alpha/2$ in the upper tail of the standard normal probability distribution, and s is the sample standard deviation.

EXAMPLE 9.3

A sample of 36 parts was assembled using a proposed production method. The results obtained are: the sample mean time to assemble a part is $\bar{x} = 15.3$ minutes, and the sample standard deviation is $s = 1.3$ minutes. The objective is to develop a 98% confidence interval estimate for the population mean time to assemble a part.

In this example, the population standard deviation σ is unknown. The sample standard deviation, $s = 1.3$, is used as an estimate of σ. For 98% confidence, $\alpha = .02$; thus we have $z_{\alpha/2} = z_{.01} = 2.33$. The resulting 98% confidence interval estimate of the population mean is calculated using (9.5):

$$\bar{x} \pm 2.33 \frac{s}{\sqrt{n}} = 15.3 \pm 2.33 \frac{1.3}{\sqrt{36}}$$

$$= 15.3 \pm .5$$

Thus we are 98% confident that the interval from 14.8 minutes to 15.8 minutes contains the mean assembly time per part for the population.

• • •

Determining the Size of the Sample

The general expression for an interval estimate of a population mean as given in (9.3) is as follows:

$$\bar{x} \pm z_{\alpha/2} \frac{\sigma}{\sqrt{n}}$$

Whenever the value of a sample mean $\bar{x}$ is used as a point estimate of a population mean μ, the *difference* between the sample mean and the population mean $(\bar{x} - \mu)$ is called the *sampling error*. In developing an interval estimate of μ, the term $\pm z_{\alpha/2}$ $(\sigma/\sqrt{n})$ provides a bound on the value of the sampling error. For example, whenever we develop a 95% confidence interval with $z_{\alpha/2} = 1.96$, we are 95% confident that the sampling error is $1.96(\sigma/\sqrt{n})$ *or less*. Letting e denote this bound, or maximum value, for the sampling error, we can write

$$e = z_{\alpha/2} \frac{\sigma}{\sqrt{n}} \qquad (9.6)$$

Since the sample size n appears in the denominator of this expression, increasing the sample size reduces the value of the sampling error. In effect, by increasing the sample size, we are improving the precision of the estimate. A key question in designing samples is: How large a simple random sample needs to be taken in order to achieve a desired degree of precision?

The answer to the sample-size question depends upon the specific sampling application. However, by obtaining three pieces of information about the application, the appropriate sample size can be determined. The three pieces of information needed are:

1. The maximum sampling error, e, that the user of the sampling results is willing to tolerate.

2. The confidence level desired for the interval estimate.
3. A planning value for the population standard deviation σ. Sometimes this planning value can be obtained from past data for the population; otherwise, either a preliminary sample* must be taken or an educated guess of the value of σ must be made.

   With the above three pieces of information, we can use (9.6) to solve for the desired sample size n. This is done by first multiplying both sides of (9.6) by $\sqrt{n}$ and then dividing both sides of the resulting expression by e. These computational steps provide the following:

$$\sqrt{n}e = z_{\alpha/2}\,\sigma$$

$$\sqrt{n} = \frac{z_{\alpha/2}\,\sigma}{e}$$

Squaring both sides of this expression, we obtain the formula for the sample size:

Sample Size for Interval Estimation of a Population Mean

$$n = \frac{(z_{\alpha/2})^2\sigma^2}{e^2} \tag{9.7}$$

EXAMPLE 9.4

In Example 9.1, 100 sixth-grade students were given a mathematics achievement test. Using the sample mean of $\bar{x} = 72.5$ and a 95% confidence level, the maximum sampling error was found to be 2.35. Recall that the population standard deviation was assumed to be $\sigma = 12$. What sample size should be used in order to reduce the sampling error to $e = 1.5$?

   Using (9.7), we have

$$n = \frac{(z_{.025})^2\sigma^2}{e^2} = \frac{(1.96)^2(12)^2}{(1.5)^2} = 245.9$$

Since we cannot sample a fraction of a student, we round up to a recommended sample size of 246 students. This sample size will provide a 95% confidence level that the sampling error is 1.5 or less.

$$\bullet \quad \bullet \quad \bullet$$

EXAMPLE 9.5

The manager of a fast-food restaurant would like to estimate the mean dollar amount spent per customer. Suppose that the manager wants to develop an estimate with a

---

*A preliminary sample is a sample taken to learn about the population and, in particular, its standard deviation σ. The preliminary sample is usually much smaller than the final sample size. However, it is helpful in providing the information needed to determine the final sample size.

maximum sampling error of \$.50. The sample standard deviation for a previous sample of 32 customers is $s = \$2.00$. How many customers should be included in the sample if a 99% confidence interval estimate of the population mean is desired?

For this information, $e = .50$, $z_{.005} = 2.58$, and a planning value for σ is 2. Thus

$$n = \frac{(z_{.005})^2 \sigma^2}{e^2} = \frac{(2.58)^2(2)^2}{(.50)^2} = 106.5$$

Hence, for a 99% confidence level, the sampling error will be \$.50 or less if we use a simple random sample of $n = 107$ customers.

• • •

EXERCISES

1. On a final examination for a chemistry course at a large university, 36 randomly selected papers were graded shortly after the exam was over. Based on the previous year, the standard deviation of examination scores for the population was assumed to be $\sigma = 15$. If the sample mean for the 36 papers was $\overline{x} = 72$, provide 90% and 95% confidence intervals for the mean examination score for the population.

2. Data were collected on the golf-ball driving distances by professional golfers in a recent tournament. Using data on the first drives of 30 randomly selected golfers, it was found that the sample mean distance was 250 yards and the sample standard deviation was 10 yards. Develop a 95% confidence interval for the mean driving distance for the population.

3. E. Lynn and Associates is an energy-research firm that provides estimates of monthly heating costs for new homes based on style of the house, square footage, insulation and so on. The firm's service is used both by builders and potential buyers of new homes who wish advance information on heating costs. For winter months, the standard deviation in the home heating bills for residential homes in a certain area is believed to be \$50. Assume that a sample of 30 homes in a particular subdivision will be used to estimate the mean monthly heating bills for all homes in this type of subdivision. If the sample mean is $\overline{x} = \$196.50$, provide a 98% confidence interval for the mean monthly heating bill.

4. A sample of 64 customers at Ron and Ted's Service Station shows a mean number of gallons of gasoline purchased per customer of 13.6 gallons. If the population standard deviation is 3.0 gallons, what is the 95% confidence interval estimate of the mean number of gallons purchased per customer?

5. During a water shortage a water company randomly sampled residential water meters in order to monitor daily water consumption. On one particular day, a sample of 50 meters showed a sample mean of $\overline{x} = 240$ gallons and a sample standard deviation of $s = 45$ gallons. Provide a 90% confidence interval estimate of the mean water consumption for the population.

6. The Benson Property Management firm located in St. Louis would like to estimate the mean cost of repairing damages in apartments that are vacated by tenants. A sample of 36 vacated apartments resulted in a sample mean repair cost of $86.00, with a sample standard deviation of $12.25. Develop a 95% confidence interval to estimate the mean repair cost for the population of apartments.

7. A simple random sample of 35 Metro buses shows a sample mean of 225 passengers carried per day per bus. The sample standard deviation is computed to be 60 passengers. Provide a 98% confidence interval estimate of the mean number of passengers carried per bus during a 1-day period.

8. Miles-per-gallon rating tests are being conducted for a particular model of automobile. If it is desired to estimate the mean miles per gallon rating to within ± 1 mile per gallon at a 95% level of confidence, how many automobiles should be used in the sample? Assume that preliminary mileage tests indicate a standard deviation in miles per gallon for the automobiles to be 2.9.

9. An educational innovation at the high-school level will enable students to earn credit by completing a self-study workbook on English literature. In order to judge the credit given to the student fairly, educators would like an estimate of the mean number of hours it will take a student to complete the self-study program. In an evaluation study, a sample of students will be monitored during the self-study process. How many students should be in the sample if it is desired to estimate the mean number of hours required to complete the self-study to within ± 3 hours at a 95% level of confidence? For planning purposes, it is estimated that the standard deviation of completion times will be approximately 10 hours.

10. Starting annual salaries for college graduates are believed to have a standard deviation of approximately $2000.00. Assume that a 95% confidence interval estimate of the mean annual starting salary is desired. How large a sample size should be taken if the size of the sampling error in the precision statement is to be
    a. $500.00
    b. $200.00
    c. $100.00

11. In developing patient appointment schedules, a medical center desires to estimate the mean time that a staff member spends with each patient. How large a sample should be taken if the precision of the estimate is to be ± 2 minutes at a 95% level of confidence? How large a sample for a 99% level of confidence? Use a planning value for the population standard deviation of 8 minutes.

12. A national survey research firm has past data that indicate that the interview time for a consumer opinion study has a standard deviation of 6 minutes.
    a. How large a sample should be taken if the firm desires a .98 probability of estimating the mean interview time to within 2 minutes or less?

b. Assume that the simple random sample you recommended in (a) is taken and that the mean interview time for the sample is 32 minutes. What is the 98% confidence interval estimate for the mean interview time for the population of interviews?

9.2 HYPOTHESIS TESTS ABOUT A POPULATION MEAN

Statistical inference is the process of drawing conclusions about a population parameter based on information contained in a sample. Point and interval estimation of a population mean, as introduced in Section 9.1, are forms of statistical inference. Another form of statistical inference is *hypothesis testing*. In developing a hypothesis test about a population mean μ, the person performing the test selects a hypothesized value for μ. Then a sample is selected from the population and the sample mean, $\bar{x}$, is computed. By comparing the value of the sample mean to the hypothesized value for the population mean, a conclusion can be reached about whether or not to reject the hypothesis.

The situation in hypothesis testing is similar to the situation found in criminal trials. The hypothesis in the trial setting is that the defendant is innocent. Under the assumption that the hypothesis of innocence is true, the trial proceeds with the presentation of sample evidence in the form of testimony. If the sample evidence or testimony is sufficient to reject the hypothesis of innocence, the defendant will be found guilty. However, if the sample evidence is *not* sufficient to reject the hypothesis, the assumption of innocence will stand and any action taken by the court will be based on this assumption.

The hypothesis tests that we consider in this section are large-sample tests concerning the value of the population mean. Recall that the large-sample condition exists whenever the sample contains at least 30 items. In this situation, the central limit theorem enables the use of a normal probability distribution to approximate the sampling distribution of $\bar{x}$. This sampling distribution is shown in Figure 9.3.

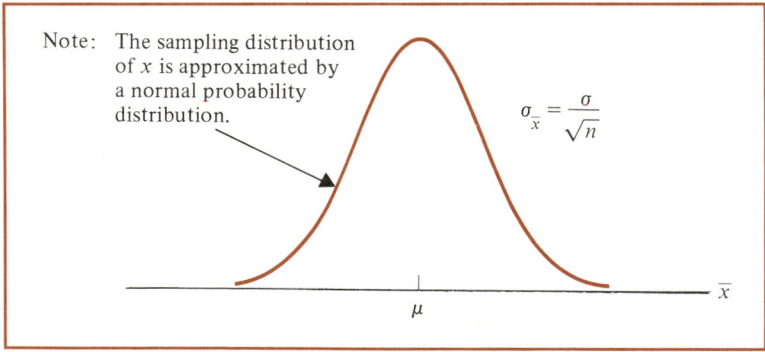

FIGURE 9.3
The Sampling Distribution of $\bar{x}$ for Simple Random Samples of Size 30 or More

EXAMPLE 9.6

KMGM is a rock radio station in Seattle, Washington. Advertising spots are sold to customers on the basis that KMGM's target listening audience has a mean age of 21 years. A particular concern of KMGM's station manager is whether or not the station is reaching this target audience. In order to answer this question, KMGM has hired an independent survey firm that uses a sampling procedure to determine characteristics of a station's listening audience. If the survey shows that KMGM is not reaching a listening audience with a mean age of $\mu = 21$ years, programming changes will be implemented by the station. The station manager would like to develop a hypothesis test about the mean age of the audience population that will help conclude whether or not the station is reaching the desired audience.

Let us begin by taking the position that the station is reaching the desired audience with a mean age of $\mu = 21$ years. A *null hypothesis,* denoted by H_0, will be formulated under this assumption. An *alternative hypothesis,* denoted by H_1, will be formulated for the case where the null hypothesis is rejected. In this situation, the alternative hypothesis is that the station is *not* reaching the desired audience. The null and alternative hypotheses and their implied conclusions and actions are as follows:

Hypothesis	Conclusion and Action
$H_0: \mu = 21$	KMGM is reaching the desired target audience; no action is necessary.
$H_1: \mu \neq 21$	KMGM is not reaching the desired target audience; consider programming modification or other corrective action.

During the audience survey, a sample of listeners will be identified and the sample mean age, $\bar{x}$, computed. Under the assumption that $H_0: \mu = 21$ is true, we are assuming that the sample mean comes from the sampling distribution shown in Figure 9.3 with the mean of this sampling distribution located at $\mu = 21$. As with all sampling distributions of $\bar{x}$, we cannot expect the value of the sample mean obtained during the survey to be exactly equal to the population mean. That is, even if $\mu = 21$ is true, we do not expect the sample mean $\bar{x}$ to be exactly equal to 21. However, if the null hypothesis $\mu = 21$ is true, we do expect the sample mean to be close to 21.

For a given value of the sample mean $\bar{x}$, we can compute a corresponding z value, which determines how many standard deviations $\bar{x}$ is from the assumed value for the population mean μ. The expression for the z value is as follows:

$$z = \frac{\bar{x} - \mu}{\sigma / \sqrt{n}} \tag{9.8}$$

In this case, z is the standard normal random variable discussed in Chapter 7. Figure 9.4 shows the distribution of z when the null hypothesis H_0 is true.

If the z value using (9.8) is close to zero, it indicates that the value of the sample mean $\bar{x}$ must be close to the hypothesized value for μ. While this condition does not

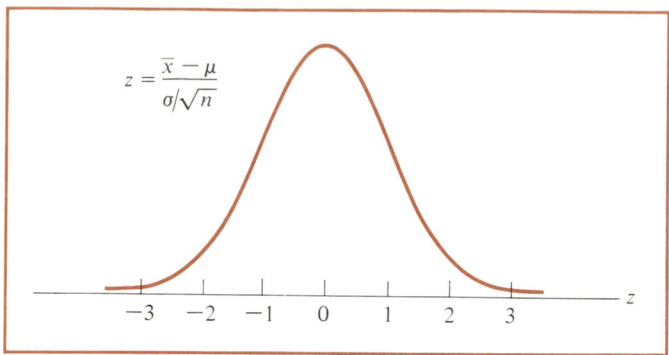

$$z = \frac{\bar{x} - \mu}{\sigma/\sqrt{n}}$$

FIGURE 9.4
Distribution of the Standard Normal Random Variable
z when the Null Hypothesis H_0 is True

prove that the hypothesized value of μ is true, it does indicate that the value of the sample mean $\bar{x}$ is consistent with the hypothesized value of μ. Hence if action is required at this point, it should be based on the assumption that the hypothesized value for μ is true. On the other hand, if the z value using (9.8) is substantially different from 0, perhaps $z = 4$, we have an indication that the sample mean is far from the hypothesized value for the population mean. In fact, if $z = 4$, the observed value of $\bar{x}$ is so far from the hypothesized value of μ that we would probably doubt that the sample mean $\bar{x}$ actually came from a sampling distribution with the hypothesized value of $\mu = 21$. In this case, we conclude that the sample results are inconsistent with the null hypothesis. As a result, the null hypothesis would be rejected and any action taken would be based on the alternative hypothesis of $\mu \neq 21$. Later in this section we discuss decision rules based on the value of z that tell us whether or not to reject the null hypothesis.

• • •

Errors Involved in Hypothesis Testing

In the hypothesis-testing procedure, we begin by tentatively assuming that the null hypothesis, H_0, is true. Our goal is to use the sample information either to confirm this assumption (accept H_0) or to reject this assumption (reject H_0). Ideally, we would like the hypothesis-testing procedure to lead us to accept H_0 when H_0 is true and to reject H_0 when H_0 is false. However, since our conclusions are based upon the results of a sample, this ideal situation cannot be guaranteed. As a result, we must consider the possibility of drawing incorrect hypothesis-testing conclusions.

First, consider the case where the null hypothesis H_0 is true. Although we hope that the testing procedure will lead to the acceptance of H_0, there is a possibility that the procedure will lead us to the incorrect conclusion of rejecting H_0. This type of error, referred to as a *type I error,* is defined as follows.

Definition of a Type I Error

A *type I error* is rejecting H_0 when it is true.

A second type of error occurs whenever the null hypothesis, H_0, is false, but the hypothesis-testing procedure leads to the incorrect conclusion of accepting H_0. This type of error, referred to as a *type II error,* is defined as follows.

Definition of a Type II Error

A *type II error* is accepting H_0 when it is false.

EXAMPLE 9.6 (continued)

The hypothesis test for the KMGM radio station in Example 9.6 is as follows:

$$H_0: \mu = 21$$

$$H_1: \mu \neq 21$$

The null hypothesis reflects the situation where the radio station is meeting the objective of having a target audience with a mean age of 21 years. The alternative hypothesis reflects the situation where the station is not meeting its target-audience age objective. If the null hypothesis is rejected, the station manager will consider programming modifications in an attempt to achieve the goal of a 21-year-old listening audience. Define the type I and type II errors for this hypothesis-testing situation.

If H_0 were true, a type I error would be made by incorrectly rejecting this null hypothesis. For KMGM this would mean concluding that the target audience is not being reached when, in fact, it is; in this case programming modifications would be undertaken when they were unnecessary. On the other hand, a type II error would be made if H_0 were false and the hypothesis-testing procedure resulted in the conclusion to accept the null hypothesis incorrectly. In this case, the station would conclude that the target audience mean age of 21 years was being reached when, in fact, it was not. As a result of this error, the station would not consider programming modifications when such modifications would be desirable.

• • •

At first you might be disappointed to learn that hypothesis testing can lead to errors. However, it is helpful to know that the probabilities of making errors can be determined and controlled. By controlling the probabilities of the two errors, we can achieve a reasonable degree of confidence that the hypothesis-testing procedure will provide the correct conclusion.

The Greek letters α (alpha) and β (beta) are used to denote the probabilities of the hypothesis-testing errors; their definitions are as follows.

$$\alpha = \text{the probability of making a type I error}$$

$$\beta = \text{the probability of making a type II error}$$

Figure 9.5 summarizes the conditions under which the type I and type II errors can be made for the KMGM hypothesis test. It also shows the situations where the hypothesis-testing procedure leads to correct decisions. Note that the box in the upper left-hand corner in Figure 9.5 corresponds to the case when H_0 is true and the hypothesis test leads to the decision to accept H_0; in this case a correct conclusion is made. The box in the lower left-hand corner corresponds to the case when the null hypothesis is true, but the procedure leads us to reject it; in this case a type I error is made. Simlarly, the box in the upper right-hand corner corresponds to making a type II error, and the box in the lower right-hand corner corresponds to the correct decision of rejecting the null hypothesis when it is false.

Hypotheses:
$H_0: \mu = 21$
$H_1: \mu \neq 21$

		Situation in the Population	
		H_0 True ($\mu = 21$)	H_0 False ($\mu \neq 21$)
Accept H_0		Correct Decision	Type II Error
Reject H_0		Type I Error	Correct Decision

Conclusion

FIGURE 9.5
Type I and Type II Errors in Hypothesis Testing

Developing Decision Rules for Hypothesis Testing

Thus far we have discussed the formulation of the null and alternative hypotheses as well as the type I and type II hypothesis-testing errors. In addition, we showed that the standard normal random variable z, where

$$z = \frac{\bar{x} - \mu}{\sigma / \sqrt{n}} \qquad (9.9)$$

is used to determine the number of standard deviations the sample mean $\bar{x}$ is from the hypothesized value of the population mean μ. The computation of the z value in (9.9) is based on the following information:

$\bar{x}$ = the value of the sample mean

μ = the value of the population mean under the assumption the null hypothesis is true

σ = the value of the population standard deviation (which may have to be estimated using the sample standard deviation, s)

n = the sample size used in the computation of the sample mean.

The distribution of z under the assumption that H_0 is true is shown in Figure 9.6. We now need to establish a decision rule based on the value of z that will enable us to

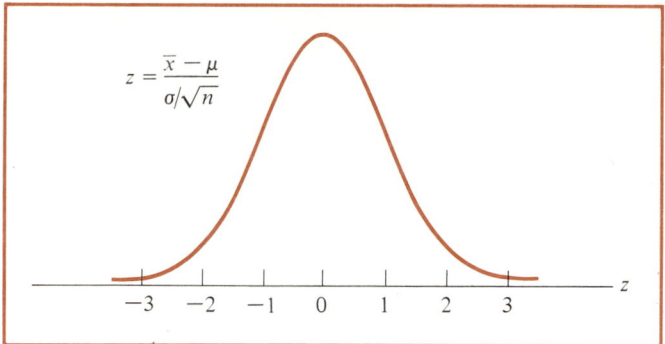

FIGURE 9.6
Distribution of the Standard Normal Random Variable z when the Null Hypothesis H_0 Is True

determine whether or not to reject the null hypothesis. Since the conclusion will be based on the value of z, z as given in (9.9) is referred to as the *test statistic* for the hypothesis test.

The key to establishing a decision rule is to specify an *allowable probability of making a type I error*. That is, under the assumption that the null hypothesis is true, the designer of the hypothesis test must specify an allowable probability of making the error of incorrectly rejecting the null hypothesis. This probability of the type I error, denoted by α, is referred to as the *level of significance* for the test. Commonly used values for the level of significance are .10, .05, .02, or .01, all of which indicate a relatively low probability of making a type I error. Let us show how to develop a decision rule by reconsidering the KMGM presented in Example 9.6.

EXAMPLE 9.6 (continued)

The hypotheses for the KMGM audience age study are written as follows:

$$H_0: \mu = 21$$

$$H_1: \mu \neq 21$$

From past experience, it is believed that $\sigma = 5$ years is the population standard deviation. If a sample of 100 listeners was found to have a sample mean age of $\bar{x} = 20.3$ years, what is the hypothesis-testing conclusion and what recommendation is appropriate? Let us see how we can develop a decision rule that will determine whether or not to reject H_0 using a .05 level of significance.

The decision rule for the hypothesis test is shown in Figure 9.7. The figure shows the distribution of z values under the assumption that $H_0: \mu = 21$ is true. Note that 95% of the z values correspond to the "Accept H_0" portion of the z axis. However, with a .025 area, or probability, in the "Reject H_0" regions in the two tails of the distribution, there is a .05 probability of having a z value in a *rejection region* of the distribution even though $H_0: \mu = 21$ is true. Thus the test is being designed with a .05 probability of making a type I error, and the desired level of significance of $\alpha = .05$ is being satisfied. The $\alpha = .05$ was split between the two tails of the distribution because the alternative

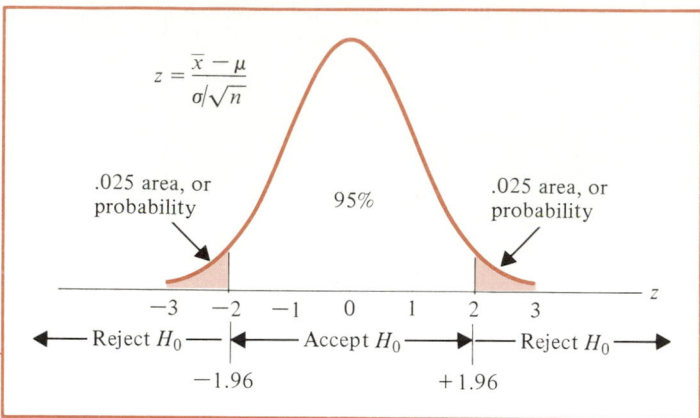

FIGURE 9.7
Decision Rule for the Null Hypothesis H_0: $\mu = 21$
with an $\alpha = .05$ Level of Significance

hypothesis shows that we want to reject H_0 if the true value of μ is *less than* 21 (lower tail) or if the true value of μ is *greater than* 21 (upper tail). Whenever a hypothesis test has rejection regions in both tails of the distribution, the hypothesis test is referred to as a *two-tailed test*. Regardless of the value specified for the level of significance α, each tail in a two-tailed test has a rejection region area, or probability, of $\alpha/2$.

In Figure 9.7 the tables of areas for the standard normal probability distribution were used to find the z values that determine the boundaries of the acceptance and rejection regions; in this case, the values are $z = -1.96$ and $z = 1.96$. In general, these z values are referred to as the *critical values* for the test. The decision rule that provides the .05 level of significance can be written as follows:

$$\text{Accept } H_0 \text{ If } -1.96 \leq z \leq +1.96$$

$$\text{Reject } H_0 \text{ Otherwise}$$

Given this decision rule, we can now turn our attention to determining whether or not to reject H_0 by computing the z value corresponding to the sample mean of $\bar{x} = 20.3$ years. Using (9.9) and assuming that the null hypothesis H_0: $\mu = 21$ is true, we have the following value for the test statistic:

$$z = \frac{\bar{x} - \mu}{\sigma/\sqrt{n}} = \frac{20.3 - 21}{5/\sqrt{100}} = -1.40$$

Referring to the decision rule, the sample mean of $\bar{x} = 20.3$ leads us to accept the null hypothesis H_0: $\mu = 21$. We are careful not to say that we have *proven* that $\mu = 21$. Rather, we are saying that with the sample mean $\bar{x} = 20.3$, we are unable to reject the null hypothesis of a mean audience age of 21 years at a .05 level of significance. Being unable to reject the null hypothesis, we would recommend that the radio station take action under the assumption that H_0 is true. Thus, in this case, the hypothesis test indicates no reason for the station to undertake programming changes.

$\bullet \ \bullet \ \bullet$

The Use of p Values

We now introduce another criterion that can be used to establish a decision rule for a hypothesis test. This criterion is called a *p value*. We show how a *p* value can be computed for an observed sample mean $\bar{x}$ and how the *p* value can be used to make the decision of whether or not to reject the null hypothesis.

The *p* value is the probability, when the null hypothesis is true, of obtaining a difference between a sample mean $\bar{x}$ and the hypothesized value of the population mean μ that is *larger* than the difference actually observed. As an illustration of the *p* value computational procedure, let us return to Example 9.6.

EXAMPLE 9.6 (continued)

The observed value of the sample mean is $\bar{x} = 20.3$ years. With the hypothesized value of $\mu = 21$ years, the observed difference between $\bar{x}$ and μ is $21 - 20.3 = .7$ years. The *p* value is the probability, when $H_0: = 21$ is true, of obtaining a difference between a sample mean $\bar{x}$ and $\mu = 21$ that is larger than .7 years.

In Example 9.6 we found the *z* value corresponding to a sample mean of $\bar{x} = 20.3$ was $z = -1.40$. Using the tables of areas for the standard normal probability distribution, we find that the area in the *left-hand tail* of the distribution at $z = -1.40$ is given by $.5000 - .4192 = .0808$. This tells us that there is a .0808 probability of obtaining a difference between $\bar{x}$ and $\mu = 21$ that is larger than .7 years by having a value of the sample mean $\bar{x}$ in the left-hand tail of the distribution. However, since we have a two-tailed hypothesis test, we must also consider the probability of having a difference between $\bar{x}$ and $\mu = 21$ that is larger than .7 years by having a value of the sample mean $\bar{x}$ in the *right-hand tail* of the distribution. Since the normal probability distribution is symmetric, this probability is given by the area in the right-hand tail at $z = +1.40$. This area is also .0808. Thus, for the observed sample result of $\bar{x} = 20.3$ years, the *p* value is $2 \times .0808 = .1616$. This indicates that there is a .1616 probability, when $H_0: \mu = 21$ is true, of obtaining a difference between a sample mean $\bar{x}$ and $\mu = 21$ that is larger than the .7 years difference actually observed.

The *p* value can now be used to make the hypothesis-testing conclusion by noting that if the *p value is greater than or equal to* α, the value of the test statistic, *z* must be in the *acceptance region*. Similarly, if the *p value is less than* α, the value of the test statistic *z* must be in the *rejection region*. Thus, the *p* value $= .1616$ is greater than $\alpha = .05$, which tells us to accept the null hypothesis $H_0: \mu = 21$.

$\bullet \quad \bullet \quad \bullet$

The *p* value and the test statistic *z* will always provide the same hypothesis-testing conclusion at a specified level of significance α. However, the *p* value carries the interpretation of being the probability of obtaining a difference between $\bar{x}$ and the hypothesized value for μ that is *larger than* the difference actually observed. Under this interpretation, a relatively large *p* value is an indication that the observed difference between $\bar{x}$ and the hypothesized value of μ is not unusual. Thus the sample result would be judged consistent with the null hypothesis and H_0 would be accepted. However, a small *p* value indicates that the difference between $\bar{x}$ and the hypothesized value for μ is so unusual that the null hypothesis should be rejected. For a given level of significance α, the decision rule for hypothesis testing is as follows.

> **_p_ Value Criterion for Hypothesis Testing**
>
> Accept H_0 If p Value $\geq \alpha$
>
> Reject H_0 If p Value $< \alpha$

EXAMPLE 9.7

Statistics have been compiled for the selling prices of condominiums in Dade County, Florida. During the previous year the mean selling price of condominiums was $81,000. A sample of condominium sales during the current year is planned in order to learn about current condominium selling prices.

a. Formulate null and alternative hypotheses that can be used to test for any change in the mean selling price of condominiums during the 1-year period.
b. Using a .10 level of significance, develop a decision rule for the values of z that will determine whether or not to reject the null hypothesis.
c. A sample of 50 current condominium sales showed a sample mean selling price of $\bar{x}$ = $79,500, with a sample standard deviation of s = $5000. What is the hypothesis-testing conclusion?
d. What is the p value associated with the sample mean of $\bar{x}$ = $79,500? What is the hypothesis-testing conclusion based on the p value?

The answer to (a) provides the following hypothesis test:

Hypothesis	Conclusion
$H_0: \mu = 81,000$	Mean sales prices are stable, showing no significant change over the 1-year period.
$H_1: \mu \neq 81,000$	Mean sales prices are not stable; the mean prices are different from the previous year.

Using an α = .10 level of significance with a .05 area or probability in each tail of the distribution, the tables of areas for the standard normal random variable can be used to obtain the following decision rule:

Accept H_0 If $-1.645 \leq z \leq +1.645$

Reject H_0 Otherwise

The acceptance and rejection regions for this two-tailed test are shown in Figure 9.8.
The hypothesis-testing decision can be based upon the value of the test statistic z

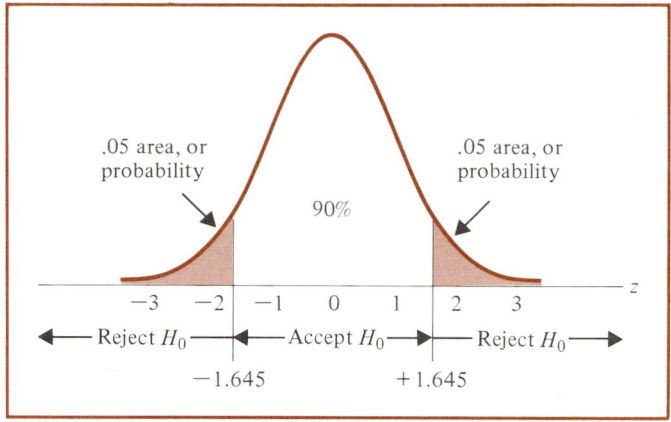

FIGURE 9.8
Decision Rule for the Null Hypothesis H_0: $\mu = 81{,}000$ with an $\alpha = .10$ Level of Significance

for the observed sample mean of $\bar{x} = \$81{,}000$. Using the information given in (c) and using the sample standard deviation s as the estimate of the population standard deviation σ, we have

$$z = \frac{\bar{x} - \mu}{\sigma / \sqrt{n}} = \frac{79{,}500 - 81{,}000}{5{,}000 / \sqrt{50}} = -2.12$$

Using the decision rule, $z = -2.12$ indicates that the null hypothesis should be rejected. The sample evidence indicates that there has been a change in the mean selling price of condominiums over the 1-year period. Although this is the primary hypothesis-testing conclusion, it is interesting to note that since the rejection of H_0 occurred in the lower tail of the distribution, the sample evidence indicates that the change in mean selling price is downward, with the current mean selling price below the mean selling price of the previous year.

Using the table of areas for the standard normal random variable, the area in the left-hand tail from $z = -2.12$ is $.5000 - .4830 = .0170$. With the two-tailed test, we double this probability to obtain a p value of $.0340$. With $.0340$ less than $\alpha = .10$, the p value criterion provides the conclusion that the null hypothesis should be rejected.

· · ·

One-Tailed Hypothesis Tests

The hypothesis tests we have been discussing have rejection regions with areas of $\alpha/2$ in both tails of the distribution. This enabled us to reject H_0 if the value of the sample mean were significantly above or below the hypothesized value of the population mean μ. These tests were referred to as two-tailed hypothesis tests.

There are actually three forms of the null and alternative hypotheses that can be formulated in tests about a population mean. Using μ_0 to denote the hypothesized value for μ, the three different forms are as follows.

Forms of Hypotheses Tests About a Population Mean

Form A Two-Tailed Test	Form B One-Tailed Test	Form C One-Tailed Test
$H_0: \mu = \mu_0$ $H_1: \mu \neq \mu_0$	$H_0: \mu = \mu_0$ $H_1: \mu < \mu_0$	$H_0: \mu = \mu_0$ $H_1: \mu > \mu_0$

Form A is the two-tailed test that we have been discussing. Forms B and C are called *one-tailed tests* because the null hypothesis will be rejected whenever the sample mean $\bar{x}$ is in one particular tail of the sampling distribution. For example, in Form B, H_0 will be rejected only if $\bar{x}$ is in the lower tail of the sampling distribution, indicating that μ is less than the hypothesized value μ_0. In Form C, H_0 will be rejected only if $\bar{x}$ is in the upper tail of the sampling distribution, indicating that μ is greater than the hypothesized value μ_0.

The form of the hypothesis test that is selected for a particular problem depends upon the situation and the objectives of the person designing the test. Nonetheless, note that the null hypothesis is always shown as an *equality*. Thus the choice of the form of the hypothesis test should be based on careful consideration of the conclusion that is reached whenever the null hypothesis is rejected.

EXAMPLE 9.8

A school administrator has developed an individualized reading-comprehension program for eighth-grade students. To evaluate this new program, a random sample of 45 eighth-grade students was selected; these students participated in the new reading program for one semester and then took a standard reading-comprehension examination. The mean test score for the population of students who had taken this test in the past was 76.

a. Determine the null and alternative hypotheses that will enable the administrator to determine whether or not the mean test score for students that take the individualized reading program is greater than the historical mean test score of $\mu = 76$.
b. Develop a decision rule for the hypothesis test using a .05 level of significance.
c. The sample results for the 45 students provided a sample mean of $\bar{x} = 79$ and a sample standard deviation of $s = 8$; what is the appropriate conclusion about the proposed individualized reading program?

We begin with the null hypothesis $H_0: \mu = 76$. Since the administrator is interested in determining if the new program will provide an *increase* in the mean test score, the alternative hypothesis is $H_1: \mu > 76$. Thus we have a one-tailed hypothesis test (Form C) with the hypotheses and the corresponding conclusions and actions as shown:

Hypothesis	Conclusion and Action
$H_0: \mu = 76$	The new program shows no improvement over the previous reading program; stay with the previous program.
$H_1: \mu > 76$	The new program shows a mean test score greater than the previous reading program; consider taking steps to implement the new program.

In this example the rejection of the null hypothesis is not necessarily a "bad" conclusion. In fact, the administrator would probably be delighted to learn that H_0 can be rejected, leading to the conclusion that the new program is better. Thus, in general, rejection of a null hypothesis is not to be interpreted as a failure. Rather, it simply points to a conclusion and action that differs from the one associated with the null hypothesis.

But what about the possibility that the new individualized program actually reduces the mean test score, that is, $\mu < 76$. This situation was not explicitly considered in the null and alternative hypotheses just shown. It does not have to be explicitly considered because the *action* implied by $\mu < 76$ is identical to the action implied by $\mu = 76$. In both cases, the new program does not show an improvement and would not be implemented; as a result, the null hypothesis can be written as either $H_0: \mu \leq 76$ or $H_0: \mu = 76$. We adopt the convention of always writing the null hypothesis with an equality. Remember that if the decision rule leads us to accept H_0, we still cannot conclude that we have proven that $\mu = 76$. Thus accepting H_0 simply indicates we do not have sufficient evidence to claim that the new program is better; in this case, the recommendation is to stay with the current program.

The hypothesis test for Example 9.9 can be made by referring to the acceptance and rejection regions shown in Figure 9.9. Note that with a .05 level of significance, the rejection region with an area or probability of .05 is located in the upper tail of the distribution. Using the table of areas for the standard normal probability distribution, we find that the critical z value for the test is 1.645. The decision rule for this one-tailed test is written as follows:

$$\text{Accept } H_0 \text{ If } z \leq 1.645$$

$$\text{Reject } H_0 \text{ Otherwise}$$

From the information in (c) we know that $n = 45$ and $\bar{x} = 79$. The sample standard deviation $s = 8$ can be used as the estimate of the population standard deviation σ. With the null hypothesis $H_0: \mu = 76$, the value of the test statistic z is as follows:

$$z = \frac{\bar{x} - \mu}{\sigma/\sqrt{n}} = \frac{79 - 76}{8/\sqrt{45}} = +2.52$$

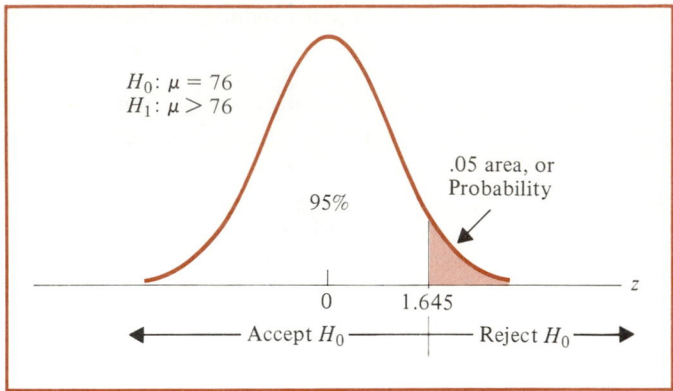

FIGURE 9.9
Decision Rule for a One-Tailed Test of the Null Hypothesis
$H_0: \mu = 76$ **with a .05 Level of Significance**

Using the decision rule, we reject H_0 and conclude that the proposed reading program provides a significant improvement in the mean test scores for the students.

A p value can also be computed for this example. Since we have a one-tailed test, we have to compute the area only in the *one rejection tail* for the z value of $+2.52$. Using the table of areas for the standard normal probability distribution, this area is $.5000 - .4941 = .0059$. This is the p value for the one-tailed test. Since the p value of $.0059$ is less than the $.05$ level of significance, the decision is to reject the null hypothesis and conclude the proposed reading program is better.

• • •

EXAMPLE 9.9

The Federal Trade Commission (FTC) has decided to select a simple random sample of 36 cans of coffee in order to check a manufacturer's claim regarding the amount of coffee in the cans. If the population mean weight is 3 or more pounds of coffee per can, no action will be taken by the FTC. However, if the sample data leads the FTC to believe that the company produces coffee with a mean weight of less than 3 pounds per can, the FTC will claim a violation exists and take action against the manufacturer.

a. Develop the null and alternative hypotheses for this test.
b. Provide the decision rule if the FTC is willing to tolerate a .01 probability of making the error of claiming the manufacturer is in violation when the manufacturer is actually filling cans with an acceptable mean of 3 pounds per can.
c. From past data, the population standard deviation of weights is $\sigma = .18$ pounds. If the sample of 36 items shows a sample mean of $\overline{x} = 2.96$ pounds, what is your conclusion and what action should the FTC take?

The hypothesis test can be written as follows:

$$H_0: \mu = 3$$

$$H_1: \mu < 3$$

This is a one-tailed test with the rejection of H_0 occurring only if the sample results indicate that the population mean filling weight is less than 3 pounds per can.

The probability of the type I error, or level of significance, is specified to be $\alpha = .01$. With the rejection region in the lower tail of the distribution, the table of areas for the standard normal probability distribution shows that the critical z value is -2.33. Thus the decision rule is written:

$$\text{Accept } H_0 \text{ if } z \geq -2.33$$

$$\text{Reject } H_0 \text{ otherwise}$$

The sample mean of $\bar{x} = 2.96$ and the hypothesized value of $\mu = 3$ provide the following value for the test statistic:

$$z = \frac{\bar{x} - \mu}{\sigma/\sqrt{n}} = \frac{2.96 - 3.00}{.18/\sqrt{36}} = -1.33$$

Since $z = -1.33$, the decision rule leads us to accept H_0. Thus, although we have not proven that the mean weight of coffee per can is $\mu = 3$, the sample evidence does not indicate the null hypothesis H_0: $\mu = 3$ should be rejected. As a result, the action taken by the FTC should be based on the assumption that the manufacturer is meeting the weight requirement.

$$\bullet \quad \bullet \quad \bullet$$

EXAMPLE 9.10

In order to test whether or not the mean highway driving speeds on a particular highway are within the posted speed limit of 55 miles per hour, the speeds for a random sample of 40 vehicles were recorded; the sample results show a mean of $\bar{x} = 57.6$ miles per hour and a standard deviation of $s = 8$ miles per hour. Using a .10 level of significance, calculate the p value that can be used to determine whether or not the mean highway driving speeds are within the 55 mile-per-hour speed limit.

The null and alternative hypotheses for this one-tailed test are

$$H_0: \mu = 55$$

$$H_1: \mu > 55$$

Thus the z value is

$$z = \frac{\bar{x} - \mu}{\sigma/\sqrt{n}} = \frac{57.6 - 55}{8/\sqrt{40}} = +2.06$$

Using the table of areas for the standard normal probability distribution and $z = +2.06$, we find an area in the rejection tail (p value) equal to $.5000 - .4803 = .0197$. At an $\alpha = .10$ level of significance, we see that the p value is less than α; thus, we reject H_0. The conclusion is that the mean driving speed on the highway exceeds the posted 55 mile-per-hour speed limit.

$$\bullet \quad \bullet \quad \bullet$$

A Summary of Decision Rules

Using the notation

$$\alpha = \text{the level of significance}$$

$$z_\alpha = \text{the } z \text{ value with an area of } \alpha \text{ in the upper tail}$$
$$\text{of the standard normal probability distribution}$$

the three forms of hypothesis tests about a population mean and the corresponding decision rules are summarized in Figure 9.10. Regardless of the form of the hypothesis test, the null hypothesis will be rejected if the p value is *less than* α.

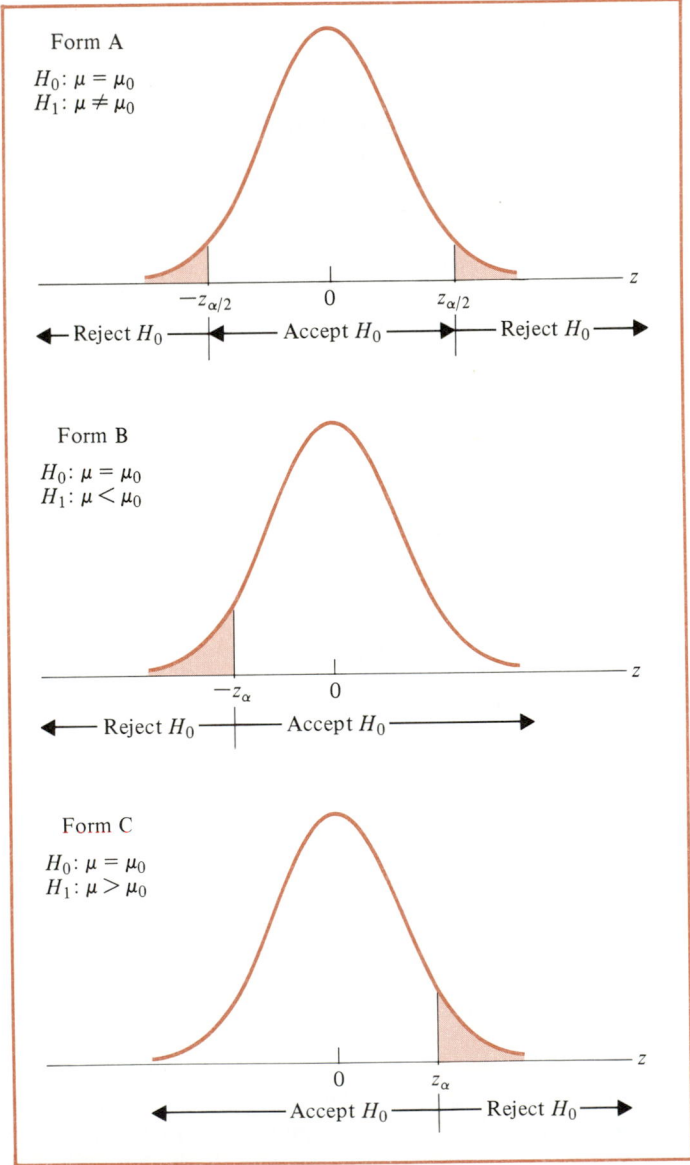

FIGURE 9.10

Summary of Decision Rules for Hypothesis Tests About a Population Mean

EXERCISES

13. A quality control inspector at Morgan Manufacturing Company tests part dimensions for a machining operation. The desired mean part diameter is two inches. A sample of parts is periodically selected and the size of each part is checked. If the sample leads the quality control inspector to believe that the mean part diameter for the population is either too large or too small, the machine will be shut down and readjusted. Which of the following hypothesis tests should be used to determine whether or not the machine should be shutdown?

A	B	C
$H_0: \mu = 2$	$H_0: \mu = 2$	$H_0: \mu = 2$
$H_1: \mu \neq 2$	$H_1: \mu < 2$	$H_1: \mu > 2$

14. What are the type I and type II errors in hypothesis testing? Define the consequences of making these errors for the hypothesis test in Exercise 13.

15. The manager of an automobile dealership is considering a new bonus plan that may increase sales volume for the dealership. Historically, the mean sales volume per sales employee has been four automobiles per month. A sample of 32 sales employees will be allowed to sell under the new bonus plan for a trial period. The manager is interested in implementing the new plan throughout the company if the sample evidence indicates the plan increases the mean sales volume. What hypotheses are appropriate for determining whether or not the new bonus plan should be implemented?

A	B	C
$H_0: \mu = 4$	$H_0: \mu = 4$	$H_0: \mu = 4$
$H_1: \mu \neq 4$	$H_1: \mu < 4$	$H_1: \mu > 4$

16. Spread Easy paint is labeled as having a mean coverage of 400 square feet per gallon. A mean coverage of more than 400 square feet is unsatisfactory and indicates that the paint is too thin. A mean coverage of less than 400 square feet is unsatisfactory and indicates that the paint is too thick. Assume $\sigma = 25$ square feet per gallon. A sample of 30 gallons of paint will be selected and used to test the coverage of the paint.
    a. What are the null and alternative hypotheses for this test?
    b. If the sample mean is $\bar{x} = 380$ square feet, what is your conclusion about the paint coverage? Use a .05 level of significance.

17. At Western University the mean scholarship examination score for entering students has been 900, with a standard deviation of 80. Each year a sample of applications is taken to see if the mean examination scores are at the same level or are changing. The null and alternative hypotheses are $H_0: \mu = 900$ and $H_1: \mu \neq 900$. A random sample of 60 students in this year's class provides a sample mean examination score of $\bar{x} = 924$.

a. Using a .05 level of significance, what conclusion should be made about any change in the examination scores?

b. What is the p value?

18. An automobile assembly-line operation has a scheduled mean completion time of 12.2 minutes. Because of the effect of completion time on both earlier and later assembly operations, it is important to maintain the 12.2-minute standard.

    a. Define the null and alternative hypotheses and determine the decision rule for conducting the test at a .02 level of significance.

    b. If a random sample of 45 completion times show $\bar{x} = 12.39$ and $s = 1.20$ minutes, what is your conclusion?

    c. What is the p value?

19. A study of the operation of a city-owned parking garage shows a historical mean parking time of 220 minutes per car. The garage area has recently been remodeled and the parking charges have been increased. The city manager would like to know if these changes have had any effect on the mean parking time of the garage customers. Test the hypotheses $H_0: \mu = 220$ and $H_1: \mu \neq 220$ at a .05 level of significance.

    a. What is your conclusion if a sample of 50 cars showed $\bar{x} = 208$ and $s = 80$?

    b. What is the p value?

20. The president of Fightmaster and Associates Real Estate, Inc., claims that the mean selling time of a residential home is 40 days or less after it is listed with the company. A sample of 50 recently sold residential homes shows a sample mean selling time of 45 days and a sample standard deviation of 20 days. Use a .02 level of significance and test the president's claim. What is the p value?

21. Fowle Marketing Research, Inc., estimates costs for a client on the assumption that a telephone interview can be completed with a mean time of 15 minutes or less. If a greater mean time is required, the client is charged a premium rate. Does a sample of 35 surveys that results in a sample mean of 17 minutes and a sample standard deviation of 4 minutes justify the premium rate? Test at a .01 level of significance. What is the p value?

22. The manager of the Danvers-Hilton Hotel believes that the mean guest bill is $250 or more. Assume that we wish to test the manager's claim with the following hypotheses: $H_0: \mu = 250$, $H_1: \mu < 250$. In this case, the manager's claim will be accepted unless H_0 is rejected. If $\sigma = 50$ and if a simple random sample of 60 billings shows a sample mean of $\bar{x} = \$235$, what conclusion would you draw? Use a .05 level of significance.

23. The manager of the Keeton Department Store believes that the mean annual income of the store's credit-card customers is at least $18,000 per year. A sample of 58 credit-card customers shows a sample mean of $17,200 and a sample standard deviation of $3000. At the .05 level of significance, can the manager's claim be rejected?

24. The Chamber of Commerce of a Florida Gulf Coast community advertises area residential lots available at a mean cost of $40,000 or less. Use a .10 level of

significance and test this claim for a sample of 32 properties that resulted in a sample mean of $41,000 with a sample standard deviation of $2500. What is the *p* value?

25. A federal funding program is available to low-income neighborhoods. To qualify for the funding, a neighborhood must have a mean household income of less than $7000 per year. Funding decisions are based on a sample of residents in a neighborhood. The hypothesis test involves $H_0: \mu = 7000$ and $H_1: \mu < 7000$. Federal funding is granted only if H_0 is rejected.

   a. Assume that a sample of 36 households in a particular neighborhood shows $\bar{x} = \$6600$ and $s = \$1200$. At a .05 level of significance, does this neighborhood qualify for federal funding?

   b. In a sample of 36 households, what is the maximum sample mean income level that will enable the neighborhood to still qualify for federal funding?

26. A company currently pays an average wage of $10.00 per hour for its production employees. The company is planning to build a new factory, and several locations are being considered. The availability of labor at a rate less than $10.00 per hour is a major factor in the location decision. For one location, a sample of 40 workers showed a current mean hourly wage of $\bar{x} = \$9.50$ and a sample standard deviation of $s = \$.60$.

   a. Using a .10 level of significance, does the sample data indicate that the location has a mean wage rate significantly below the $10.00 per hour rate?

   b. What is the *p* value?

9.3 INFERENCES USING SMALL SAMPLES

The methods of estimation and hypothesis testing that we have discussed thus far have required sample sizes of at least 30 items. The reason for this is that in the large-sample situation ($n \geq 30$), the central limit theorem can be used to approximate the sampling distribution of $\bar{x}$ with a normal probability distribution. Thus the random variable

$$z = \frac{\bar{x} - \mu}{\sigma/\sqrt{n}} \qquad (9.9)$$

is a standard normal random variable. Recall that z values were used in both interval-estimation and hypothesis-testing procedures.

   If the sample size is small (that is, $n < 30$), the central limit theorem can no longer be used to specify the sampling distribution of $\bar{x}$; in these cases we have to consider other methods of estimation and hypothesis testing.

The t Distribution

W. S. Gosset, who published his writings under the pen name 'Student,' found that if the population being sampled has a *normal probability distribution* and if the sample standard deviation s is used as an estimate of the population standard deviation σ,

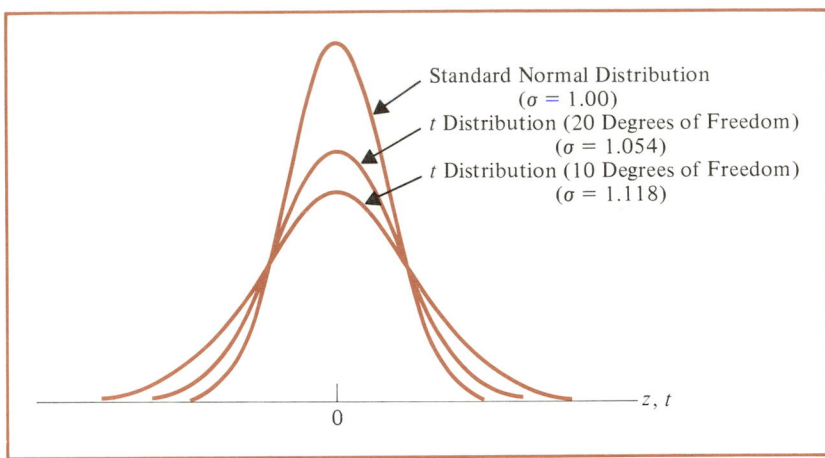

statistical inferences about a population mean can be based on the random variable

$$t = \frac{\overline{x} - \mu}{s/\sqrt{n}} \tag{9.10}$$

The sampling distribution of t is called the Student t probability distribution, or simply the t *distribution*.

It is important to realize that if the population has a normal distribution and if s is used as an estimator of σ, the t distribution is applicable for *any sample size*. In fact, as the sample size increases, the t distribution approaches the standard normal distribution. As a result, in the large-sample case, interval estimation and hypothesis tests based on z values and t values provide similar conclusions. However, in Sections 9.1 and 9.2, we already have presented methods for interval estimation and hypothesis tests about a population mean with large samples. Thus we do not need to draw upon the t distribution and its required assumption of a normally distributed population until we encounter a small-sample case.

The t distribution is actually a family of similar probability distributions. Each specific t distribution depends upon a parameter known as its *degrees of freedom*;* there is a unique t distribution with 1 degree of freedom, a unique t distribution with 2 degrees of freedom, a unique t distribution with 3 degrees of freedom, and so on. As the number of degrees of freedom increases, the difference between the t distribution and the standard normal distribution becomes smaller and smaller. Figure 9.11 shows t distributions with 10 and 20 degrees of freedom and their relationships to the standard normal probability distribution.

*The reason the terminology degrees of freedom is used is that the computation of s is based on the sum of the squared deviations $\Sigma(x_i - \overline{x})^2$ and only $n - 1$ of the deviations, $(x_i - \overline{x})$, are independent. That is, if we know $n - 1$ of the $(x_i - \overline{x})$ values, the remaining deviation can be determined, since for any sample $\Sigma(x_i - \overline{x}) = 0$. In this case, we have $n - 1$ degrees of freedom.

TABLE 9.1

t Distribution Table for Areas in the Upper Tail.
Example: With 10 Degrees of Freedom $t_{.025} = 2.228$

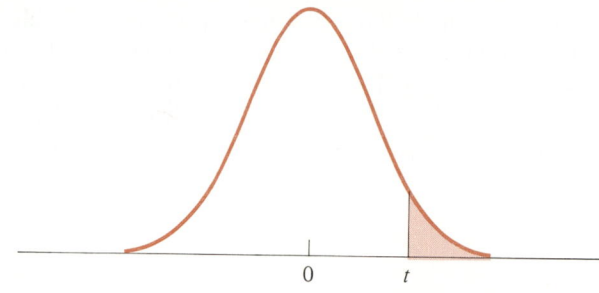

Degrees of Freedom	Upper-Tail Areas (Shaded)				
	.10	.05	.025	.01	.005
1	3.078	6.314	12.706	31.821	63.657
2	1.886	2.920	4.303	6.965	9.925
3	1.638	2.353	3.182	4.541	5.841
4	1.533	2.132	2.776	3.747	4.604
5	1.476	2.015	2.571	3.365	4.032
6	1.440	1.943	2.447	3.143	3.707
7	1.415	1.895	2.365	2.998	3.499
8	1.397	1.860	2.306	2.896	3.355
9	1.383	1.833	2.262	2.821	3.250
10	1.372	1.812	2.228	2.764	3.169
11	1.363	1.796	2.201	2.718	3.106
12	1.356	1.782	2.179	2.681	3.055
13	1.350	1.771	2.160	2.650	3.012
14	1.345	1.761	2.145	2.624	2.977
15	1.341	1.753	2.131	2.602	2.947
16	1.337	1.746	2.120	2.583	2.921
17	1.333	1.740	2.110	2.567	2.898
18	1.330	1.734	2.101	2.552	2.878
19	1.328	1.729	2.093	2.539	2.861
20	1.325	1.725	2.086	2.528	2.845
21	1.323	1.721	2.080	2.518	2.831
22	1.321	1.717	2.074	2.508	2.819
23	1.319	1.714	2.069	2.500	2.807
24	1.318	1.711	2.064	2.492	2.797
25	1.316	1.708	2.060	2.485	2.787
26	1.315	1.706	2.056	2.479	2.779
27	1.314	1.703	2.052	2.473	2.771
28	1.313	1.701	2.048	2.467	2.763
29	1.311	1.699	2.045	2.462	2.756
30	1.310	1.697	2.042	2.457	2.750
40	1.303	1.684	2.021	2.423	2.704
60	1.296	1.671	2.000	2.390	2.660
120	1.289	1.658	1.980	2.358	2.617
∞	1.282	1.645	1.960	2.326	2.576

We use a subscript for t to indicate the area in the *upper tail* of the t distribution. For example, we use $t_{.025}$ to indicate a t value with a .025 area in the upper tail of the distribution. A table for the t distribution is provided in Table 2 of Appendix B. This table is also shown in Table 9.1. Note, for example, that for a t distribution with 10 degrees of freedom, $t_{.025} = 2.228$. Similarly, for a t distribution with 20 degrees of freedom, $t_{.025} = 2.086$. Now that we have an idea of what the t distribution is, let us show how it is used to make statistical inferences about a population mean.

Estimation

Assuming that the population has a normal probability distribution and that the sample standard deviation s is used as an estimate of the population standard deviation σ, interval estimation of a population mean is similar to the approach we used in Section 9.1; the exception is that a t value is used instead of a z value. Using $1 - \alpha$ to denote the confidence coefficient, the following formula provides the general procedure for an interval estimate of a population mean.

Interval Estimation of a Population Mean (Small-Sample Case)

$$\bar{x} \pm t_{\alpha/2} \frac{s}{\sqrt{n}} \qquad (9.11)$$

where $1 - \alpha$ is the confidence coefficient and $t_{\alpha/2}$ is the t value providing an area of $\alpha/2$ in the upper tail of a t distribution with $n - 1$ *degrees of freedom.*

EXAMPLE 9.11

Assume that the duration times of long-distance telephone calls are normally distributed. A sample of 20 telephone calls shows a sample mean of $\bar{x} = 12.5$ minutes and a sample standard deviation of $s = 5$ minutes. Develop 95% and 98% confidence interval estimates of the mean duration time for the population of long-distance telephone calls.

At 95% confidence, $1 - \alpha = .95$ and $\alpha = .05$. Using (9.11), we obtain

$$\bar{x} \pm t_{.025} \frac{s}{\sqrt{n}}$$

The $t_{.025}$ value is based on $n - 1 = 20 - 1 = 19$ degrees of freedom. From Table 9.1, $t_{.025} = 2.093$. Thus we have

$$12.5 \pm 2.093 \frac{5}{\sqrt{20}} = 12.5 \pm 2.34$$

Thus the 95% confidence interval for the population mean is 10.16 to 14.84 minutes per telephone call.

At 98% confidence, $1 - \alpha = .98$ and $\alpha = .02$. Using (9.11), we have

$$\bar{x} \pm t_{.01} \frac{s}{\sqrt{n}}$$

or

$$12.5 \pm 2.539 \frac{5}{\sqrt{20}} = 12.5 \pm 2.84$$

This result provides a 98% confidence interval for the population mean from 9.66 to 15.34 minutes per telephone call.

• • •

Hypothesis Testing

The small-sample procedure for hypothesis tests about a population mean follows the general hypothesis-testing procedure of Section 9.2; the exception is that the t distribution is used to determine the critical value(s) for the test and t given in (9.10) is the test statistic.

EXAMPLE 9.12

The heights of a particular plant are normally distributed with a mean of 28 inches. A new plant food is tested on a sample of 12 plants. Results of the sample show a sample mean height of 29.4 inches and a sample standard deviation of 3 inches. Using a .10 level of significance, is there reason to believe that the new plant food *increases* plant growth?

The hypothesis test is as follows:

$$H_0: \mu = 28$$

$$H_1: \mu > 28$$

The rejection region is located in the upper tail of the sampling distribution. With $n - 1 = 12 - 1 = 11$ degrees of freedom, we have $t_{.10} = 1.363$ as the critical value for the test. Thus the decision rule is to compute the test statistic

$$t = \frac{\bar{x} - \mu}{s/\sqrt{n}}$$

and if $t \leq 1.363$, accept the null hypothesis; otherwise H_0 is rejected. Using the sample results, we have the following value for the test statistic:

$$t = \frac{\bar{x} - \mu}{s/\sqrt{n}} = \frac{29.4 - 28}{3/\sqrt{12}} = 1.62$$

Since 1.62 is greater than 1.363, the null hypothesis is rejected. At a .10 level of significance it can be concluded that the mean plant height exceeds 28 inches when the new plant food is used.

• • •

The Use of p Values

The p value for Example 9.12 can be determined by using the observed t value of 1.62. However, due to the format of the t distribution table, p values are slightly more difficult to determine than they were in Section 9.2. For example, the t distribution in Example 9.12 has 11 degrees of freedom. Referring to the row for 11 degrees of freedom, in Table 9.1, we see that 1.62 is between 1.363, occurring at a p value of .10, and 1.796, occurring at a p value of .05. Interpolation shows that the value 1.62 corresponds to a p value of approximately .08. With a p value of .08, which is less than $\alpha = .10$, we reject H_0 and conclude that the new plant food will increase the mean plant height.

EXAMPLE 9.13

A production process is designed to fill containers with a mean filling weight of $\mu = 16$ ounces. An undesirable condition exists if the process is underfilling containers and the consumer is not receiving the amount of product indicated on the container label. In addition, an equally undesirable condition exists if the process is overfilling containers; in this case the firm is losing money since the process is placing more product in the container than is required. Suppose that quality assurance personnel periodically select a simple random sample of eight containers and test the following two-tailed hypothesis:

$$H_0: \mu = 16$$

$$H_1: \mu \neq 16$$

If H_0 is rejected, the production manager will request that the production process be stopped and that the mechanism for regulating filling weights be readjusted to provide a mean filling weight of 16 ounces. If the sample provides data values of 16.02, 16.22, 15.82, 15.92, 16.22, 16.32, 16.12, and 15.92 ounces, what action should be taken at a .05 level of significance? Assume that the population of filling weights is normally distributed.

Since the data have not been summarized, we must first compute the sample mean and sample standard deviation. Doing so provides the following results:

$$\bar{x} = \frac{\Sigma x_i}{n} = \frac{128.56}{8} = 16.07 \text{ ounces}$$

and

$$s = \sqrt{\frac{\Sigma(x_i - \bar{x})^2}{n - 1}} = \sqrt{\frac{.22}{7}} = .18 \text{ ounces}$$

With a two-tailed test and $\alpha = .05$, the critical values are $-t_{.025}$ and $+t_{.025}$ from a t distribution with $n - 1 = 8 - 1 = 7$ degrees of freedom. Using Table 9.1, we obtain -2.365 and $+2.365$ as the critical values. Thus H_0 will be accepted if the value of the test statistic, t, is between -2.365 and $+2.365$; otherwise H_0 will be rejected.

Using $\bar{x} = 16.07$ and $s = .18$, we have

$$t = \frac{\bar{x} - \mu}{s/\sqrt{n}} = \frac{16.07 - 16.00}{.18/\sqrt{8}} = 1.10$$

Thus, since t is in the acceptance region, the null hypothesis $\mu = 16$ ounces cannot be rejected. The conclusion is that the production process should continue to run.

Using Table 9.1 and the row for 7 degrees of freedom, we see that the computed t value of 1.10 has an upper tail area of more than .10. While the format of the table prevents us from being more specific, we can at least conclude that the two-tailed p value is greater than .20. Since this is greater than the .05 level of significance, we see that the p value leads to the same conclusion; that is, accept H_0 and continue the production process.

• • •

EXERCISES

27. For a t distribution with 12 degrees of freedom, find the area, or probability, that lies in each region.
   a. to the left of 1.782
   b. to the right of -1.356
   c. to the right of 2.681
   d. to the left of -1.782
   e. between -2.179 and $+2.179$
   f. between -1.356 and $+1.782$

28. Find the t value for each of the following.
   a. upper tail area of .05 with 18 degrees of freedom
   b. lower tail area of .10 with 22 degrees of freedom
   c. upper tail area of .01 with 5 degrees of freedom
   d. 90% of the area is between these two t values with 14 degrees of freedom
   e. 95% of the area is between these two t values with 28 degrees of freedom

29. A test is made on the breaking strength of a new synthetic fishing line. The breaking strength of six sections of the line showed a sample mean of $\bar{x} = 20$ pounds and a sample standard deviation of $s = 2.3$ pounds. Develop a 95% confidence interval for the mean breaking strength of the new line under the assumption that the population of breaking strengths is normally distributed.

30. A sample of 12 cab fares in New York City shows a sample mean of $\bar{x} = \$8.50$ and a sample standard deviation of $s = \$2.40$. Develop a 90% confidence interval estimate of the mean cab fares in New York City.

31. A simple random sample of five people provided the following data on ages: 21, 25, 20, 18, and 21. Develop a 95% confidence interval for the mean age of the population being sampled.

32. In the testing of a new production method, 18 employees were randomly selected and asked to try the new method. The sample mean production rate for the 18 employees was 80 parts per hour. The sample standard deviation was 10 parts per hour. Provide 90% and 95% confidence interval estimates for the mean production rate for the new method.

33. The directors of a university computer center have been studying how many of the center's 30 computer terminals are in use at 9:00 P.M. on Friday evenings. A sample of 5 weeks resulted in the following data.

Week	1	2	3	4	5
Number in use	12	18	21	15	9

Treat the data as being from a simple random sample and develop a 95% confidence interval estimate for the mean number of terminals in use on Friday evenings at 9:00 P.M.

34. Shown below are the duration (in minutes) for a sample of 20 telephone flight reservations:

2.1	4.8	5.5	10.4
3.3	3.5	4.8	5.8
5.3	5.5	2.8	3.6
5.9	6.6	7.8	10.5
7.5	6.0	4.5	4.8

a. What is the point estimate of the population mean time for flight-reservation phone calls?
b. Develop a 95% confidence interval estimate of the population mean time.

35. A bath soap manufacturing process is designed with the expectation that each batch prepared in the mixture department will produce a mean of 120 bars of soap per batch. A mean over or under this standard is undesirable. A sample of ten batches shows the following numbers of bars of soap.

108, 118, 120, 122, 119, 113, 124, 122, 120, 123

a. Use a .05 level of significance and test to see if the null hypothesis H_0: $\mu = 120$ should be rejected.
b. What is the p value?

36. It is estimated that, on the average, a housewife with a husband and two children works 55 hours or less per week on household related activities. Shown below are the hours worked during a week for a sample of eight housewives.

58, 52, 64, 63, 59, 62, 62, 55

a. Use $\alpha = .05$ to test the hypotheses $H_0: \mu = 55$, $H_1: \mu > 55$. What is your conclusion about the mean number of hours worked per week?

b. What is the p value?

37. A study of a drug designed to reduce blood pressure used a sample of 25 men between the ages of 45 and 55. With μ indicating the mean change in blood pressure for the population of men receiving the drug, the hypotheses in the study were written: $H_0: \mu = 0$ and $H_1: \mu < 0$. Rejection of H_0 shows that the mean change is negative, indicating that the drug is effective in lowering blood pressure.

a. At a .05 level of significance, what conclusion should be made if $\bar{x} = -10$ and $s = 15$?

b. What is the p value?

38. Last year the number of lunches served at an elementary school cafeteria was normally distributed with a mean of 300 lunches per day. At the beginning of the current year, the price of a lunch was raised by 25¢. A sample of 6 days during the months of September, October and November provided the following number of children being served lunches: 290, 275, 305, 260, 270, and 275. Do these data indicate that the mean number of lunches per day has dropped compared to last year? Test the hypothesis $H_0: \mu = 300$ against the alternative hypothesis $H_1: \mu < 300$ at a .10 level of significance.

Summary

In this chapter we discussed procedures for making statistical inferences about a population mean. In particular, we described both the methods of interval estimation and hypothesis testing. In the large-sample case ($n \geq 30$), the central limit theorem enables us to approximate the sampling distribution of $\bar{x}$ with a normal probability distribution. As a result, the standard normal random variable z was used in both the large-sample interval-estimation and hypothesis-testing procedures.

We indicated that an interval estimate is developed for a stated confidence level. For a given sample size, higher confidence levels require wider intervals. We showed that given information regarding the maximum acceptable sampling error, the desired level of confidence, and a planning value for the population standard deviation σ, a formula can be used for determining the sample size that would meet the desired level of precision for the estimation of the population mean.

We then showed how to conduct hypothesis tests about the value of a population mean. The purpose of hypothesis testing is to make a statistical conclusion about a hypothesized value of the population mean. A null hypothesis H_0 and an alternative hypothesis H_1 must be formulated for the test. The type I error is the error of rejecting H_0 when it is true, and the type II error is the error of accepting H_0 when it is false. By specifying a maximum allowable probability of making a type I error, called the level of significance, a decision rule can be established for determining whether or not the null hypothesis should be rejected.

The p value was introduced as another criterion for determining whether or not to reject the null hypothesis. The p value indicates the probability of obtaining a difference between the observed sample mean and the hypothesized value of the population mean that is larger than the difference actually observed. The decision rule indicates the null hypothesis should be rejected only if the p value is less than the level of significance for the test.

Finally, we introduced the t distribution as a special probability distribution that can be used to make inferences about a population mean whenever the population is normal and the sample standard deviation s is used as an estimate of σ. The t distribution is particularly helpful in making inferences about a population mean whenever the sample size is small ($n < 30$). The specific steps of using the t distribution parallel the estimation and hypothesis-testing procedures for the large-sample case.

Statistics in Practice

CONFIDENCE INTERVALS FOR INVENTORY VALUATION

Thriftway, Inc., is a supermarket chain that serves communities throughout southwestern Ohio and northern Kentucky. Being in an inventory-intense business with over 25,000 supermarket and nonfood items, Thriftway made the decision to adopt the last in–first out (LIFO) method of inventory valuation. Under this accounting practice, the inventory on hand at the close of an accounting period is valued at the first price paid regardless of any fluctuations that may have affected the actual cost of the inventory. While we do not want to go into the complete details of the LIFO method of inventory valuation, the Internal Revenue Service requires a company using LIFO to establish an annual index for the valuation of its inventory.

The approach Thriftway uses to determine its annual LIFO index employs standard sampling procedures. First, a random sample consisting of 500 items is selected from the population of 25,000 items carried in inventory. The physical inventory counts for the sample items are taken during the last week of December. A clerical employee identifies the current cost per unit and the prior-year cost per unit for each item in the sample. Component 1 of the LIFO index is the dollar value of the sample inventory using the current costs, while component 2 is the dollar value of the sample inventory using the prior-year costs. The ratio of the sample current costs to the sample prior-year costs provides the point estimate for the company's LIFO index.

For example, for a given year the sample LIFO index was 1.045. This index value estimates that the company's inventory at current costs contains a 4.5% increase in value due to the increases in costs that have occurred during the one-year period. However, the sample LIFO index is only an estimate of the population LIFO index. Statements about the sampling error and the associated confidence interval estimate are essential in determining the goodness of the sample index. Using the sample results, the standard error was computed to be .01. With a 95% confidence level, the maximum sampling error was approximately .02. Thus the interval of 1.025 to 1.065 provided the 95% confidence interval estimate for the actual LIFO index.

The sample of 500 items (2% of all items in the population) provided the time and cost savings that made the election of the LIFO inventory policy acceptable for Thriftway. The 95% confidence interval was reported to Internal Revenue Service as support for the precision of the inventory valuation data.

Glossary

Point estimate—A single numerical value used as the estimate of the value of a population parameter. In this chapter, the value of the sample mean $\bar{x}$ provided the point estimate of the population mean μ.

Interval estimate—An estimate of a population parameter that provides an interval of values believed to contain the value of the parameter.

Confidence coefficient—The confidence that an interval estimate contains the value of the parameter of interest. For example, if an interval-estimation procedure provides intervals such that 95% of the time the value of the population mean μ is included, an interval estimate is said to be constructed at the 95% confidence level, and .95 is referred to as the confidence coefficient.

Sampling error—The difference between a point estimate and the value of the population parameter. In the case of the mean, the sampling error can be written $\bar{x} - \mu$.

Null hypothesis—The hypothesis of the form H_0: $\mu = \mu_0$. This hypothesis is assumed true, with the sample results indicating whether or not the null hypothesis should be rejected.

Alternative hypothesis—The hypothesis concluded if the null hypothesis is rejected.

Type I error—The error of rejecting H_0 when it is true.

Type II error—The error of accepting H_0 when it is false.

Test statistic—A random variable such as z or t that is used to reach a hypothesis-testing conclusion. Comparison of the value of the test statistic to the critical value for the test provides the acceptance or rejection conclusion.

Level of significance—The maximum probability of making a type I error that the user will tolerate in the hypothesis-testing procedure.

Two-tailed test—A hypothesis test in which rejection of the null hypothesis occurs in either tail of the sampling distribution.

One-tailed test—A hypothesis test in which rejection of the null hypothesis occurs in only one tail of the sampling distribution.

Critical value—The value of z or t that separates the acceptance and rejection regions for a hypothesis test. The decision rule for the test can be stated in terms of the critical value or values.

p value—The probability of obtaining a difference between the value of the sample mean $\bar{x}$ and the hypothesized value of μ that is greater than or equal to the difference actually observed. If the p value is less than the level of significance for the test, the null hypothesis should be rejected.

t distribution—A family of probability distributions that can be used to make inferences about a population mean whenever the population is normal and the sample standard deviation s is used as an estimate of the population standard deviation σ.

Degrees of freedom—A parameter of the t distribution that specifies the t distribution of interest. Whenever the t distribution is used to make inferences about a population mean, the appropriate t distribution has $n - 1$ degrees of freedom, where n is the size of the simple random sample.

Key Formulas

Interval Estimation of a Population Mean

$$\bar{x} \pm z_{\alpha/2} \frac{\sigma}{\sqrt{n}} \tag{9.3}$$

Sample Standard Deviation

$$s = \sqrt{\frac{\Sigma(x_i - \bar{x})^2}{n - 1}} \tag{9.4}$$

Interval Estimation of a Population Mean—σ Unknown

$$\bar{x} \pm z_{\alpha/2} \frac{s}{\sqrt{n}} \tag{9.5}$$

Sample Size for Interval Estimation of a Population Mean

$$n = \frac{(z_{\alpha/2})^2 \sigma^2}{e^2} \tag{9.7}$$

Standard Normal Random Variable

$$z = \frac{\bar{x} - \mu}{\sigma/\sqrt{n}} \tag{9.9}$$

t Random Variable

$$t = \frac{\bar{x} - \mu}{s/\sqrt{n}} \tag{9.10}$$

Interval Estimation of a Population Mean (Small-Sample Case)

$$\bar{x} \pm t_{\alpha/2} \frac{s}{\sqrt{n}} \tag{9.11}$$

Review Quiz

TRUE/FALSE

1. Whenever the central limit theorem is applicable, the sampling error will be zero when using $\bar{x}$ as a point estimator of μ.
2. Interval estimates provide information on how close the value of the sample mean $\bar{x}$ is to the population mean μ.
3. A point estimate does not contain information about the size of the sampling error.
4. An interval estimate contains information about the size of the sampling error.
5. In order to determine the appropriate sample size, the user must specify a maximum allowable sampling error.
6. In hypothesis testing, we do not assume the null hypothesis true until after the sample results have been analyzed.
7. In hypothesis testing, the z value can be interpreted as the number of standard deviations the sample mean is from the hypothesized value of μ.
8. The type I error is the error of accepting H_0 when it is false.
9. If the level of significance, α, is made smaller, the rejection region becomes larger.
10. Decision rules based on critical z values and p values will always provide the same hypothesis-testing conclusions.
11. In order to use the t distribution, we must be willing to assume that the population has a normal or near-normal distribution.
12. When using the t distribution to form a 95% confidence interval, the interval will be smaller than if a z value from a standard normal probability distribution is used.
13. The t distribution can never be used if the sample size is 30 or more.

MULTIPLE CHOICE

Use the following information for questions 14–16. A random sample of 81 automobile tires has a mean tread life of 36,000 miles. It is known that the standard deviation of tread life of tires is $\sigma = 4500$ miles.

14. A 95% confidence interval for the population mean is
    a. 35,500 to 36,500
    b. 35,177.5 to 36,822.5
    c. 35,020 to 36,980
    d. none of the above
15. If the sample mean of 36,000 had been from a random sample size 50, the 95% confidence interval would have been
    a. the same
    b. a wider interval

    c. a narrower interval

    d. none of the above

16. A 90% confidence interval for the population mean is

    a. 35,000 to 37,000

    b. 35,177.5 to 36,822.5

    c. 35,020 to 36,980

    d. none of the above

17. The useful life of a certain type of light bulb is known to have a standard deviation of $\sigma = 40$ hours. How large a sample should be taken if it is desired to have an error of 10 hours or less at a 95% level of confidence?

    a. 62

    b. 44

    c. 37

    d. 8

18. If a hypothesis test leads to the rejection of the null hypothesis

    a. a type I error is always committed

    b. a type II error is always committed

    c. a type I error may have been committed

    d. a type II error may have been committed

Use the following information for questions 19–21. The ABC Electronics Company claims that the batteries it produces have a useful life of at least 100 hours. It is known that the standard deviation is 20 hours. A test is undertaken to check the validity of this claim.

19. With the level of significance set at .05, the critical value or values for the test based on a sample of 49 batteries is

    a. $z = -1.645$

    b. $z = -1.435$

    c. $z = -1.96$ and $+1.96$

    d. $z = +1.645$

20. If the tested random sample of 49 batteries resulted in an average life of 96 hours, can the manufacturer's claim be rejected at the .05 level of significance?

    a. Yes, the null hypothesis can be rejected.

    b. No, do not reject the null hypothesis.

21. What is the p value associated with the sample mean of 96 hours?

    a. .042

    b. .081

    c. .419

    d. .96

22. If the level of significance of a hypothesis test is raised from .01 to .05, the probability of a type II error

    a. will also be increased from .01 to .05

    b. will not be changed

    c. will be decreased

    d. not enough information is given to answer this question

Supplementary Exercises

39. Sales personnel for Skillings Distributors are required to submit weekly reports listing the customer contacts made during the week. A sample of 60 weekly contact reports shows a mean of 22.4 customer contacts per week for the sales personnel. The sample standard deviation was 5 contacts. Compute a 95% confidence interval for the mean number of weekly customer contacts for the population of sales personnel.

40. The North Carolina Savings and Loan Association would like to develop an estimate of the mean size of home improvement loans granted by its member institutions. A sample of 100 loans granted by member institutions resulted in a sample mean of $3,400 and a sample standard deviation of $650. With these data develop a 98% confidence interval for the mean dollar amount of home improvement loans.

41. In a test of phone utilization, a firm recorded the length of time for phone calls handled by its main switchboard. A sample of 50 phone calls provided a sample mean of 8.9 minutes and a sample standard deviation of 5 minutes.
    a. What is a 90% confidence interval estimate for the mean phone call duration?
    b. What is a 99% confidence interval estimate?

42. A utility company found that a sample of 100 delinquent accounts yielded an average amount owed of $131.44, with a sample standard deviation of $16.19. Develop a 90% confidence interval for the population mean amount owed.

43. In Exercise 42 a utility company sampled 100 delinquent accounts in order to estimate the mean amount owed by these accounts. The sample standard deviation was $16.19. How large a sample should be taken if the company wants to be 90% confident that the estimate of the population mean will have a sampling error of $1.50 or less?

44. The owner of a pay-fishing lake wishes to determine the mean weight of the fish caught by the patrons. A preliminary sample of 10 fish caught shows a sample standard deviation of 1.6 pounds. How many fish should be included in the sample if we would like to estimate the mean weight of the fish to within ± .25 pounds at a 95% level of confidence?

45. A gasoline service station shows a standard deviation of $6.25 for the charges made by the credit-card customers. Assume that the station's management would like an estimate of the mean gasoline bill for its credit-card customers to within ±$1.00 of the actual population mean. For a 95% confidence level, how large a sample would be necessary?

46. Because of production changeover time and costs, a director of manufacturing must convince management that a proposed manufacturing method reduces costs before the new method can be implemented. The current production line operates with a mean cost of $220 per hour. A new production line has been proposed and a sample production period specified. What hypothesis test should be formulated in order to test whether or not the company should convert to the new production line?

47. The monthly rent for a two-bedroom apartment in a particular city is reported to

average \$350. Assume that we would like to test the hypothesis H_0: $\mu = 350$ versus H_1: $\mu \neq 350$. A sample of 36 two-bedroom apartments is selected. The sample mean turns out to be $\bar{x} = \$338$, with a sample standard deviation of $s = \$40$.

a. Conduct this hypothesis test with a .05 level of significance.

b. Use the sample results to construct a 95% confidence interval for the population mean.

48. A long-distance trucking firm believes that its mean weekly loss due to damaged shipments is \$2000 or less. A sample of 35 weeks of operations shows a sample mean weekly loss of \$2200, with a sample standard deviation of \$500. Use a .05 level of significance and test the trucking firm's claim that the mean weekly loss is \$2,000 or less. (*Note:* H_0: $\mu = 2000$ and H_1: $\mu > 2000$.)

49. New tires manufactured by a company in Findlay, Ohio, are designed to provide a mean of at least 28,000 miles. Tests with 40 tires show a sample mean of 27,200 miles with a sample standard deviation of 1000 miles. Use a .01 level of significance and test for whether or not there is sufficient evidence to reject the claim of a mean of at least 28,000 miles. What is the p value?

50. In making bids on building projects, Sonneborn Builders, Inc. assumes construction workers are idle on the average no more than 15% of the time. For a normal 8-hour shift, the mean idle time per worker should be 72 minutes or less per day. A sample of 30 construction workers found a mean idle time of 80 minutes per day. The sample standard deviation was 20 minutes.

a. Formulate the null and alternative hypotheses so that the claim of 72 minutes or less per day will be rejected if H_0 is rejected.

b. What is the p value associated with the sample result?

c. Using a .05 level of significance and the p value, what is your conclusion?

51. Stout Electric Company operates a fleet of trucks for its electrical service to the construction industry. Monthly mean maintenance costs have been \$75 per truck. A random sample of 40 trucks shows a sample mean maintenance cost of \$82.50 per month, with a sample standard deviation of \$30. Management would like a test to determine whether or not the mean monthly maintenance cost has increased.

a. What is your conclusion based on the sample mean of \$82.50? Use a .05 level of significance.

b. What is the p value associated with this sample result? What is your conclusion based on the p value?

52. The lifetimes of a certain battery are normally distributed. In a random sample of 25 batteries, the sample results showed that $\bar{x} = 60$ hours and $s = 10$ hours. Provide a 90% confidence interval for the mean lifetime of this type of battery.

53. Sample assembly times for a particular manufactured part were 8, 10, 10, 12, 15, and 17 minutes. If the mean of the sample is to be used to estimate the mean of the population of assembly times, provide a point estimate and a 90% confidence interval estimate of the population mean.

54. Dailey Paints, Inc. implemented a long-term painting test study designed to check the wear resistance of its major brand of paint. The test consisted of painting 8 houses in various parts of the United States and observing the number of months until signs of peeling were observed. The following data were obtained.

House	1	2	3	4	5	6	7	8
Time Until Signs of Peeling (months)	60	51	64	45	48	62	54	56

    a. What is a point estimate of the mean number of months until signs of peeling are observed?

    b. Develop a 95% confidence interval to estimate the population mean number of months until signs of peeling are observed.

    c. Develop a 99% confidence interval for the population mean.

55. Joan's Nursery specializes in custom designed landscaping for residential areas. The labor cost associated with a particular landscaping proposal is estimated based on the number of plantings of trees, shrubs, and so on associated with the project. For labor cost estimating purposes, management allows a maximum of 2 hours of labor time for the planting of a medium-size tree. Actual times from a sample of 10 plantings during the past month are as follows (times in hours).

1.9, 2.1, 2.8, 3.0, 2.6, 2.5, 2.8, 3.2, 2.2, 2.5

Using a .05 level of significance, test the claim that the mean tree planting time is 2 hours or less. What is your conclusion, and what recommendations would you consider making to management? What is the p value?

Computer Exercise

A consumer research organization has been studying the repair history of diesel automobiles produced by a major Detroit manufacturer. Of particular concern has been the performance of diesel engines used in the manufacturer's full-sized cars. Preliminary evidence shows that owners of these cars have experienced relatively early failures in the car's transmission system. Part of the consumer research organization's study has uncovered the fact that the transmission used by the manufacturer may be too small for the diesel engines. To aid in the investigation, the research organization sent questionnaires to owners of the diesel automobiles. Data were collected from 50 owners who had experienced a transmission failure. The following data shows the number of miles the vehicle had been driven at the time of the failure.

85,092	32,609	59,465	77,437	32,534	64,090	32,464	59,902
39,323	89,641	94,219	116,803	92,857	63,436	65,605	85,861
64,342	61,978	67,998	59,817	101,769	95,774	121,352	69,568
74,276	66,998	40,001	72,069	25,066	77,098	69,922	35,662
74,425	67,202	118,444	53,550	79,294	64,544	86,813	116,269
37,831	89,341	73,341	85,288	138,114	53,402	85,586	82,256
77,539	88,798						

QUESTIONS

1.   Use appropriate descriptive statistics to summarize this data.
2.   Develop a 95% confidence interval for the mean number of miles driven until transmission failure for the population of vehicles that have experienced transmission failure.
3.   Discuss the implications of your statistical findings in terms of the claim that some owners of the diesel automobiles have been experiencing early transmission failures.

CHAPTER 10

Inferences About a Population Proportion

What You Will Learn in This Chapter

- the process of using a sample proportion to make inferences about a population proportion

- the characteristics of the sampling distribution of the sample proportion

- how to construct and interpret an interval estimate of a population proportion

- how to determine the sample size for estimation of a population proportion

- how to conduct hypothesis tests about a population proportion

Contents

Statistics in the News

DOES THE MISS AMERICA PAGEANT EXPLOIT WOMEN?

According to Albert Marks, chairman of the Miss America Pageant, the greatest spectator sport in America is not football or baseball—it's girl-watching. While some individuals claim the Miss America pageant is sexist, silly, destructive, and exploits women, others argue that the pageant is simply entertainment and that the claims of sexism are unfounded. Pageant participants apparently do not feel exploited; many point to the pageant as a high point in their lives.

A recent survey of 1764 people was conducted by Valley Forge Information Service of Pennsylvania in order to estimate the proportion of the American population that find the Miss America pageant acceptable. Although the results show a variety of opinions, the survey still found much support for the crown jewel of beauty pageants. Specific estimates from the survey show that 48.2% of the adults with daughters old enough to participate in the pageant would encourage participation; 40.1% said they would prefer their daughters not participate. In addition, 50.4% of the people surveyed did not believe that the 63-year-old pageant exploits women, whereas 36.9% did believe that it exploits women. Apparently the majority of Americans find the pageant fun and enjoy watching it on television. The television audience of this year's pageant is expected to reach 70 million; 75% of the viewers will be female.

Contestants participating in a recent Miss America Pageant.

Beauty pageants are an American tradition. About 750,000 beauty contests are held annually, with the total number of contestants running into the millions. Eighty thousand young women compete in the Miss America event each year. Pageant officials are quick to point out that the Miss America contest is a nonprofit operation and that pageant participants share over $4 million in college scholarships each year. While the debate over the pageant continues, millions of Americans are supportive of it and are satisfied that it is a pleasant American tradition that harms no one.

Based on "Pageant's Harmless and Fun for Millions," *USA Today* (September 14, 1984).

In Chapters 8 and 9 we showed how a sample mean can be used to estimate and conduct hypothesis tests about a population mean. In some statistical applications, however, we find that a population proportion is of more interest than a population mean. The "Statistics in the News" article on the Miss America pageant described a survey that was conducted in order to estimate the *proportion* of the people in the United States who support the pageant. Several proportions were reported in the article, including the proportion who would encourage their daughters to participate, the proportion who did not believe the pageant exploits women, the proportion of the television audience who are women, and so on. Other situations where the interest is on a population proportion include the proportion of voters who prefer a particular political candidate, the proportion of adults who have a particular disease, the proportion of college students who are in-state residents, and so on. In situations such as these, we use the sample proportion $\bar{p}$ to make statistical inferences about the population proportion p. This statistical process is depicted in Figure 10.1.

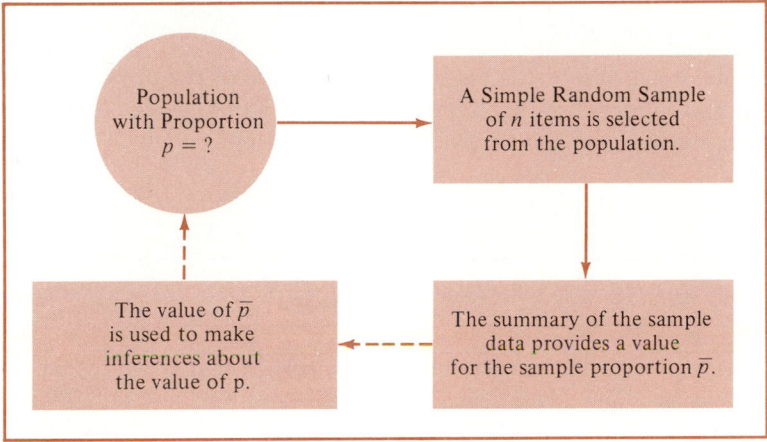

FIGURE 10.1
The Statistical Process of Using a Sample Proportion to Make Inferences About a Population Proportion

EXAMPLE 10.1

A sample of 200 married couples selected from throughout the United States showed 84 couples in which both the husband and the wife hold full-time jobs. Use the sample results to estimate the proportion of all married couples in the United States with both the husband and wife holding full-time jobs. Let

$$x = \text{the number of couples in the sample with both the}$$
$$\text{husband and the wife holding full-time jobs}$$

$$n = \text{the number of couples in the sample}$$

The sample proportion is computed as follows:*

$$\bar{p} = \frac{x}{n} \tag{10.1}$$

*We use the notation $\bar{p}$ to denote the sample proportion because it is analogous to the sample mean $\bar{x}$. If we define a random variable

$$x_i = \begin{cases} 1 & \text{if the } i\text{th couple in the sample both have full-time jobs} \\ 0 & \text{otherwise,} \end{cases}$$

then the sample proportion $\bar{p} = \Sigma x_i / n$ is also a sample mean.

Using the given data, we have

$$\bar{p} = \frac{x}{n} = \frac{84}{200} = .42$$

This sample proportion $\bar{p} = .42$ is the *point estimate* of the population proportion p. The value $\bar{p} = .42$ provides the estimate that for 42% of married couples in the United States both the husband and the wife hold full-time jobs.

10.1 THE SAMPLING DISTRIBUTION OF $\bar{p}$

As was the case with the sample mean $\bar{x}$, we realize that if we repeat the sampling process shown in Figure 10.1 several times, we can anticipate obtaining a variety of different values for the sample proportion $\bar{p}$. Since the different samples consist of different items from the population, we expect the different samples to provide different values for the sample statistic $\bar{p}$. This situation would be observed in Example 10.1 if we randomly selected a new sample of 200 married couples. Perhaps this time we would find 96 couples with both the husband and the wife holding full-time jobs. The sample proportion for this sample would be $\bar{p} = 96/200 = .48$.

Because different simple random samples can provide different values for the sample statistic, whenever we make inferences about a population proportion p, we are interested in the probability distribution of all possible values of the sample proportion $\bar{p}$. This probability distribution is referred to as the *sampling distribution of $\bar{p}$*. If we can identify the properties of the sampling distribution of $\bar{p}$, we can use this distribution to make inferences about a population proportion p in *exactly* the same way that we used the sampling distribution of $\bar{x}$ to make inferences about the population mean μ. Let us begin by defining what is known about the the properties of the sampling distribution of $\bar{p}$.

Expected Value of $\bar{p}$

The *expected value* of $\bar{p}$ is the mean value of all possible $\bar{p}$ values that can be observed when simple random samples are selected from a population with a proportion denoted by p. Letting $E(\bar{p})$ denote the expected value of $\bar{p}$, or simply the mean of all possible $\bar{p}$ values, it can be shown that

---

Expected Value of $\bar{p}$

$$E(\bar{p}) = p \tag{10.2}$$

Standard Deviation of $\bar{p}$

The *standard deviation* of $\bar{p}$ is used as the measure of dispersion in the possible $\bar{p}$ values. When simple random samples of size n are selected from a population with proportion p, the expression for the standard deviation of $\bar{p}$ is as follows.*

Standard Deviation of $\bar{p}$

$$\sigma_{\bar{p}} = \sqrt{\frac{p(1-p)}{n}} \qquad\qquad (10.3)$$

The standard deviation of $\bar{p}$ is called the *standard error of the proportion.*

Form of the Sampling Distribution of $\bar{p}$

In Chapter 8 we saw that the central limit theorem enabled us to conclude that the sampling distribution of $\bar{x}$ can be approximated by a normal probability distribution whenever the sample size is large. Applying the same central limit theorem as it relates to the sample proportion $\bar{p}$, we have the following statement about the form of the sampling distribution of $\bar{p}$.

In selecting simple random samples of size n from a population with proportion p, the probability distribution of the sample proportion $\bar{p}$ approaches a normal distribution with mean p and standard deviation $\sqrt{p(1-p)/n}$ as the sample size becomes large.

Thus we can also use a normal probability distribution to approximate the sampling distribution of $\bar{p}$ provided the sample size is large. With proportions, the sample size can be considered large whenever the following two conditions are satisfied:

$$np \geq 5$$

$$n(1-p) \geq 5$$

To understand the rationale behind this rule of thumb for a large sample size, first note that the population proportion p is equivalent to the probability of success associated with the binomial probability distribution. In fact, the exact sampling

---

*As we indicated in Chapter 8, throughout the text we assume that the population is *large* relative to the sample size, with $n/N \leq .05$. In this situation the finite population correction factor, $\sqrt{(N-n)/(N-1)}$ is approximately 1 and is not needed in the formulas for the standard deviation or standard error. Thus in the problems and examples in this chapter, the populations will be considered large and (10.3) may be used to compute $\sigma_{\bar{p}}$. If the population size N and the sample size n are values such that $n/N > .05$, the finite population correction factor should appear in (10.3) with $\sigma_{\bar{p}} = \sqrt{(N-n)/(N-1)} \sqrt{p(1-p)/n}$.

distribution of $\bar{p}$ can be determined by using the binomial probability distribution. However, as we saw in Chapter 7, whenever n is large it is computationally convenient to use the normal distribution to approximate the binomial distribution. The given rule of thumb indicates when the normal approximation of the binomial distribution is appropriate and thus also when the normal approximation is appropriate for the sampling distribution of $\bar{p}$.

In the following examples, we determine the sampling distribution of $\bar{p}$ and then show how the sampling distribution can be used to make probability statements about the value of the sample proportion $\bar{p}$ that will be found from a simple random sample.

EXAMPLE 10.2

Consider the population of fourth-grade students in the Cleveland, Ohio, public school system. Assume that the proportion of students in this population who wear eyeglasses is $p = .25$. If a random sample of 50 students is to be selected, define the characteristics of the sampling distribution of $\bar{p}$, where $\bar{p}$ is the proportion of students in the sample who wear eyeglasses.

From (10.2), we know the mean of the sampling distribution is

$$E(\bar{p}) = p = .25$$

From (10.3), we know that the standard deviation is

$$\sigma_{\bar{p}} = \sqrt{\frac{p(1-p)}{n}} = \sqrt{\frac{.25(1-.25)}{50}} = .0612$$

Finally, for the sample size of $n = 50$, we have $np = 50(.25) = 12.5$ and $n(1-p) = 50(1-.25) = 37.5$; thus the large sample conditions are satisfied and the sampling distribution of $\bar{p}$ can be approximated by a normal probability distribution. The complete sampling distribution of $\bar{p}$ for this example is shown in Figure 10.2.

• • •

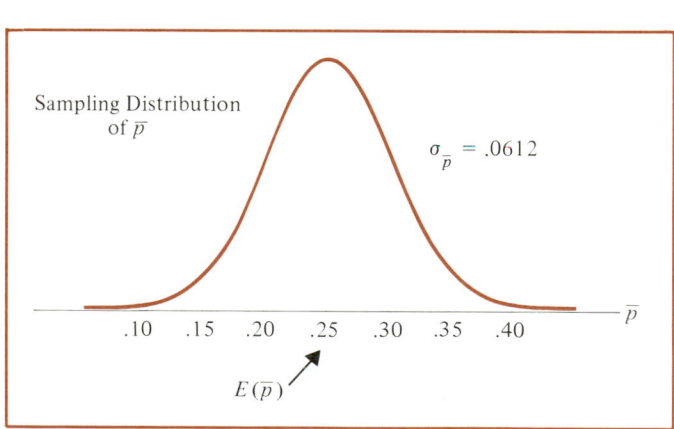

FIGURE 10.2
Sampling Distribution of $\bar{p}$ for the Sample Proportion of Students Wearing Eyeglasses

EXAMPLE 10.3

Use the sampling distribution of $\bar{p}$ in Figure 10.2 to compute the probability of selecting a simple random sample of 50 students and finding a sample proportion wearing eyeglasses between .20 and .30. The sampling distribution of $\bar{p}$ and the area showing the probability of $.20 \leq \bar{p} \leq .30$ is shown in Figure 10.3.

Since the sampling distribution of $\bar{p}$ can be approximated by a normal distribution, we can use the areas under the curve of the standard normal distribution to answer the probability question. With the normal distribution, we need to compute a z value, where

$$z = \frac{\bar{p} - p}{\sigma_{\bar{p}}} \tag{10.4}$$

The above value is identical to the z value we have always used, with the exception of the notation $\bar{p}$ and $\sigma_{\bar{p}}$, which is used to remind us we are dealing with the sampling distribution of $\bar{p}$.

$$\text{At } \bar{p} = .30, \qquad z = \frac{\bar{p} - p}{\sigma_{\bar{p}}} = \frac{.30 - .25}{.0612} = +.82$$

Using $z = .82$ in the table of areas for the standard normal distribution, we find an area between .25 and .30 of .2939.

$$\text{At } \bar{p} = .20, \qquad z = \frac{\bar{p} - p}{\sigma_{\bar{p}}} = \frac{.20 - .25}{.0612} = -.82$$

This z value shows that the area between .20 and .25 must also be .2939. Thus the total probability of selecting a sample with a sample proportion $\bar{p}$ between .20 and .30 must be $.2939 + .2939 = .5878$.

• • •

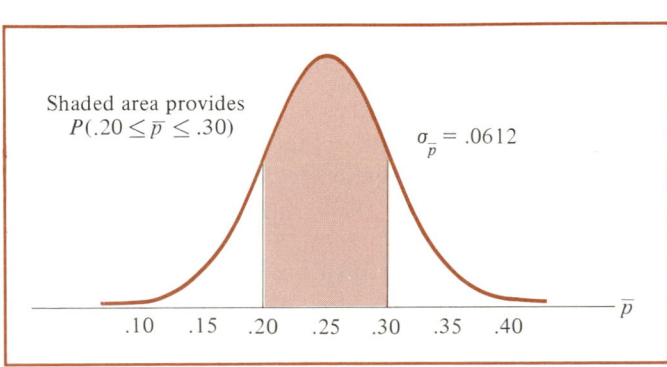

FIGURE 10.3
Sampling Distribution of $\bar{p}$ Showing $P(.20 \leq \bar{p} \leq .30)$ for Example 10.3

EXAMPLE 10.4

An experiment consists of a sample of 100 flips of a coin, where the probability of a head is .50. What is the probability that the proportion of heads in the sample will be .60 or more?

Letting $\bar{p}$ denote the proportion of heads in the sample and noting that $p = .50$, the sampling distribution of $\bar{p}$ for this example is shown in Figure 10.4.

$$\text{At } \bar{p} = .60, \qquad z = \frac{\bar{p} - p}{\sigma_{\bar{p}}} = \frac{.60 - .50}{.05} = 2.0$$

Using the table of areas for the standard normal distribution, we find that at $z = 2.0$, the area between .50 and .60 must be .4772. Thus the probability of a sample proportion $\bar{p} = .60$ or more is $.5000 - .4772 = .0228$.

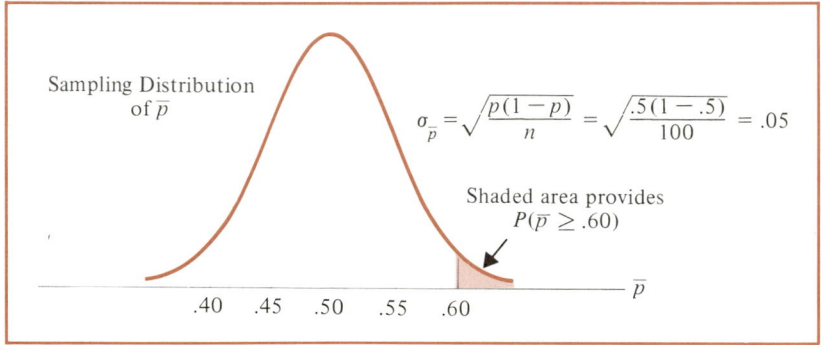

FIGURE 10.4

Sampling Distribution of $\bar{p}$ for the Proportion of Heads Occurring in 100 Flips of A Coin

• • •

EXAMPLE 10.5

At a large university, 40% of all undergraduate women students are members of a sorority. Thus, for the population of undergraduate women, the population proportion for being a member of a sorority is $p = .40$. If simple random samples of size 80 are to be taken from this population, we expect different samples to consist of different students and, therefore, to provide a variety of values for $\bar{p}$, the proportion of women in the sample who are a member of a sorority. Find a range of values for $\bar{p}$ that will include 95% of all possible sample proportions that can be observed.

First we show the sampling distribution for the sample proportion of women students belonging to a sorority. Using $p = .40$, this sampling distribution is shown in Figure 10.5.

For the standard normal probability distribution, we know that 95% of the area under the curve is between $z = -1.96$ and $z = 1.96$. Thus the range containing 95% of all possible $\bar{p}$ values must be from

$$p - 1.96\sigma_{\bar{p}} = .40 - 1.96\sigma_{\bar{p}} = .40 - 1.96(.0548) = .29$$

$$p + 1.96\sigma_{\bar{p}} = .40 + 1.96\sigma_{\bar{p}} = .40 + 1.96(.0548) = .51$$

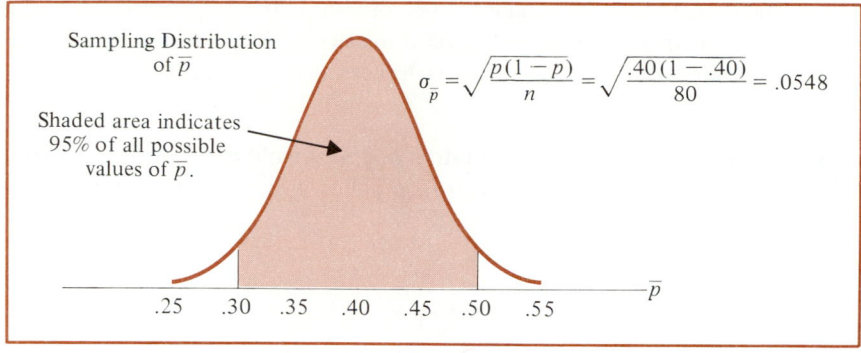

Sampling Distribution of $\bar{p}$

Shaded area indicates 95% of all possible values of $\bar{p}$.

$$\sigma_{\bar{p}} = \sqrt{\frac{p(1-p)}{n}} = \sqrt{\frac{.40(1-.40)}{80}} = .0548$$

FIGURE 10.5

Sampling Distribution of $\bar{p}$ for the Proportion of Women Students Belonging to a Sorority

Thus, there is a .95 probability of selecting a simple random sample of 80 women students and obtaining a sample proportion between .29 and .51.

• • •

EXERCISES

1. Nine nurses selected at random from a large hospital were asked if they believed that the hospital nursing department is understaffed. The responses are as shown.

Nurse	1	2	3	4	5	6	7	8	9
Response	No	Yes	Yes	No	No	Yes	No	No	No

Develop a point estimate of the proportion of all hospital nurses that believe that the hospital nursing department is understaffed.

2. A survey of 400 women college students was conducted to determine future plans concerning career, marriage, and family. The following results were recorded:

   310 women answered yes to the question, Do you plan to begin a full-time career immediately following graduation?
   225 women answered yes to the question, Do you plan to marry before the age of 30?
   175 women answered yes to the question, Do you plan to have children?

   Use the survey results to provide point estimates of each of the following.
   a. the proportion of college women planning to begin full-time careers immediately following graduation
   b. the proportion of college women planning to marry before the age of 30
   c. the proportion of college women planning to have children

3. Develop a point estimate of the proportion of the pages in this text that contain a figure or a table. That is, if a randomly selected page has at least one figure or at

least one table, the response is yes; otherwise the response is no. Use the random numbers in Table 6 of Appendix B to select a simple random sample of 20 pages. Compare your point estimate with those of others in the class.

4. Consider the following population of 25-year-old males, where Yes indicates that the individual has a life insurance policy and No indicates that the individual does not have such a policy.

Individual	Response
1	Yes
2	No
3	No
4	Yes
5	No
6	Yes

a. Selecting simple random samples of size 4 provides a total of 15 possible samples. List the 15 samples.
b. Compute the proportion of Yes responses for each sample, and show a histogram of the sampling distribution of $\bar{p}$.

5. A family has five children, three girls and two boys. A random sample of three children will be selected in order to estimate the proportion of girls in the family where $p = 3/5 = .60$.
a. List the 10 possible samples.
b. Compute the sample proportion $\bar{p}$ for each possible sample.
c. Show a graphical representation of the probability distribution of the $\bar{p}$ values. This is the graph of the sampling distribution of $\bar{p}$.
d. Compute the mean of the 10 values of $\bar{p}$ and compare this to $E(\bar{p})$ given in Equation (10.2).

6. A research finding published by a medical institute indicates that 8% of the population have chronic headache problems and frequently experience more than one headache per month. A sample of 50 individuals will be taken to verify these results.
a. Assuming $p = .08$ is true, state as much as you can about the characteristics of the probability distribution of the various values of $\bar{p}$ that can be observed.
b. Assuming $p = .08$ is true and that the sample size is increased to $n = 100$, state as much as you can about the characteristics of the probability distribution of the various values of $\bar{p}$ that can be observed.
c. What information about the sampling distribution of $\bar{p}$ were you able to provide in (b) that you could not provide in (a)? Why was this possible?

7. For a population where $p = .35$ is the proportion of persons having a college degree:
   a. Explain how the sampling distribution of $\bar{p}$ results from random samples of size 80 being used to estimate the proportion of individuals having a college degree.
   b. Show the sampling distribution for $\bar{p}$ in this case.
   c. If the sample size is increased to 200, what happens to the sampling distribution of $\bar{p}$? Compare the standard error for the $n = 80$ and $n = 200$ alternatives.

8. The president of Doerman Distributors, Inc. believes that 30% of the firm's orders come from new, or first-time, customers. A simple random sample of 100 orders will be used to estimate the proportion of new customers. The results of the sample will be used to verify the president's claim of $p = .30$.
   a. Assume that the president is correct with $p = .30$. What is the sampling distribution of $\bar{p}$ for this study?
   b. What is the probability that the sample proportion $\bar{p}$ will be between .20 and .40?
   c. What is the probability that the sample proportion will be within plus or minus .05 of the population proportion $p = .30$?

9. A particular county in West Virginia has a 9% unemployment rate. A monthly survey of 800 individuals is conducted by a state agency. This study provides the basis for monitoring the unemployment rate of the county.
   a. Assume that $p = .09$. What is the sampling distribution of $\bar{p}$ when a sample of size 800 is used?
   b. What is the probability that a sample proportion $\bar{p}$ of at least .08 will be observed?

10. A doctor believes that 80% of all patients having a particular disease will be fully recovered within 3 days after receiving a new drug.
    a. A simple random sample of 20 medical records will be used to develop an estimate of the proportion of patients who were fully recovered within 3 days after receiving the drug. If a data analyst suggests using a normal probability distribution approximation for the sampling distribution of $\bar{p}$, what would you say? Explain.
    b. If the sample of patient records is increased to 60, what is the probability that the sample proportion will be within $\pm.10$ of the population proportion? (Assume that the population proportion is .80.)

11. The proportion of individuals insured by the All-Driver Automobile Insurance Company who have received at least one traffic ticket during a 5-year period is .15.
    a. Show the sampling distribution of $\bar{p}$ if a random sample of 150 insured individuals is used to estimate the proportion having received at least one ticket.
    b. What is the probability that the sample proportion will be within $\pm.03$ of the population proportion?

12. Historical records show that .50 of all orders placed at Big Burger Fast Food restaurants include a soft drink. With a simple random sample of 40 orders, what is the probability that between .45 and .55 of the sampled orders will include a soft drink?

13. Lori Jeffrey is a successful sales representative for a major publisher of college textbooks. Historically, Lori obtains a book adoption on 25% of her sales calls. Viewing her sales calls for one month as a sample of all possible sales calls, a statistical analysis of the data yields a standard error of the proportion of .0625.
    a. How large was the sample used in this analysis? That is, how many sales calls did Lori make during the month?
    b. Let $\bar{p}$ indicate the sample proportion of book adoptions obtained during the month. Show the sampling distribution $\bar{p}$.
    c. Use the sampling distribution of $\bar{p}$. What is the probability that Lori will obtain book adoptions on 30% or more of her sales calls during the one month period?

10.2 ESTIMATION OF A POPULATION PROPORTION

In Chapter 9 we used the value of the sample mean $\bar{x}$ to estimate the population mean μ. The confidence interval estimation procedure was based upon knowledge of the characteristics of the sampling distribution of $\bar{x}$. In this chapter, we are using the value of the population proportion $\bar{p}$ to estimate the population proportion p. Since the characteristics of the sampling distribution of $\bar{p}$ are known, we can use the logic we used to estimate μ in order to develop an interval estimate of p.

Based on the fact that the sampling distribution of $\bar{x}$ can be approximated by a normal probability distribution, the following expression was used to obtain an interval estimate of a population mean:

$$\bar{x} \pm z_{\alpha/2} \frac{\sigma}{\sqrt{n}} \tag{10.5}$$

Recall that $1 - \alpha$ denotes the confidence coefficient for the interval. $z_{\alpha/2}$ is the z value providing an area of $\alpha/2$ in the upper tail of the standard normal probability distribution. The term $\sigma/\sqrt{n}$ is the standard error of the mean.

Since the sampling distribution of $\bar{p}$ can also be approximated by a normal probability distribution, the logic behind the development of (10.5) can be used to develop the following expression for an interval estimate of a population proportion:

$$\bar{p} \pm z_{\alpha/2} \sqrt{\frac{p(1-p)}{n}} \tag{10.6}$$

Note the similarity in (10.5) and (10.6). Specifically, (10.6) uses the sample proportion $\bar{p}$ in place of the sample mean $\bar{x}$. The $z_{\alpha/2}$ term is used in both due to the fact that both sampling distributions are approximately normal. In addition, in (10.6) the standard error of the proportion, $\sqrt{p(1-p)/n}$, is substituted for the standard error of the mean in (10.5). Although the two expressions use different notation, they are based on the same logic, and each provides the confidence interval estimate of its corresponding population parameter.

However, note that in using (10.6) to develop an interval estimate of a population proportion p, the value of p would have to be *known*. Since the value of p is what we are trying to estimate and is *unknown*, we simply substitute the sample proportion $\bar{p}$ for p. As a result, the general expression for a confidence interval estimate of a population proportion is as follows:*

Interval Estimation of a Population Proportion

$$\bar{p} \pm z_{\alpha/2} \sqrt{\frac{\bar{p}(1-\bar{p})}{n}} \qquad (10.7)$$

where $1-\alpha$ is the confidence coefficient and $z_{\alpha/2}$ is the z value providing an area of $\alpha/2$ in the upper tail of the standard normal probability distribution.

By choosing different values of z, we can obtain interval estimates with different levels of confidence. Some of the common confidence levels and their corresponding z values are as follows:

z Values for Selected Confidence Levels

90% Confidence:	$z = 1.645$
95% Confidence:	$z = 1.96$
98% Confidence:	$z = 2.33$
99% Confidence:	$z = 2.58$

EXAMPLE 10.6

Based upon a sample of 250 credit-card holders, a department store found that 185 card holders incurred a monthly interest charge on an unpaid balance. Develop 95% and 90% confidence interval estimates of the proportion of the population of credit-card holders who incur a monthly interest charge.

The point estimate of the population proportion p is $\bar{p} = 185/250 = .74$. For a 95% confidence interval, we have $\alpha = .05$ and $z_{\alpha/2} = z_{.025} = 1.96$. Using (10.7) provides

$$\bar{p} \pm z_{.025} \sqrt{\frac{\bar{p}(1-\bar{p})}{n}} = .74 \pm 1.96 \sqrt{\frac{.74(1-.74)}{250}}$$

$$= .74 \pm .054$$

---

*An unbiased estimate of the standard error of the proportion is given by $\sqrt{\bar{p}(1-\bar{p})/(n-1)}$. The bias introduced by using n in the denominator does not cause any difficulty because large samples are used in making estimates concerning population proportions and the numerical difference between the results using n and $n-1$ is negligible.

Thus the 95% confidence interval estimate of the population proportion is .686 to .794.

At 90% confidence, $z_{\alpha/2} = z_{.05} = 1.645$. Using (10.7) provides

$$.74 \pm 1.645 \sqrt{\frac{.74(1 - .74)}{250}} = .74 \pm .046$$

Thus we are 90% confident that the population proportion p is between .694 and .786.

• • •

EXAMPLE 10.7

A survey conducted by the United State Department of Labor found that 48 out of 500 heads of households were unemployed. Develop a 98% confidence interval estimate of the proportion of unemployed heads of households in the population.

The point estimate of the population proportion is $\bar{p} = 48/500 = .096$. At a 98% confidence level, $z_{.01} = 2.33$. Using (10.7), the confidence interval becomes

$$.096 \pm 2.33 \sqrt{\frac{.096(1 - .096)}{500}} = .096 \pm .031$$

Thus we are 98% confident that the population proportion of unemployed heads of households is between .065 and .127.

• • •

Determining the Size of the Sample

We can now discuss the issue of determining how large a sample should be taken whenever a sample proportion is to be used to estimate a population proportion. Referring to the expression for interval estimation of a population proportion as it was written in (10.6), we see that the term $z_{\alpha/2} \sqrt{p(1 - p)/n}$ provides a bound on the sampling error whenever $\bar{p}$ is used as an estimator of p. Since the sample size, n, appears in the denominator, the bound on the sampling error can be decreased by increasing the sample size. Our goal is to determine the sample size necessary to estimate p with a reasonable degree of precision.

Generally, the answer to the sample-size question depends upon the specific sampling application. However, by obtaining the following three pieces of information about the application, the appropriate sample size can be determined:

1. The maximum sampling error, e, that the user of the sampling results is willing to tolerate.
2. The confidence level desired for the interval estimate. The value of $z_{\alpha/2}$ will be based on the confidence level specified.
3. A planning value for the population proportion p. This planning value can be obtained from past data, from a preliminary sample, or from an educated guess.

Using e as the maximum value for the sampling error, we can write

$$e = z_{\alpha/2} \sqrt{\frac{p(1 - p)}{n}}$$

Squaring both sides of this equation provides

$$e^2 = z_{\alpha/2}^2 \frac{p(1 - p)}{n}$$

Multiplying both sides by n gives

$$ne^2 = z_{\alpha/2}^2 p(1 - p)$$

Finally, dividing by e^2 provides the following formula for the sample size when estimating a population proportion:

Sample Size for Interval Estimation of a Population Proportion

$$n = \frac{z_{\alpha/2}^2 p(1 - p)}{e^2} \tag{10.8}$$

This expression shows that in order to determine the sample size, we must have a preliminary idea of the value of the population proportion, p. Past data, a preliminary sample, or an educated guess are suggested ways for obtaining the necessary planning value for p. However, in some cases it may be difficult or impossible to specify the planning value. In order to handle these situations, note that the numerator of (10.8) shows that the sample size is proportional to the quantity $p(1 - p)$. In Table 10.1 we show some possible values for this quantity. To be on the conservative side, we need to consider the largest possible value for $p(1 - p)$, since this will provide the largest recommended sample size. As the values in Table 10.1 suggest, if an appropriate planning value for p cannot be obtained, use $p = .50$. This planning value will provide the largest possible recommended sample size. If the population proportion is different than the .50 planning value, the sample proportion will have more precision than necessary. However, in using $p = .50$ to determine n, we have guaranteed that the required precision will be obtained.

TABLE 10.1

Possible Values for the Quantity $p(1 - p)$

If p	Then $p(1 - p)$	
.10	$(.10)(.90) = .09$	
.30	$(.30)(.70) = .21$	
.40	$(.40)(.60) = .24$	
.50	$(.50)(.50) = .25$	←Largest value
.60	$(.60)(.40) = .24$	
.70	$(.70)(.30) = .21$	
.90	$(.90)(.10) = .09$	

EXAMPLE 10.8

A medical experiment is being conducted to determine the recovery rate of patients given a new drug. In particular, the researcher would like to estimate the proportion of patients who fully recover within two weeks from when they begin taking the drug. The desired precision of the estimate is expressed as a 98% confidence that the maximum sampling error will be .04 or less. For planning purposes, the researcher anticipates that .80 of the patients receiving the new drug will fully recover within the 2-week period. How large should the sample size be in this study?

Using (10.8) with $z = 2.33$ for 98% confidence, a planning value of $p = .80$, and a maximum sampling error $e = .04$, we have

$$n = \frac{z_{\alpha/2}^2 \, p(1 - p)}{e^2} = \frac{(2.33)^2 (.80)(1 - .80)}{(.04)^2} = 542.9 \text{ patients}$$

As usual, we round up to a sample size of 543 to ensure that the bounds on the sampling error are maintained.

• • •

EXAMPLE 10.9

A national survey of registered voters is being conducted to determine the proportion of voters who favor a particular candidate. Assume that the desired confidence level is 95% and that the desired maximum sampling error is .02.

a. How large a sample is needed if it is believed that approximately 35% of the population currently support the candidate?
b. How large a sample is needed if no information is available on the proportion of voters currently supporting the candidate?

i. The planning value for p is .35. With $z_{.025} = 1.96$ and $e = .02$, the necessary sample size is

$$n = \frac{z_{\alpha/2}^2 p(1 - p)}{e^2} = \frac{(1.96)^2 (.35)(1 - .35)}{(.02)^2} = 2184.9$$

Rounding up, we recommend a sample of 2185 voters.
ii. No planning value is available for p. Using the conservative approach discussed earlier, we base the sample size on a planning value of $p = .50$. Doing so provides

$$n = \frac{z_{\alpha/2}^2 p(1 - p)}{e^2} = \frac{(1.96)^2 (.50)(1 - .50)}{(.02)^2} = 2401$$

Note that the recommended sample size of 2401 voters is larger than the sample size of 2185 voters recommended in (a). This larger sample size should have been anticipated due to the fact that $p = .50$ is a conservative planning value and guarantees that the sample size will be large enough to satisfy the precision requirement regardless of the actual value of p.

• • •

EXERCISES

14. A professor of sociology selected a random sample of 120 students who were attending a large eastern university. The professor asked each student to state a preference in donating his or her own money to one of the following: (1) an agency providing food for the people of India, or (2) a fund for a copier, which will be made available for student use. The students responded to the question anonymously by writing their choices on pieces of paper. If 92 of 120 students selected alternative 2, develop a 95% confidence interval for the proportion of students who would prefer donating money to the copier fund.

15. A simple random sample of 100 residents of Watkins Glen, New York, resulted in 65 individuals stating that they would support a newly proposed water-treatment facility. Develop a 95% confidence interval estimate of the proportion of all Watkins Glen residents that would support the new water-treatment facility.

16. In a telephone follow-up survey of a new advertising campaign, 45 of 150 individuals contacted could recall the new advertising slogan associated with the product. Develop a 90% confidence interval estimate of the proportion in the population that will recall the advertising slogan.

17. A sample of 90 students at a particular college showed that 27 students favor pass-fail grades for elective courses.
 a. What is the point estimate of the proportion of all students who would favor pass-fail grades for elective courses?
 b. Provide a 90% confidence interval estimate of the proportion of the population of students who would favor pass-fail grades for elective courses.

18. The New Orleans Beverage Company has been experiencing problems with the automatic machine that places labels on bottles. The company desires an estimate of the percentage of bottles that have improperly applied labels. A simple random sample of 400 bottles resulted in 18 bottles with improperly applied labels. Using these data, develop a 90% confidence interval estimate of the population proportion of bottles with improperly applied labels.

19. A sample of 200 people were asked to identify their major source of news information; 110 stated that their major source was television news coverage.
 a. Construct a 95% confidence interval for the proportion of the people in the population that consider television news their major source.
 b. How large a sample would be necessary to estimate the population proportion with a sampling error of .05 or less at a 95% confidence level?

20. A survey is to be taken to estimate the proportion of high school graduates in a particular school district that plan to attend college. How large a sample of students should be selected if the survey is to provide a 95% confidence of reporting a sample proportion that is within $\pm.025$ of the population proportion? Use $p = .35$ as a planning value for the proportion of high school students who plan to attend college.

21. A new cheese product is to be test-marketed by giving a free sample to randomly selected customers and asking them to state whether or not they like the product. With a 98% confidence level and a target sampling error of .05 or less, what sample size would you recommend in each case?
    a. Preliminary estimates are that approximately 35% of the individuals in the population will like the product.
    b. No information is available about the proportion in the population that will like the product.

22. In an election campaign, a campaign manager requests that a sample of voters be polled to determine the support for the candidate. From a sample of 120 voters, 64 express plans to support the candidate.
    a. What is the point estimate of the proportion of the voters in the population who will support the candidate?
    b. Develop and interpret the 95% confidence interval for the proportion of voters in the population who will support the candidate.
    c. From the result obtained in (b), is the campaign manager justified in feeling confident that the candidate has the support of at least 50% of the voters? Explain.
    d. How many voters should be sampled if we desired to estimate the population proportion with a sampling error of 5% or less? Continue to use the 95% confidence level.

10.3 HYPOTHESIS TESTS ABOUT A POPULATION PROPORTION

Hypothesis tests about a population proportion p are based on knowledge of the sampling distribution of $\bar{p}$. The mechanics of conducting the test follow the procedure used in Chapter 9 to make hypothesis tests about a population mean μ. The only difference is that in this section we use the sample proportion $\bar{p}$ and the sampling distribution of $\bar{p}$ in the analysis.

We begin by formulating a null hypothesis and an alternative hypothesis about the value of the population proportion. We then consider the sampling distribution of $\bar{p}$ under the assumption that the null hypothesis is true. Based on whether the hypotheses are one-tailed or two-tailed and the level of significance, a critical value(s) is selected for z. Then, using the value of the sample proportion $\bar{p}$, we compute a value for the test statistic z. The comparison of this value of z to the critical value enables us to determine whether or not the null hypothesis should be rejected.

EXAMPLE 10.10

A newspaper article contains the statement that, nationwide, 60% of all college seniors have a job prior to graduation. The director of a college placement office at a large university is interested in testing this claim for her university. The hypotheses about the proportion of the students having a job prior to graduation are as follows:

$$H_0: p = .60$$

$$H_1: p \neq .60$$

If a random sample shows that 40 of 75 recent graduates had a job prior to graduation, what conclusion can be drawn? Use a .05 level of significance.

Under the assumption that the null hypothesis is true, we have $np = 75(.60) = 45$ and $n(1 - p) = 75(.40) = 30$. Since both of these values exceed 5, the sampling distribution of $\bar{p}$ can be approximated by a normal probability distribution. Again assuming $p = .60$, the standard error of the proportion is computed to be

$$\sigma_{\bar{p}} = \sqrt{\frac{p(1 - p)}{n}} = \sqrt{\frac{.60(1 - .60)}{75}} = .0566$$

The quantity

$$z = \frac{\bar{p} - p}{\sigma_{\bar{p}}} \tag{10.9}$$

is a standard normal random variable that can be used to determine the number of standard deviations (or standard errors) an observed value of a sample proportion $\bar{p}$ is from the hypothesized value for the population proportion p. Using a .05 level of significance, the critical values for the two-tailed hypothesis test are shown in Figure 10.6. The decision rule is

$$\text{Accept } H_0 \text{ If } -1.96 \leq z \leq 1.96$$

$$\text{Reject } H_0 \text{ Otherwise}$$

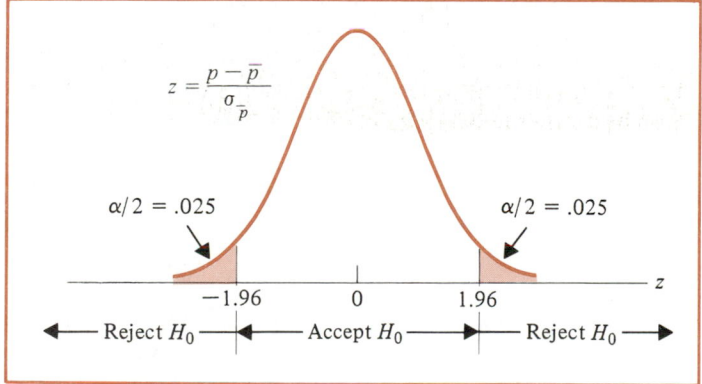

FIGURE 10.6
Acceptance and Rejection Regions for a Two-Tailed Test with $\alpha = .05$

With the sample proportion $\bar{p} = 40/75 = .533$ and the hypothesized value for the population proportion $p = .60$, using (10.9) we have the following value for the test statistic:

$$z = \frac{\bar{p} - p}{\sigma_{\bar{p}}} = \frac{.533 - .60}{.0566} = -1.18$$

This value of z is in the acceptance region of Figure 10.6. Thus, the claim that 60% of the graduating seniors have a job prior to graduation cannot be rejected.

Using the table of areas for the standard normal probability distribution, the area in the tail of the distribution at $z = -1.18$ is $.5000 - .3810 = .1190$. For a two-tailed test, we have a *p value* of $2 \times .1190 = .2380$. With $.2380 > \alpha$, we see that the *p* value criterion also indicates that the null hypothesis cannot be rejected. The computation and interpretation of *p* values for tests about a population proportion is the same as we used for tests about a population mean.

$$\bullet \quad \bullet \quad \bullet$$

EXAMPLE 10.11

The manager of an Italian restaurant is considering opening a carryout food service. However, the manager is concerned that not all individuals placing orders by phone actually pick up the order. If 90% or less of the phone orders are picked up, the restaurant will not have a profitable carryout operation. However, if it can be concluded that more than 90% of the phone orders are picked up, the carryout operation will be a worthwhile addition for the restaurant. The hypothesis test of interest is

$$H_0: p = .90$$

$$H_1: p > .90$$

This one-tailed test indicates that the restaurant should implement the carryout operation if H_0 can be rejected. During a 2-week test period, 234 orders in a sample of 250 phone-in orders were picked up. Using a .05 level of significance, what conclusion should be made?

With the one-tailed test and $\alpha = .05$, rejection of the null hypothesis will occur in the upper tail of the distribution with a critical value of $z = 1.645$. Thus, if the data show $z \leq 1.645$, the null hypothesis should be accepted; otherwise it should be rejected.

Given the hypothesized value of $p = .90$ and the sample proportion $\bar{p} = 234/250 = .936$, the value of the test statistic is as follows:

$$z = \frac{\bar{p} - p}{\sigma_{\bar{p}}} = \frac{\bar{p} - p}{\sqrt{p(1-p)/n}} = \frac{.936 - .900}{\sqrt{.90(1 - .90)/250}} = 1.90$$

Since this value is greater than 1.645, the null hypothesis is rejected. The manager of the restaurant should be safe in concluding that the proportion of carryout orders picked up for the population exceeds .90. Thus it is recommended the restaurant begin the carryout food service. For $z = 1.90$, the corresponding p value is $.5000 - .4713 = .0287$.

• • •

EXAMPLE 10.12

During a water shortage in Florida, restaurants were asked not to serve water with meals unless requested to do so by the customers. In the initial 3-month period, 45% of the customers served at a particular restaurant requested water with their meal. Recently the restaurant placed a card at each table describing the water-shortage problem and pointing out that considering the drinking water, the ice, and the water to wash the glass, it requires 24 ounces of water for every 8 ounces of water served. The restaurant would like to use a statistical test to determine if placing their cards at each table significantly decreases the proportion of customers requesting water with their meal. Use a .02 level of significance. If a sample of 150 customers showed 53 customers ordering water, what is your conclusion?

The hypothesis test is as follows:

$$H_0: p = .45$$

$$H_1: p < .45$$

Rejection of H_0 will occur in the lower tail of the distribution. With an area of $\alpha = .02$ in the lower tail, the table of areas for the standard normal distribution shows the critical z value to be -2.05. Thus, the decision rule is to reject H_0 if $z < -2.05$.

Using the hypothesized value $p = .45$ and the sample proportion $\bar{p} = 53/150 = .353$, the value of the test statistic is

$$z = \frac{\bar{p} - p}{\sigma_{\bar{p}}} = \frac{\bar{p} - p}{\sqrt{p(1-p)/n}} = \frac{.353 - .450}{\sqrt{.45(1 - .45)/150}} = -2.39$$

Since $z = -2.39$ is in the rejection region, the null hypothesis $H_0: p = .45$ is rejected. The restaurant can conclude that the cards have helped reduce the proportion of customers who order water with their meal.

• • •

EXERCISES

23. A magazine claims that 25% of its readers are college students. A random sample of 200 readers is taken. It is found that 42 of these readers are college students.
   a. Use a .10 level of significance to test the $H_0: p = .25$ versus $H_0: p \neq .25$.
   b. Using the sample results, develop a 90% confidence interval for the proportion of the population of readers that are college students.

24. A new television series must prove that it has more than 25% of the viewing audience after its initial 13-week run in order to be judged successful. Assume that in a sample of 400 households, 112 were watching the series.

a. At a .10 level of significance, can the series be judged successful based on the sample information?

b. What is the *p* value for the sample results? What is your hypothesis-testing conclusion?

25. An accountant believes that the company's cash flow problems are a direct result of the slow collection of accounts receivable. The accountant claims that at least 70% of the current accounts receivable are over 2 months old. A sample of 120 accounts receivable yielded 78 accounts that were over 2 months old. Test the accountant's claim at accounts that were the .05 level of significance. What is the *p* value?

26. A supplier claims that at least 96% of the parts it supplies meet the product specifications. In a sample of 500 parts received over the past 6 months, 36 were defective. Use the *p* value to test the supplier's claim at a .05 level of significance.

27. If 20% or more of the population have a negative reaction to a new drug, the drug will not be marketed.

a. Define the hypotheses such that the drug will be marketed only if H_0 is rejected.

b. In a sample of 80 patients, 14 patients experience a negative reaction to the drug. Test the hypotheses in (a) at a .02 level of significance. What is your conclusion?

28. The manager of K-Mark Supermarkets estimates that at least 30% of the Saturday customers purchase the price-reduced special advertised in the Friday newspaper. Use $\alpha = .05$ and test the manager's claim if a sample of 250 Saturday customers show that 60 purchased the advertised special. What is the *p* value for the sample results?

29. The filling machine for a production operation must be adjusted if more than 8% of the items being produced are underfilled. A random sample of 80 items from the day's production contained 9 underfilled items. Does the sample evidence indicate that the filling machine should be adjusted? Use $\alpha = .02$.

30. A radio station in a major resort area announced that at least 90% of the hotels and motels would be full for the Memorial Day weekend. The station went on to advise listeners to make reservations in advance if they planned to be in the resort over the weekend. On Saturday night a sample of 58 hotels and motels showed 49 with a no-vacancy sign and 9 with vacancies. What is your reaction to the radio station's claim based on the sample evidence? Use $\alpha = .05$ in making this statistical test. What is the *p* value for the sample results?

Summary

In this chapter we discussed the procedures for making statistical inferences about a population proportion. After describing the properties of the sampling distribution of $\bar{p}$, we presented the methods of interval estimation and hypothesis testing for a population proportion. In the large-sample case where both $np \geq 5$ and $n(1 - p) \geq 5$, the central limit theorem enables us to use the normal probability distribution to approximate the sampling distribution of $\bar{p}$.

The procedure for developing a confidence interval estimate of a population proportion uses the same logic as the procedure for developing a confidence interval estimate of a population mean. The essential difference is that the sampling distribution of $\bar{p}$ is used as the basis for developing the confidence interval instead of the sampling distribution of $\bar{x}$.

A procedure was presented for determining the sample size that will meet a desired precision when estimating a population proportion. This procedure requires the person conducting the test to specify a planning value for the population proportion p. If p is unknown, using a planning value of $p = .50$ provides a sample size that will satisfy the precision requirements regardless of the actual value of the population proportion.

The procedure for testing hypotheses about a population proportion follows the logic used for testing hypotheses about a population mean. Examples of both one-tail and two-tail hypothesis tests for a population proportion were presented.

Statistics in Practice

TESTING FOR PROPORTION DEFECTIVE

The RF Communications Division of Harris Corporation in Melbourne, Florida is a major manufacturer of radio communications equipment. One of the higher-volume items at the catalog products factory is an assembly called an RF deck. Each assembly consists of 16 electronic components soldered to a machined casting, which forms the plated surface of the deck. During a manufacturing run, a problem developed in the soldering process; the flow of solder onto the deck was not working properly.

It was initially believed that the soldering problem was due to defective platings being shipped by one of the firm's suppliers. An experiment was designed to test this hypothesis. Letting p denote the proportion of defective platings in the current Harris inventory and p_0 denote the proportion of defective platings allowable from the supplier, the following hypothesis test was conducted:

$$H_0: p = p_0$$

$$H_1: p > p_0$$

If H_0 could not be rejected, the test would support the conclusion that the proportion of defective platings in inventory was within the level required from the supplier. However, if H_0 could be rejected, the test would indicate that the Harris inventory had a defective plating proportion that exceeded the supplier's allowed defective proportion.

Production tests were made on a sample of platings taken from the Harris inventory. The sample proportion defective was .15. This proportion defective resulted in the rejection of H_0. The conclusion reached was that the current inventory had a higher defective proportion than would exist if the supplier's operation conformed to specifications. As a result, pressure was applied to purchasing management to have the plating supplier held responsible for both rejected parts and damages.

After reflecting on the experimental process, the plant manager and materials manager concluded that the hypothesis test correctly indicated an undesirably high proportion of defective platings in the Harris inventory. However, the test did not prove that defective platings from the supplier were the cause of the problem. Additional investigation ultimately revealed that shelf contamination received while the platings were stored in inventory was causing the high proportion of defectives. Thus, although the statistical test of the proportion defective pointed to the high defective proportion in the inventory, management judgment was essential in determining that the defective problem was not caused by the supplier.

Glossary

Sampling distribution of $\bar{p}$ — The probability distribution showing all possible values of the sample proportion $\bar{p}$.

Standard error of the proportion — The standard deviation of all possible values of the sample proportion $\bar{p}$.

Key Formulas

Sample Proportion

$$\bar{p} = \frac{x}{n} \tag{10.1}$$

Expected Value of $\bar{p}$

$$E(\bar{p}) = p \tag{10.2}$$

Standard Deviation of $\bar{p}$

$$\sigma_{\bar{p}} = \sqrt{\frac{p(1-p)}{n}} \tag{10.3}$$

z Value

$$z = \frac{\bar{p} - p}{\sigma_{\bar{p}}} \tag{10.4}$$

Interval Estimation of a Population Proportion

$$\bar{p} \pm z_{\alpha/2} \sqrt{\frac{\bar{p}(1-\bar{p})}{n}} \tag{10.7}$$

Sample Size for Interval Estimation of a Population Proportion

$$n = \frac{z_{\alpha/2}^2 p(1-p)}{e^2} \tag{10.8}$$

Review Quiz

TRUE/FALSE

1. The sample proportion $\bar{p}$ is not a point estimator of the population proportion p.
2. The central limit theorem cannot be applied to the sampling distribution of $\bar{p}$.
3. The standard error of the proportion, $\sigma_{\bar{p}}$, depends upon the value of the population proportion p.
4. In computing confidence intervals for a population proportion, the sample proportion $\bar{p}$ can be used to obtain an estimate of the standard error of the proportion, $\sigma_{\bar{p}}$.
5. In determining the sample size, the larger the planning value for p, the larger the sample size.
6. In conducting hypothesis tests about a population proportion, the value of p specified in the null hypothesis is used to compute the standard error of the proportion, $\sigma_{\bar{p}}$.

MULTIPLE CHOICE

Use the following information for questions 7–9. A random sample of 300 voters showed .47 in favor of a certain ballot proposal.

7. Which of the following best describes the form of the sampling distribution of the sample proportion?
   a. When standardized, it is exactly the standard normal distribution.
   b. When standardized, it is the t distribution.
   c. It is approximately normal because $n > 30$.
   d. It is approximately normal because $n\bar{p} \geq 5$ and $n(1 - \bar{p}) \geq 5$.
8. An estimate of the standard deviation of the sampling distribution of the sample proportion is
   a. .016
   b. .025
   c. .029
   d. .035

9. A 90% confidence interval estimate for the true proportion of voters favoring the proposal is
    a. .38 to .56
    b. .401 to .539
    c. .413 to .527
    d. none of the above

10. In choosing a sample size for a public-opinion survey, what hypothesized value of the population proportion will lead to the largest sample size when the confidence level and error allowance are given?
    a. $p = .1$
    b. $p = .5$
    c. $p = .99$
    d. The confidence level must be known before an answer can be given.

Supplementary Exercises

31. Consider a population of six people, four of whom are college graduates.
    a. Identify the 15 different possible samples of four people that can be selected from this population.
    b. Compute the sample proportion of college graduates for each sample.
    c. Show the sampling distribution of $\bar{p}$ by showing the probability distribution of the values of $\bar{p}$ found in (b).

32. A survey of 350 teenagers, conducted in the spring of 1984, found Michael Jackson to be the favorite entertainer among teenagers. If 160 of these teenagers voted for Michael Jackson as their first choice, develop a 99% confidence interval for the proportion of all teenagers who would select Michael Jackson as their favorite entertainer.

33. Assume that 60% of the management staff in a large corporation has completed the company's special management-training program on improving communication skills. What is the probability that a sample of managers will provide a sample proportion $\bar{p}$ that is within $\pm.05$ of the population proportion of $p = .60$? Answer this question for sample sizes of 30, 60, and 100. What is the advantage of the larger sample size?

34. Assume that 15% of the items produced in an assembly-line operation are defective but that the firm's production manager is not aware of this situation. Assume further that 50 parts are tested by the quality assurance department in order to determine the quality of the assembly operation. Let $\bar{p}$ be the sample proportion defective found by the quality assurance test.
    a. Show the sampling distribution for $\bar{p}$.
    b. What is the probability that the sample proportion will be within $\pm.03$ of the population proportion defective?

35. A market research firm conducts telephone surveys with a 40% historical response rate. What is the probability that in a new sample of 400 telephone numbers at least 150 individuals will cooperate and respond to the questions? In

other words, what is the probability of a sample proportion $\bar{p} \geq 150/400 = .375$?

36. A production run is not acceptable for shipment to customers if a sample of 100 items contains at least 5% defective. If a production run has a population proportion defective of $p = .10$, what is the probability that $\bar{p}$ will be at least .05?

37. H. G. Forester and Company is a distributor of lumber supplies throughout the southwest United States. Management of H. G. Forester would like to check a shipment of over one million pine boards in order to determine if excessive warpage exists for the boards. A sample of 50 boards resulted in the identification of 7 boards with excessive warpage. With these data, develop a 95% confidence interval estimate of the proportion of boards defective in the whole shipment.

38. Consider the H. G. Forester and Company problem represented in Exercise 37. How large a sample would be required to estimate the proportion of boards with warpage to within $\pm .01$ at a 95% confidence level?

39. The Tourism Institute for the State of Florida plans to sample visitors at major beaches throughout the state in order to estimate the proportion of beach visitors that are residents of states other than Florida. Preliminary estimates are that 55% of the beach visitors are not residents of Florida.

    a. How large a sample should be taken to estimate the proportion of out-of-state visitors to within $\pm 3\%$ of the actual value at a 95% confidence level?

    b. How large a sample should be taken if the acceptable error is increased to $\pm 6\%$?

40. A fast-food restaurant plans to initiate a special offer, which will enable customers to purchase specially designed drink glasses featuring well-known cartoon characters. If 15% or fewer of the customers will purchase the glasses, the special offer will not be initiated. If a preliminary test is set up at several locations and 88 of 500 customers purchase the glasses, should the special glass offer be made? Test the hypothesis that will support your decision. Use a .01 level of significance. What is your recommendation?

41. It has been hypothesized that 5% of the students at one college make blood donations during a given year. If, in a random sample, 10 of 250 students have given blood during the past year, test $H_0{:}p = .05$ versus $H_1{:}p \neq .05$. Use a .05 level of significance in reaching your conclusion. What is the p value?

42. It has been hypothesized that at least 90% of juvenile first-time criminals are given probation upon the admission of guilt. Test this hypothesis at a .02 level of significance, if a sample of 92 juvenile criminal convictions shows 78 juveniles receiving probation. What is the p value?

Computer Exercise

One of the critical issues facing the Congress of the United States deals with whether or not budget cuts should be made in the area of federal aid to higher education. In order to learn about the views of constituents, a congresswoman from an eastern state requested a survey be conducted to obtain a more accurate picture of public opinion on the issue. Part of the data collected during the survey showed whether or not the respondent's household had college-age children and whether or not the respondent opposed budget cuts in federal aid to higher education. Results for the first 70 responses received during the survey are as follows:

Household with Children	Opposes Budget Cuts	Household with Children	Opposes Budget Cuts
Yes	Yes	No	No
Yes	Yes	No	Yes
No	No	Yes	Yes
No	Yes	No	Yes
Yes	No	No	No
Yes	No	Yes	Yes
No	No	No	No
No	Yes	Yes	No
No	No	Yes	Yes
No	Yes	No	Yes
No	No	No	No
Yes	Yes	Yes	Yes
Yes	No	Yes	Yes
No	Yes	Yes	Yes
Yes	Yes	Yes	No
No	Yes	No	No
Yes	No	Yes	No
No	No	Yes	No
No	No	No	Yes
No	Yes	Yes	No
Yes	Yes	Yes	Yes
Yes	Yes	No	No
No	Yes	Yes	No
No	No	No	Yes
No	No	No	Yes
Yes	Yes	Yes	Yes
No	No	No	No
Yes	No	No	No
Yes	No	No	No
No	No	Yes	Yes
No	Yes	Yes	Yes
No	No	Yes	Yes
No	No	No	No
Yes	Yes	Yes	Yes
No	Yes	Yes	No

QUESTIONS

1. Use descriptive statistics and a tabular format to summarize the data.
2. Provide a 95% confidence interval for the proportion in the population that have households with college-age children.

3. Provide a 95% confidence interval for the proportion in the population that oppose budget cuts in the federal-aid program for higher education.

4. Provide 95% confidence intervals for the proportion opposing budget cuts for the population of households with college-age children and the proportion opposing budget cuts for the population of households without college-age children. Does there appear to be a difference between the opinions of the households with and without college-age children? Comment on any differences observed.

CHAPTER 11

Inferences About Means and Proportions with Two Populations

What You Will Learn in This Chapter

- how to construct interval estimates and conduct hypothesis tests about the difference between the means of two populations

- when and how to use the t distribution to conduct inferences about the difference between the means of two populations

- the difference between independent and matched samples

- how to compute a pooled variance estimate

- how to construct interval estimates and conduct hypothesis tests about the difference between the proportions of two populations

Contents

369

Statistics in the News

MEN VERSUS WOMEN: THE PAY DIFFERENTIAL CONTINUES

More and more women are working outside the home, but odds are that they are making less money than their male counterparts. As a result of the growing concern over the persistence of earning differences between men and women, policymakers, researchers, and others have become increasingly interested in obtaining earnings data for both the male population and the female population. Wide-ranging information on these two populations is collected routinely through the federal government's Current Population Survey (CPS).

Not surprisingly, the most highly paid occupations for men are in the professional and managerial fields. Engineers, economists, lawyers, physicians, health administrators, school administrators, and financial managers all have jobs that rank in the 20 top-paying occupations for men. Using 1981 CPS data, these occupations paid men median incomes of from $507 to $619 per week. Much like the situation for men, the most highly paid occupations for women are in the professional and managerial fields. However, while the 20-top paying occupations for women were similar to the top 20 occupations for men, these occupations paid women median incomes ranging from only $318 to $422 per week.

It is interesting to note that the highly paid occupations for women are in the male-dominated professional and managerial fields. The occupations that are female-dominated (e.g., 90% of the clerical workers are female) are not included in

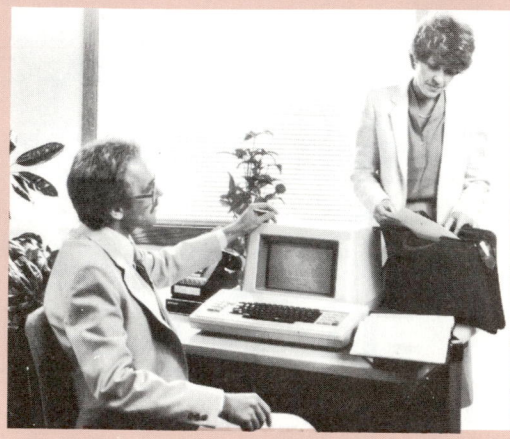

Men and women in similar jobs may still experience a salary differential with men more often making a higher salary.

the top-paying occupations for either men or women.

On an annual basis, the median income for the population of men is estimated to be $19,000, compared to a median income of $11,500 for the population of women. This is a pay differential for the two populations of $7500 per year. For individuals with college degrees, the median incomes for men ($24,400) and women ($15,100) show an estimated pay differential of $9300 per year. Considering median income levels for the two populations, women are currently earning at a rate of about 60% of that for men.

Based on "Earnings of Men and Women: A Look At Specific Occupations," *Monthly Labor Review* (April 1982).

In the preceding three chapters we have developed statistical methodology for interval estimation and hypothesis tests for population means and population proportions. However, the statistical procedures we have discussed thus far have considered only single-population situations. In this chapter we expand our discussion to cases where *two populations* are involved. Specifically, we will be selecting random samples and performing statistical analyses that will enable us to draw conclusions about the difference between the means and/or the proportions for two populations. An example of this type of situation appeared in "Statistics in the News" where pay differentials between the population of men and the population of women were reported. We begin the study of two-population situations by showing how to estimate the difference between the means of two populations.

11.1 ESTIMATION OF THE DIFFERENCE BETWEEN THE MEANS OF TWO POPULATIONS—INDEPENDENT SAMPLES

In some statistical applications we are faced with two populations where the difference between the means of the two populations is of prime importance. We know from Chapters 9 and 10 that we can take a simple random sample from a single population and use the sample mean $\bar{x}$ as an estimator of the population mean. In the two-population case we will select two independent simple random samples, one from

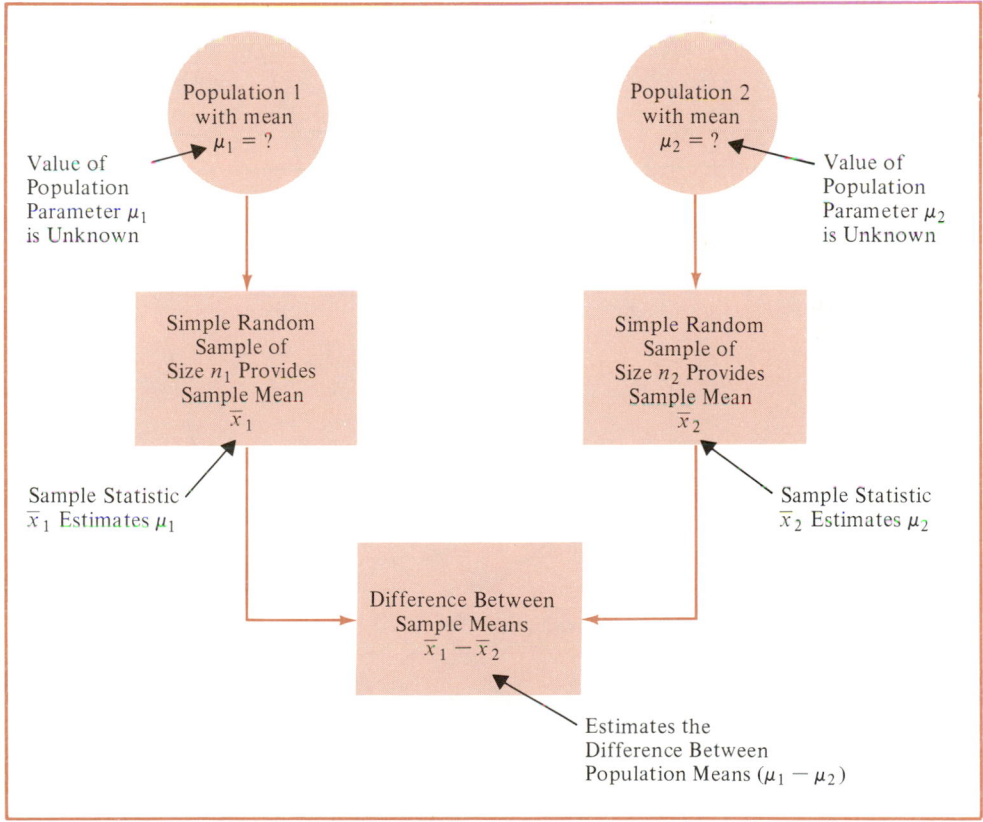

FIGURE 11.1

Statistical Process of Using the Difference Between Sample Means to Estimate the Difference Between Population Means

population 1 and another from population 2. Let

μ_1 = mean of population 1

μ_2 = mean of population 2

$\bar{x}_1$ = sample mean of the simple random sample taken from population 1 (i.e., the estimator of μ_1)

$\bar{x}_2$ = sample mean of the simple random sample taken from population 2 (i.e., the estimator of μ_2)

The difference between the two population means is given by $\mu_1 - \mu_2$. A point estimator of this difference is given by $\bar{x}_1 - \bar{x}_2$. Thus we see that an estimate of the difference between two population means is given by the difference between the two sample means. The situation is depicted in Figure 11.1.

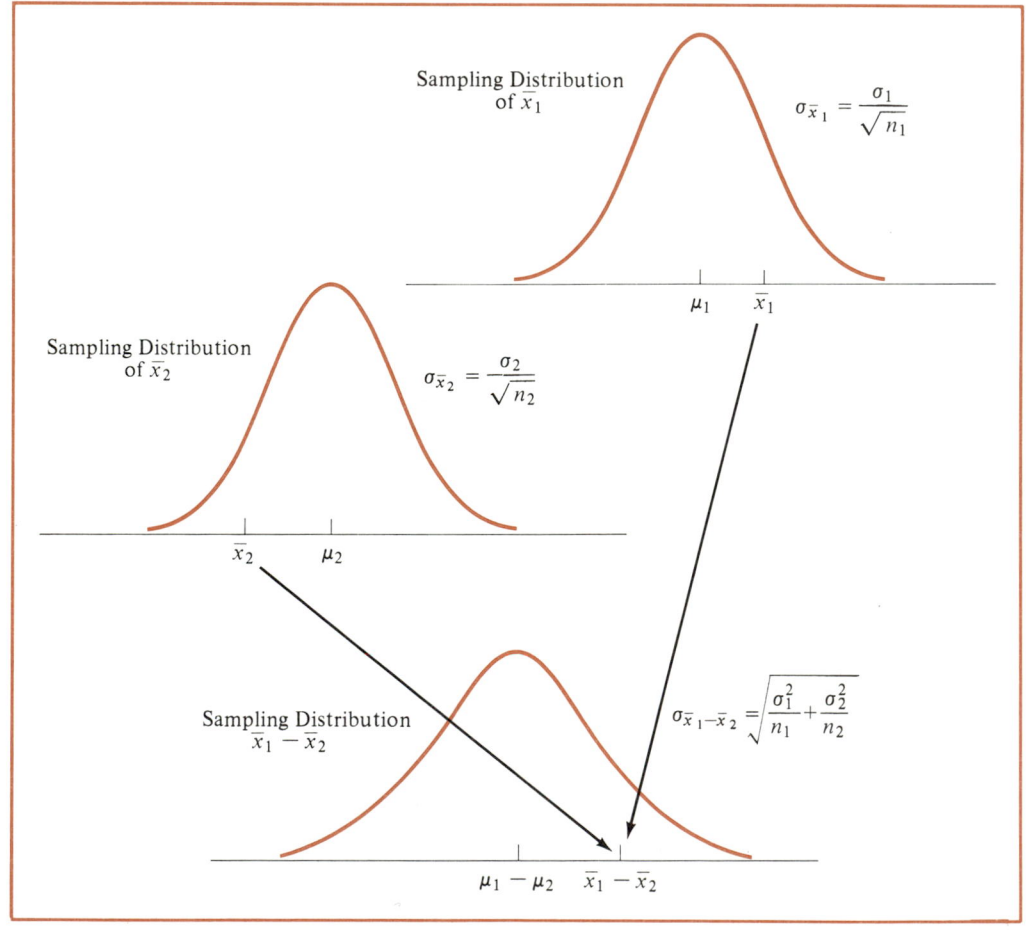

FIGURE 11.2

Sampling Distribution of $\bar{x}_1 - \bar{x}_2$ and Its Relationship to the Individual Sampling Distributions of $\bar{x}_1$ and $\bar{x}_2$

Sampling Distribution of $\bar{x}_1 - \bar{x}_2$

In statistical inferences about the difference between the means of two populations, $\bar{x}_1 - \bar{x}_2$ is the statistic of interest. In Chapter 8 we showed that different simple random samples taken from the same population provide a variety of different values for the sample mean. Accordingly, we can anticipate that both $\bar{x}_1$ and $\bar{x}_2$ will take on different values for different simple random samples. Thus $\bar{x}_1 - \bar{x}_2$, just like other sample statistics, is subject to variability. The probability distribution showing all possible values of $\bar{x}_1 - \bar{x}_2$ is called the *sampling distribution of $\bar{x}_1 - \bar{x}_2$.*

If we can identify the properties of the sampling distribution of $\bar{x}_1 - \bar{x}_2$, we can develop interval estimates and conduct hypothesis tests about $\mu_1 - \mu_2$ in much the same way we used the sampling distribution of $\bar{x}$ to make inferences about the single population mean μ. The properties of the sampling distribution of $\bar{x}_1 - \bar{x}_2$ are as follows:

Sampling Distribution of $\bar{x}_1 - \bar{x}_2$

Expected Value:
$$E(\bar{x}_1 - \bar{x}_2) = \mu_1 - \mu_2 \tag{11.1}$$

Standard Deviation:
$$\sigma_{\bar{x}_1 - \bar{x}_2} = \sqrt{\frac{\sigma_1^2}{n_1} + \frac{\sigma_2^2}{n_2}} \tag{11.2}$$

where

σ_1 = standard deviation of population 1

σ_2 = standard deviation of population 2

n_1 = sample size for the simple random sample selected from population 1

n_2 = sample size for the simple random sample selected from population 2

Distribution Form: Based on the central limit theorem, if both sample sizes are large ($n_1 \geq 30$ and $n_2 \geq 30$), the sampling distribution of $\bar{x}_1 - \bar{x}_2$ can be approximated by a normal probability distribution.

Figure 11.2 shows the sampling distribution of $\bar{x}_1 - \bar{x}_2$ and its relationship to the individual sampling distributions of $\bar{x}_1$ and $\bar{x}_2$. We will use the sampling distribution of $\bar{x}_1 - \bar{x}_2$ to make inferences about the difference between the means of two populations. We first show the method for large sample sizes (both $n_1 \geq 30$ and $n_2 \geq 30$). Then we show the method for situations where one or both samples are small ($n_1 < 30$ and/or $n_2 < 30$).

Interval Estimation of $\mu_1 - \mu_2$: Large Sample Case

Using the known properties of the sampling distribution of $\bar{x}_1 - \bar{x}_2$, we can follow the same procedure for interval estimation that we used in previous chapters. Using

$(1 - \alpha)$ as the confidence coefficient, the general statement for a confidence interval estimate of the difference between two population means is as follows:

Interval Estimation of the Difference Between the Means of Two Populations—(Large-Sample Case with $n_1 \geq 30$ and $n_2 \geq 30$)

$$\bar{x}_1 - \bar{x}_2 \pm z_{\alpha/2} \sqrt{\frac{\sigma_1^2}{n_1} + \frac{\sigma_2^2}{n_2}} \qquad (11.3)$$

If the two population variances σ_1^2 and σ_2^2 are unknown, the sample variances s_1^2 and s_2^2 can be substituted for the population variances in (11.3) to provide the interval estimate.

EXAMPLE 11.1

In a national study of the effects of a college education on the wage-earning potential of women, samples were collected from the following two populations:

Population 1: Women who graduated from high school and later obtained a bachelor's degree in college

Population 2: Women who graduated from high school and did not attend college

With μ_1 denoting the mean hourly wage of women who obtained a college degree (population 1) and μ_2 denoting the mean hourly wage of women who did not attend college (population 2), an objective of the study was to estimate the difference between the two population means 7 years after graduation from high school.

A sample of 200 women from population 1 provided a sample mean of $\bar{x}_1 = \$6.06$ per hour and a sample standard deviation of $s_1 = \$1.20$. A sample of 250 women from population 2 provided a sample mean of $\bar{x}_2 = \$4.57$ per hour and a sample standard deviation of $s_2 = \$.75$. Using a 95% confidence interval, estimate the difference between the mean hourly wages of the population of women who obtained a college degree and the population of women who did not attend college.

The point estimate of the difference in mean hourly wages is $\bar{x}_1 - \bar{x}_2 = \$6.06 - \$4.57 = \$1.49$ per hour. Thus, we estimate that women with a college degree earn an average of $1.49 more per hour 7 years after high school graduation than women who did not attend college. The 95% confidence interval estimate for the difference in mean wages can be computed using (11.3):

$$\bar{x}_1 - \bar{x}_2 \pm z_{\alpha/2} \sqrt{\frac{\sigma_1^2}{n_1} + \frac{\sigma_2^2}{n_2}} = 6.06 - 4.57 \pm 1.96 \sqrt{\frac{(1.20)^2}{200} + \frac{(.75)^2}{250}}$$

$$= 1.49 \pm 1.96(.097)$$

$$= 1.49 \pm .19$$

Thus an interval estimate of $1.49 \pm .19$ or 1.30 to 1.68 tells us we can be 95% confident that the mean hourly wage of women with college degrees is between $1.30 and $1.68 greater than the mean hourly wage of women who did not attend college. Using $1.49 as the point estimate of the difference in hourly wages and assuming 40 hours per week for 52 weeks a year, the earning power of the college women averages $(\$1.49)(40)(52) = \$3099$ more per year.

· · ·

EXAMPLE 11.2

A firm with department stores in Atlanta, Georgia has some stores located in the inner city and some stores located in suburban shopping centers. In a study designed to learn about the different characteristics of the inner-city and suburban customer populations, a sample of 60 inner-city customers and a sample of 80 suburban customers was taken. The data obtained on customer ages is summarized below:

Store Type	Sample Size	Sample Mean Age	Sample Standard Deviation
Inner city	60	$\bar{x}_1 = 40$ years	$s_1 = 9$ years
Suburban	80	$\bar{x}_2 = 35$ years	$s_2 = 10$ years

Using 90% and 99% confidence intervals, estimate the difference between the mean ages of the two populations of customers.

At 90%, $z_{\alpha/2} = z_{.05} = 1.645$. Using (11.3), we have

$$40 - 35 \pm 1.645 \sqrt{\frac{(9)^2}{60} + \frac{(10)^2}{80}} = 5 \pm 1.645(1.61)$$

$$= 5 \pm 2.65$$

Thus, we are 90% confident that the difference between the mean ages of the two populations is between 2.35 and 7.65 years with the inner-city stores, on the average, having the older customers.

At 99%, $z_{\alpha/2} = z_{.005} = 2.58$. Using (11.3), we have

$$40 - 35 \pm 2.58 \sqrt{\frac{(9)^2}{60} + \frac{(10)^2}{80}} = 5 \pm 2.58(1.61)$$

$$= 5 \pm 4.15$$

Hence, the 99% confidence interval for the difference between the mean ages is .85 to 9.15 years. As expected, the interval estimate corresponding to a 99% confidence level results in a wider interval than the interval estimate with a 90% confidence level.

· · ·

Interval Estimation of $\mu_1 - \mu_2$: Small Sample Case

The method we have been discussing for interval estimation of the difference between two population means has used large samples ($n_1 \geq 30$ and $n_2 \geq 30$). As a result the central limit theorem enables us to conclude that the sampling distribution of $\bar{x}_1 - \bar{x}_2$ can be approximated by a normal probability distribution. Thus $z_{\alpha/2}$ is used in (11.3) to compute the interval estimate of $\mu_1 - \mu_2$.

If either or both of the sample sizes are small ($n_1 < 30$ and/or $n_2 < 30$), the central limit theorem cannot be used to conclude that the sampling distribution of $\bar{x}_1 - \bar{x}_2$ can be approximated by a normal distribution. However, as we saw in Chapter 9, whenever the population has a normal probability distribution and the sample standard deviation is used to estimate the population standard deviation, the t distribution can be used to make inferences about a population mean.

In order to use the t distribution to develop interval estimates of the difference between two population means, the following two assumptions must be satisfied.

1. Both populations must have normal probability distributions.
2. The variances of the two populations must be equal, i.e., $\sigma_1^2 = \sigma_2^2 = \sigma^2$, where σ^2 is the variance for both populations.

Because of the equal variance assumption, the expression for the standard deviation of $\bar{x}_1 - \bar{x}_2$ as given by (11.2) can be written

$$\sigma_{\bar{x}_1 - \bar{x}_2} = \sqrt{\frac{\sigma^2}{n_1} + \frac{\sigma^2}{n_2}} = \sqrt{\sigma^2 \left(\frac{1}{n_1} + \frac{1}{n_2}\right)} \qquad (11.4)$$

Generally, the value of σ^2 will be unknown. However, since the two sample variances s_1^2 and s_2^2 both provide estimates of σ^2, we can *combine* s_1^2 and s_2^2 to obtain an estimate of σ^2. The process of combining the results of the two independent samples to provide one estimate of σ^2 is referred to as *pooling*. The *pooled variance estimate* of σ^2 is as follows.

Pooled Variance Estimate

$$s_p^2 = \frac{(n_1 - 1)s_1^2 + (n_2 - 1)s_2^2}{(n_1 + n_2 - 2)} \qquad (11.5)$$

Substituting s_p^2 for σ^2 in (11.4), we obtain the following estimate of the standard deviation of $\bar{x}_1 - \bar{x}_2$:

$$s_{\bar{x}_1 - \bar{x}_2} = \sqrt{s_p^2 \left(\frac{1}{n_1} + \frac{1}{n_2}\right)} \qquad (11.6)$$

The notation $s_{\bar{x}_1 - \bar{x}_2}$ is used to indicate that (11.6) provides an estimate of $\sigma_{\bar{x}_1 - \bar{x}_2}$.

With normally distributed populations and with the common variance σ^2, estimated by s_p^2, the t distribution can be used to compute a confidence interval for the difference between the means of two populations. The confidence interval for this difference is given by the following expression.

> **Interval Estimation of the Difference Between the Means of Two Populations—(Small-Sample Case with $n_1 < 30$ and/or $n_2 < 30$)**
>
> $$\bar{x}_1 - \bar{x}_2 \pm t_{\alpha/2} \sqrt{s_p^2 \left(\frac{1}{n_1} + \frac{1}{n_2} \right)} \qquad (11.7)$$
>
> where the t value is based on a t distribution with $n_1 + n_2 - 2$ degrees of freedom.

The $t_{\alpha/2}$ values in (11.7) are found in the table for the t distribution (see Table 2 in Appendix B). As we saw in Chapter 9, $t_{\alpha/2}$ cuts off an area of $\alpha/2$ in the upper tail of the t distribution and thus corresponds to an interval with a confidence coefficient of $1 - \alpha$.

EXAMPLE 11.3

An urban-planning group is interested in estimating the difference between the mean household incomes for two neighborhoods in a large metropolitan area. Independent random samples of households in the neighborhoods provided the following results:

	Neighborhood 1	Neighborhood 2
Sample size	$n_1 = 8$ households	$n_2 = 12$ households
Sample mean	$\bar{x}_1 = \$15,700$	$\bar{x}_2 = \$13,500$
Sample standard deviation	$s_1 = \$700$	$s_2 = \$850$

The point estimate of the difference in mean household incomes for the two neighborhoods is

$$\bar{x}_1 - \bar{x}_2 = \$15,700 - \$13,500 = \$2,200$$

Making the assumptions that the incomes are normally distributed in both neighborhoods and that the population variances are equal (11.7) can be used to develop a 95% confidence interval for the difference between the mean incomes in the two neighborhoods. First, we use (11.5) to compute the pooled estimate of σ^2. This computation is as follows:

$$s_p^2 = \frac{(n_1 - 1)s_1^2 + (n_2 - 1)s_2^2}{n_1 + n_2 - 2}$$

$$= \frac{(8 - 1)(700)^2 + (12 - 1)(850)^2}{8 + 12 - 2}$$

$$= \frac{7(490,000) + 11(722,500)}{18} = 632,083$$

With $n_1 + n_2 - 2 = 18$, we use the t distribution with 18 degrees of freedom to find $t_{.025} = 2.101$. (See Table 2 in Appendix B.) Thus, using (11.7), we obtain the 95% confidence interval of

$$\bar{x}_1 - \bar{x}_2 \pm t_{.025} \sqrt{s_p^2 \left(\frac{1}{n_1} + \frac{1}{n_2} \right)} = 2200 \pm (2.101) \sqrt{(632{,}083)\left(\frac{1}{8} + \frac{1}{12} \right)}$$

$$= 2200 \pm (2.101)(362.883)$$

$$= 2200 \pm 762.42$$

Subtracting and adding 762.42 to the estimate of 2200 provides the 95% confidence interval of $1437.58 to $2962.42. Thus we are 95% confident that the difference between the mean household incomes of the two neighborhoods is in this interval.

• • •

EXAMPLE 11.4

A sociology class project involves the study of dating practices on a major college campus. One aspect of the study compared the frequency of dating by freshmen women and freshmen men. A sample of 15 freshman women showed a mean of 8.2 dates per month with a standard deviation of 2.5. A sample of 10 freshman men showed a mean of 6.2 dates per month with a standard deviation of 2.2. What is the 90% confidence interval estimate of the difference between the mean number of dates per month for freshmen women and freshmen men?

With the small-sample case, we make the assumptions that the populations are normally distributed with equal variances. The pooled estimate of the common variance is

$$s_p^2 = \frac{(n_1 - 1)s_1^2 + (n_2 - 1)s_2^2}{n_1 + n_2 - 2}$$

$$= \frac{(14)(2.5)^2 + 9(2.2)^2}{15 + 10 - 2} = 5.698$$

At 90% confidence with $n_1 + n_2 - 2 = 15 + 10 - 2 = 23$ degrees of freedom, the t distribution table shows $t_{.05} = 1.714$. Thus we have

$$\bar{x}_1 - \bar{x}_2 \pm t_{.05} \sqrt{s_p^2 \left(\frac{1}{n_1} + \frac{1}{n_2} \right)} = 8.2 - 6.2 \pm (1.714) \sqrt{5.698 \left(\frac{1}{15} + \frac{1}{10} \right)}$$

$$= 2.0 \pm 1.714(.975)$$

$$= 2.0 \pm 1.67$$

Thus we are 90% confident that freshman women students average from .33 to 3.67 more dates per month than do freshman men.

• • •

As a final comment, in Chapter 9 we pointed out that the t distribution is not restricted to the small-sample situation. Anytime we are interested in the difference between two population means and the populations are normally distributed with equal variances, the t distribution can be used to develop the appropriate confidence interval. However, (11.3) shows how to compute confidence intervals when the sample sizes are large. In this situation the use of the t distribution and its corresponding assumptions are not required. Thus, we do not need to refer to the t distribution unless we have a small-sample-size situation.

EXERCISES

1. In the evaluation of an eighth-grade reading comprehension program, a standardized test is to be given to a sample of eighth graders and to a sample of seventh graders from the same school. Let the population of eighth graders be denoted as population 1, with test scores having characteristics $\mu_1 = 77$ and $\sigma_1 = 12$. Let the population of seventh graders be denoted as population 2, with test scores having characteristics $\mu_2 = 72$ and $\sigma_2 = 14$. Answer the following questions if a random sample of 40 eighth graders and a random sample of 40 seventh graders will be used in the study.
   a. Show the sampling distribution of $\bar{x}_1 - \bar{x}_2$.
   b. What is the probability that the eighth graders in the study will show a mean test score between 3 and 7 points greater than the seventh graders?
   c. What is the probability of $\bar{x}_1 - \bar{x}_2$ being negative, which would mean that the seventh graders did better on the test than the eighth graders?

2. A college admissions board is interested in estimating the difference between the mean grade point averages of students from two high schools. Independent simple random samples of students at the two high schools provide the following results.

Mt. Washington	Country Day
$n_1 = 46$	$n_2 = 33$
$\bar{x}_1 = 3.02$	$\bar{x}_2 = 2.72$
$s_1 = .38$	$s_2 = .45$

   a. What is the point estimate of the difference between the means of the two populations?
   b. Develop an interval estimate of the difference between the two population means with a confidence coefficient of .90.
   c. Answer (b) using a .95 confidence coefficient.

3. The Butler County Bank and Trust Company is interested in estimating the difference between the mean credit card balances at two of its branch banks. Independent samples of credit card customers provide the following results.

Branch 1	Branch 2
$n_1 = 32$	$n_2 = 36$
$\bar{x}_1 = \$500$	$\bar{x}_2 = \$375$
$s_1 = \$150$	$s_2 = \$130$

    a. Develop a point estimate of the difference between the mean balances at the two branches.

    b. Develop an interval estimate of the difference between the mean balances. Use a confidence coefficient of .99.

4. Starting annual salaries for individuals with master's and bachelor's degrees were obtained from two independent random samples. Use the data shown below to provide a 90% confidence interval estimate of the increase in mean starting salary that can be expected upon completion of the master's degree.

Master's Degree	Bachelor's Degree
$n_1 = 60$	$n_2 = 80$
$\bar{x}_1 = \$21,000$	$\bar{x}_2 = \$19,000$
$s_1 = \$\ 2,500$	$s_2 = \$\ 2,000$

5. A sample of 15 recently released prisoners who had convictions for armed robbery showed that the time in prison had a sample mean of 5.2 years, with a sample standard deviation of 1.4 years. A sample of 12 recently released prisoners who had convictions for assault showed a sample mean time in prison of 2.7 years, with a sample standard deviation of 1.1 years.

    a. What is the pooled estimate of the population variance σ^2?

    b. Using a 95% confidence interval, estimate the difference in mean time in prison for armed robbery and assault convictions.

11.2 HYPOTHESIS TESTS ABOUT THE DIFFERENCE BETWEEN THE MEANS OF TWO POPULATIONS— INDEPENDENT SAMPLES

In this section we continue our discussion of inferences about the difference between the means of two populations based upon two independent simple random samples. We consider the statistical procedure of using the sample information to conduct hypothesis tests about the difference between the population means.

Hypothesis tests about the difference between the means of two populations use the same logic as the hypothesis tests in Chapters 9 and 10, except that the tests are based on the sample statistic $\bar{x}_1 - \bar{x}_2$. Thus we use the sampling distribution of $\bar{x}_1 - \bar{x}_2$

to conduct the test. In the large-sample case, the sampling distribution can be approximated by a normal probability distribution. Thus the following quantity is a standard normal random variable z and is the test statistic for the hypothesis test.

Observed Value of $\bar{x}_1 - \bar{x}_2$ 　　　　　　　　 Hypothesized Value of $\mu_1 - \mu_2$

$$z = \frac{(\bar{x}_1 - \bar{x}_2) - (\mu_1 - \mu_2)}{\sqrt{\dfrac{\sigma_1^2}{n_1} + \dfrac{\sigma_2^2}{n_2}}} \qquad (11.8)$$

Standard Deviation of $\bar{x}_1 - \bar{x}_2$.

In this large-sample case, if the population variances σ_1^2 and σ_2^2 are unknown, (11.8) can be used with the sample variances s_1^2 and s_2^2 substituted for σ_1^2 and σ_2^2.

EXAMPLE 11.5

A medical research study was conducted to determine if there is a difference between the effectiveness of two pain-relief medicines used for headaches. Over a 6-month period of time, a sample of individuals used one of the medicines, whereas another sample of individuals used the other medicine. Data collected during the study showed the time required to receive pain relief. Letting

$$\mu_1 = \text{mean pain-relief time for medicine 1}$$

$$\mu_2 = \text{mean pain-relief time for medicine 2}$$

the hypothesis test is expressed as follows:

$$H_0: \mu_1 - \mu_2 = 0$$

$$H_1: \mu_1 - \mu_2 \neq 0$$

If the null hypothesis is rejected, the test will have shown that the two medicines differ in terms of pain-relief speed.

Using the following data, conduct the test and draw a conclusion comparing the two medicines. Use $\alpha = .05$.

	Individuals Using Medicine 1	Individuals Using Medicine 2
Sample size	$n_1 = 248$	$n_2 = 225$
Sample mean	$\bar{x}_1 = 24.8$ minutes	$\bar{x}_2 = 26.1$ minutes
Sample standard deviation	$s_1 = 3.3$ minutes	$s_2 = 4.2$ minutes

With $\alpha = .05$ and a two-tailed test, the critical z values are -1.96 and $+1.96$. The decision rule can be stated as follows:

$$\text{Accept } H_0 \text{ If } -1.96 \le z \le 1.96$$

$$\text{Reject } H_0 \text{ Otherwise}$$

Using the test statistic of (11.8), we have

$$z = \frac{(\bar{x}_1 - \bar{x}_2) - (\mu_1 - \mu_2)}{\sqrt{\dfrac{\sigma_1^2}{n_1} + \dfrac{\sigma_2^2}{n_2}}} = \frac{(24.8 - 26.1) - 0}{\sqrt{\dfrac{(3.3)^2}{248} + \dfrac{(4.2)^2}{225}}}$$

$$= \frac{-1.3}{.35} = 3.71$$

With this value of z, we reject H_0. The conclusion is that there is a significant difference in the pain-relief effectiveness of the two medicines. Medicine 1 provides speedier pain relief.

Using the p value criteria and the table of areas in the standard normal probability distribution, $z = -3.71$ has an area of less than .001 in the tail of the distribution. For the two-tailed hypothesis test, the p value would be less than .001 $\times$ 2 = .002. Thus using the p value, we would reject H_0 at the .05 level of significance.

EXAMPLE 11.6

It has been suggested that college students learn more and obtain higher grades in small classes (40 students or less) when compared to large classes (150 students or more). To test this claim, a university assigned a professor to teach a small class and a large class of the same course. At the end of the course students from the two classes were given the same final exam. Final grade differences for the two classes would provide a basis for testing the difference between the small-class and large-class situations.

Letting μ_1 denote the mean exam score for the population of students taking a small class and μ_2 denote the mean exam score for the population of students taking a large class, the hypothesis test is as follows:

$$H_0: \mu_1 - \mu_2 = 0$$

$$H_1: \mu_1 - \mu_2 > 0$$

Rejecting H_0 will lead to the conclusion that the mean exam score is greater in the small class. However, if the test is unable to reject H_0 the conclusion will be that the small class does not show a significantly higher grade performance.

Viewing the students actually taking the courses as samples from the populations of students in small and large classes, the following data were obtained.

	Individuals Taking Small Class	Individuals Taking Large Class
Sample Size	$n_1 = 35$	$n_2 = 170$
Sample Mean Exam Score	$\bar{x}_1 = 74.2$	$\bar{x}_2 = 71.7$
Sample Standard Deviation	$s_1 = 14$	$s_2 = 13$

Test the hypothesis and draw a conclusion about the small and large classes based on the given data. Use $\alpha = .05$.

For the one-tailed test, the critical z value is 1.645. The corresponding decision rule becomes

$$\text{Accept } H_0 \text{ If } z \leq 1.645$$

$$\text{Reject } H_0 \text{ Otherwise}$$

The point estimate of $\mu_1 - \mu_2$ is given by $\bar{x}_1 - \bar{x}_2 = 74.2 - 71.7 = 2.5$; thus the small class group has a sample mean that is 2.5 points greater than that for the large class. Using (11.8), the value of the test statistic becomes

$$z = \frac{(\bar{x}_1 - \bar{x}_2) - (\mu_1 - \mu_2)}{\sqrt{\dfrac{\sigma_1^2}{n_1} + \dfrac{\sigma_2^2}{n_2}}} = \frac{(2.5) - 0}{\sqrt{\dfrac{14^2}{35} + \dfrac{13^2}{170}}}$$

$$= \frac{2.5}{2.57} = .97$$

Since $z < 1.645$, we are unable to reject H_0. As a result, we are unable to conclude that for this situation small classes enable students to obtain higher grades. With $z = .97$, the p value for the test result is $.5000 - .3340 = .1660$.

Our analysis of the data indicate that student performance on the final exam is essentially the same whether the course was taken in a small or large class. We must add, however, that these results were based upon an experiment with only one professor. While it does not appear that student performance improves in smaller classes with this professor, further testing should be done before arriving at the same conclusion for other professors.

● ● ●

In the preceding examples we have shown how hypothesis tests can be conducted for situations involving the means of two populations using *large* samples. Similar hypothesis tests can be made for the small-sample case ($n_1 < 30$ and/or $n_2 < 30$) by assuming that the populations have normal probability distributions with equal variances. In such cases the t distribution is used; the test statistic t is given by

$$t = \frac{(\bar{x}_1 - \bar{x}_2) - (\mu_1 - \mu_2)}{\sqrt{s_p^2 \left(\dfrac{1}{n_1} + \dfrac{1}{n_2}\right)}} \tag{11.9}$$

The t distribution corresponding to (11.9) has $n_1 + n_2 - 2$ degrees of freedom. Comparing (11.8) and (11.9), we note that the only difference is in the standard error term in the denominator. The form of the denominator in (11.9) provides an estimate of the standard deviation of $\bar{x}_1 - \bar{x}_2$ based on the pooled variance estimate of σ^2 denoted by s_p^2, where

$$s_p^2 = \frac{(n_1 - 1)s_1^2 + (n_2 - 1)s_2^2}{n_1 + n_2 - 2}$$

EXAMPLE 11.7

Automobile gasoline mileage tests were conducted for similar-sized foreign and domestic automobiles. Test the hypothesis that the mean number of miles per gallon is the same for foreign and domestic automobiles based on the following sample results. Use $\alpha = .05$.

	Foreign Automobiles	Domestic Automobiles
Sample size	$n_1 = 8$	$n_2 = 10$
Sample mean	$\bar{x}_1 = 36.5$	$\bar{x}_2 = 32.4$
Sample standard deviation	$s_1 = 2.3$	$s_2 = 2.8$

The null and alternative hypotheses for this test are

$$H_0: \mu_1 - \mu_2 = 0$$

$$H_1: \mu_1 - \mu_2 \neq 0$$

With $n_1 + n_2 - 2 = 8 + 10 - 2 = 16$ degrees of freedom, the table for the t distribution shows $t_{.025} = 2.12$. The decision rule for the hypothesis test is

$$\text{Accept } H_0 \text{ If } -2.12 \leq t \leq 2.12$$

$$\text{Reject } H_0 \text{ Otherwise}$$

The calculations made with the sample data are as follows:

$$s_p^2 = \frac{(n_1 - 1)s_1^2 + (n_2 - 1)s_2^2}{n_1 + n_2 - 2} = \frac{7(2.3)^2 + 9(2.8)^2}{8 + 10 - 2} = 6.72$$

Then using (11.9) as the test statistic we have

$$t = \frac{(36.5 - 32.4) - 0}{\sqrt{6.72\left(\dfrac{1}{8} + \dfrac{1}{10}\right)}} = \frac{4.1}{1.23} = 3.33$$

Since $3.33 > 2.12$, we reject H_0 and conclude that there is a significant difference between the mean number of miles per gallon achieved by foreign and domestic automobiles.

• • •

EXERCISES

6. Sample weights of babies (in pounds) born in two different countries show the following:

Country 1	Country 2
$\bar{x}_1 = 7.1$	$\bar{x}_2 = 6.5$
$s_1 = .7$	$s_2 = .4$
$n_1 = 125$	$n_2 = 100$

Using a .05 level of significance, can these data support the conclusion that there is a difference in the weights of babies born in the two countries? What is the p value?

7. In a wage discrimination case involving male and female employees, independent samples of male and female employees with 5 years or more experience show the following hourly wage results:

Male Employees	Female Employees
$n_1 = 44$	$n_2 = 32$
$\bar{x}_1 = \$6.25$	$\bar{x}_2 = \$5.70$
$s_1 = \$1.00$	$s_2 = \$.80$

The null hypothesis is stated such that male employees have a mean hourly wage less than or equal to that of the female employees. Rejection of H_0 leads to the conclusion that male employees have a mean hourly wage exceeding the female employee wages. Test the hypothesis with $\alpha = .01$. Does wage discrimination appear to exist in this case?

8. Safegate Foods, Inc. is redesigning the check-out lanes in its supermarkets throughout the country. Two designs have been suggested. Tests on customer

check-out times (in minutes) have been collected at two stores where the two new systems have been installed. The sample data are as follows:

Times for Check-out System A	Times for Check-out System B
$n_1 = 120$	$n_2 = 100$
$\bar{x}_1 = 4.1$	$\bar{x}_2 = 3.3$
$s_1 = 2.2$	$s_2 = 1.5$

Test at the .05 level of significance to determine if there is a difference in the mean check-out times for the two systems. Which system is preferred? What is the p value?

9. Samples of final examination scores for two statistics classes with different instructors showed the following results:

Instructor A's Class	Instructor B's Class
$n_1 = 12$	$n_2 = 15$
$\bar{x}_1 = 72$	$\bar{x}_2 = 78$
$s_1 = 8$	$s_2 = 10$

With $\alpha = .05$, use the p value to test whether or not these data are sufficient to conclude that the mean grades differ for the two classes.

10. A firm is studying the delivery times (in days) for two raw material suppliers. The firm is basically satisfied with its current supplier, referred to as supplier A, and will stay with this supplier provided that the mean delivery times are the same as or less than those of supplier B. However, if the firm finds that the mean delivery times from supplier B are less than those of supplier A, it will begin making raw material purchases from supplier B.
    a. What are the null and alternative hypotheses for this situation?
    b. Assume that independent samples show the following delivery time characteristics for the two suppliers:

Supplier A	Supplier B
$n_1 = 50$	$n_2 = 30$
$\bar{x}_1 = 14$	$\bar{x}_2 = 12.5$
$s_1 = 3$	$s_2 = 2$

Show the sampling distribution of $\bar{x}_1 - \bar{x}_2$ for the hypothesis test.
    c. For $\alpha = .05$, what is the decision rule for the test?

d. What is your conclusion for the hypotheses from part (a)? What action do you recommend in terms of supplier selection?

11.3 INFERENCES ABOUT THE DIFFERENCE BETWEEN THE MEANS OF TWO POPULATIONS—MATCHED SAMPLES

Suppose that a manufacturing company has two methods available for employees to perform a certain production task. In order to maximize production output, the company would like to identify the method with the smallest mean completion time per unit. If a difference between mean completion times exists, the company will implement the method with the smaller mean completion time. If no difference between means can be detected, the choice between the two production methods will be based on a criterion other than completion time. The hypotheses to be tested are stated as follows:

Hypothesis	Conclusion
H_0: $\mu_1 - \mu_2 = 0$	No difference exists between the mean completion times of the two methods
H_1: $\mu_1 - \mu_2 \neq 0$	A difference exists between the mean completion times; select the method with the smaller mean completion time

In designing the sampling procedure that will be used to collect production time data and test the above hypotheses, we consider two alternative designs. One is based on *independent samples,* and the other is based on *matched samples.* The designs are described as follows:

Independent-sample design—A random sample of workers is selected and uses method 1. A second and independent random sample of workers is selected and uses method 2. The test of the difference between means is based on the procedures of Section 11.2.

Matched-sample design—One random sample of workers is selected with the workers first using one method and then using the other method. The order of the two methods is assigned randomly to the workers, with some workers performing method 1 first and others performing method 2 first. Each worker provides a pair of data values, one value for method 1 and another value for method 2.

Our interest in the matched-sample design is that since both production methods are tested under similar conditions (i.e., same workers), this design often leads to a smaller sampling error than the independent sample design. The primary reason for this is that each worker in a matched sample design provides data first under one method and then under the other method. Thus variation between workers is eliminated as a source of the sampling error. This variation between workers cannot be eliminated when the independent-sample design is used.

The advantage of the matched-sample design is that it allows the experimenter to control some factors in order to obtain a better measure of others. In this situation, completion times are affected by two factors: method used and individual differences in worker speed. Only the effect of method used is of interest here, so the matched-sample design is useful in allowing us to eliminate (control for) the effect of differences in individual worker speed.

EXAMPLE 11.8

A matched-sample design was used for two production methods. A sample of six workers was taken; each worker performed the task for both production methods. The data on completion times for the six workers are as follows.

Worker	Completion Time, Method 1 (minutes)	Completion Time, Method 2 (minutes)	Difference in Completion Times (d_i)
1	6.0	5.4	.6
2	5.0	5.2	−.2
3	7.0	6.5	.5
4	6.2	5.9	.3
5	6.0	6.0	.0
6	6.4	5.8	.6

Note that each worker provides a pair of data values, one for each production method. Also note that the last column contains the difference (d_i) in completion times for each worker in the sample. For example, $d_i = .6$ shows that worker 1 required .6 minutes more time for method 1. A negative d_i value indicates worker 1 required more time for method 2.

The key to analyzing a matched-sample design is to use the difference data only. Doing so converts the situation to a single sample containing six values. Then the procedures introduced in Chapter 9 can be used to make an inference about the mean difference for the population of all possible differences.

Let μ_d = the mean of the *difference* values for the population of workers. With this notation the null and alternative hypotheses are rewritten as follows:

Hypothesis	Conclusion
$H_0: \mu_d = 0$	No difference exists in the mean completion time of the two methods
$H_1: \mu_d \neq 0$	A difference exists in the mean completion time; select the method with the smaller mean completion time

The sample mean and sample standard deviation for the six difference values are as follows:

$$\bar{d} = \frac{\Sigma d_i}{n} = \frac{[.6 + (-.2) + .5 + .3 + .0 + .6]}{6} = \frac{1.8}{6} = .3$$

$$s_d = \sqrt{\frac{\Sigma (d_i - \bar{d})^2}{n - 1}}$$

$$= \sqrt{\frac{(.3)^2 + (-.5)^2 + (.2)^2 + (0)^2 + (-.3)^2 + (.3)^2}{6 - 1}}$$

$$= \sqrt{\frac{.56}{5}} = .335$$

In Chapter 9, we found that if the population has a normal probability distribution and if s is used as an estimate of σ, the t distribution with $n - 1$ degrees of freedom can be used to test a hypothesis about the population mean. The test statistic t was given by

$$t = \frac{\bar{x} - \mu}{s/\sqrt{n}}$$

Assuming the population of difference values (d_i) has a normal probability distribution and using the sample standard deviation of differences, s_d, to estimate the population standard deviation, this same formula can be used to test the hypothesis $H_0: \mu_d = 0$. To indicate that we are using the d_i values of the matched-sample design, we write the test statistic t as follows:

$$t = \frac{\bar{d} - \mu_d}{s_d/\sqrt{n}} \tag{11.10}$$

With a sample of six workers (differences), we have $n - 1 = 5$ degrees of freedom. Using $\alpha = .05$, we find that $t_{.025} = 2.571$. The decision rule is to accept H_0 if $-2.571 \le t \le 2.571$; otherwise, H_0 is rejected. Using the sample results, we have

$$t = \frac{\bar{d} - \mu_d}{s_d/\sqrt{n}} = \frac{.3 - 0}{.335/\sqrt{6}} = 2.19$$

Since t is between -2.571 and 2.571, H_0 cannot be rejected. The sample data do not provide sufficient evidence to reject the assumption of no difference between the mean completion times of the two methods.

Using the sample results, we could also develop an interval estimate of the difference between the means of the two populations using the methodology of interval estimation for one population, as introduced in Chapter 9. Using this approach, we have

$$\bar{d} \pm t_{.025} \frac{s_d}{\sqrt{n}} = .3 \pm 2.571 \frac{.335}{\sqrt{6}} = .3 \pm .35 \tag{11.11}$$

Thus the 95% confidence interval estimate of the difference in the means of the two production methods is $-.05$ to $.65$ minutes.

• • •

In the last example, workers performed the production task using first one method and then the other method. This is an example of a matched-sample design, where each sampled item (worker) provides a pair of data values. Although this is often the procedure used in the matched-samples analysis, it is possible to use different but similar items to provide the pair of data values. In this sense, a worker at one location could be matched with a similar worker at another location (similarity based on age, education, sex, experience, etc.). The pairs of workers would provide the difference data that could be used in the matched-sample analysis.

Since a matched-sample procedure for inferences about two population means generally provides a better precision than the two-independent-samples approach, it is the recommended design. However, in some applications the matching cannot be achieved, or perhaps the time and cost associated with matching is excessive. In these cases the independent-sample design should be used.

The example presented in this section employed a sample size of six workers. As such, the small-sample case existed, and the t distribution was used in both the test of hypothesis and interval estimation computations. If the sample size is large ($n \geq 30$), the statistical computations can be based on the z values of the standard normal probability distribution.

EXERCISES

11. A manufacturer produces both a deluxe and a standard model automatic sander designed for home use. Selling prices obtained from a sample of retail outlets are as follows.

Retail Outlet	Price, Deluxe Model	Price, Standard Model
1	$39	$27
2	39	28
3	45	35
4	38	30
5	40	30
6	39	34
7	35	29

The manufacturer's suggested retail prices for the two models show a $10 difference in prices. Use a .05 level of significance and test to see if the mean difference between prices of the two models is at the $10 value. What is the 95% confidence interval estimate of the difference between mean prices for the two models?

12. Figure Perfect, Inc. is a women's figure salon that specializes in weight-reduction programs. Weights (pounds) for a sample of clients before and after a 6-week introductory program are as follows.

Client	Weight Before	Weight After
1	140	132
2	160	158
3	210	195
4	148	152
5	190	180
6	170	164

Use $\alpha = .05$ and test to determine if the introductory program provides a weight loss. What is the p value?

13. The pulse rates of patients before and after being given a certain tranquilizer are as follows.

Before	After
81	77
80	79
82	75
79	80
84	78
80	74

a. Use a .05 level of significance to test for the ability of the tranquilizer to reduce the pulse rate of the patients.

b. Provide a 95% confidence interval estimate of the mean decrease in pulse rate attributable to the tranquilizer.

14. Word processing systems are often justified on the basis of improved efficiencies for a secretarial staff. Given are typing rates in words per minute for seven secretaries who previously used electronic typewriters and who are now using computer-based word processors. Test at the .05 level of significance to see if there has been any change in the mean typing rate due to the word processor system.

Secretary	Electronic Typewriter	Word Processor
1	72	75
2	68	66
3	55	60
4	58	64
5	52	55
6	55	57
7	64	64

11.4 INFERENCES ABOUT THE DIFFERENCE BETWEEN THE PROPORTIONS OF TWO POPULATIONS

We now consider the case where two populations are involved and we are interested in making inferences about the difference between the proportions of the two populations. Let

p_1 = proportion for population 1

p_2 = proportion for population 2

$\bar{p}_1$ = sample proportion for a simple random sample taken from population 1 (i.e., the point estimator of p_1)

$\bar{p}_2$ = sample proportion for a simple random sample taken from population 2 (i.e., the point estimator of p_2)

The difference between the two population proportions is given by $p_1 - p_2$. The point estimator of this difference is:

$$\bar{p}_1 - \bar{p}_2 \tag{11.12}$$

Thus we see that the estimator of the difference between two population proportions is the difference between the two sample proportions.

Sampling Distribution of $\bar{p}_1 - \bar{p}_2$

In the study of the difference between two population proportions, $\bar{p}_1 - \bar{p}_2$ is the sample statistic of interest. As we have seen in several previous cases, the sampling distribution of the sample statistic is a key factor in developing confidence interval estimates and in testing hypotheses about the parameters of interest. The properties of

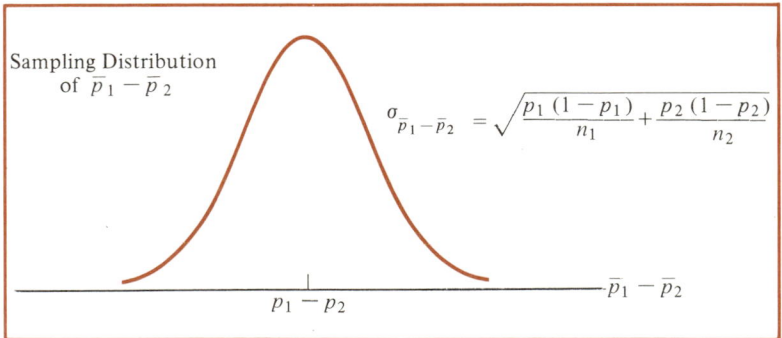

FIGURE 11.3
Sampling Distribution of $\bar{p}_1 - \bar{p}_2$

the sampling distribution of $\bar{p}_1 - \bar{p}_2$ are as follows:

Sampling Distribution of $\bar{p}_1 - \bar{p}_2$

Expected Value: $\qquad E(\bar{p}_1 - \bar{p}_2) = p_1 - p_2 \qquad\qquad$ (11.13)

Standard Deviation: $\qquad \sigma_{\bar{p}_1 - \bar{p}_2} = \sqrt{\dfrac{p_1(1 - p_1)}{n_1} + \dfrac{p_2(1 - p_2)}{n_2}} \qquad$ (11.14)

where

n_1 = sample size for the simple random sample selected from
population 1

n_2 = sample size for the simple random sample selected from
population 2

Distribution Form: Provided that the sample sizes are large (that is, $n_1 p_1$, $n_1(1 - p_1)$, $n_2 p_2$, and $n_2(1 - p_2)$ are all greater than or equal to 5), the sampling distribution of $\bar{p}_1 - \bar{p}_2$ can be approximated by a normal probability distribution.

Figure 11.3 shows the sampling distribution of $\bar{p}_1 - \bar{p}_2$.

Note that the formula for the standard deviation of $\bar{p}_1 - \bar{p}_2$ as given by (11.14) requires that we know the actual values for the population proportions p_1 and p_2. Since p_1 and p_2 are unknown, we cannot use (11.14) to compute $\sigma_{\bar{p}_1 - \bar{p}_2}$. However, using $\bar{p}_1$ as the estimate of p_1 and $\bar{p}_2$ as the estimate of p_2, we can estimate $\sigma_{\bar{p}_1 - \bar{p}_2}$ as follows:

Estimate of $\sigma_{\bar{p}_1 - \bar{p}_2}$

$$s_{\bar{p}_1 - \bar{p}_2} = \sqrt{\dfrac{\bar{p}_1(1 - \bar{p}_1)}{n_1} + \dfrac{\bar{p}_2(1 - \bar{p}_2)}{n_2}} \qquad (11.15)$$

Interval Estimation of $p_1 - p_2$

With the sampling distribution of $\bar{p}_1 - \bar{p}_2$ known and with $s_{\bar{p}_1 - \bar{p}_2}$ providing an estimate of $\sigma_{\bar{p}_1 - \bar{p}_2}$, we follow the same procedure for interval estimation that we have used previously. We have the following expression for a confidence interval:

Interval Estimation of the Difference Between the Proportions of Two Populations—(Applies with $n_1 p_1$, $n_1(1 - p_1)$, $n_2 p_2$, $n_2(1 - p_2)$ all greater than or equal 5)

$$\bar{p}_1 - \bar{p}_2 \pm z_{\alpha/2} \sqrt{\dfrac{\bar{p}_1(1 - \bar{p}_1)}{n_1} + \dfrac{\bar{p}_2(1 - \bar{p}_2)}{n_2}} \qquad (11.16)$$

Note that as we have done with previous interval estimations, we can change the $z_{\alpha/2}$ value to reflect different levels of confidence.

EXAMPLE 11.9

A firm that specializes in preparing tax returns for its clients is interested in comparing the quality of work at two of its regional offices. A random sample of tax returns prepared at each office is selected and verified for accuracy. Of concern to the firm is the proportion of erroneous returns prepared at each office. In particular, the firm would like an estimate of the difference between the proportions of erroneous returns prepared at the two offices.

Sample results showed that of 250 returns verified at office 1, 35 were in error and that of 300 returns verified at office 2, 27 were in error. Develop 95% and 90% confidence interval estimates for the difference between the two population proportions.

The sample proportions at the two offices are as follows.

$$\bar{p}_1 = \frac{35}{250} = .14$$

$$\bar{p}_2 = \frac{27}{300} = .09$$

Using (11.12), we estimate the difference between the proportion of erroneous tax returns for the two populations to be $\bar{p}_1 - \bar{p}_2 = .14 - .09 = .05$. Specifically, we are led to believe that office 1 possesses a 5% greater error rate than office 2. Using (11.16), the 95% confidence interval for the difference between the two population proportions becomes

$$\bar{p}_1 - \bar{p}_2 \pm z_{.025} \sqrt{\frac{\bar{p}_1(1 - \bar{p}_1)}{n_1} + \frac{\bar{p}_2(1 - \bar{p}_2)}{n_2}} =$$

$$.14 - .09 \pm 1.96 \sqrt{\frac{.14(1 - .14)}{250} + \frac{.09(1 - .09)}{300}} =$$

$$.05 \pm 1.96(.0275) = .05 \pm .054$$

Thus the 95% confidence interval estimate of the difference in error rates at the two offices is $-.004$ to $.104$.

The 90% confidence interval uses $z_{.05} = 1.645$. Thus, replacing the 1.96 with 1.645 in (11.16) provides the 90% confidence interval estimate. This substitution will result in a confidence interval of $.05 \pm 1.645(.0275)$, or $.05 \pm .045$. Thus we can be 90% confident that the difference in the error rates at the two offices is in the interval $.005$ to $.095$.

● ● ●

EXAMPLE 11.10

In a study of household television-viewing habits, individuals in sampled households were asked to participate by keeping a daily diary of television viewing. The letters requesting participation in the study were sent to two groups. The first group received a letter requesting participation and $1.00 as a token of appreciation for their participation. The second group received only the letter requesting participation in the study. Of 300 letters sent to the first group, 141 households participated in the study. Of 500 letters sent to the second group, 150 households participated in the study. Provide a 95% confidence interval estimate for the difference in participation rate that exists if $1.00 is included with the request for participation.

Using the sample results, we have

$$\bar{p}_1 = \frac{141}{300} = .47$$

$$\bar{p}_2 = \frac{150}{500} = .30$$

Using $z_{.025} = 1.96$ and (11.16) provides the confidence interval estimate of

$$.47 - .30 \pm 1.96 \sqrt{\frac{.47(1 - .47)}{300} + \frac{.30(1 - .30)}{500}} = .17 \pm 1.96(.0354)$$

$$= .17 \pm .069$$

Thus, we can be 95% confident that including $1.00 with the letter increases the participation rate between .101 and .239.

$$\bullet \quad \bullet \quad \bullet$$

Hypothesis Tests About $p_1 - p_2$

Let us now consider statistical inferences involving hypothesis tests about the difference between the proportions of two populations. As with other hypothesis-testing situations, we use the standardized normal random variable z to compare the value of the sample statistic with the hypothesized value for the population parameter. For hypothesis tests concerning the proportions of two populations, the test statistic takes the form

$$z = \frac{(\bar{p}_1 - \bar{p}_2) - (p_1 - p_2)}{\sigma_{\bar{p}_1 - \bar{p}_2}} \tag{11.17}$$

In this expression, $\sigma_{\bar{p}_1 - \bar{p}_2}$ is the standard deviation of the $\bar{p}_1 - \bar{p}_2$ statistic and is defined in (11.14).

We may be tempted to use (11.15) to compute $s_{\bar{p}_1 - \bar{p}_2}$ and use this value as an estimate of $\sigma_{\bar{p}_1 - \bar{p}_2}$. However, whenever the null hypothesis for a hypothesis test about the difference between two population proportions is $H_0: p_1 - p_2 = 0$, the hypothesis-testing procedure requires us to assume H_0 is true and $p_1 - p_2 = 0$. Whenever we make this assumption there is no need to use the individual values of $\bar{p}_1$ and $\bar{p}_2$ to estimate $\sigma_{\bar{p}_1 - \bar{p}_2}$. Using $p_1 = p_2 = p$, (11.14) can be written

$$\sigma_{\bar{p}_1 - \bar{p}_2} = \sqrt{\frac{p(1-p)}{n_1} + \frac{p(1-p)}{n_2}} = \sqrt{p(1-p)\left(\frac{1}{n_1} + \frac{1}{n_2}\right)} \qquad (11.18)$$

With (11.18), we see that we need an estimate of p in order to estimate $\sigma_{\bar{p}_1 - \bar{p}_2}$. The estimate of p is provided by

$$\bar{p} = \frac{n_1 \bar{p}_1 + n_2 \bar{p}_2}{n_1 + n_2} \qquad (11.19)$$

In effect, (11.19) provides a combined, or pooled estimate of the population proportion p. The following expression for $s_{\bar{p}_1 - \bar{p}_2}$ can then be used to estimate $\sigma_{\bar{p}_1 - \bar{p}_2}$ in (11.17).

> **Estimate of $\sigma_{\bar{p}_1 - \bar{p}_2}$ when $H_0: p_1 - p_2 = 0$**
>
> $$s_{\bar{p}_1 - \bar{p}_2} = \sqrt{\bar{p}(1 - \bar{p})\left(\frac{1}{n_1} + \frac{1}{n_2}\right)} \qquad (11.20)$$

EXAMPLE 11.11

A sample of driving records over a 2-year period showed that 16 of 400 adults had received traffic citations, whereas 24 of 300 teen-age drivers had received traffic citations. Test the hypothesis that there is no difference between the traffic citation rate for adult and teen-age drivers.

Letting p_1 = the population proportion of adult drivers receiving traffic citations and p_2 = the population proportion of teen-age drivers receiving traffic citations, we want to test the hypotheses

$$H_0: p_1 - p_2 = 0$$

$$H_1: p_1 - p_2 \neq 0$$

With $\alpha = .05$, we accept H_0 if $-1.96 \leq z \leq 1.96$. Otherwise, we reject H_0.
Sample results show the following values for the sample proportions:

$$\text{Adults:} \qquad \bar{p}_1 = \frac{16}{400} = .04$$

$$\text{Teen-Agers:} \qquad \bar{p}_2 = \frac{24}{300} = .08$$

Using (11.19), we have

$$\bar{p} = \frac{n_1 \bar{p}_1 + n_2 \bar{p}_2}{n_1 + n_2} = \frac{400(.04) + 300(.08)}{400 + 300} = .057$$

as the pooled estimate of p. The value of the test statistic is then computed as follows:

$$z = \frac{(\bar{p}_1 - \bar{p}_2) - (p_1 - p_2)}{\sqrt{\bar{p}(1 - \bar{p})\left(\dfrac{1}{n_1} + \dfrac{1}{n_2}\right)}} = \frac{(.04 - .08) - 0}{\sqrt{.057(1 - .057)\left(\dfrac{1}{400} + \dfrac{1}{300}\right)}}$$

$$= \frac{-.04}{.0177} = -2.26$$

With this value for z, we reject H_0 and conclude that there is a significant difference in the traffic-citation rate for adult and teen-age drivers.

$$\bullet \quad \bullet \quad \bullet$$

EXAMPLE 11.12

In the validation of examination questions used in a physics course at a large university, an instructor would like to compare the proportions of A students and F students answering the questions correctly. A particular examination question is judged to be a good discriminator if the proportion of A students (p_1) answering the question correctly is greater than the proportion of F students (p_2) answering the question correctly. That is, rejecting H_0 in the following hypothesis test will support the conclusion that the examination question is a good discriminator between A and F students.

$$H_0: p_1 - p_2 = 0$$

$$H_1: p_1 - p_2 > 0$$

Using $\alpha = .05$, H_0 will be accepted if $z \leq 1.645$; otherwise H_0 will be rejected.

At the end of the term, sample results showed that 85 of the 110 students who received an A in the course had answered the examination question correctly, whereas 50 of the 95 students who received an F in the course answered the examination question correctly. The computations required to conduct the test of hypothesis are as follows:

$$\text{A Students:} \quad \bar{p}_1 = \frac{85}{110} = .773$$

$$\text{F Students:} \quad \bar{p}_2 = \frac{50}{95} = .526$$

$$\text{Overall:} \quad \bar{p} = \frac{110(.773) + 95(.526)}{110 + 95} = .659$$

$$z = \frac{(\bar{p}_1 - \bar{p}_2) - (p_1 - p_2)}{\sqrt{\bar{p}(1 - \bar{p})\left(\dfrac{1}{n_1} + \dfrac{1}{n_2}\right)}} = \frac{(.773 - .526) - 0}{\sqrt{.659(1 - .659)\left(\dfrac{1}{110} + \dfrac{1}{95}\right)}}$$

$$= \frac{.247}{.066} = 3.74$$

Thus we reject H_0 and conclude that A students do significantly better than F students on the examination question.

• • •

EXERCISES

15. A sample of 400 items produced by supplier A contained 70 defective items. A sample of 300 items produced by supplier B contained 40 defective items. Compute a 90% confidence interval estimate of the difference in the proportion defective from the two suppliers.

16. During the primary elections a particular presidential candidate showed the following pre-election voter support in Wisconsin and Illinois:

State	Voters Surveyed	Voters Favoring the Candidate
Wisconsin	500	270
Illinois	360	162

Compute a 95% confidence interval estimate for the difference between the proportion of voters favoring the candidate in the two states.

17. In a study of the eating habits of teen-agers, it was found that more girls than boys did not eat breakfast on a regular basis. Provide a 90% confidence interval estimate of the difference between the proportion of girls and proportion of boys that do not eat breakfast on a regular basis if sample data show that 85 of 210 girls and 48 of 200 boys do not have breakfast on a regular basis.

18. In a study of coffee-drinking habits, 50 of 240 men and 55 of 180 women expressed a preference for decaffeinated coffee. Using a .05 level of significance, test for a difference between the proportion of men and the proportion of women who prefer decaffeinated coffee. What is your conclusion? What is the p value?

19. A survey firm conducts door-to-door surveys on a variety of issues. Some individuals cooperate with the interviewer and complete the interview questionnaire, while others do not. The following sample data are available (showing the response data for men and women).

	Sample Size	Number Completing the Survey
Men	200	110
Women	300	210

a. Use the p value and $\alpha = .05$ to test the hypothesis that the response rate is the same for both men and women.
b. Compute the 95% confidence interval estimate for the difference between the proportion of men and the proportion of women that cooperate with the survey.

20. In a test of the quality of two television commercials, each commercial was shown in a separate test area six times over a one-week period. The following week a telephone survey was conducted to identify individuals who had seen the commercials. The individuals who had seen the commercials were asked to state the primary message in the commercial. The following results were recorded.

	Number Reporting Having Seen the Commercial	Number Recalling Primary Message
Commercial A	150	63
Commercial B	200	60

a. Use $\alpha = .05$ to test the hypothesis that there is no difference in the recall proportions for the two commercials.
b. Compute a 95% confidence interval for the difference between the recall proportions for the two populations.

21. A political opinion survey shows that of 200 Republicans surveyed, 80 opposed the building of power plants using fission processes (nuclear energy). Similar results for a sample of 300 Democrats showed that 150 opposed building nuclear power plants. Do these results indicate that there is a significant difference in the attitudes of Republicans and Democrats on this issue? Use $\alpha = .05$.

Summary

In this chapter we have discussed procedures for interval estimation and hypothesis testing involving two populations. Specifically, we showed how to make inferences about the difference between the means of two populations when independent simple random samples are selected. Two cases were considered. In the first case, where the sample sizes were large, the z values from the standard normal probability distribution are used for inferences about the difference between two population means. In the second case, where the populations are normally distributed with equal variances, the t distribution is used.

Inferences about the difference between the means of two populations were discussed for the matched-sample design. In the matched-sample design each data value from one sample is matched with a data value in the other sample. The difference in the pair of data values is then used in the statistical analysis. The matched-sample design is generally preferred over the independent-sample design because the matched-sample procedure tends to reduce the sampling error and thus improves the precision of the estimate.

Finally, interval estimation and hypothesis testing involving the difference between two population proportions were discussed. Statistical procedures for analyzing the difference between two population proportions are similar to the procedures for analyzing the difference between two population means.

Statistics in Practice

PRIVATE VERSUS PUBLIC SCHOOLS

Are students better prepared academically by attending private or public schools? Researchers Julius Sassenrath, Michelle Croce, and Manuel Penaloza of the University of California, at Davis, investigated this question by undertaking a two-population study with seniors in private high schools compared academically to seniors in public high schools. Two samples were taken, with 49 students chosen from each of the two populations. In an attempt to eliminate other factors that may have affected academic achievement, the students were matched on age, ethnicity, gender, socioeconomic status, and IQ data obtained 10 years earlier when all the students in the study had attended public elementary schools. With this matching process, the primary difference between the two groups of high school seniors was that one group obtained education by switching from public to private schools, whereas the other group obtained education by remaining in public schools.

During their senior years in high school, the 98 students in the study were given the same reading and mathematics tests. The mean test scores were as follows:

Test	Private School	Public School	Difference
Reading	33.10	33.10	.00
Mathematics	30.12	30.00	.12

Tests for differences between means showed that the differences observed were not statistically significant. The researchers concluded that students with similar characteristics such as age, ethnicity, gender, socioeconomic status, and IQ will have similar reading and mathematical abilities regardless of whether they receive their education from private or public schools. From this study, it appears that private and public schools have about the same influence on academic achievement.

While this study indicates no difference between the academic achievement of students from private and public schools, the researchers point out that there are often other considerations that affect the choice of a school. For instance the social environment provided by a particular school may be an important factor in the selection process for many parents. However, the research study indicates that the choice of a private or public school should not be based on the general claim that private schools lead to higher academic achievement.

"Private and Public School Students: Longitudinal Achievement Differences?" by Julius Sassenrath, Michelle Croce, and Manuel Penaloza, *American Educational Research Journal* (Fall 1984).

Glossary

Pooled variance estimate—An estimate of the variance of a population based on the combination of two (or more) sample results. The pooled variance estimate is appropriate whenever the variances of two (or more) populations are assumed equal. For the two-population case, the pooled estimate of the variance is computed as follows:

$$s_p^2 = \frac{(n_1 - 1)s_1^2 + (n_2 - 1)s_2^2}{n_1 + n_2 - 2}.$$

Independent samples—Samples selected from two (or more) populations where the elements making up one sample are chosen independently of the elements making up the other sample(s).

Matched samples—Samples where each data value is a difference between matched or paired observations.

Key Equations

Expected Value of $\bar{x}_1 - \bar{x}_2$

$$E(\bar{x}_1 - \bar{x}_2) = \mu_1 - \mu_2 \tag{11.1}$$

Standard Deviation of $\bar{x}_1 - \bar{x}_2$

$$\sigma_{\bar{x}_1 - \bar{x}_2} = \sqrt{\frac{\sigma_1^2}{n_1} + \frac{\sigma_2^2}{n_2}} \tag{11.2}$$

Interval Estimation of the Difference Between the Means of Two Populations (Large Sample Case with $n_1 \geq 30$ and $n_2 \geq 30$)

$$\bar{x}_1 - \bar{x}_2 \pm z_{\alpha/2} \sqrt{\frac{\sigma_1^2}{n_1} + \frac{\sigma_2^2}{n_2}} \tag{11.3}$$

Pooled Variance Estimate

$$s_p^2 = \frac{(n_1 - 1)s_1^2 + (n_2 - 1)s_2^2}{(n_1 + n_2 - 2)} \tag{11.5}$$

Interval Estimation of the Difference Between the Means of Two Populations (Small Sample Case with $n_1 < 30$ and/or $n_2 < 30$)

$$\bar{x}_1 - \bar{x}_2 \pm t_{\alpha/2} \sqrt{s_p^2 \left(\frac{1}{n_1} + \frac{1}{n_2} \right)} \tag{11.7}$$

Standard Random Variable z

$$z = \frac{(\bar{x}_1 - \bar{x}_2) - (\mu_1 - \mu_2)}{\sqrt{\frac{\sigma_1^2}{n_1} + \frac{\sigma_2^2}{n_2}}} \tag{11.8}$$

Random Variable t

$$t = \frac{(\bar{x}_1 - \bar{x}_2) - (\mu_1 - \mu_2)}{\sqrt{s_p^2 \left(\frac{1}{n_1} + \frac{1}{n_2} \right)}} \tag{11.9}$$

Random Variable t for Matched Samples

$$t = \frac{\bar{d} - \mu_d}{s_d / \sqrt{n}} \tag{11.10}$$

Expected Value of $\bar{p}_1 - \bar{p}_2$

$$E(\bar{p}_1 - \bar{p}_2) = p_1 - p_2 \tag{11.13}$$

Standard Deviation of $\bar{p}_1 - \bar{p}_2$

$$\sigma_{\bar{p}_1 - \bar{p}_2} = \sqrt{\frac{p_1(1 - p_1)}{n_1} + \frac{p_2(1 - p_2)}{n_2}} \tag{11.14}$$

Interval Estimation of the Difference Between the Proportions of Two Populations

$$\bar{p}_1 - \bar{p}_2 \pm z_{\alpha/2} \sqrt{\frac{\bar{p}_1(1 - \bar{p}_1)}{n_1} + \frac{\bar{p}_2(1 - \bar{p}_2)}{n_2}} \tag{11.16}$$

Estimate of $\sigma_{\bar{p}_1 - \bar{p}_2}$ when $H_0: p_1 - p_2 = 0$

$$\bar{p} = \frac{n_1\bar{p}_1 + n_2\bar{p}_2}{n_1 + n_2} \tag{11.19}$$

and

$$s_{\bar{p}_1 - \bar{p}_2} = \sqrt{\bar{p}(1 - \bar{p})\left(\frac{1}{n_1} + \frac{1}{n_2}\right)} \tag{11.20}$$

Review Quiz

TRUE/FALSE

1. A point estimator of the difference between two population means is the corresponding difference between the two sample means.
2. The central limit theorem cannot be applied to the sampling distribution of $\bar{x}_1 - \bar{x}_2$.
3. The pooled estimate of σ^2 is a weighted average of the two sample variances.
4. The only assumption necessary in using the t distribution for interval estimation of the difference between population means is that the variances of the two populations are equal.
5. The normal probability distribution cannot be used for large-sample hypothesis tests concerning the difference between population means.
6. If sampling n_1 items from one population and n_2 items from a second population, the large-sample case is applicable as long as $n_1 + n_2 \geq 60$.
7. When the matched-sample approach is used for inferences about the difference between population means each item in the sample provides two data values.
8. The advantage of the matched-sample design is that it allows the experimenter to control some factors in order to obtain a sharper measure of others.
9. The sampling distribution of $\bar{p}_1 - \bar{p}_2$ can be approximated by a normal distribution, provided that one of the sample sizes is large.
10. When conducting the hypothesis test $H_0: p_1 - p_2 = 0$, a pooled estimate of the population proportion is computed from the two sample proportions.

MULTIPLE CHOICE

11. Independent samples are obtained from two normal populations with equal variances in order to construct a confidence interval estimate for the difference

between the population means. If the first sample contains 16 items and the second sample contain 36 items, the correct form to use for the sampling distribution is:
a. normal distribution
b. t distribution with 15 degrees of freedom
c. t distribution with 35 degrees of freedom
d. t distribution with 50 degrees of freedom

12. Information about s_1^2 and s_2^2 can be used to estimate
a. if both $n_1 \geq 30$ and $n_2 \geq 30$
b. if either $n_1 \geq 30$ or $n_2 \geq 30$
c. always
d. only when both populations are normal

Use the following for questions 13–16. A testing company is checking to see if there is any significant difference in the coverage of two different brands of paint for a hardware store chain. The results are summarized below.

	Amazon Paint	Coverup Paint
Mean coverage (in square feet)	305	295
Standard deviation	20	25
Sample size	31	41

13. A point estimate of the difference between the population means is
a. -5
b. 0
c. 5
d. 10

14. A point estimate of the standard deviation of the difference between the sample means is
a. -5
b. 5.3
c. 28.1
d. 32

15. The form of the sampling distribution of the difference between the sample means is the
a. normal distribution, approximately
b. t distribution with 30 degrees of freedom
c. t distribution with 35 degrees of freedom
d. t distribution with 40 degrees of freedom

16. If a two-tailed test is used with a .05 level of significance, the critical values are
a. -1.96 and $+1.96$
b. $-.4$ and $+20.4$
c. -10.4 and $+10.4$
d. none of the above

Supplementary Exercises

22. Samples of dinner and luncheon receipts at a major downtown restaurant show the following results.

Dinner Receipts	Luncheon Receipts
$n_1 = 70$	$n_2 = 55$
$\bar{x}_1 = \$32.65$	$\bar{x}_2 = \$12.80$
$s_1 = \$\ 7.20$	$s_2 = \$\ 3.60$

Provide 90% and 98% confidence interval estimates of the difference between mean receipt amounts for the two meals.

23. Production quantities for two assembly-line workers are shown. Each data value indicates the amount produced during a randomly selected 1-hour period.

Worker 1	Worker 2
20	22
18	18
21	20
22	23
20	24

a. Develop a point estimate of the difference between the mean hourly production rates of the two workers. Which worker appears to have the higher mean production rate?

b. Develop a 90% confidence interval estimate for the difference between the mean production rates of the two workers. Consider the confidence interval estimate. Does the result provide conclusive evidence that the worker having the higher sample mean production rate is actually the worker with the overall higher production rate? Explain.

24. A realtor is interested in estimating the difference between the mean selling prices of new homes in two sections of the city. Assume that the standard deviations of the selling prices are approximately $12,000 for both areas. How large a sample should be taken in each area to have a 95% confidence that the sampling error for the difference between mean prices will be $5000 or less? Use the same sample size for both sections of the city.

25. In a study of job attitudes and job satisfaction, a sample of 50 men and 50 women were asked to rate their overall job satisfaction on a 1 to 10 scale. A high rating indicates a higher degree of job satisfaction. Using the sample results shown, does there appear to be a significant difference in the level of job satisfaction of men and women? Use $\alpha = .05$. What is the p value?

Men	Women
$\bar{x}_1 = 7.2$	$\bar{x}_2 = 6.8$
$s_1^2 = 2.8$	$s_2^2 = 1.8$

26. A production line is designed on the assumption that the difference in mean assembly times for two operations is 5 minutes. Independent tests for the two assembly operations show the following results.

Operation A	Operation B
$n_1 = 100$	$n_2 = 50$
$\bar{x}_1 = 14.8$ minutes	$\bar{x}_2 = 10.4$ minutes
$s_1 = .8$ minutes	$s_2 = .6$ minutes

Using $\alpha = .02$, test the hypothesis that the difference between the mean assembly times is 5 minutes.

27. Salary surveys of chemistry and physics majors show the following starting annual salary data:

Chemistry Majors	Physics Majors
$n_1 = 14$	$n_2 = 16$
$\bar{x}_1 = \$19,800$	$\bar{x}_2 = \$19,300$
$s_1 = \$\ 1,000$	$s_2 = \$\ 1,400$

Consider the test of the hypothesis that the mean annual salaries are the same for both majors.

a. What assumptions must be made in order to test the hypothesis?

b. Assume that these assumptions are appropriate. What is the pooled estimate of the population variance?

c. Using $\alpha = .05$, can you conclude that a difference exists in the mean annual salary for the two majors?

d. What is the p value?

28. A market research firm used a sample of individuals to rate their potential to purchase a particular product before and after they saw a new television commercial about the product. The potential-to-purchase ratings were based on a 0 to 10 scale, with higher values indicating a higher potential-to-purchase. The null hypothesis stated that the mean rating after seeing the commercial would be less than or equal to the mean rating before. Rejection of this hypothesis would provide the conclusion that the commercial improved the mean potential-to-purchase rating. Use $\alpha = .05$ and the following data to test the hypothesis. Comment on the value of the commercial.

Individual	Purchase Rating Before Seeing Commercial	Purchase Rating After Seeing Commercial
1	5	6
2	4	6
3	7	7
4	3	4
5	5	3
6	8	9
7	5	7
8	6	6

29. A company wants to evaluate the potential for a new bonus plan by having a random sample of five salespersons use the bonus plan for a trial period. The weekly sales volumes (units) before and after implementing the bonus plan are shown below:

Salesperson	Weekly Sales Before	Weekly Sales After
1	15	18
2	12	14
3	18	19
4	15	18
5	16	18

a. Use $\alpha = .05$ and test to see if it can be concluded that the bonus plan will result in an increase in the mean weekly sales.

b. Provide a 90% confidence interval estimate for the mean *increase* in weekly sales that can be expected if a new bonus plan is implemented.

30. A cable television firm is considering submitting bids for rights to operate in two regions of the state of Florida. Surveys of the two regions provide the following data on customer acceptance of the cable television service.

Region I	Region II
$n_1 = 500$	$n_2 = 800$
Number indicating	Number indicating
likely to purchase = 175	likely to purchase = 360

Develop a 99% confidence interval estimate of the difference between the population proportions of likely-to-purchase customers in the two regions.

31. A large automobile insurance company selected samples of single and married male policyholders and recorded the number who had made an insurance claim over the past 3-year period:

Single Policyholders	Married Policyholders
$n_1 = 400$ Number making claims $= 76$	$n_2 = 900$ Number making claims $= 90$

    a. Test using $\alpha = .05$ to determine if the claim rates differ between single and married male policyholders. What is the p value?

    b. Provide a 95% confidence interval estimate of the difference between the claim proportions for the two populations.

32. Two loan officers at the North Ridge National Bank show the following data for defaults on loans that they have approved (the data are based on samples of loans granted over the past 5 years).

	Number of Loans Reviewed	Number of Defaulted Loans
Loan officer A	60	9
Loan officer B	80	6

Use $\alpha = .05$ to test the hypothesis that the default rates are the same for the two loan officers. What is the p value?

Computer Exercise

Par, Inc. is a major manufacturer of golf equipment. The research group at Par has been investigating a new golf ball designed to resist cuts and yet still offer good driving distances. In tests with the new golf balls, 40 balls of the new model and 40 balls of the current model were subjected to distance tests. The testing was performed with a mechanical hitting machine in ideal weather conditions, so that if a difference existed between the mean distance of the two models it could be attributed to a real difference in design performance. The results of the test are shown. The distance data are measured to the nearest yard.

Current Model	New Model	Current Model	New Model
242	274	248	269
239	266	265	256
245	260	267	261
250	263	258	277
236	259	250	271
261	248	253	278
236	259	243	273
244	286	238	266
237	283	256	265
248	261	253	259
241	271	259	280
242	263	252	247
262	259	251	250
241	268	241	257
238	257	253	267
261	278	245	260
233	247	257	258
250	260	252	252
244	275	254	260
246	261	240	276

QUESTIONS

1. Develop numerical and graphical measures to summarize the given data.
2. Develop confidence interval estimates for the mean distance traveled for both types of balls.
3. What statistical conclusion can you reach regarding the mean distances for the two models? What are your recommendations?

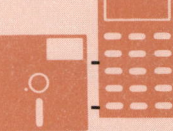

CHAPTER 12

Inferences About Population Variances

What You Will Learn in This Chapter

- situations where inferences about variances are needed

- how the chi-square distribution is used to make inferences about a population variance

- how the *F* distribution is used for hypothesis tests concerning the equality of two population variances

Contents

Statistics in the News

PRODUCT QUALITY: THE CHALLENGE FROM JAPAN

The Japanese emphasis on quality has pushed many Japanese products to the top of the competitive ladder. In recent years, U.S. companies have been striving to improve product quality and recapture their market leadership. In many cases, the quality of U.S. products is still lagging behind Japanese products, and indications are that U.S. companies still have a long way to go to meet the Japanese challenge. One wonders how the Japanese obtained the knowledge that enabled them to leap past the United States in product quality.

In tracing the history of the Japanese emphasis on quality, many point to the year 1950, when an American statistician by the name of W. Edwards Deming began to talk with Japanese managers about how to build quality products. Deming's efforts were part of Japan's postwar recovery program. Eventually his contributions to Japanese product quality and productivity resulted in the Emperor giving him a medal and naming a prestigious prize after him.

The heart of Deming's method for achieving high quality is statistical in nature. Every process, whether it be on the factory floor or in the office, has variations from the ideal. These variations can be measured statistically with variances and standard deviations. Deming shows clients systematic methods for identifying the causes of variations and then reducing them. The result is a steady improvement in the production process and, ultimately, product quality.

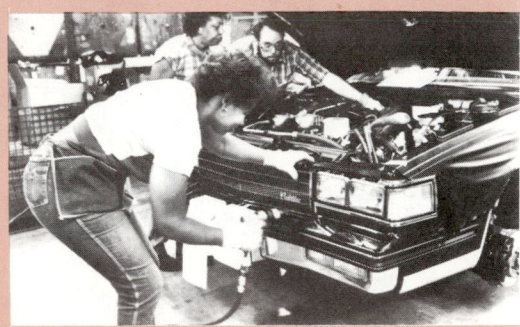

Adding the grille and making final engine adjustments, these workers send a new Cadillac on its way at the General Motors' Orion assembly line.

Thirty years after the Japanese started listening to Deming, American companies began to listen. Firms such as Ford and A.T.&T. pay consulting fees as high as $5000 per day for Deming's keys to product quality. Deming's message emphasizes improving the product by reducing the variance. Ford has adopted Deming's statistical process controls throughout the company.

At the age of 83, Deming still feels compelled to deliver his message wherever possible. He believes the United States is still falling behind Japan, and—as a result—the American standard of living is getting lower. Deming estimates that it will take another 30 years for the United States to catch up with Japan.

Based on the "The Curmudgeon Who Talks Tough on Quality," *Fortune* (June 25, 1984).

In the previous four chapters we focused our attention on methods of statistical inference involving means and proportions. In this chapter we continue the discussion of statistical inference by considering methods for making inferences about variances.

The discussion of Japan's emphasis on product quality in "Statistics in the News" focuses on the variance as an important consideration in quality control. As another example of where variance is important, consider the production process of filling containers with a liquid detergent product. The filling mechanism for the process is adjusted so that the mean filling weight is 16 ounces per container. Although a mean of 16 ounces is desired, the variance of the fillings is also critical. That is, even with the filling mechanism properly adjusted for the mean of 16 ounces, we cannot expect every container to have exactly 16 ounces. By selecting a sample of containers, we can compute a sample variance for the filling weights. This value serves as an estimate of the variance for the population of containers being filled by the production process. If the sample variance is modest, the production process is continued. However, if the sample variance is excessive, overfilling and underfilling can create problems even though the mean filling weight is correct (16 ounces). In this case, the filling mechanism must be readjusted to reduce the filling variance for the containers.

In the following section we consider methods for making inferences about the variance of a single population. Later we discuss procedures for making inferences about the variances of two populations. Since the standard deviation is the square root of the variance, the methods introduced in this chapter can also be used to make inferences about population standard deviations.

12.1 INFERENCES ABOUT A POPULATION VARIANCE

Let us consider a population with an unknown variance denoted by σ^2. The sample variance s^2 will be used to make inferences about the population variance σ^2. The formula for computing a sample variance is restated.

Sample Variance

$$s^2 = \frac{\Sigma(x_i - \bar{x})^2}{n - 1} \tag{12.1}$$

where

$\bar{x}$ = sample mean

n = sample size

The process of using s^2 to make inferences about σ^2 is shown in Figure 12.1.

The Sampling Distribution of $(n - 1)s^2/\sigma^2$

In previous chapters we have shown that knowledge of a sampling distribution is essential for computing interval estimates and conducting hypothesis tests about a population mean or a population proportion. Thus it should not be surprising that in order to make inferences about a population variance, we also must use a sampling distribution. For inferences about a population variance, we find it is appropriate to work with the sampling distribution of the term $(n - 1)s^2/\sigma^2$ rather than the sampling distribution of s^2 directly. The reason for this is that for normally distributed populations, the sampling distribution of the quantity $(n - 1)s^2/\sigma^2$ is known to have a special probability distribution referred to as a *chi-square distribution*. Since tables of

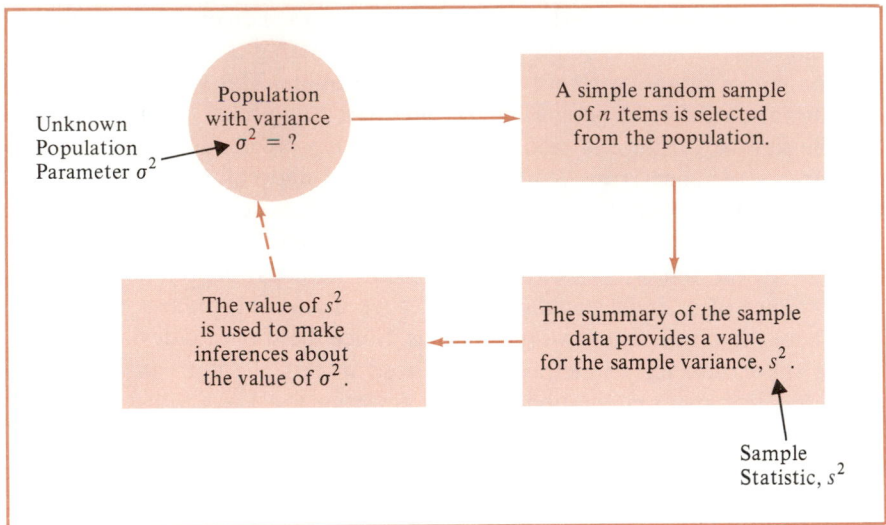

FIGURE 12.1
**The Statistical Process of Using a Sample Variance
to Make Inferences About a Population Variance**

probabilities are available for the chi-square distribution, it becomes relatively easy to use the chi-square distribution to make interval estimates and hypothesis tests about the value of a population variance.

The sampling distribution of $(n - 1)s^2/\sigma^2$ is described as follows:

Sampling Distribution of $(n - 1)s^2/\sigma^2$

Whenever a random sample of size n is selected from a *normally distributed population,* the quantity

$$\frac{(n - 1)s^2}{\sigma^2} \tag{12.2}$$

has a *chi-square distribution* with $n - 1$ degrees of freedom, where s^2 is the sample variance and σ^2 is the population variance.

Typical graphs of the sampling distributions of $(n - 1)s^2/\sigma^2$ are shown in Figure 12.2. The graphs show the sampling distributions resulting when random samples of size 3, 6, and 11 are taken from a normally distributed population with variance σ^2.

The Chi-Square Distribution

Like the t distribution, the chi-square distribution is a family of similar probability distributions. Each specific chi-square distribution depends upon a *degrees of freedom* parameter. That is, there is a chi-square distribution with 1 degree of freedom, a chi-square distribution with 2 degrees of freedom and so on. Figure 12.2 is thus a graph

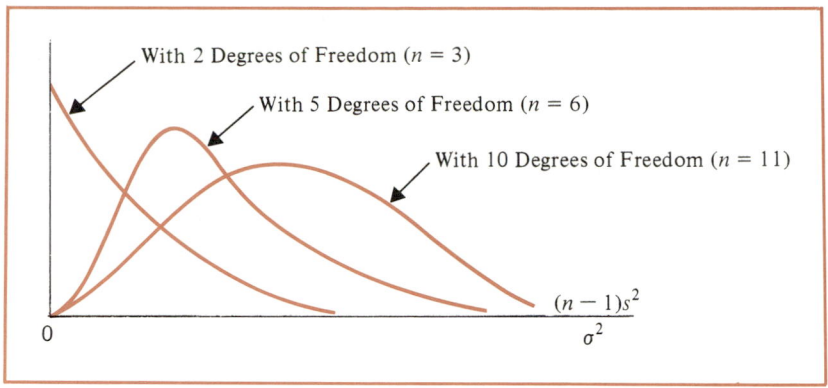

FIGURE 12.2

Examples of the Sampling Distribution of $(n - 1)s^2/\sigma^2$
(a Chi-Square Distribution) with 2, 5, and 10 Degrees of Freedom

of the chi-square distributions for 2, 5, and 10 degrees of freedom. In using a sample of size n to make inferences about a population variance, we find that the appropriate chi-square distribution has $n - 1$ degrees of freedom.

To obtain a better feel for the chi-square distribution and to see how tables of chi-square values can be used to make inferences about a population variance, let us consider a situation where the sample size is 20. The degrees of freedom in this case would be $n - 1 = 20 - 1 = 19$. A graph of the chi-square distribution with 19 degrees of freedom is shown in Figure 12.3. Using the symbol χ^2 to refer to the chi-square value, note that the distribution shows 95% of the possible χ^2 values must be between 8.90655 and 32.8523. That is, when a random sample is selected from a normal population, there is a .95 probability that the chi-square value (i.e., $\chi^2 = (n - 1)s^2/\sigma^2$) will be between 8.90655 and 32.8523.

Table 12.1 contains a table of values for selected chi-square distributions. In referring to the chi-square distribution table, we use a subscript on χ^2 to denote the area or probability under the curve to the *right* of the stated χ^2 value. For example, the area under the curve to the right of $\chi^2_{.025}$ is .025. Thus $\chi^2_{.025}$ corresponds to a chi-square value in the upper tail of the distribution. Similarly, $\chi^2_{.975}$ (97.5% of the chi-square

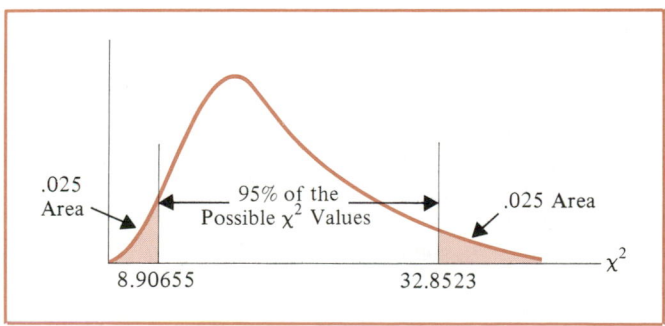

FIGURE 12.3

Example of a Chi-Square Distribution with 19 Degrees of Freedom

TABLE 12.1

Chi-Square Distribution Table

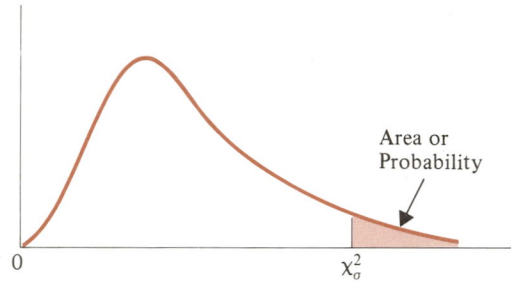

Degrees of Freedom	.99	.975	.95	.05	.025	.01
1	$157{,}088 \times 10^{-9}$	$982{,}069 \times 10^{-9}$	$393{,}214 \times 10^{-8}$	3.84146	5.02389	6.63490
2	.0201007	.0506356	.102587	5.99147	7.37776	9.21034
3	.114832	.215795	.351846	7.81473	9.34840	11.3449
4	.297110	.484419	.710721	9.48773	11.1433	13.2767
5	.554300	.831211	1.145476	11.0705	12.8325	15.0863
6	.872085	1.237347	1.63539	12.5916	14.4494	16.8119
7	1.239043	1.68987	2.16735	14.0671	16.0128	18.4753
8	1.646482	2.17973	2.73264	15.5073	17.5346	20.0902
9	2.087912	2.70039	3.32511	16.9190	19.0228	21.6660
10	2.55821	3.24697	3.94030	18.3070	20.4831	23.2093
11	3.05347	3.81575	4.57481	19.6751	21.9200	24.7250
12	3.57056	4.40379	5.22603	21.0261	23.3367	26.2170
13	4.10691	5.00874	5.89186	22.3621	24.7356	27.6883
14	4.66043	5.62872	6.57063	23.6848	26.1190	29.1413
15	5.22935	6.26214	7.26094	24.9958	27.4884	30.5779
16	5.81221	6.90766	7.96164	26.2962	28.8454	31.9999
17	6.40776	7.56418	8.67176	27.5871	30.1910	33.4087
18	7.01491	8.23075	9.39046	28.8693	31.5264	34.8053
19	7.63273	8.90655	10.1170	30.1435	32.8523	36.1908
20	8.26040	9.59083	10.8508	31.4104	34.1696	37.5662
21	8.89720	10.28293	11.5913	32.6705	35.4789	38.9321
22	9.54249	10.9823	12.3380	33.9244	36.7807	40.2894
23	10.19567	11.6885	13.0905	35.1725	38.0757	41.6384
24	10.8564	12.4011	13.8484	36.4151	39.3641	42.9798
25	11.5240	13.1197	14.6114	37.6525	40.6465	44.3141
26	12.1981	13.8439	15.3791	38.8852	41.9232	45.6417
27	12.8786	14.5733	16.1513	40.1133	43.1944	46.9630
28	13.5648	15.3079	16.9279	41.3372	44.4607	48.2782
29	14.2565	16.0471	17.7083	42.5569	45.7222	49.5879
30	14.9535	16.7908	18.4926	43.7729	46.9792	50.8922
40	22.1643	24.4331	26.5093	55.7585	59.3417	63.6907
50	29.7067	32.3574	34.7642	67.5048	71.4202	76.1539
60	37.4848	40.4817	43.1879	79.0819	83.2976	88.3794
70	45.4418	48.7576	51.7393	90.5312	95.0231	100.425
80	53.5400	57.1532	60.3915	101.879	106.629	112.329
90	61.7541	65.6466	69.1260	113.145	118.136	124.116
100	70.0648	74.2219	77.9295	124.342	129.561	135.807

Additional values of chi-square can be found in Table 3 of Appendix B.

values are to the right of $\chi^2_{.975}$) corresponds to a chi-square value in the lower tail of the distribution.

　Now let us use the chi-square distribution information in Table 12.1 to show how the values of χ^2 in Figure 12.3 were obtained. Since the chi-square distribution in Figure 12.3 is based on 19 degrees of freedom, we see that $\chi^2_{.975} = 8.90655$ and $\chi^2_{.025} = 32.8523$ are found in row 19 and the .975 and .025 columns of Table 12.1. Thus there is a .95 probability that a chi-square value will be between $\chi^2_{.975}$ and $\chi^2_{.025}$. This statement holds for all chi-square distributions. However, the numerical values of $\chi^2_{.975}$ and $\chi^2_{.025}$ change depending upon the number of degrees of freedom. If we had wanted an interval containing 90% of the chi-square values, we could have used the interval from $\chi^2_{.95}$ to $\chi^2_{.05}$. With 19 degrees of freedom, we see from Table 12.1 that this interval is from 10.1170 to 30.1435.

EXAMPLE 12.1

Consider a chi-square distribution with 15 degrees of freedom. Find the value of χ^2 that cuts off an area of .01 in the upper tail of the distribution.

　Referring to Table 12.1, we find that $\chi^2_{.01} = 30.5779$ cuts off an area of .01 in the upper tail.

•　•　•

EXAMPLE 12.2

Consider a chi-square distribution with 9 degrees of freedom. Find the value of χ^2 that cuts off an area of .05 in the lower tail of the distribution.

　Referring to Table 12.1, we find that $\chi^2_{.95} = 3.32511$ cuts off an area of .05 in the lower tail of the distribution.

•　•　•

EXAMPLE 12.3

Consider a chi-square distribution with 50 degrees of freedom. Find values of χ^2 that cut off an area of .01 in each tail of the distribution.

　Referring to Table 12.1, we find that $\chi^2_{.99} = 29.7067$ cuts off an area of .01 in the lower tail and $\chi^2_{.01} = 76.1539$ cuts off an area of .01 in the upper tail of the distribution.

•　•　•

Interval Estimation of σ^2

Let us now show how the chi-square distribution can be used to develop a confidence interval estimate of a population variance σ^2 for a normal population. Based upon our previous discussion of the chi-square distribution, we can conclude that there is a $1 - \alpha$ probability that χ^2 will be between $\chi^2_{1-\alpha/2}$ and $\chi^2_{\alpha/2}$. That is, there is a $1 - \alpha$

probability $\chi^2_{1-\alpha/2} \leq \chi^2 \leq \chi^2_{\alpha/2}$. From (12.2) we also know the quantity $(n-1)s^2/\sigma^2$ follows the chi-square distribution. Therefore, there must be a $1-\alpha$ probability

$$\chi^2_{1-\alpha/2} \leq \frac{(n-1)s^2}{\sigma^2} \leq \chi^2_{\alpha/2} \tag{12.3}$$

Working with the right-hand side of (12.3), we have

$$\frac{(n-1)s^2}{\sigma^2} \leq \chi^2_{\alpha/2} \tag{12.4}$$

After taking a sample, the sample size n and the sample variance will be known. In addition, the value of $\chi^2_{\alpha/2}$ can be found by using the chi-square distribution table. Thus the population variance σ^2 is the only unknown quantity in (12.4). Multiplying (12.4) by σ^2 provides the following inequality:

$$(n-1)s^2 \leq \sigma^2 \chi^2_{\alpha/2}$$

Dividing both sides of this inequality by $\chi^2_{\alpha/2}$, we obtain (12.5):

$$\frac{(n-1)s^2}{\chi^2_{\alpha/2}} \leq \sigma^2 \tag{12.5}$$

With n, s^2, and $\chi^2_{\alpha/2}$ known, we can use (12.5) to compute a lower limit for the value of the population variance σ^2. Using a similar approach with the left-hand inequality in (12.3), we obtain an upper limit for the value of the σ^2 expressed in terms of n, s^2, and $\chi^2_{1-\alpha/2}$.

Based on these results the following general procedure can be used to find $(1-\alpha)\%$ confidence interval estimate of a population variance. (Recall that the population is assumed to be normally distributed.)

$(1-\alpha)\%$ Confidence Interval for a Population Variance

$$\frac{(n-1)s^2}{\chi^2_{\alpha/2}} \leq \sigma^2 \leq \frac{(n-1)s^2}{\chi^2_{1-\alpha/2}} \tag{12.6}$$

where the values of χ^2 are based on the chi-square distribution with $n-1$ degrees of freedom.

By altering the values of α in (12.6), we can obtain the appropriate level of confidence. For example, $\chi^2_{.025}$ and $\chi^2_{.975}$ will provide a 95% confidence interval, $\chi^2_{.05}$ and $\chi^2_{.95}$ will provide a 90% confidence interval, and $\chi^2_{.005}$ and $\chi^2_{.995}$ will provide a 99% confidence interval.

EXAMPLE 12.4

A production process is designed to fill 16-ounce containers with liquid detergent. The production manager is concerned about the variance in filling weights. A sample of 20 containers provides a sample variance of $s^2 = .0025$. Develop a 95% confidence interval for the population variance, σ^2, of the filling weights.

With $n - 1 = 19$ degrees of freedom, we have $\chi^2_{.025} = 32.8523$ and $\chi^2_{.975} = 8.90655$. Using (12.6), the 95% confidence interval becomes

$$\frac{(20 - 1)(.0025)}{32.8523} \leq \sigma^2 \leq \frac{(20 - 1)(.0025)}{8.90655}$$

or

$$.0014 \leq \sigma^2 \leq .0053$$

By taking the square root of these terms, we can also find the 95% confidence interval for the population standard deviation σ:

$$.037 \leq \sigma \leq .073$$

Thus, we are 95% confident the population variance is between .0014 and .0053 and the population standard deviation is between .037 and .073.

• • •

EXAMPLE 12.5

Twenty-eight children were given a language test with the scores recorded on a scale of 0 to 100. The variance in the test scores serves as a measure of the homogeneity in language skills for the children. The sample standard deviation of the test scores was found to be $s = 12$. Provide a 90% confidence interval estimate of the variance and standard deviation of the test scores for the population of children.

With $n - 1 = 27$ degrees of freedom, Table 12.1 shows $\chi^2_{.95} = 16.1513$ and $\chi^2_{.05} = 40.1133$. Using the sample variance of $s^2 = (12)^2 = 144$ in (12.6), we have

$$\frac{(28 - 1)(144)}{40.1133} \leq \sigma^2 \leq \frac{(28 - 1)(144)}{16.1513}$$

or

$$96.93 \leq \sigma^2 \leq 240.72$$

This shows a 90% confidence interval for the population variance of 96.93 to 240.72. Taking the square root of the above values provides a 90% confidence interval for the population standard deviation of 9.85 to 15.52.

The variability in language skills found through this procedure might be used as a justification for grouping the children in separate, more homogeneous, sections to tailor the instruction more closely to student skills.

• • •

Hypothesis Tests About σ^2

Hypothesis tests about the value of a population variance are based on the value of $(n-1)s^2/\sigma^2$ and the chi-square distribution with $n-1$ degrees of freedom. The value of σ^2 specified in the null hypothesis is used in the denominator of $(n-1)s^2/\sigma^2$. The decision rule for accepting or rejecting the hypothesis is based on the value of χ^2 in a manner similar to the use of z and t values in previous hypothesis-testing applications.

EXAMPLE 12.6

The St. Louis Metro Bus Company has recently made a concerted effort to improve reliability by encouraging its drivers to maintain consistent schedules. As a standard policy, the company expects arrival times at a bus stop to have low variability. In terms of the variance of arrival times, the company specifies an arrival time standard deviation of 2 minutes or less. This is a variance of 4 for the arrival times. A sample of 10 arrival times shows a sample variance of 5. Using a .05 level of significance, should the company reject the hypothesis that the arrival time variance for the population is within the allowable variance of 4?

The hypotheses for the study are

$$H_0: \sigma^2 = 4$$

$$H_1: \sigma^2 > 4$$

Rejection of H_0 will imply the company is not meeting the reliability guideline.

With $\alpha = .05$, the one-tailed test decision rule is based on the upper-tail chi-square value of $\chi^2_{.05}$. With $n - 1 = 9$, Table 12.1 shows $\chi^2_{.05} = 16.919$. Thus if $(n - 1)s^2/\sigma^2$ is less than or equal to 16.919, we will not reject H_0. The chi-square distribution with the accept and reject regions is shown in Figure 12.4.

With the sample variance of $s^2 = 5$, do we have sufficient evidence to reject H_0 and conclude that the buses are not meeting the arrival variance guideline? The test

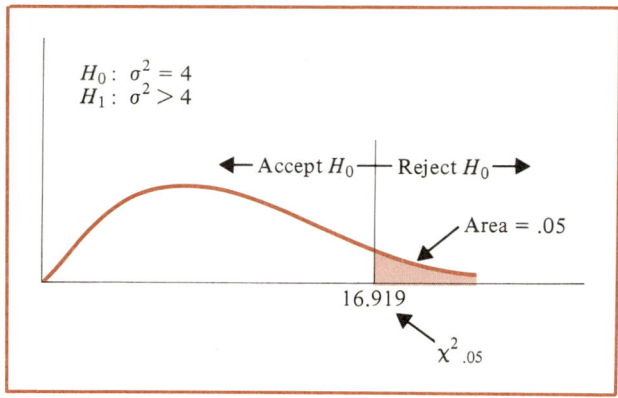

FIGURE 12.4
Chi-Square Distribution with 9 Degrees of Freedom

statistic χ^2 is computed as follows:

$$\chi^2 = \frac{(n-1)s^2}{\sigma^2} = \frac{(10-1)(5)}{4} = 11.25$$

Comparing 11.25 to the $\chi^2_{.05} = 16.919$ value, we see that the null hypothesis cannot be rejected. Thus the sample of 10 bus arrivals provides insufficient evidence to conclude that the bus arrival time variance is not meeting the company standard.

• • •

EXAMPLE 12.7

Historically, the variance in test scores for individuals applying for drivers' licenses has been $\sigma^2 = 100$. A new examination has been designed. However, motor vehicle administrators believe that it is desirable for the variance in the test scores to remain at the historical level. A sample of 30 individuals are given the new version of the driver's examination; the sample variance is 64. Is there reason to believe the variance of the test scores has changed? Use $\alpha = .05$.

The hypothesis test is

$$H_0: \sigma^2 = 100$$

$$H_1: \sigma^2 \neq 100$$

The two-tailed test with $\alpha = .05$ requires the use of $\chi^2_{.975}$ and $\chi^2_{.025}$. With $n - 1 = 29$ degrees of freedom, Table 12.1 shows $\chi^2_{.975} = 16.0471$ and $\chi^2_{.025} = 45.7222$. Thus H_0 cannot be rejected if the value of the test statistic is between 16.0471 and 45.7222.

With $\sigma^2 = 100$ from H_0, the value of χ^2 is computed to be

$$\chi^2 = \frac{(n-1)s^2}{\sigma^2} = \frac{29(64)}{100} = 18.56$$

With this value of χ^2, we conclude the new examination does not appear to cause a significantly different variance in the test scores.

• • •

EXERCISES

1. The scores for a biology examination are normally distributed with a population standard deviation of $\sigma = 10$. A sample of 12 examinations will be selected and the sample standard deviation computed. With different samples possible, it is expected that the sample standard deviation will vary based on the sample selected.

   a. What is the sampling distribution of the quantity $(n - 1)s^2/\sigma^2$? Show a graph of this sampling distribution

   b. Using Table 12.1, what is the probability that the sample results will show $(n - 1)s^2/\sigma^2$ greater than or equal to 4.57481? Greater than or equal to 19.6751?

   c. Use the results from (b) to show that there is a .90 probability that the sample standard deviation s will be between 6.45 and 13.37.

2. For 20 randomly selected days, the standard deviation of the number of inmates in a county jail was 5.2. Place a 95% confidence interval on the population standard deviation of number of inmates in the jail.

3. The variance in unit weights is very critical in the pharmaceutical industry. For a specific drug, with unit weights measured in grams, a sample of 18 units provided a sample variance of $s^2 = .36$.

   a. Construct a 90% confidence interval estimate for the population variance for the weights of this drug.

   b. Construct a 90% confidence interval estimate for the population standard deviation.

4. A sample of cans of soups produced by Carle Foods shows the following weights, measured in ounces.

12.2	11.9	12.0	12.2
11.7	11.6	11.9	12.0
12.1	12.3	11.8	11.9

   Provide 95% confidence interval estimates for both the variance and the standard deviation of the population.

5. A certain part must be machined to very close tolerances or it is not acceptable to customers. Production specifications call for a maximum standard deviation in the lengths of the parts of .02 inches. The sample variance for 30 parts turns out to be $s^2 = .0005$. Using $\alpha = .05$, test to see if the production specifications are being violated.

6. City Trucking, Inc. claims consistent delivery times for its routine customer deliveries. A sample of 22 truck deliveries shows a sample variance of 1.5. Test to

determine if the company can justifiably claim that the standard deviation in its delivery times is 1 hour or less. Use $\alpha = .10$.

7. The variance in the filling amounts for cups of soft drink from an automatic drink machine is an important consideration to the owner of the soft-drink service. If the variance is too large, overfilling and underfilling of cups will cause customer dissatisfaction with the service. An acceptable variance in filling amounts is $\sigma^2 = .25$ when filling amounts are measured in ounces. In a test of filling amounts for a particular machine, a sample of 18 cups showed a sample variance of .40.
   a. Use a .05 level of significance. Do the sample results indicate that the filling mechanism on the machine should be replaced due to a large variance in filling amounts?
   b. Provide a 90% confidence interval for the variance in the filling amounts for this machine.

8. From a sample of 9 days over the past 6 months, a dentist has seen the following number of patients: 22, 25, 20, 18, 15, 22, 24, 19, and 26. The number of patients seen per day appears to be normally distributed. Would analysis of this sample data support the hypothesis that the variance in the number of patients seen per day is 10? Use a .10 level of significance. What is your conclusion?

12.2 INFERENCES COMPARING THE VARIANCES OF TWO POPULATIONS

In some statistical applications it is desirable to compare the variances of two populations. For instance, we might want to compare the variability in product quality resulting from two different production processes, the variability in assembly times for two assembly methods, or the variability in temperatures for two heating devices. In addition, recall that in Chapter 11 we developed a pooled variance estimate based on the assumption that two populations had equal variances. Thus we might want to compare the variances of two populations to determine if the equal variance assumption, and thus pooling, can be justified.

The Sampling Distribution of s_1^2/s_2^2

In making comparisons about the variances of two normally distributed populations, we use data collected from two independent random samples, one from population 1 and another from population 2. The sample variances s_1^2 and s_2^2 serve as the estimates of the corresponding population variances, σ_1^2 and σ_2^2. The statistic of interest is the ratio of the two sample variances, s_1^2/s_2^2. If the two populations involved both have normal probability distributions with equal variances, the ratio s_1^2/s_2^2 is known to have a special probability distribution known as an *F distribution*. Tables of areas or probabilities are available for this distribution, and they can be used to make inferences about the variances of the populations.

The sampling distribution of s_1^2/s_2^2 is described as follows:

> **Sampling Distribution of s_1^2/s_2^2**
>
> Whenever independent random samples of sizes n_1 and n_2 are selected from normal populations with equal variances, the ratio
>
> $$F = \frac{s_1^2}{s_2^2} \qquad (12.7)$$
>
> has an F distribution with $n_1 - 1$ degrees of freedom for the numerator and $n_2 - 1$ degrees of freedom for the denominator, where s_1^2 is the sample variance for the random sample of n_1 items from population 1 and s_2^2 is the sample variance for the random sample of n_2 items from population 2.

A graph of the sampling distribution of s_1^2/s_2^2 with 20 degrees of freedom associated with s_1^2 and 20 degrees of freedom associated with s_2^2 is shown in Figure 12.5. This distribution, which is an F distribution, shows what happens to the ratio s_1^2/s_2^2 as random samples of size $n_1 = 21$ and $n_2 = 21$ are taken from two normal populations with equal variances.

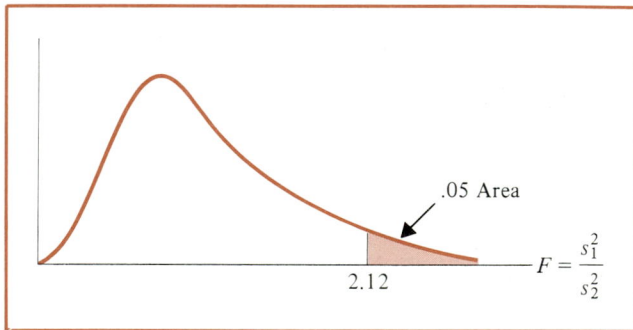

FIGURE 12.5
F **Distribution with 20 Degrees of Freedom for the Numerator and 20 Degrees of Freedom for the Denominator**

The F Distribution

With two sample variances involved in computing the F value of s_1^2/s_2^2, it should not be surprising to find that the F distribution requires the specification of two degrees of freedom values. Each specific F distribution depends upon the number of degrees of freedom associated with the numerator and the number of degrees of freedom associated with the denominator.

We use F_α to denote the point on the horizontal axis of the F distribution that cuts off an area of α in the upper tail of the F distribution. For instance, $F_{.05}$ cuts off an area in the upper tail of .05. Similarly, $F_{1-\alpha}$ cuts off an area of α in the lower tail of the F distribution. Thus $F_{.95}$ cuts off an area of .05 in the lower tail of the F distribution.

TABLE 12.2

Values of $F_{.05}$ for the F Distribution

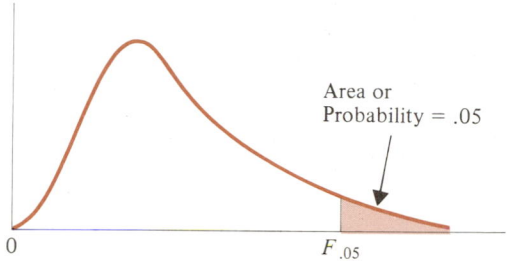

Area or Probability = .05

0 $F_{.05}$

Entries in the table give $F_{.05}$ values, where .05 is the area or probability in the upper tail of the F distribution. For example, with 12 numerator degrees of freedom and 15 denominator degrees of freedom, $F_{.05} = 2.48$.

Table of $F_{.05}$ Values

Denominator Degrees of Freedom	\multicolumn{17}{c}{Numerator Degrees of Freedom}

Denominator Degrees of Freedom	1	2	3	4	5	6	7	8	9	10	12	15	20	24	30	40
1	161.4	199.5	215.7	224.6	230.2	234.0	236.8	238.9	240.5	241.9	243.9	245.9	248.0	249.1	250.1	251.1
2	18.51	19.00	19.16	19.25	19.30	19.33	19.35	19.37	19.38	19.40	19.41	19.43	19.45	19.45	19.46	19.47
3	10.13	9.55	9.28	9.12	9.01	8.94	8.89	8.85	8.81	8.79	8.74	8.70	8.66	8.64	8.62	8.59
4	7.71	6.94	6.59	6.39	6.26	6.16	6.09	6.04	6.00	5.96	5.91	5.86	5.80	5.77	5.75	5.72
5	6.61	5.79	5.41	5.19	5.05	4.95	4.88	4.82	4.77	4.74	4.68	4.62	4.56	4.53	4.50	4.46
6	5.99	5.14	4.76	4.53	4.39	4.28	4.21	4.15	4.10	4.06	4.00	3.94	3.87	3.84	3.81	3.77
7	5.59	4.74	4.35	4.12	3.97	3.87	3.79	3.73	3.68	3.64	3.57	3.51	3.44	3.41	3.38	3.34
8	5.32	4.46	4.07	3.84	3.69	3.58	3.50	3.44	3.39	3.35	3.28	3.22	3.15	3.12	3.08	3.04
9	5.12	4.26	3.86	3.63	3.48	3.37	3.29	3.23	3.18	3.14	3.07	3.01	2.94	2.90	2.86	2.83
10	4.96	4.10	3.71	3.48	3.33	3.22	3.14	3.07	3.02	2.98	2.91	2.85	2.77	2.74	2.70	2.66
11	4.84	3.98	3.59	3.36	3.20	3.09	3.01	2.95	2.90	2.85	2.79	2.72	2.65	2.61	2.57	2.53
12	4.75	3.89	3.49	3.26	3.11	3.00	2.91	2.85	2.80	2.75	2.69	2.62	2.54	2.51	2.47	2.43
13	4.67	3.81	3.41	3.18	3.03	2.92	2.83	2.77	2.71	2.67	2.60	2.53	2.46	2.42	2.38	2.34
14	4.60	3.74	3.34	3.11	2.96	2.85	2.76	2.70	2.65	2.60	2.53	2.46	2.39	2.35	2.31	2.27
15	4.54	3.68	3.29	3.06	2.90	2.79	2.71	2.64	2.59	2.54	2.48	2.40	2.33	2.29	2.25	2.20
16	4.49	3.63	3.24	3.01	2.85	2.74	2.66	2.59	2.54	2.49	2.42	2.35	2.28	2.24	2.19	2.15
17	4.45	3.59	3.20	2.96	2.81	2.70	2.61	2.55	2.49	2.45	2.38	2.31	2.23	2.19	2.15	2.10
18	4.41	3.55	3.16	2.93	2.77	2.66	2.58	2.51	2.46	2.41	2.34	2.27	2.19	2.15	2.11	2.06
19	4.38	3.52	3.13	2.90	2.74	2.63	2.54	2.48	2.42	2.38	2.31	2.23	2.16	2.11	2.07	2.03
20	4.35	3.49	3.10	2.87	2.71	2.60	2.51	2.45	2.39	2.35	2.28	2.20	2.12	2.08	2.04	1.99
21	4.32	3.47	3.07	2.84	2.68	2.57	2.49	2.42	2.37	2.32	2.25	2.18	2.10	2.05	2.01	1.96
22	4.30	3.44	3.05	2.82	2.66	2.55	2.46	2.40	2.34	2.30	2.23	2.15	2.07	2.03	1.98	1.94
23	4.28	3.42	3.03	2.80	2.64	2.53	2.44	2.37	2.32	2.27	2.20	2.13	2.05	2.01	1.96	1.91
24	4.26	3.40	3.01	2.78	2.62	2.51	2.42	2.36	2.30	2.25	2.18	2.11	2.03	1.98	1.94	1.89
25	4.24	3.39	2.99	2.76	2.60	2.49	2.40	2.34	2.28	2.24	2.16	2.09	2.01	1.96	1.92	1.87
26	4.23	3.37	2.98	2.74	2.59	2.47	2.39	2.32	2.27	2.22	2.15	2.07	1.99	1.95	1.90	1.85
27	4.21	3.35	2.96	2.73	2.57	2.46	2.37	2.31	2.25	2.20	2.13	2.06	1.97	1.93	1.88	1.84
28	4.20	3.34	2.95	2.71	2.56	2.45	2.36	2.29	2.24	2.19	2.12	2.04	1.96	1.91	1.87	1.82
29	4.18	3.33	2.93	2.70	2.55	2.43	2.35	2.28	2.22	2.18	2.10	2.03	1.94	1.90	1.85	1.81
30	4.17	3.32	2.92	2.69	2.53	2.42	2.33	2.27	2.21	2.16	2.09	2.01	1.93	1.89	1.84	1.79
40	4.08	3.23	2.84	2.61	2.45	2.34	2.25	2.18	2.12	2.08	2.00	1.92	1.84	1.79	1.74	1.69
60	4.00	3.15	2.76	2.53	2.37	2.25	2.17	2.10	2.04	1.99	1.92	1.84	1.75	1.70	1.65	1.59
120	3.92	3.07	2.68	2.45	2.29	2.17	2.09	2.02	1.96	1.91	1.83	1.75	1.66	1.61	1.55	1.50
∞	3.84	3.00	2.60	2.37	2.21	2.10	2.01	1.94	1.88	1.83	1.75	1.67	1.57	1.52	1.46	1.39

Additional values for the F distribution can be found in Table 4 of Appendix B.

Table 12.2 contains a table of $F_{.05}$ values for various numerator and denominator degrees of freedom. A more complete table for the F distribution is provided in Table 4 of Appendix B.

EXAMPLE 12.8

Consider a sample of size 21 from population 1 and a sample of size 21 from population 2. Find $F_{.05}$ for the F distribution corresponding to $F = s_1^2/s_2^2$.

Referring to Table 12.2, we see that with 20 degrees of freedom in the numerator and 20 degrees of freedom for the denominator, $F_{.05} = 2.12$. This value is shown on the graph in Figure 12.5.

• • •

EXAMPLE 12.9

Find $F_{.01}$ for an F distribution with 15 degrees of freedom in the numerator and 24 degrees of freedom in the denominator.

Referring to Table 4 of Appendix B we find that the column corresponding to 15 numerator degrees of freedom and the row corresponding to 24 denominator degrees of freedom provides $F_{.01} = 2.89$.

• • •

The tables for the F distribution provide F values that cut off only areas of .05, .025, and .01 in the upper tail of the F distribution. However, F values that cut off these same areas in the lower tail of the F distribution can be easily computed from the inverse relationship of the F distribution.

Inverse Relationship of F Statistic

$$F_{(1-\alpha)} = \frac{1}{F_\alpha} \qquad (12.8)$$

where $F_{(1-\alpha)}$ is from an F distribution with n_1 degrees of freedom in the numerator and n_2 degrees of freedom in the denominator and F_α is from an F distribution with n_2 degrees of freedom in the *numerator* and n_1 degrees of freedom in the *denominator.*

EXAMPLE 12.10

Consider a case in which a sample of size 25 has been taken from population 1 and a sample of size 10 has been taken from population 2. Find $F_{.05}$ and $F_{.95}$.

Refer to Table 12.2 in the column corresponding to 24 degrees of freedom and the row corresponding to 9 degrees of freedom. We find $F_{.05} = 2.90$.

To find $F_{.95}$ we must employ the inverse relationship. From (12.8) we have

$$F_{.95} = \frac{1}{F_{.05}}$$

where the numerator and denominator degrees of freedom for $F_{.95}$ and $F_{.05}$ are reversed. Using Table 12.2, we find that for 9 numerator degrees of freedom and 24 denominator degrees of freedom, $F_{.05} = 2.30$. Therefore

$$F_{.95} = \frac{1}{F_{.05}} = \frac{1}{2.30} = .435$$

Here $F_{.95}$ has 24 numerator and 9 denominator degrees of freedom, respectively.

• • •

Hypothesis Tests About σ_1^2 and σ_2^2

Let us now see how the F distribution can be used for hypothesis tests concerning the variances of two normally distributed populations. Hypothesis tests about the variances of two populations are based on the value of s_1^2/s_2^2. The decision rule for accepting or rejecting the hypothesis is based on the F value in a manner similar to how z, t, and χ^2 values have been used in previous hypothesis-testing applications.

EXAMPLE 12.11

Dullus County Schools is renewing its school bus service contract for the coming year and must select one of the two bus companies, the Milbank Company or the Gulf Park Company. We will be interested in using the variance of the arrival or pickup/delivery times as a primary measure of the quality of the bus service. Low variance values will indicate the more consistent and higher-quality service. With equality of variances for the populations of arrival times associated with the two services, Dullus School administrators will select the company offering the better financial terms. However, if sample data on bus arrival times for the two companies indicate that a significant difference exists between the variances, the administrators may want to give special consideration to the company with the better or lower-variance service. The appropriate hypotheses and their associated conclusions and actions are as follows:

Hypothesis	Conclusion and Action
$H_0: \sigma_1^2 = \sigma_2^2$	Equal quality of service; base service selection decision on financial terms
$H_1: \sigma_1^2 \neq \sigma_2^2$	Unequal quality of service; give special consideration to the low-variance service

Assume that the above hypothesis test will be conducted with $\alpha = .10$. Furthermore, assume that we obtain samples of arrival times from school systems currently using the two school bus services. A sample of 25 arrival times is available for the Milbank service (population 1) and a sample of 16 arrival times is available for the Gulf Park service (population 2). The graph of the F distribution with $n_1 - 1 = 24$ degrees of freedom and $n_2 - 1 = 15$ degrees of freedom is shown in Figure 12.6. Note that for $\alpha = .10$, the two-tailed rejection regions are indicated by the critical values at $F_{.95}$ and $F_{.05}$.

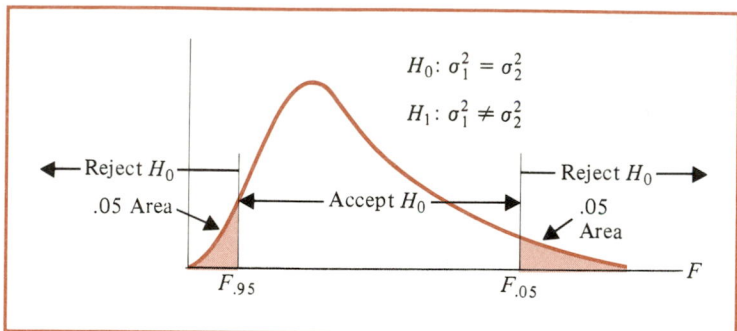

FIGURE 12.6
F Distribution with 24 Degrees of Freedom for the
Numerator and 15 Degrees of Freedom for the Denominator

Using Table 12.2, we find that with 24 and 15 degrees of freedom, $F_{.05} = 2.29$. While the $F_{.95}$ value is not available in this table, it can be found from the inverse relationship of (12.8). With 15 and 24 numerator and denominator degrees of freedom, respectively, Table 12.2 shows $F_{.05} = 2.11$. Substituting into (12.8), we find that with 24 and 15 numerator and denominator degrees of freedom, respectively,

$$F_{.95} = \frac{1}{2.11} = .474$$

Thus the decision rule for our hypothesis test is

Accept H_0 If $.474 \leq F \leq 2.29$

Reject H_0 Otherwise

Suppose that the two samples of arrival times show sample variances of $s_1^2 = 48$ for Milbank and $s_2^2 = 20$ for Gulf Park. The corresponding F value is

$$F = \frac{s_1^2}{s_2^2} = \frac{48}{20} = 2.4$$

Since $F = 2.40 > 2.29$, we reject H_0. The bus services appear to differ in terms of pickup and delivery-time variances. The recommendation is that the Dullus school administrators give special consideration to the better, or lower-variance, service offered by the Gulf Park Company.

• • •

One-tailed tests involving two population variances are also possible. Again the F distribution is used, with the one-tailed rejection region enabling us to conclude whether or not one population variance is significantly greater or significantly less than the other. Only upper-tail F values are needed. For any one-tailed test we set up the null hypothesis so that the rejection region is in the upper tail. This can be accomplished by a judicious choice of which population is labeled population 1. For example, one-tailed hypothesis tests about two population variances should be stated in the following form

$$H_0: \sigma_1^2 = \sigma_2^2$$

$$H_1: \sigma_1^2 > \sigma_2^2$$

In this form, rejection of H_0 will only occur when the value of s_1^2/s_2^2 is in the upper tail of the F distribution. Thus the population with the variance that involves "greater than" in the alternative hypothesis should be labeled population 1.

EXAMPLE 12.12

A sample of men and women were selected to participate in a study designed to measure attitudes about current political issues. The attitude scores of 31 men showed a sample variance of 80. The attitude scores of 41 women showed a sample variance of 120. From this evidence, can it be concluded that women demonstrate a greater variation in attitude on political issues? Use $\alpha = .05$.

In formulating the hypothesis to be tested, we note that for women to demonstrate greater variance, women must show larger variance in the alternative hypothesis. Thus the hypothesis test should be stated as

$$H_0: \sigma^2_{women} = \sigma^2_{men}$$

$$H_1: \sigma^2_{women} > \sigma^2_{men}$$

In our computations for the F test, women will be population 1 and men will be population 2. Thus the corresponding F distribution has 40 numerator degrees of freedom and 30 denominator degrees of freedom. Table 12.2 shows that with these degrees of freedom, $F_{.05} = 1.79$. Computing the F value from the sample results, we have

$$F = \frac{s_1^2}{s_2^2} = \frac{120}{80} = 1.5$$

Since 1.5 is less than $F_{.05} = 1.79$, H_0 cannot be rejected. Thus the sample results do not support the position that women have significantly greater variation in attitudes on political issues.

• • •

EXERCISES

9. Assume that we have two automobile mechanics that are equal in ability and that the variances of their completion times for automobile repair jobs are the same. Suppose that we have a sample of six jobs performed by the first mechanic and eight jobs performed by the second mechanic. Furthermore, assume that we will

compute the sample variances s_1^2 and s_2^2 for the completion times of the two mechanics.

a.   What is the sampling distribution of s_1^2/s_2^2? Show a graph of this sampling distribution.

b.   Use Table 12.2 to determine the probability that the ratio of the two sample variances is less than or equal to 3.97.

10.   Two secretaries are each given eight typing assignments of equal difficulty. The sample standard deviations of the completion times were 3.8 minutes and 5.2 minutes, respectively. Do the data suggest that there is a difference in the variability of completion times for the two secretaries? Test the hypothesis at a .10 level of significance.

11.   It is assumed that the variance of stopping distances of automobiles on wet pavement is substantially greater than the variance of stopping distances of automobiles on dry pavement. In testing this assumption, 16 automobiles traveling at the same speeds are tested with respect to stopping distances on wet pavement and stopping distances on dry pavement. On wet pavement, the standard deviation of stopping distances was 32 feet. On dry pavement, the standard deviation was 16 feet.

a.   At a .05 level of significance, do the sample data provide a clear indication that the variability in stopping distances on wet pavement is greater than the variability in stopping distances on dry pavement?

b.   What are the implications of your statistical conclusions in terms of driving safety recommendations?

12.   Assume that ball bearings are produced on two different shifts. Sample results concerning the standard deviation of bearing sizes (in inches) from the two shifts are as follows:

First Shift	Second Shift
$n_1 = 22$	$n_2 = 25$
$s_1 = .12$	$s_2 = .09$

Are the sample results sufficient to conclude that the variances for the ball bearings differ for the two shifts? Test with $\alpha = .10$.

13.   Consider the problem of estimating the difference between the mean checking account balances at two branches of the Clearview National Bank. The data collected from two independent random samples are as follows.

Branch Bank	Number of Checking Accounts	Sample Mean Balance	Sample Standard Deviation
Cherry Grove	12	$\bar{x}_1 = \$1000$	$s_1 = \$150$
Beechmont	10	$\bar{x}_2 = \$ 920$	$s_2 = \$120$

In using the t distribution to estimate the difference between means we assume that the variances of the two populations are equal. This assumption is the basis for developing a pooled variance estimate. Use $\alpha = .10$ and conduct a test for the equality of the population variances. Do your results justify the use of a pooled variance estimate?

14. Two new assembly methods are tested with the following variances in assembly times.

Method	Sample Size	Sample Variance
A	31	$s_1^2 = 25$
B	25	$s_2^2 = 12$

Use $\alpha = .10$ and test for equality of the two population variances.

15. Independent random samples of parts manufactured by two suppliers show the following results.

Supplier	Sample Size	Sample Variance of Part Sizes
Durham Electric	41	$s_1^2 = 3.8$
Raleigh Electronics	31	$s_2^2 = 2.0$

A firm currently using the Durham supplier will continue to do so unless a hypothesis test shows that the Raleigh supplier provides a significantly lower variance in part sizes. Use $\alpha = .05$ and conduct the statistical test that will help the firm decide whether or not to change suppliers.

Summary

In this chapter we have presented statistical procedures that can be used to make inferences about population variances. In the process we introduced two new probability distributions: the chi-square distribution and the F distribution. The chi-square distribution can be used as the basis for interval estimation and hypothesis tests concerning the variance of a normally distributed population. In particular, we showed that with random samples of size n selected from a normally distributed population, the quantity $(n - 1)s^2/\sigma^2$ has a chi-square distribution with $n - 1$ degrees of freedom.

We illustrated the use of the F distribution in making hypothesis tests concerning the variances of two normally distributed populations. In particular, we showed that with independent random samples of sizes n_1 and n_2 selected from two normal populations with equal variances, $\sigma_1^2 = \sigma_2^2$, the sampling distribution of the ratio of the two sample variances, s_1^2/s_2^2, has an F distribution with $n_1 - 1$ degrees of freedom for the numerator and $n_2 - 1$ degrees of freedom for the denominator.

Statistics in Practice

ENVIRONMENTAL PROTECTION

As part of a program established by the United States Department of Interior to clean up the nation's rivers and lakes, environmental protection grants were made to a number of small cities scattered throughout the country. Congress asked the United States General Accounting Office (GAO), an independent and nonpolitical audit organization in the legislative branch of the Federal government, to determine how effectively the program was operating. To make this determination, GAO examined records and conducted site visits at several sewage treatment plants.

One objective of these audits was to ensure that the effluent (treated sewage) at the plants met certain standards before entering rivers and/or streams. Among other things, the following characteristics of the effluents were examined:

1. Oxygen content
2. pH level
3. Amount of suspended solids
4. Amount of soluble solids

GAO's investigation of one particular plant began with an examination of a sample of the records that were periodically sent to the state engineering department. Initially, GAO auditors were concerned with determining whether or not various characteristics of the effluent were within acceptable limits. In this case the measurements were within acceptable limits, but the auditors noted that there was very little variance in the data over several periods of time.

For example, the pH level of water is 7. A certain variance is normal and expected for samples taken from different sources at different times. However, the apparent low variance in the pH level data at this plant caused the auditors to conduct the following hypothesis test concerning the variance in pH level data:

$$H_0: \sigma^2 = \text{standard}$$

$$H_1: \sigma^2 < \text{standard}$$

The *standard* variance in this case was taken to be the variance in pH level found at plants known to be functioning properly. The hypothesis test led to the rejection of H_0 and the conclusion that the variance in the pH level at the plant in question was significantly less than the variance at properly functioning plants.

The auditors then made a plant visit to examine the measuring equipment and discuss their findings with the plant operator. The auditors found that the measuring equipment at the plant was not being used because the plant operator did not know how to operate it. Instead, the operator was told by an engineer what an acceptable pH level was and had simply recorded numbers within the acceptable range each day without actually conducting the required tests. The low variance in the pH level data had alerted the auditors to the potential violation. The plant was ordered to correct its procedures and to begin using the measuring equipment. Further study led to a GAO recommendation that cities receiving funds from the program be required to establish training programs for plant operators in order to ensure that proper testing procedures were being conducted.

Key Formulas

$(1 - \alpha)\%$ Confidence Interval for a Population Variance

$$\frac{(n - 1)s^2}{\chi^2_{\alpha/2}} \leq \sigma^2 \leq \frac{(n - 1)s^2}{\chi^2_{1-\alpha/2}} \tag{12.6}$$

Sampling Distribution of s_1^2/s_2^2 when $\sigma_1^2 = \sigma_2^2$

$$F = \frac{s_1^2}{s_2^2} \tag{12.7}$$

Review Quiz

TRUE/FALSE

1. The sample variance should not be used as a point estimator of the population variance.
2. The sampling distribution of $(n - 1)s^2/\sigma^2$ is a normal probability distribution.
3. The random variable for the chi-square probability distribution may not assume negative values.
4. The chi-square probability distribution is symmetric.
5. In order to determine the appropriate number of degrees of freedom for a chi-square distribution, one must know the sample size.
6. When applying the chi-square distribution, we must be able to assume the population being sampled follows a normal probability distribution.
7. For the F distribution the number of degrees of freedom in the numerator must be greater than or equal to the number of degrees of freedom in the denominator.
8. An F test can be used for a hypothesis test concerning the equality of two population variances.

MULTIPLE CHOICE

9. The sampling distribution of the quantity $(n - 1)s^2/\sigma^2$ is the
   a. chi-square distribution
   b. normal distribution
   c. F distribution
   d. t distribution
10. The sampling distribution of the ratio of independent sample variances extracted from two normally distributed populations with equal variances is the
    a. chi-square distribution
    b. normal distribution
    c. F distribution
    d. t distribution
11. When a sample variance of 25 is obtained from a sample of 10 items, the 90% confidence interval for a population variance is
    a. 12.3 to 57.1
    b. 13.3 to 67.7
    c. 14.1 to 46.25
    d. 15.3 to 53.98
12. Compared to a 95% confidence interval, a 99% confidence interval would be
    a. a narrower interval
    b. a wider interval
    c. the same width because you use the same data
    d. either narrower or wider

Use the following information for questions 13–15. These sample results were obtained for independent random samples from two normally distributed populations with equal variances.

	Sample 1	Sample 2
Sample size	10	16
Sample variance	25	20

13. Using a .05 level of significance, one critical value for an F test on the equality of the respective population variances is:
    a. 2.59
    b. 3.01
    c. 3.12
    d. 3.89
14. The test statistic F is
    a. .13
    b. .5
    c. .8
    d. 1.25

15. The conclusion reached would be which of the following?
    a. There is a statistically significant difference between the variances of the two populations.
    b. There is no statistically significant difference between the variances of the two populations.

Supplementary Exercises

16. Because of staffing decisions, management of the Gibson-Marimont Hotel is interested in the variability for the number of rooms occupied per day during a particular season of the year. A sample of 20 days of operation shows a sample mean of 290 rooms occupied per day and a sample standard deviation of 30 rooms.
    a. What is the point estimate of the population variance?
    b. Provide a 90% confidence interval estimate for the population variance.
    c. Provide a 90% confidence interval estimate for the population standard deviation.
17. A random sample of 30 days of sales for United Mufflers, Inc., shows a sample mean of 22.5 mufflers sold per day, with a sample standard deviation of 6. Provide 95% confidence interval estimates of both the population variance and the population standard deviation for the muffler sales data.
18. Historical delivery times for Buffalo Trucking, Inc., have had a mean of 3 hours and a standard deviation of .5 hours. A sample of 22 deliveries over the past month provides a sample mean of 3.1 hours and a sample standard deviation of .75 hours.
    a. Use a test of hypothesis about a population variance, and test to determine if the sample results support the historical delivery variability of $\sigma^2 = (.5)^2 = .25$. Use $\alpha = .05$.
    b. Provide a 95% confidence interval estimate for both the population variance and the population standard deviation.
19. Part variability is very critical in the manufacturing of ball bearings. Large variances in the size of the ball bearings cause bearing failure and rapid wearout. Production standards call for a maximum variance of .0001 when the bearing sizes are measured in inches. A sample of 15 bearings shows a sample standard deviation of .014 inches.
    a. Test with $\alpha = .10$ to determine if the sample bearings were taken from a population having a variance of .0001 or less.
    b. Provide a 90% confidence interval estimate for the variance of the ball bearings in the population.
20. The filling variance for boxes of cereal is designed to be .02 or less. A sample of 41 boxes of cereal shows a sample standard deviation of .16 ounces. Test with $\alpha = .05$ to determine if the variance in the cereal box fillings is exceeding the standard.
21. A sample standard deviation for the number of passengers taking a particular airline flight is 8. A 95% confidence interval estimate of the standard deviation is 5.86 to 12.62.

    a. Was a sample size of 10 or 15 used in the above statistical analysis?

    b. If the sample standard deviation of $s = 8$ had been based on a sample of 25 flights, what change would you expect in the confidence interval for the population standard deviation? Compute a 95% confidence interval for σ if a sample size 25 had been used.

22. The following sample data have been collected from two independent random samples:

Population	Sample Size	Sample Mean	Sample Variance
A	$n_A = 25$	$\bar{x}_A = 40$	$s_A^2 = 5$
B	$n_B = 21$	$\bar{x}_B = 50$	$s_B^2 = 11$

In a test for the difference between the two population means, the statistical analyst is considering using a pooled estimate of the population variance based on the assumption the variances of the two populations are equal. Is pooling appropriate in this case? Use $\alpha = .10$ for your test.

23. A firm gives a mechanical aptitude test to all job applicants. A sample of 20 male applicants shows a sample variance of 80 for the test scores. A sample of 16 female applicants shows a sample variance of 220. Test with $\alpha = .05$ to determine if the test score variances differ for male and female job applicants. If a difference in variances exist, which group has the higher variance in mechanical aptitude?

24. The accounting department analyzes the variances of the weekly unit costs reported by two production departments. A sample of 16 cost reports for each of the two departments shows cost variances of 2.3 and 5.4, respectively. Is this sample sufficient to conclude that the two production departments differ in terms of unit cost variances? Use $\alpha = .10$.

25. In using the t distribution to estimate the difference between two population means an analyst is interested in computing a pooled estimate of the variance of the populations. Pooling is justified only if it appears reasonable to assume that the two populations have equal variances. Use the following data to determine if pooling is appropriate for this situation (test with $\alpha = .10$):

      Sample 1   80, 72, 75, 90, 78, 75, 72, 85.
      Sample 2   50, 48, 45, 60, 65, 66, 70, 54.

If pooling is appropriate, what is the pooled estimate of the population variances?

Computer Exercise

An Air Force introductory course in electronics is currently taught using computer-assisted instruction, with each student in the course working individually at a computer terminal. It has been proposed that a better approach to teaching the course would be to have a pair of students work together at each computer terminal. In addition to the fact that a greater number of students could be taught at the same time, the proposed method may have the positive effect of reducing overall training time due to the fact that the students can help each other learn. In order to test the proposed method, an entering class of 120 students was randomly assigned to two groups of 60 students each. One group of 60 was taught using the current method; the second group was taught using the new method. The time in hours was recorded for each student in the sample. The data are shown.

Proposed Method	Current Method	Proposed Method	Current Method
75	73	75	79
76	77	71	79
75	72	75	74
73	76	72	79
74	65	77	72
73	72	78	75
75	71	74	75
73	73	76	71
74	67	75	74
76	82	77	80
77	68	76	82
77	72	71	76
76	69	77	76
72	76	74	77
77	79	82	81
75	76	75	74
76	80	73	82
77	73	72	75
72	79	76	72
75	77	73	76
74	75	77	63
75	77	75	79
70	70	73	75
74	73	78	74
78	78	75	76
80	73	74	77
72	76	72	74
76	72	77	70
76	61	76	72
73	76	78	78

QUESTIONS

1. Develop numerical and graphical measures to summarize the data.
2. Does there appear to be any difference between the mean training times for the two methods? Explain.
3. Does variance in the training times appear to be a significant factor in terms of the difference between the two methods?
4. What conclusion can you reach about any differences in the two methods? What is your recommendation? Explain.

CHAPTER 13

Tests of Goodness of Fit and Independence

What You Will Learn in This Chapter

- what a goodness of fit test is

- what a contingency table is

- how to use the chi-square distribution to conduct tests of goodness of fit and independence

Contents

Statistics in the News

CONCERN FOR WAR ON THE RISE

What do you think is the most important problem facing this country today? Questions of this form are periodically asked in public opinion polls across the United States in an attempt to learn about the concerns of the American people.

A November 1983 Gallup Poll found that, for the first time since the Vietnam war of the late 1960s, the American people were seeing international tensions and the threat of war as the most important problem facing the United States. The Gallup Poll findings were based on personal interviews of 1504 adults, 18 and older, conducted in more than 300 scientifically selected localities across the nation. Shown below are the number of people responding in each category for the Gallup Poll surveys of October 1982 and November 1983. A statistical procedure known as a chi-square test reveals that the two surveys show a statistically significant change in public opinion. In particular, the threat of war, which ranked a distant third (6%) in October 1982, had risen to the number one concern (37%) by November 1983.

The 1982 and 1983 surveys bracketed the ter-

The awesome power of nuclear weapons poses a real threat to the future of the world.

rorist bombings of the Marine barracks in Beirut and the U.S. operation in Grenada. In addition, the 1983 survey may also have reflected the publicity surrounding the televised nuclear war movie "The Day After," which was viewed by roughly half of all U.S. adults. With incidents such as these, it is perhaps not surprising that for the first time in 15 years, the American people were seeing war as their number one concern.

| | Most Important Problem in United States Today | | | | |
	Threat of War	Unemployment	Inflation	Moral Decline	Other
October 1982 survey	91	932	271	45	165
November 1983 survey	557	481	165	90	211

Based on "War Concerns Rise," *Tampa Tribune* (December 9, 1983).

In Chapter 12 we introduced the chi-square distribution and illustrated how it could be used in interval estimation and hypothesis testing. In this chapter we introduce two more hypothesis-testing procedures that are based on the use of the chi-square distribution. As with other hypothesis-testing procedures, these tests compare observed sample results with those that are expected when the null hypothesis is true. The acceptance or rejection of the null hypothesis is based upon how "close" the sample or observed results are to the expected results.

In the following section we introduce a goodness-of-fit test involving a multinomial population. Later we discuss a test for independence using contingency tables.

13.1 GOODNESS OF FIT TEST—A MULTINOMIAL POPULATION

In this section we consider the situation in which each element of a population is assigned to one and only one of several classes or categories. Such a population is described by a *multinomial probability distribution*. Example 13.1 describes a multinomial distribution with three classes, or categories.

EXAMPLE 13.1

Patients that arrive for treatment at the emergency room of a large metropolitan hospital are assigned to one of the following three categories based upon the seriousness of their condition.

> Category 1: Patient condition is stable; immediate treatment by a physician is not required.
> Category 2: Patient condition is serious; immediate treatment is not required, but patient should be monitored for vital signs until a physician is available.
> Category 3: Patient condition is critical; the patient's life will be endangered without immediate treatment.

In this example the population of interest is a multinomial population, since the condition of each patient is classified into one, and only one, category. We have a multinomial population with three classifications or categories: stable, serious, and critical.

• • •

In general, a multinomial population involves k categories. A goodness-of-fit test can be used to determine if a particular multinomial distribution provides a good description of the population. To perform such a test we must first formulate a null hypothesis concerning the particular multinomial probability distribution.

EXAMPLE 13.1 (continued)

Over the past year the hospital's records show that 50% of the patients that arrived for treatment were classified as stable, 30% were classified as serious, and 20% were classified as critical. The hospital's reputation for providing superior emergency room treatment, however, has resulted in an increased volume for the emergency room. The director is concerned that the percentage of patients classified as having stable, serious, or critical conditions may have also changed. Let us define the following notation:

$$p_1 = \text{percentage of patients classified as stable}$$

$$p_2 = \text{percentage of patients classified as serious}$$

$$p_3 = \text{percentage of patients classified as critical}$$

Based upon the assumption that the increase in volume for the emergency room has not altered the distribution of patients among the categories, the null and alternative hypotheses would be stated as follows:

$$H_0\text{: } p_1 = .50, p_2 = .30, \text{ and } p_3 = .20$$

$$H_1\text{: The population proportions are not}$$

$$p_1 = .50, p_2 = .30, \text{ and } p_3 = .20$$

$$\bullet \quad \bullet \quad \bullet$$

Once the hypotheses have been formulated we must obtain a simple random sample of n items from the population in order to conduct the test. Using the sample of n items, we record the observed frequencies for each of the k classes or categories. Then, given the usual hypothesis-testing assumption that the null hypothesis is true, we determine the expected frequency for each category. The expected frequency for each category is found by multiplying the sample size n by the proportion assumed to be in that category under the null hypothesis.

EXAMPLE 13.1 (continued)

Suppose the hospital has selected a sample of 200 patients who have been treated since the volume increased in the emergency room. The observed frequencies for this group are summarized as follows.

Stable Condition	Serious Condition	Critical Condition
98	48	54

The next step is to compute the expected frequencies for the 200 patients under the null hypothesis assumption that $p_1 = .50$, $p_2 = .30$, and $p_3 = .20$. Doing so provides the following expected frequencies:

Stable Condition	Serious Condition	Critical Condition
200(.50) = 100	200(.30) = 60	200(.20) = 40

Note that the expected frequency for each category is found by multiplying the sample size of 200 by the hypothesized proportion for the category.

$$\bullet \quad \bullet \quad \bullet$$

The goodness-of-fit test now focuses on the differences between the observed and expected frequencies. One or more large differences between observed and expected frequencies casts doubt on the assumption that the hypothesized proportions are correct. However, small differences between observed and expected frequencies do not provide sufficient evidence to reject the null hypothesis. In the case where observed and expected frequencies are equal, the sample data provide a perfect fit to the hypothesized distribution.

Let

O_i = observed frequency for category i

E_i = expected frequency for category i based on the assumption that the null hypothesis is true,

k = the number of categories.

If the null hypothesis is true and if the sample size is large, then the quantity

$$\chi^2 = \sum_{i=1}^{k} \frac{(O_i - E_i)^2}{E_i} \tag{13.1}$$

has a chi-square distribution. From the numerator of (13.1), we see that larger differences between observed and expected frequencies cause larger values of χ^2, and vice versa. For the case where the null hypothesis involves proportions for k categories of a multinomial population, the appropriate chi-square distribution has $k - 1$ degrees of freedom. The requirement of a *large* sample size is satisfied whenever the expected frequency for each category is 5 or more.

EXAMPLE 13.1 (continued)

With $E_i \geq 5$ for all three categories, the sample size is large, and we can proceed with the computation of the chi-square (χ^2) value in (13.1), as follows:

$$\chi^2 = \frac{(98 - 100)^2}{100} + \frac{(48 - 60)^2}{60} + \frac{(54 - 40)^2}{40}$$

$$= \frac{4}{100} + \frac{144}{60} + \frac{196}{40}$$

$$= .04 + 2.40 + 4.90 = 7.34$$

Suppose we test the null hypothesis that the multinomial population has the proportions of $p_1 = .50$, $p_2 = .30$, and $p_3 = .20$ at the $\alpha = .05$ level of significance. We will reject the null hypothesis only if the differences between observed and expected cell frequencies are *large*. Thus the larger the value of χ^2, the more likely it is we will reject the null hypothesis. Using $\alpha = .05$, we will place a rejection area of .05 in the upper tail of the chi-square distribution. Checking the chi-square distribution table (Table 3 of the Appendix B), we find that with $k - 1 = 3 - 1 = 2$ degrees of freedom $\chi^2_{.05} = 5.99147$. Thus, as with similar one-tailed tests, we will reject H_0 if the computed chi-square value exceeds the critical value of 5.99147.

From our previous calculations we have computed $\chi^2 = 7.34$. Since this value is larger than the critical value of 5.99147, we reject the null hypothesis. In rejecting H_0 we conclude that the increase in volume for the emergency room has altered the percentages of patients whose conditions are stable, serious, or critical. While the goodness-of-fit test itself permits no further conclusions, we can informally compare the observed and expected frequencies to obtain an idea of where the significant differences are. For instance, considering the critical-condition category, we find that the observed frequency of 54 is larger than the expected frequency of 40. Since the expected frequency was based upon the historical percentage observed, the larger observed frequency suggests that associated with the increase in volume for the emergency room has been an increase in the percentage of patients whose conditions are classified as critical. Similarly, there has been a decrease in the percentage of patients whose conditions are classified as serious.

• • •

As illustrated in Example 13.1, the goodness-of-fit test uses the chi-square distribution to determine if a hypothesized probability distribution for a population provides a good fit. Acceptance or rejection of the hypothesized population distribution is based upon differences between observed frequencies in a sample and the expected frequencies based on the assumed population distribution. Let us outline the general steps that can be used to conduct a goodness-of-fit test for any hypothesized population distribution:

1. Formulate a null hypothesis indicating a hypothesized distribution for k classes or categories of a population.
2. Use a simple random sample of n items and record the observed frequencies for each of the k classes or categories.
3. Use the assumption that the null hypothesis is true and determine the expected frequencies for each category.
4. Use the observed and expected frequencies and (13.1) to compute a value of χ^2 for the test.
5. Decision rule:

$$\text{Accept } H_0 \text{ If } \chi^2 \leq \chi^2_\alpha$$

$$\text{Reject } H_0 \text{ If } \chi^2 > \chi^2_\alpha$$

where α is the level of significance for the test.

EXERCISES

1. During the first 13 weeks of the television season, the Saturday evening 8:00 P.M. to 9:00 P.M. audience proportions were recorded as: ABC, 29%; CBS, 28%; NBC,

25%; and independents 18%. A sample of 300 homes two weeks after a Saturday night schedule revision showed the following viewing audience data: ABC 95 homes, CBS 70 homes, NBC 89 homes, and independents 46 homes. Test with $\alpha = .05$ to determine if the viewing audience proportions have changed.

2. A new container design has been adopted by a manufacturer. Color preferences indicated in a sample of 150 individuals are as follows:

Red	Blue	Green
40	64	46

Test using $\alpha = .10$ to see if the color preferences are the same. (*Hint:* Formulate the null hypothesis as $H_0: p_1 = p_2 = p_3 = \frac{1}{3}$.)

3. Grade distribution guidelines for a statistics course at a major university are as follows: 10% A, 30% B, 40% C, 15% D, and 5% F. A sample of 120 statistics grades at the end of a semester showed 18 A's, 30 B's, 40 C's, 22 D's, and 10 F's. Use $\alpha = .05$ and test to see if the actual grades differ significantly from the grade distribution guidelines.

4. Consumer panel preferences for three proposed store displays are as follows:

Display A	Display B	Display C
43	53	39

Use $\alpha = .05$ and test to see if there is a preference among the three display designs

5. At Ontario University entering freshmen have historically selected the following colleges:

College	Percentage
Business	15%
Education	20%
Engineering	30%
Liberal Arts	25%
Science	10%

Data obtained for the most recent class show that 73 students selected business, 105 selected education, 150 selected engineering, 124 chose liberal arts, and 47 selected science. Use $\alpha = .10$ and test whether or not the historical percentages have changed.

13.2 TEST OF INDEPENDENCE—CONTINGENCY TABLES

Another important application of the chi-square distribution involves using sample data to test for the independence of two variables. To illustrate the test of independence, we consider the study conducted by Alber's Brewery.

EXAMPLE 13.2

Alber's Brewery manufactures and distributes three types of beers: a low-calorie light beer, a regular beer, and a dark beer. In an analysis of the market segments for the three beers, the firm's market research group has raised the question of whether or not preferences for the three beers differ between male and female beer drinkers. If beer preference is independent of the sex of the beer drinker, one advertising campaign will be initiated for all of Alber's beers. However, if beer preference depends upon the sex of the beer drinker, the firm will tailor its promotions toward different target markets.

A test of independence addresses the question of whether or not the beer preference (light, regular, or dark) is independent of the sex of the beer drinker (male, female). The hypotheses for this test of independence are as follows:

H_0: Beer preference is independent of the sex of the beer drinker

H_1: Beer preference is not independent of the sex of the beer drinker (i.e., males and females differ in their preferences)

Table 13.1 can be used to describe the situation being studied. By identifying the population as all male and female beer drinkers, a sample can be selected and each individual asked to state his or her preference for the three Alber's beers. Every individual in the sample will be placed in one of the six cells in the table. For example, an individual may be a male preferring regular beer (cell 2), a female preferring light beer (cell 4), a female preferring dark beer (cell 6), and so on. Since we have listed all possible combinations of beer preference and sex—or, in other words, listed all possible contingencies—Table 13.1 is called a *contingency table*. The test of independence

TABLE 13.1

Contingency Table—Beer Preference and Sex of Beer Drinkers

	Beer Preference		
Sex	Light	Regular	Dark
Male	(Cell 1)	(Cell 2)	(Cell 3)
Female	(Cell 4)	(Cell 5)	(Cell 6)

makes use of the contingency table format and for this reason is sometimes referred to as a *contingency table test*.

Suppose that a simple random sample of 150 beer drinkers has been selected. After taste-testing the three beers, the individuals in the sample are asked to state their preference, or first choice. The contingency table in Table 13.2 summarizes the responses to the study. As we see in the contingency table, the data for the test of

TABLE 13.2

Sample Results of Beer Preferences for Male and Female Beer Drinkers (Observed Frequencies)

Sex	Beer Preference			Total
	Light	Regular	Dark	
Male	20	40	20	80
Female	30	30	10	70
Totals	50	70	30	150

independence are collected in terms of counts, or frequencies, for each cell or category. Thus of the 150 individuals in the sample, 20 were men who favored light beer, 40 were men who favored regular beer, 20 were men who favored dark beer, and so on.

The data in Table 13.2 contain the sample, or observed, frequencies for each of six classes or categories. If we can determine the expected frequencies under the assumption of independence between beer preference and sex of the beer drinker, we can use the chi-square distribution, just as we did in the previous section, to determine whether or not there is a significant difference between observed and expected frequencies.

Expected frequencies for the cells of the contingency table are based on the following rationale: First, we assume that the null hypothesis of independence between beer preference and sex of the beer drinker is true. Then we note that the sample of 150 beer drinkers showed a total of 50 preferring light beer, 70 preferring regular beer, and 30 preferring dark beer. In terms of fractions, we conclude that $50/150 = 1/3$ of the beer drinkers prefer light beer, $70/150 = 7/15$ prefer regular beer, and $30/150 = 1/5$ prefer dark beer. If the *independence* assumption is valid, we argue that these same fractions must be applicable to both male and female beer drinkers. Thus under the assumption of independence, we would expect the 80 male beer drinkers to show that $(1/3)80 = 26.67$ prefer light beer, $(7/15)80 = 37.33$ prefer regular beer, and $(1/5)80 = 16$ prefer dark beer. Application of these same fractions to the 70 female beer drinkers provides the expected frequencies as shown in Table 13.3.

Let E_{ij} stand for the expected frequency for the contingency table category in row i and column j. With this notation let us reconsider the expected frequency calculation for males (row $i = 1$) who prefer regular beer (column $j = 2$)—that is, expected frequency E_{12}. Following our previous argument for the computation of expected

TABLE 13.3

Expected Frequencies If Beer Preference Is Independent of the Sex of the Beer Drinker

Sex	Beer Preference			Total
	Light	Regular	Dark	
Male	26.67	37.33	16.00	80
Female	23.33	32.67	14.00	70
Total	50	70	30	150

frequencies, we showed that

$$E_{12} = \left(\frac{7}{15}\right)80 = 37.33$$

Writing this slightly differently, we find

$$E_{12} = \left(\frac{7}{15}\right)80 = \left(\frac{70}{150}\right)80 = \frac{(80)(70)}{150} = 37.33$$

Note that the 80 in the above expression is the total number of males (row 1), the 70 is the total number preferring regular beer (column 2), and the 150 is the total sample size. Thus we see that

$$E_{12} = \frac{(\text{Row 1 Total})(\text{Column 2 Total})}{\text{Sample Size}}$$

• • •

Generalization of this last expression shows that the following formula provides the expected frequencies for a contingency table in the test for independence.

Expected Frequencies for Contingency Tables Under the Assumption of Independence

$$E_{ij} = \frac{(\text{Row } i \text{ Total})(\text{Column } j \text{ Total})}{\text{Sample Size}} \tag{13.2}$$

The test procedure for comparing the observed frequencies of Table 13.2 with the expected frequencies of Table 13.3 is similar to the goodness-of-fit calculations made in the previous section. Specifically, the value of χ^2, based on the differences between the observed and expected frequencies, is computed as follows.

$$\chi^2 = \sum_i \sum_j \frac{(O_{ij} - E_{ij})^2}{E_{ij}} \tag{13.3}$$

where

O_{ij} = observed frequency for contingency table category in row i and column j,

E_{ij} = expected frequency for contingency table category in row i and column j.

The double summation in (13.3) indicates that the values must be summed for each cell in the contingency table.

Before we apply (13.3), we must check to see that the expected frequencies in each cell are at least 5. This is the same check that we used in the previous section to determine whether or not the sample size was large enough for the chi-square distribution assumption to be made.

EXAMPLE 13.2 (continued)

Since all expected frequencies in Table 13.3 are at least 5, we can conclude that the sample size is adequate. The resulting value of χ^2 is found as follows:

$$\chi^2 = \frac{(20 - 26.67)^2}{26.67} + \frac{(40 - 37.33)^2}{37.33} + \cdots + \frac{(10 - 14.00)^2}{14.00}$$

$$= 1.67 + .19 + \cdots + 1.14 = 6.13$$

The number of degrees of freedom for the appropriate chi-square distribution is computed by multiplying the *number of rows minus* 1 times the *number of columns minus* 1. With two rows and three columns, we have $(2 - 1)(3 - 1) = (1)(2) = 2$ degrees of freedom for the test of independence of beer preference and the sex of the beer drinker. Using $\alpha = .05$ for the level of significance of the test, Table 3 of the Appendix shows an upper-tail value of $\chi^2_{.05} = 5.99147$. Here again we are using the upper-tail value because we will reject the null hypothesis only if the differences in observed and expected frequencies provide a large value of χ^2. In our example, $\chi^2 = 6.13$ is greater than the critical value of $\chi^2_{.05} = 5.99147$. Thus we reject the null hypothesis of independence and conclude that the preference for the beers is not independent of the sex of the beer drinkers.

Although the test for independence allows only the above conclusion, we can again informally compare the observed and expected frequencies in order to obtain an idea of how the dependence between the beer preference and sex of the beer drinker comes about. (Refer to Tables 13.2 and 13.3.) We see that male beer drinkers have higher observed than expected frequencies for both regular and dark beers, while female beer drinkers have a higher observed than expected frequency for only the light beer. These observations give us an insight into the differing preferences between male and female beer drinkers. This information can be used by the company in targeting its promotions for the different beers.

• • •

EXERCISES

6. The number of units of three different products sold by three salespersons over a 3-month period are shown.

Salesperson	Product A	Product B	Product C
Troutman	14	12	4
Kempton	21	16	8
McChristian	15	5	10

Use $\alpha = .05$ and test for the independence of salesperson and type of product sold.

7. Starting positions for business and engineering graduates are classified by industry as shown.

Degree Major	Oil	Chemical	Industry Electrical	Computer
Business	30	15	15	40
Engineering	30	30	20	20

Use $\alpha = .01$ and test for independence of degree major and industry type.

8. A sport preference poll shows the following data for men and women.

Sex	Baseball	Favorite Sport Basketball	Football
Men	19	15	24
Women	16	18	16

Use $\alpha = .05$ and test for similar sport preferences by men and women. What is your conclusion?

9.   Three suppliers provide the following data on defective parts.

Supplier	Good	Part Quality Minor Defect	Major Defect
A	90	3	7
B	170	18	7
C	135	6	9

Use $\alpha = .05$ and test for independence between the supplier and the part quality. What does the result of your analysis tell the purchasing department?

10.   A study of educational levels of voters and their political party affiliations showed the following results.

Educational Level	Democratic	Party Affiliation Republican	Independent
Did not complete high school	40	20	10
High school diploma	30	35	15
College degree	30	45	25

Use $\alpha = .01$ and test to see if party affiliation is independent of the educational level of the voters.

Summary

In this chapter we introduced the goodness-of-fit test and the test of independence procedures, both of which are based on the chi-square distribution. The purpose of the goodness-of-fit test is to determine whether or not a hypothesized probability distribution provides a good description of a particular population of interest. The computations for conducting the goodness-of-fit test involve comparing observed frequencies from a sample with expected frequencies when the hypothesized probability distribution is assumed true. A chi-square distribution is used to determine if the differences in observed and expected frequencies are large enough to reject the hypothesized probability distribution. We illustrated the goodness-of-fit test for an assumed multinomial probability distribution.

A test of independence for two variables is a straightforward extension of the methodology employed in the goodness-of-fit test for a multinomial population. A contingency table is used to display the observed and expected frequencies. Then a chi-square value is computed. Large chi-square values, caused by large differences between observed and expected frequencies, lead to the rejection of the null hypothesis of independence.

Statistics in Practice

EVALUATING COMMUNITY ATTITUDES TOWARD CHARITIES

A local United Way agency was interested in determining community attitudes about itself, its operation, and the charities it supported. For example, one area of interest was assessing the community attitudes toward the amount of funds United Way allocated to its administrative activities. To obtain data on this and other issues, a questionnaire was developed and distributed to individuals at 18 different organizations in the local area. One of the questions asked of respondents was:

Of the funds collected, what percentage do you feel goes to United Way administrative expenses?

( ) up to 10%
( ) 11%–20%
( ) 21% and over

Each respondent was also asked to indicate his or her occupation according to the following classification:

( ) Production line/assemblers
( ) Maintenance/warehouse
( ) Craftsmen and foremen
( ) Clerical worker
( ) Sales worker
( ) Managers and administrators
( ) Professional and technical workers
( ) Other _____

Since data was collected regarding perception of administrative expenses as well as occupation, it was suggested that a chi-square test for independence be performed. In this case a test of independence addressed the question of whether or not the perception of the United Way administrative expenditures is independent of the occupation of the respondent. The hypotheses for the test of independence are as follows:

H_0: Perception of Administrative Expenditures Is
Independent of the Occupation

H_1: Perception of Administrative Expenditures Is
Not Independent of the Occupation

Using $\alpha = .05$ for the chi-square test of independence, H_0 was rejected. As a result, it was concluded that individuals in different occupational classes do not have the same perceptions of the administrative expenses associated with United Way. Statistical tests such as these have helped the United Way draw conclusions from questionnaire data and obtain ideas for making adjustments in fund-raising efforts.

Glossary

Goodness-of-fit test—A statistical test conducted to determine whether to accept or reject a hypothesized probability distribution for a population.

Contingency table—A table used to summarize observed and expected frequencies for a test of independence of two variables associated with a population.

Key Formulas

Goodness-of-Fit Test

$$\chi^2 = \sum_{i=1}^{k} \frac{(O_i - E_i)^2}{E_i} \tag{13.1}$$

Expected Frequencies for Contingency Tables under the Assumption of Independence

$$E_{ij} = \frac{(\text{Row } i \text{ Total})(\text{Column } j \text{ Total})}{\text{Sample Size}} \tag{13.2}$$

Contingency Table Test

$$\chi^2 = \sum_i \sum_j \frac{(O_{ij} - E_{ij})^2}{E_{ij}} \tag{13.3}$$

Review Quiz

TRUE/FALSE

1. A goodness-of-fit test can be used to determine if a particular multinomial distribution provides a good description of a population.
2. The chi-square probability distribution should not be used for a goodness-of-fit test.
3. In computing the expected frequencies for a goodness-of-fit test, it is not proper to assume the null hypothesis is true.
4. In conducting a goodness-of-fit test, the expected frequency for each category must be greater than or equal to 10.

5. In conducting a goodness-of-fit test, the observed frequency for one or more categories may be less than 5.
6. The chi-square distribution is used in conducting a contingency table test.
7. In a contingency table test of independence, the number of rows must be equal to the number of columns.
8. In conducting a contingency table test for independence, the expected frequency for each cell must be at least 5.
9. In conducting either a goodness-of-fit or contingency table test, the larger the differences between the observed and expected frequencies, the more likely it is that the null hypothesis will be rejected.
10. The appropriate number of degrees of freedom for a contingency table test is given by the product of the number of rows times the number of columns.

MULTIPLE CHOICE

11. The sampling distribution for a goodness of fit test is the
    a. chi-square distribution
    b. t distribution
    c. F distribution
    d. normal distribution

Use the following information for questions 12–15. A firm that manufactures kitchen appliances takes a random sample of 300 families to check whether or not there is any significant difference in color preferences for appliances. The results are as follows.

Color Preferred	Number of Families
White	65
Coppertone	89
Avocado	72
Harvest gold	74
	300

12. The critical value for the multinomial goodness-of-fit test with a .01 level of significance is
    a. 6.25139
    b. 7.77944
    c. 11.3449
    d. 13.2767
13. The expected frequency is
    a. 4 for each color
    b. 50 for each color
    c. 75 for each color
    d. different for each color

14. The calculated value of the test statistic for this chi-square goodness-of-fit test is
    a. 4.08
    b. 8.52
    c. 11.3
    d. 306
15. The result of the test at the .01 level of significance is which of the following?
    a. There is a significant difference in color preference.
    b. There is no significant difference in color preference.
16. Set up a contingency table for the survey findings in the chapter-opening "Statistics in the News" article about the important problems facing the United States. The expected frequency in the "Unemployment" column and "October 1982 Survey" row is closest to
    a. 550
    b. 700
    c. 800
    d. 900
17. Refer to the contingency table in question 16. The difference between the observed and expected frequencies for the "Threat of War" issue in the 1982 survey is closest to
    a. 230
    b. -230
    c. -500
    d. 1400
18. Refer to the contingency table in question 16. The computed value of χ^2 is closest to
    a. 3
    b. 15
    c. 45
    d. 87
19. The value of χ^2 that cuts off an area in the upper tail of .01 with 8 degrees of freedom is closest to
    a. 14
    b. 19
    c. 22
    d. 25
20. The value of χ^2 that cuts off an area in the upper tail of .025 with 70 degrees of freedom is closest to
    a. 20
    b. 50
    c. 75
    d. 100

Supplementary Exercises

11. A regional transit authority was concerned about the number of riders on one of their routes. In setting up this route, the assumption was that the number of

riders was uniformly distributed from Monday through Friday. The following historical data were obtained:

Day	Number of Riders
Monday	13
Tuesday	16
Wednesday	28
Thursday	17
Friday	16

Test using $\alpha = .05$ to determine if the transit authority's assumption appears to be justified.

12. An automobile dealer sells three models of a certain make of pickup truck. Over the most recent sales period the dealer sold 27 units of model 1, 39 units of model 2, and 30 units of model 3. Using $\alpha = .05$, test to determine whether or not consumer preferences vary among the three models.

13. In setting sales quotas, the marketing manager makes the assumption that order potentials are the same in each of four sales territories. A sample of 200 sales shows the following number of orders from each region.

Sales Territories			
I	II	III	IV
60	45	59	36

Do these data support the manager's assumptions? Use $\alpha = .05$.

14. A sample of parts provided the following contingency table data concerning part quality and production shift.

Shift	Number Good	Number Defective
First	368	32
Second	285	15
Third	176	24

Use $\alpha = .05$ and test the assumption that part quality is independent of the production shift. What is your conclusion?

15. An analysis of attendance records and performance on the final examination was made for a freshman mathematics course. The following results were obtained.

Number of Classes Missed	Grade on Final			
	80 or Above	70s	60s	Below 60
None	18	11	5	1
1–5	14	12	3	3
More than 5	3	9	13	7

a. Use $\alpha = .05$ and test for independence between number of classes missed and the grade on the final exmination.

b. Were the assumptions required for using the chi-square test for independence satisfied for this data set? Explain and, if necessary, perform any further analysis required to determine if the two classifications are independent.

c. What does the result of your analysis tell the professor in charge of the course?

16. As part of a study conducted to determine if the teacher's perceptions of the behavior of first-grade students were any different than the parents' perceptions, the following results were obtained.

Teacher's Perception	Parents' Perception		
	Shy	Typical	Outgoing
Shy	9	7	4
Typical	7	12	6
Outgoing	4	16	10

Use $\alpha = .01$ and test to determine if the teacher's perception of students is different than the parent's perception.

17. As part of the standard course evaluation, students are asked to rate the course as either poor, good, or excellent. The course evaluation form also asks students to indicate whether or not the course taken was a required part of their academic program or was taken as an elective. The dean of the college is interested in determining if the rating of the course is independent of the reason for taking the course. The following results were obtained.

Reason for Taking the Course	Rating		
	Poor	Good	Excellent
Required	16	38	16
Elective	4	10	16

How would you respond to the dean? Use $\alpha = .01$.

18. The director of placement for Southwestern University has been studying the records of engineering students who have interviewed with Computer Systems, Inc., one of the major employers over the past 5 years. One interest was to see whether or not receiving a job offer was dependent upon the student's grade-point average. The following results were obtained.

Grade Point Average	Company Decision Offer Made	No Offer Made
Below 2.5	5	15
2.5–3.5	7	8
Above 3.5	18	12

At the $\alpha = .10$ level, determine whether or not the company decision to make a job offer is independent of the student's grade-point average.

19. A lending institution shows the following data regarding loan approvals by four different loan officers.

Loan Officer	Loan Approval Decision Approved	Rejected
Miller	24	16
McMahon	17	13
Games	35	15
Runk	11	9

Use $\alpha = .05$ and test to determine if the loan approval decision is independent of the loan officer reviewing the loan application.

20. Entering freshmen at a state college are given both a mathematics achievement test and a reading rate test. The following results were obtained.

Reading Rate	Mathematics Achievement Low	Normal	High
Poor	21	14	24
Below average	14	23	15
Average	18	17	34
Above average	28	16	21
Excellent	17	12	16

Use $\alpha = .05$ and test to determine if mathematics achievement and reading rate are independent.

Computer Exercise

The computer exercise in Chapter 2 described a study conducted by Consolidated Foods, Inc. The company had taken a sample of 100 customers in order to learn about how customers were paying for their food purchases. The data collected for each customer included how much was spent on the purchase and how the customer paid for the purchase. The alternative payment methods included cash, an approved check, or a credit card. In addition, data were also collected on the sex of the customer. The data obtained for the 100 customers are shown below.

Amount Spent ($)	Sex	Method of Payment	Amount Spent ($)	Sex	Method of Payment
84.12	Male	Check	86.34	Female	Check
34.66	Male	Credit card	20.23	Female	Credit card
37.27	Female	Credit card	108.70	Female	Check
38.82	Female	Credit card	45.36	Female	Credit card
46.50	Female	Credit card	83.31	Male	Check
99.67	Female	Check	64.45	Male	Credit card
70.18	Female	Check	54.33	Female	Credit card
99.21	Male	Check	16.78	Female	Cash
138.42	Female	Check	115.96	Male	Check
93.68	Female	Check	95.83	Female	Check
120.89	Female	Check	19.76	Female	Cash
10.14	Female	Cash	35.37	Male	Cash
74.51	Male	Check	111.98	Female	Check
17.91	Male	Check	103.95	Female	Check
49.59	Male	Check	90.40	Male	Credit card
4.74	Male	Cash	6.68	Male	Cash
48.14	Male	Cash	32.09	Male	Credit card
65.67	Male	Credit card	79.70	Male	Credit card
89.66	Female	Check	96.08	Male	Credit card
96.40	Female	Check	20.60	Male	Cash
54.16	Female	Credit card	78.81	Female	Check
79.55	Female	Check	123.62	Female	Check
67.95	Female	Check	125.01	Female	Check
30.69	Male	Cash	41.58	Male	Credit card
151.89	Female	Check	36.73	Male	Credit card
130.41	Female	Check	52.07	Female	Credit card
98.80	Female	Check	19.78	Male	Cash
23.59	Female	Cash	66.44	Female	Check
104.67	Female	Check	5.08	Male	Cash
90.04	Female	Check	50.15	Male	Credit card
77.62	Female	Check	114.42	Female	Check
36.01	Male	Cash	97.26	Male	Credit card
88.17	Female	Check	22.75	Male	Cash
66.76	Female	Credit card	53.63	Female	Credit card
23.50	Male	Cash	132.31	Female	Check

Amount Spent ($)	Sex	Method of Payment	Amount Spent ($)	Sex	Method of Payment
127.34	Female	Check	105.54	Male	Check
26.02	Male	Cash	66.09	Male	Check
79.77	Male	Check	62.24	Female	Check
29.35	Male	Check	97.93	Female	Check
71.31	Female	Credit card	10.57	Female	Cash
43.57	Female	Credit card	51.21	Male	Credit card
76.18	Female	Credit card	90.17	Female	Check
59.38	Male	Credit card	24.08	Male	Credit card
72.99	Male	Credit card	42.72	Male	Cash
19.24	Male	Cash	97.72	Female	Check
80.20	Female	Check	112.67	Female	Check
55.79	Female	Cash	14.30	Female	Cash
134.27	Female	Check	28.76	Male	Credit card
64.68	Female	Credit card	81.85	Female	Check
75.54	Female	Check	56.84	Female	Credit card

QUESTIONS

1. Develop a contingency table showing the sex of the customer and the method of payment.
2. Is method of payment independent of the sex of the customer? Explain.

CHAPTER 14

Experimental Design and the Analysis of Variance

What You Will Learn in This Chapter

- how to use analysis of variance to test for equality of the means of three or more populations

- how the *F* distribution is used in analysis of variance

- what an analysis of variance table is and how it is constructed

- the assumptions of analysis of variance

- the completely randomized and randomized block experimental designs

- how to make computations and conclusions for completely randomized and randomized block experimental designs

Contents

461

Statistics in the News

HIGH SELF-ESTEEM:
A KEY TO ACADEMIC ACHIEVEMENT

Self-esteem has been the subject of a great amount of research that reveals its importance in the school setting. Students with high self-esteem are more receptive to the educational process and respond in more positive directions toward the teacher, assignments, and school in general.

Robert H. Phillips, a researcher at Fordham University, reports of a study designed to investigate ways in which teacher interaction with elementary-school students can improve student self-esteem. A sample of 30 elementary-school students was selected from 10 low-income areas of New York City. Each student was assigned to one of three groups. The first group, the experimental group, was provided an environment in which the students were given positive reenforcement by the teacher for any positive statements the students made about themselves. For example, if a student offered a legitimate positive self-statement, the teacher was instructed to respond with statements such as, "I'm proud of you too," "Yes, you did do well," "You are doing beautifully," and so on. The second group, the control group, was provided the same physical environment but the teacher made no comment if students offered positive statements about themselves. The third group, referred to as the inventory group, received no special instructions and operated under normal conditions.

All students in the research project were given self-esteem tests, at the beginning and at the end of the 7-week study. A measure of how each student's self-esteem score changed during the study was recorded. The statistical technique known as analysis of variance was used to see if

A teacher's individual attention and positive statements help students improve self esteem and academic achievement.

the three groups differed in terms of their change in self-esteem scores. The statistical conclusion was that there was a significant difference among the self-esteem scores for the three groups. The experimental group with its reinforcing teacher statements showed a significantly greater increase in self-esteem scores.

Because self-esteem is important in children's academic experiences, this research suggests that teacher responses designed to enhance student self-esteem should be part of regular educational programs. Future research investigations are planned to gain additional evidence concerning the degree to which improvements in self-esteem scores lead to corresponding improvements in academic achievement.

Based on "Increasing Positive Self-Referent Statements to Improve Self-Esteem in Low-Income Elementary School Children" by Robert H. Phillips, *The Journal of School Psychology* (Summer 1984).

In Chapter 11 we discussed how to test whether or not the means of two populations are equal. Recall that the test required the selection of an independent random sample from each of the two populations. In this chapter we present a statistical procedure for determining whether or not the means of more than two populations are equal. This technique is called *analysis of variance (ANOVA)*. In "Statistics in the News,"

ANOVA was used to test differences between mean self-esteem scores for three groups of students. In this chapter we also discuss the process of planning an experiment in order to collect the appropriate data; this process is referred to as *experimental design.*

14.1 EXPERIMENTAL DESIGN CONSIDERATIONS

In order to introduce the concepts of experimental design and analysis of variance, let us consider the situation described in the following example.

EXAMPLE 14.1

The Penn Yan School District has developed an 80-question reading competency examination in order to identify eighth-grade students with reading deficiencies. For students who answer less than 55 questions correctly on this examination, three remedial programs have been proposed:

Program 1: A review class that meets 1 hour each week for 10 weeks.
Program 2: A personalized system of instruction using microcomputers.
Program 3: A formal 10-week class that meets for 1 hour each day.

Before making a final decision about which remedial program to adopt, the board of education has requested that further study be conducted to determine how the proposed programs affect reading competency examination scores.

• • •

In Example 14.1, the reading competency remedial program is referred to as a *factor.* Since there are three remedial programs, or *levels,* corresponding to this factor, we say that there are three *treatments* associated with the experiment: one treatment corresponds to Program 1, another to Program 2, and the third to Program 3. In general, a treatment corresponds to a level of a factor.*

The random variable of interest in this experiment is the reading competency examination scores received by the students when they are retested. Our primary statistical objective is to determine whether or not the mean reading competency examination scores are the same for all three programs. In experimental design terminology, the random variable of interest is referred to as the *dependent variable,* the *response variable,* or the *response.*

Let us now turn our attention to how the experiment is actually conducted. To begin with, let us assume that a random sample of three students has been selected from the population of students who answered less than 55 questions correctly on the

---

*The term *treatment* was originally used in experimental design because many of the applications were in agriculture, where the treatments often corresponded to different types of fertilizers applied to selected agricultural plots. Today the term is used in a more general context.

original reading competency examination. In experimental design terminology, the students are referred to as *experimental units*.

The experimental design that we will use for this experiment is referred to as a *completely randomized design*. This type of design requires that each of the three remedial programs, or treatments, be randomly assigned to one of the experimental units or students. For example, Program 1 might be randomly assigned to the second student, Program 2 to the first student, and Program 3 to the third student. The concept of randomization as illustrated in this example is an important principle of all experimental designs.

Note that our experiment as described would result in only one measurement of reading competency score for each treatment. In other words, we have a sample size of 1 corresponding to each treatment. In order to obtain additional data for each preparation program, we must repeat, or *replicate*, the basic experimental process. For example, suppose that instead of selecting just three students at random, we had selected 15 students. If *each* of the three remedial programs is randomly assigned to five students, we say that five replicates have been obtained. The concept of *replication* is another important principle of experimental design. Figure 14.1 shows the completely randomized design for the reading competency remedial program study.

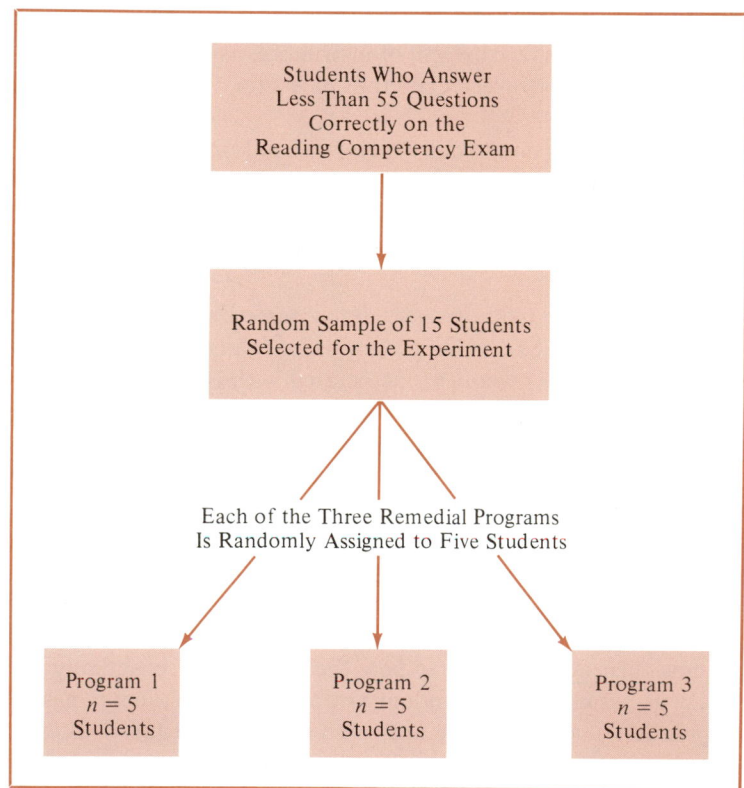

FIGURE 14.1
Completely Randomized Design for Evaluating the Remedial Programs

Data Collection

Once we are satisfied with the experimental design, we proceed by carrying out the study and collecting the data. In this case the students would receive additional reading instruction under their assigned remedial program. The students would then retake the reading competency exam, and their scores would be recorded. Let us assume that this has been done and that the scores for the 15 students in the study are as shown in Table 14.1. Using these data we calculate the sample mean score for each of the three remedial programs, as shown.

Remedial Program	Mean Exam Score
Program 1	55
Program 2	68
Program 3	57

From these data it appears that Program 2 may result in higher retest scores than either of the other methods. However, before we make any final recommendations we must remember that each of the given means is based upon the test results of just five students. Thus we are looking at three sample means drawn from the three populations representing all students who might participate in these programs. The real issue, then, is whether or not the three sample means observed are different enough for us to conclude that the means of the populations corresponding to the three remedial programs are different. Let

μ_1 = mean retest score for the population of all students assigned to Program 1

μ_2 = mean retest score for the population of all students assigned to Program 2

μ_3 = mean retest score for the population of all students assigned to Program 3.

We use $\bar{x}_1$, $\bar{x}_2$, and $\bar{x}_3$ to denote the corresponding sample means.

The experimental results for the three remedial programs yielded $\bar{x}_1 = 55$,

TABLE 14.1

**Reading Competency Examination Scores
for the 15 Students Who Were Retested**

		Observation				
	Program 1	48	54	57	54	62
Treatment	Program 2	73	63	66	64	74
	Program 3	51	63	61	54	56

$\bar{x}_2 = 68$, and $\bar{x}_3 = 57$. Although we will never know the actual values of μ_1, μ_2, and μ_3, we want to use the sample results to test the following hypotheses:

$$H_0: \mu_1 = \mu_2 = \mu_3$$

$$H_1: \text{Not All Means Are Equal}$$

In the next section we show how the ANOVA procedure can be used to determine if there is a significant difference between the means of the three populations.

14.2 THE ANALYSIS OF VARIANCE PROCEDURE FOR COMPLETELY RANDOMIZED DESIGNS

The analysis of variance (ANOVA) procedure that we describe is designed to test the following hypotheses.

Hypotheses for Analysis of Variance

$$H_0: \mu_1 = \mu_2 = \cdots = \mu_k$$

$$H_1: \text{Not All } \mu_i \text{ Are Equal}$$

where

$$k = \text{number of populations or treatments}$$

We assume that a simple random sample of size n has been selected from each of the k populations and that the sample means $\bar{x}_1, \bar{x}_2, \ldots, \bar{x}_k$ have been computed. In general, we refer to the sample mean corresponding to the ith population as $\bar{x}_i$ and the overall sample mean as $\bar{\bar{x}}$. That is,

$$\bar{x}_i = \frac{\sum\limits_{j} x_{ij}}{n} \tag{14.1}$$

$$\bar{\bar{x}} = \frac{\sum\limits_{i} \sum\limits_{j} x_{ij}}{n_T} \tag{14.2}$$

where

$$x_{ij} = j\text{th observation corresponding to the } i\text{th treatment}$$

$$n_T = \text{total sample size for the experiment}$$

Note that since the sample size is n for each of the k treatments, $n_T = kn$ and $\bar{\bar{x}} = \Sigma \bar{x}_i / k$.

EXAMPLE 14.1 (continued)

Since there are three remedial programs, or treatments, $k = 3$. Also, since each of the three programs is randomly assigned to five students, $n = 5$. Thus $n_T = (3)(5) = 15$.

$$\bullet \quad \bullet \quad \bullet$$

Assumptions for Analysis of Variance

The ANOVA procedure is based upon the following two assumptions:

1. The response variable, or variable of interest for each population, has a normal probability distribution.
2. The variance of the response variable is the same for each population.

EXAMPLE 14.1 (continued)

Assumption 1 requires that the variable of interest, the retest score, be normally distributed for each of the three programs under study. Assumption 2 requires that the variance of retest scores be the same for each program.

$$\bullet \quad \bullet \quad \bullet$$

We denote the common population variance (Assumption 2) of the response variable as σ^2. The basis for the methodology of the ANOVA procedure is the development of two independent estimates of σ^2. One estimate of σ^2 that we will develop is based upon the differences *between* the treatment means and the overall sample mean. The other estimate is based upon the differences of observations *within* each treatment from the corresponding treatment mean. By comparing these two estimates of σ^2, we will be able to determine whether or not we can reject the null hypothesis that the population means are equal.

Between-Treatments Estimate of Population Variance

The procedure for developing an estimate of σ^2 based upon the differences between the treatment means and the overall sample mean depends on the assumption that the null hypothesis is true and that the two assumptions given earlier are valid. In this case, if we let μ denote the common population mean under the null hypothesis (that is $\mu_1 = \mu_2 = \cdots = \mu_k = \mu$), then all the sample observations would represent data values drawn from the same normal probability distribution with mean μ and variance σ^2. If we let $\bar{x}$ denote the mean of a simple random sample of size n selected from this probability distribution, then the sampling distribution of $\bar{x}$ is normally distributed with mean μ and variance $\sigma_{\bar{x}}^2 = \sigma^2 / n$. Figure 14.2 illustrates such a sampling distribution. Thus, under the null hypothesis, we can think of each $\bar{x}_i$ as a value drawn

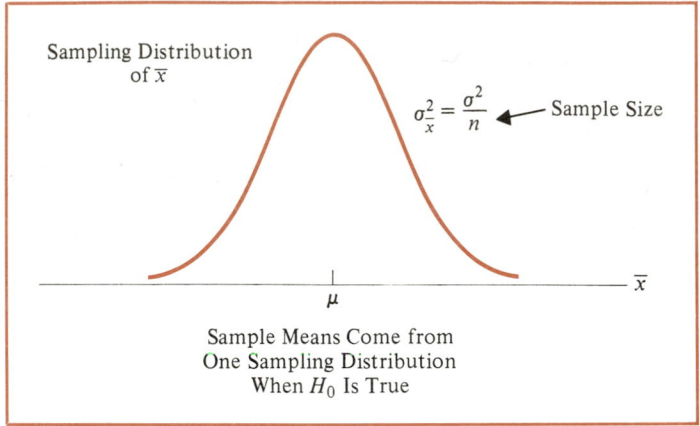

FIGURE 14.2
Sampling Distribution of $\bar{x}$ Given the Null Hypothesis
$\mu_1 = \mu_2 = \cdots = \mu_k = \mu$ **and a Sample of Size** n

at random from this sampling distribution. An estimate of the mean μ of this sampling distribution can be obtained by computing $\bar{\bar{x}}$, the overall sample mean.

EXAMPLE 14.1 (continued)

For the retest scores in Table 14.1, we use all the data to find

$$\sum_i \sum_j x_{ij} = (48 + 54 + 57 + \cdots + 54 + 56) = 900$$

Then using (14.2) with $n_T = 15$, we compute the overall sample mean,

$$\bar{\bar{x}} = \frac{\sum_i \sum_j x_{ij}}{n_T} = \frac{900}{15} = 60$$

as an estimate of μ.

• • •

To estimate the variance of the sampling distribution of $\bar{x}$ (that is, $\sigma_{\bar{x}}^2$), we can use the variance of the individual sample means about the overall sample mean, $\bar{\bar{x}}$. For the case where each sample is the same size, the estimated variance, denoted by $s_{\bar{x}}^2$, is

$$s_{\bar{x}}^2 = \frac{\sum_i (\bar{x}_i - \bar{\bar{x}})^2}{k - 1} \tag{14.3}$$

where $\bar{x}_i$ = mean of the ith sample.

EXAMPLE 14.1 (continued)

Since our previous computations of the three sample means showed $\bar{x}_1 = 55$, $\bar{x}_2 = 68$, and $\bar{x}_3 = 57$, we have

$$\sum_i (\bar{x}_i - \bar{\bar{x}})^2 = (55 - 60)^2 + (68 - 60)^2 + (57 - 60)^2$$

$$= 25 + 64 + 9 = 98$$

With $k = 3$ populations, or treatments, we have $k - 1 = 2$. Thus with (14.3), the estimate of the variance of the sampling distribution becomes

$$s_{\bar{x}}^2 = \frac{\sum_i (\bar{x}_i - \bar{\bar{x}})^2}{k - 1} = \frac{98}{2} = 49$$

$$\bullet \quad \bullet \quad \bullet$$

Since we know that $\sigma_{\bar{x}}^2 = \sigma^2/n$, solving for σ^2 gives

$$\sigma^2 = n\sigma_{\bar{x}}^2 \tag{14.4}$$

where n is the sample size involved in computing $\bar{x}$. Hence an estimate of the population variance, σ^2, can be obtained from (14.4) by multiplying n by the estimate of $\sigma_{\bar{x}}^2$.

EXAMPLE 14.1 (continued)

$$\text{Estimate of } \sigma^2 = n(\text{Estimate of } \sigma_{\bar{x}}^2) = ns_{\bar{x}}^2 \tag{14.5}$$

With all samples of size $n = 5$ and with $s_{\bar{x}}^2 = 49$, for the reading competency experiment we have

$$\text{Estimate of } \sigma^2 = 5(49) = 245$$

$$\bullet \quad \bullet \quad \bullet$$

This estimate of σ^2 is given the name *mean square between treatments* and is denoted by MSTR.

From (14.3) and (14.5) we see that MSTR can be written

$$\text{MSTR} = \frac{n\sum_i (\bar{x}_i - \bar{\bar{x}})^2}{k - 1} . \tag{14.6}$$

Tne numerator of (14.6) is called the *sum of squares between treatments* and is denoted by SSTR. The denominator, $k - 1$, represents the *degrees of freedom* corresponding to SSTR.

> **Between Treatments Estimate of Population Variance**
>
> $$\text{MSTR} = \frac{\text{SSTR}}{k-1} \qquad\qquad (14.7)$$
>
> where
>
> $$\text{SSTR} = n\sum_i (\bar{x}_i - \bar{\bar{x}})^2$$

EXAMPLE 14.1 (continued)

Using our previous calculations, we have the following values for SSTR, degrees of freedom, and MSTR:

$$\text{SSTR} = n\sum_i (\bar{x}_i - \bar{\bar{x}})^2 = 5\,(98) = 490. \qquad k - 1 = 3 - 1 = 2$$

and thus

$$\text{MSTR} = \frac{\text{SSTR}}{k-1} = \frac{490}{2} = 245$$

• • •

Within-Treatments Estimate of Population Variance

We now develop a second estimate of σ^2 that is not based on the null hypothesis assumption that the population means are equal. Instead, it is based upon the variation of the sample observations "within" each treatment and is called the *mean square due to error,* denoted by MSE. This estimate is also referred to as the *mean square within treatments.*

If each of the k samples is a simple random sample, then each of the k sample variances provides an estimate of σ^2. For each sample, the sample variance is computed in the usual fashion.

$$\text{Variance of Sample } i = s_i^2 = \frac{\displaystyle\sum_j (x_{ij} - \bar{x}_i)^2}{n-1} \qquad\qquad (14.8)$$

Since each sample variance provides an individual estimate of σ^2, it would seem reasonable that the overall best estimate of σ^2 could be obtained by pooling the individual estimates. In our case, where the sample size corresponding to each treatment is the same, this can be done by simply averaging the individual estimates of

σ^2. Thus

$$\text{MSE} = \frac{s_1^2 + s_2^2 + \cdots + s_k^2}{k}$$

Using (14.8) to substitute for each of the within-treatment variances, s_i^2, and simplifying algebraically yields the following result.

$$\text{MSE} = \frac{\sum_j (x_{1j} - \bar{x}_1)^2 + \sum_j (x_{2j} - \bar{x}_2)^2 + \cdots + \sum_j (x_{kj} - \bar{x}_k)^2}{k(n-1)}$$

$$= \frac{\sum_i \sum_j (x_{ij} - \bar{x}_i)^2}{kn - k} \tag{14.9}$$

$$= \frac{\sum_i \sum_j (x_{ij} - \bar{x}_i)^2}{n_T - k}$$

The numerator in (14.9) is given the name *sum of squares within,* or *sum of squares due to error,* and is denoted by SSE. The denominator of MSE is the number of *degrees of freedom* associated with the within-treatment variance estimate. In general, then, we can compute MSE as follows.

Within-Treatments Estimate of Population Variance

$$\text{MSE} = \frac{\text{SSE}}{n_T - k} \tag{14.10}$$

where

$$\text{SSE} = \sum_i \sum_j (x_{ij} - \bar{x}_i)^2 \tag{14.11}$$

EXAMPLE 14.1 (continued)

In considering the retest scores for the five students assigned to Program 1, we obtain the following result.

$$\sum_j (x_{1j} - \bar{x}_1)^2$$

$$= (48 - 55)^2 + (54 - 55)^2 + (57 - 55)^2 + (54 - 55)^2 + (62 - 55)^2$$

$$= 49 + 1 + 4 + 1 + 49 = 104$$

Similarly, for Programs 2 and 3 we obtain

$$\sum_j (x_{2j} - \bar{x}_2)^2 = 106 \qquad \sum_j (x_{3j} - \bar{x}_3)^2 = 98$$

Hence

$$\begin{aligned} \text{SSE} &= \sum_i \sum_j (x_{ij} - \bar{x}_i)^2 \\ &= \sum_j (x_{1j} - \bar{x}_1)^2 + \sum_j (x_{2j} - \bar{x}_2)^2 + \sum_j (x_{3j} - \bar{x}_3)^2 \\ &= 104 + 106 + 98 = 308 \end{aligned}$$

Thus, since $n_T - k = (3)(5) - 3 = 12$, using (14.10) we obtain

$$\text{MSE} = \frac{\text{SSE}}{n_T - k} = \frac{308}{12} = 25.67$$

$\bullet \quad \bullet \quad \bullet$

Comparing the Variance Estimates: The F Test

We have now developed two estimates of σ^2. The first estimate (MSTR) is based upon the variation between treatment means, and the second estimate (MSE) is based upon the variation within each treatment. Recall that in order to compute MSTR we had to assume that the null hypothesis was true (that is, $\mu_1 = \mu_2 = \cdots = \mu_k$). However, in computing MSE this assumption was not required. In fact, regardless of whether or not the means of the k populations are equal, MSE will always provide an unbiased estimate of σ^2.

It can be shown that MSTR provides an unbiased estimate of σ^2 when H_0 is true. However, if the means of the k populations are not equal (that is, H_1 is true), MSTR is not an unbiased estimate of σ^2. In fact, in this case MSTR overestimates σ^2. This is the key to the analysis of variance procedure: that is, we test the hypotheses

$$H_0: \mu_1 = \mu_2 = \cdots = \mu_k$$

$$H_1: \text{not all } \mu_i \text{ are equal}$$

by comparing the two estimates of the population variance, MSTR and MSE. If MSTR and MSE are approximately equal, such that the ratio MSTR/MSE is near 1, we cannot reject H_0; in this case our statistical evidence does not lead us to conclude that the means are not all equal. However, if MSTR is much larger than MSE, such that the ratio MSTR/MSE is much larger than 1, we reject H_0 and conclude that the means are not all equal.

To obtain a better intuitive feel for why larger values of MSTR are obtained when the null hypothesis is false, recall the formula for computing MSTR.

$$\text{MSTR} = \frac{\text{SSTR}}{k-1} = \frac{n\Sigma(\overline{x}_i - \overline{\overline{x}})^2}{k-1} \tag{14.12}$$

Note that the numerator (SSTR) is based on the dispersion of the k sample means, $\overline{x}_i$'s, around the overall sample mean $\overline{\overline{x}}$. Now consider the two situations shown in Figure 14.3. In (a) we have the sampling distribution of $\overline{x}$ under the assumption that H_0 is true with $\mu_1 = \mu_2 = \mu_3 = \mu$. In this case each sample comes from the same population, and there is only one sampling distribution. In this case MSTR as computed in (14.12) will provide an unbiased estimate of σ^2.

The second diagram, (b) of Figure 14.3, shows a situation where H_0 is false and the three means μ_1, μ_2, and μ_3 are not the same. What happens to the value of SSTR in this

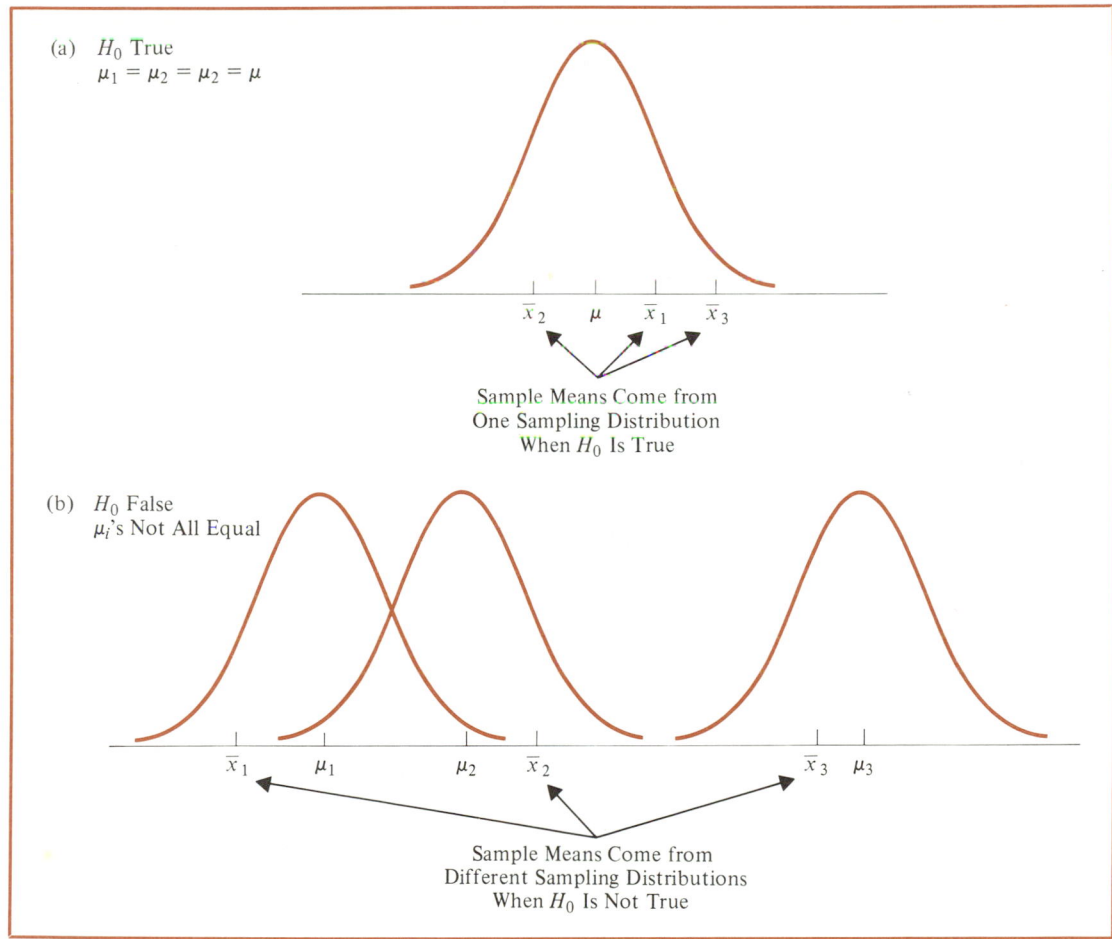

FIGURE 14.3

Examples of the Sampling Distributions of $\overline{x}$ for the Cases of H_0 True (*a*) and H_0 False (*b*)

case? Note that since the sample means $\bar{x}_1$, $\bar{x}_2$, and $\bar{x}_3$ are coming from distributions with different μ's, they have different sampling distributions and show a much larger dispersion than when H_0 is true. In this case the value of SSTR in (14.12) will be larger, causing the value of MSTR to overestimate the population variance σ^2.

Let us assume for the moment that the null hypothesis is true and that $\mu_1 = \mu_2 = \cdots = \mu_k$. In this case it can be shown that MSTR and MSE provide two independent unbiased estimates of σ^2. Recall from Chapter 12 that the sampling distribution of the ratio of two independent estimates of σ^2 for a normal population follows an F probability distribution. Thus if the null hypothesis is true and the ANOVA assumptions are valid, the sampling distribution of MSTR/MSE is an F distribution with numerator degrees of freedom equal to $k - 1$ and denominator degrees of freedom equal to $n_T - k$.

On the other hand, if the means of the k populations are not all equal, the value of MSTR/MSE will be inflated because MSTR overestimates σ^2. Hence we will reject H_0 if the resulting value of MSTR/MSE appears to be too large to have been selected at random from an F distribution with degrees of freedom $k - 1$ in the numerator and $n_T - k$ in the denominator. The value of F that will cause us to reject H_0 depends upon α, the level of significance. Once α is selected a critical value of F can be determined. Figure 14.4 shows the sampling distribution of MSTR/MSE and the rejection region associated with a level of significance equal to α. Note that F_α denotes the critical value.

EXAMPLE 14.1 (continued)

Let us finish the ANOVA procedure for our reading comprehension problem. Assume that the board of education was willing to accept a type I error of $\alpha = .05$. From Table 4 of Appendix B we can determine the critical value by locating the value corresponding to numerator degrees of freedom equal to $k - 1 = 3 - 1 = 2$ and denominator degrees of freedom equal to $n_T - k = 15 - 3 = 12$. Thus we obtain the value $F_{.05} = 3.89$. Hence

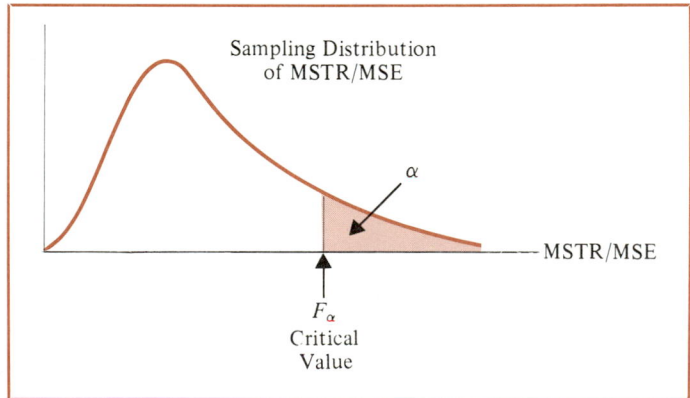

FIGURE 14.4
Sampling Distribution of MSTR/MSE; The Critical Value for Rejecting the Null Hypothesis of Equality of Means Is F_α

the appropriate decision rule for our problem is written

$$\text{If MSTR/MSE} \leq 3.89 \quad \text{Do not Reject } H_0$$

$$\text{If MSTR/MSE} > 3.89 \quad \text{Reject } H_0$$

where

$$H_0: \mu_1 = \mu_2 = \mu_3$$

$$H_1: \text{not all } \mu_i \text{ are equal}$$

Since MSTR/MSE = 245/25.67 = 9.55 is greater than the critical value $F_{.05} = 3.89$, there is sufficient statistical evidence to reject the null hypothesis that the means of the three programs are the same.

• • •

EXERCISES

1. The Jacobs Chemical Company wants to estimate the mean time (minutes) required to mix a batch of material on machines produced by three different manufacturers. In order to limit the cost of testing, four batches of material were mixed on machines produced by each of the three manufacturers. The time needed to mix the material was recorded. The times in minutes are shown.

Manufacturer 1	Manufacturer 2	Manufacturer 3
20	28	20
26	26	19
24	31	23
22	27	22

Use these data and test to see if the mean time needed to mix a batch of material is the same for each manufacturer. Use $\alpha = .05$.

2. Four different paints are advertised as having the same drying time. In order to check the manufacturers' claims, five paint samples were tested for each make of paint. The time in minutes until the paint was dry enough for a second coat to be applied was recorded. The following data were obtained.

Paint 1	Paint 2	Paint 3	Paint 4
128	144	133	150
137	133	143	142
135	142	137	135
124	146	136	140
141	130	131	153

At the $\alpha = .05$ level of significance, test to see if the mean drying time is the same for each type of paint.

3. Three different brands of radial tires were compared for wear characteristics. For each brand of tire, 10 tires were randomly selected and subjected to standard wear-testing procedures. The average mileage obtained for each brand of tire and the sample standard deviations are shown.

	Brand A	Brand B	Brand C
Average mileage	36,400	38,200	33,100
Sample standard deviation	1,650	1,800	1,500

Use these data and test to see if the mean mileage for all three brands of tires is the same. Use $\alpha = .05$.

4. Three top-of-the-line intermediate-sized automobiles manufactured in the United States have been test driven and compared on a variety of criteria by a well-known automotive magazine. In the area of gasoline mileage performance, five automobiles of each brand were each test-driven 500 miles; the miles per gallon data obtained are shown.

Miles per Gallon Data				
Automobile A	19	21	20	19	21
Automobile B	19	20	22	21	23
Automobile C	24	26	23	25	27

Use the analysis of variance procedure with $\alpha = .05$ to determine if there is a significant difference in the mean miles per gallon for the three types of automobiles.

5. Three brands of paper towels were tested for their abilities to absorb water. Equal-sized towels were used, with four sections of towels tested per brand. The absorbency rating data are below. Using a .05 level of significance, does there appear to be a difference in the ability of the brands to absorb water?

Brand	Absorbency Rating			
X	91	100	88	89
Y	99	96	94	99
Z	83	88	89	76

6. In order to study the effect of temperature upon yield in a chemical process, five batches were produced under each of three temperature levels. The results are given. Construct an analysis of variance table. Using a .05 level of significance, test to see if the temperature level appears to have an effect upon the mean yield of the process.

Temperature (°C)	Yield				
50	34	24	36	39	32
60	30	31	34	23	27
70	23	28	28	30	31

14.3 THE ANOVA TABLE AND OTHER CONSIDERATIONS

A convenient way to summarize the computations and results of the analysis of variance procedure is to develop an *analysis of variance (ANOVA) table*. Before presenting the ANOVA table for the reading comprehension study, however, let us perform some more calculations with the data collected (see Table 14.1). Treating the entire data set as one sample of 15 observations, we have already found the overall sample mean to be $\bar{\bar{x}} = 60$. Using the entire data set as one sample, let us now compute an overall sample variance.

$$s^2 = \frac{\sum_i \sum_j (x_{ij} - \bar{\bar{x}})^2}{n_T - 1} \tag{14.13}$$

The numerator of (14.13) is called the *total sum of squares about the mean* (SST), and the denominator represents the degrees of freedom associated with this total sum of squares.

Total Sum of Squares About the Mean

$$SST = \sum_i \sum_j (x_{ij} - \bar{\bar{x}})^2$$

where

$$\frac{\text{Degrees of Freedom}}{\text{for Total Sum of Squares}} = n_T - 1$$

EXAMPLE 14.1 (continued)

Check for yourself that the 15 values in Table 14.1 result in the following values for SST and $n_T - 1$.

$$
\begin{aligned}
\text{SST} &= \sum_i \sum_j (x_{ij} - \bar{\bar{x}})^2 \\
&= (48 - 60)^2 + (54 - 60)^2 + \cdots + (54 - 60)^2 + (56 - 60)^2 \\
&= 798
\end{aligned}
$$

$$ n_T - 1 = 15 - 1 = 14 $$

• • •

We can now develop the ANOVA table for a completely randomized design. Table 14.2 shows a general ANOVA table, and Table 14.3 shows the ANOVA table for the reading comprehension problem.

Note that the rows of the table contain information concerning the two sources of variation: between treatments, which refers to the between-group variation, and error, which refers to the within-group variation. The "Sum of Squares" column and the "Degrees of Freedom" column provide the corresponding values as defined in the previous section. The column labeled "Mean Square" is simply the sum of the squares divided by the corresponding degrees of freedom. In Section 14.2 we showed that when we divided the sum of squares by the corresponding degrees of freedom, we obtained two estimates of the population variance (MSTR and MSE). The ANOVA table simply summarizes these values in the "Mean Square" column.

Finally, the last column in the table contains the F value corresponding to MSTR/MSE. Since the variance estimates are found in the "Mean Square" column, the F value is computed by dividing the mean square treatments (MSTR) by the mean square error (MSE). That is, $F = \text{MSTR/MSE}$.

What observation can you make from the "Sum of Squares" columns in Tables

TABLE 14.2

ANOVA Table for a Completely Randomized Design

Source of Variation	Sum of Squares	Degrees of Freedom	Mean Square	F
Between treatments	SSTR	$k - 1$	$\text{MSTR} = \dfrac{\text{SSTR}}{k - 1}$	$\dfrac{\text{MSTR}}{\text{MSE}}$
Error (within treatments)	SSE	$n_T - k$	$\text{MSE} = \dfrac{\text{SSE}}{n_T - k}$	
Total	SST	$n_T - 1$		

TABLE 14.3

ANOVA Table for the Reading Comprehension Experiment

Source of Variation	Sum of Squares	Degrees of Freedom	Mean Square	F
Between treatments	490	2	245	9.55
Error (within treatments)	308	12	25.667	
Total	798	14		

14.2 and 14.3? Note in particular that the following condition holds.

$$\text{SST} = \text{SSTR} + \text{SSE} \tag{14.14}$$

or

$$\sum_i \sum_j (x_{ij} - \bar{\bar{x}})^2 = n \sum_i (\bar{x}_i - \bar{\bar{x}})^2 + \sum_i \sum_j (x_{ij} - \bar{x}_i)^2. \tag{14.15}$$

Thus we see that SST can be partitioned into two sums—one called the sum of squares between treatments and one called the sum of squares due to error. This is known as *partitioning* the sum of squares.

Note also that the degrees of freedom corresponding to SST ($n_T - 1 = 14$) can be partitioned into the degrees of freedom corresponding to SSTR ($k - 1 = 2$) and the degrees of freedom corresponding to SSE ($n_T - k = 12$). The analysis of variance procedure can be viewed as the process of partitioning the sum of squares and degrees of freedom into their corresponding sources: treatments and error. Dividing the sum of squares by the appropriate degrees of freedom provides the variance estimates and the F value used to test the hypothesis of equal population or treatment means.

Unbalanced Designs

Equation (14.7) provides SSTR for *balanced designs*. Balanced designs are ones in which the sample size is the same for each treatment. Any experimental design for which the sample size is not the same for each treatment is said to be *unbalanced*. Although we would prefer a balanced design, in some cases we must work with unequal sample sizes.* For unbalanced designs the analysis of variance procedure described above may be used with the following modification of the formula for SSTR.

$$\text{SSTR} = \sum_i n_i (\bar{x}_i - \bar{\bar{x}})^2 \tag{14.16}$$

*With a balanced design the distribution of the F statistic is less sensitive to small departures from the assumption that σ^2 is the same for each population. Also, choosing equal sample sizes maximizes the power of the test.

where

$$n_i = \text{sample size for the } i\text{th treatment.}$$

In such cases

$$n_T = n_1 + n_2 + \cdots + n_k$$

$$= \sum_i n_i.$$

Note that (14.16) yields the same formula we used earlier when the sample size is the same for each treatment, that is, when $n_i = n$.

Easing the Computational Burden

The computational aspect of the analysis variance procedure is devoted primarily to computing the appropriate sums of squares. Once the sums of squares are determined, the other computations in the ANOVA table are rather straightforward. The following step-by-step procedure is designed to ease the burden in computing the appropriate sums of squares. The formulas shown below can be applied to both balanced and unbalanced designs.

Computational Steps for Computing the Sums of Squares for a Completely Randomized Design

$$x_{ij} = \text{value of the } j\text{th observation under treatment } i$$

$$T_i = \text{sum of all observations in treatment } i$$

$$T = \sum_i T_i, \text{ the sum of all observations}$$

$$n_i = \text{sample size for the } i\text{th treatment}$$

$$n_T = \text{total sample size for the experiment}$$

STEP 1　Compute $T_i = \sum_j x_{ij}$ for $i = 1, 2, \ldots, k$, and $T = \sum_i T_i$.

STEP 2　Compute $\sum_i \sum_j x_{ij}^2$.

STEP 3　Compute the sum of squares about the mean (SST).

$$\text{SST} = \sum_i \sum_j (x_{ij} - \bar{\bar{x}})^2 = \sum_i \sum_j x_{ij}^2 - \frac{T^2}{n_T} \qquad (14.17)$$

STEP 4   Compute the sum of squares due to treatments (SSTR).

$$\text{SSTR} = \sum_i n_i(\bar{x}_i - \bar{\bar{x}})^2 = \sum_i \frac{T_i^2}{n_i} - \frac{T^2}{n_T} \qquad (14.18)$$

STEP 5   Compute the sum of squares due to error (SSE).

$$\text{SSE} = \text{SST} - \text{SSTR} \qquad (14.19)$$

EXAMPLE 14.1 (continued)

Using this computational procedure with retest score data in Table 14.1 we obtain the following results.

STEP 1 $T_1 = 48 + 54 + 57 + 54 + 62 = 275$, $T_2 = 340$, $T_3 = 285$, and $T = 275 + 340 + 285 = 900$

STEP 2 $\sum_i \sum_j x_{ij}^2 = 48^2 + 54^2 + \cdots + 54^2 + 56^2 = 54{,}798$

STEP 3 $\text{SST} = 54{,}798 - (900)^2/15 = 798$

STEP 4 $\text{SSTR} = (275)^2/5 + (340)^2/5 + (285)^2/5 - (900)^2/15 = 490$

STEP 5 $\text{SSE} = \text{SST} - \text{SSTR} = 798 - 490 = 308$

$\bullet \ \ \bullet \ \ \bullet$

Computer Results for Analysis of Variance

As you can see, the computation of the sums of squares as required by the analysis of variance procedure can be quite a job even when a hand calculator is used in conjunction with the step-by-step procedure. Furthermore, the difficulty increases as the sample size and/or the number of treatments increases. Because of the computational burden, computer software packages are often used for ANOVA computations.

```
ANALYSIS OF VARIANCE
SOURCE       DF         SS         MS          F
FACTOR        2      490.0      245.0       9.55
ERROR        12      308.0       25.7
TOTAL        14      798.0

LEVEL         N       MEAN      STDEV
1             5      55.00       5.10
2             5      68.00       5.15
3             5      57.00       4.95
```

FIGURE 14.5
Computer Ouput for Reading Comprehension Problem

Figure 14.5 shows part of the output for the reading comprehension problem using the ANOVA procedure from the Minitab computer package. We see that the computer output contains the familiar ANOVA table format. Comparing Figure 14.5 with Table 14.3, we see that the same information is available.

EXERCISES

7. Solve Exercise 1 again, this time using the computational step-by-step procedure described in this section. Summarize your computations and results by setting up the ANOVA table.

8. Solve Exercise 4 again by using the computational step-by-step procedure. Set up the ANOVA table for this problem.

9. Three different methods for assembling a product were proposed by an industrial engineer. To investigate the number of units assembled correctly using each method, 30 employees were randomly selected and randomly assigned to the three proposed methods, such that 10 workers were associated with each method. The number of units assembled correctly was recorded, and the analysis of variance procedure was applied to the resulting data set. The following results were obtained: SST = 10,800, SSTR = 4560.
   a. Set up the ANOVA table for this problem.
   b. Using $\alpha = .05$, test for any significant difference in the means for the three assembly methods.

10. In a completely randomized experimental design, seven experimental units were used for each of the five levels of the factor. Complete the analysis of variance table shown.

Source of Variation	Sum of Squares	Degrees of Freedom	Mean Squares	F
Between treatments	300			
Error				
Total	480			

11. In an experiment designed to test the breaking strength of four types of cables, the following results were obtained: SST = 85.05, SSTR = 61.64, $n_T = 24$. Set up the ANOVA table and test for any significant difference in the mean breaking strength of the four cables. Use $\alpha = .05$.

12. To test for any significant difference in the time between breakdowns for four machines, the following data were obtained.

	Time (hours) Between Breakdowns					
Machine 1	6.4	7.8	5.3	7.4	8.4	7.3
Machine 2	8.7	7.4	9.4	10.1	9.2	9.8
Machine 3	11.1	10.3	9.7	10.3	9.2	8.8
Machine 4	9.9	12.8	12.1	10.8	11.3	11.5

Use the computational step-by-step procedure to develop the ANOVA table for this problem. At the $\alpha = .05$ level of significance, is there any difference in the mean time between breakdowns among the four machines?

13. In an unbalanced experimental design, 12 experimental units were used for the first treatment, 15 experimental units were used for the second treatment, 20 experimental units were used for the third treatment, and 10 experimental units were used for the fourth treatment. Complete the analysis of variance table shown. Using a .05 level of significance, is there a significant difference between the treatments?

Source	Sum of Squares	Degrees of Freedom	Mean Square	F
Between treatments	1200			
Error				
Total	1800			

14. Develop the analysis of variance computations for the following unbalanced experimental design. Using $\alpha = .05$, is there a significant difference between the treatment means?

Treatment									
A	136	120	113	107	131	114	129	102		
B	107	114	120	104	107	109	97	114	104	94
C	92	82	85	101	89	117	110	120	98	106

14.4 RANDOMIZED BLOCK DESIGN

Thus far we have considered the completely randomized experimental design. Recall that in order to test for a difference in means, we computed an F value using the ratio

$$F = \frac{\text{MSTR}}{\text{MSE}}$$

(14.20)

A problem can arise whenever differences due to extraneous factors (ones not considered) cause the MSE term in this ratio to become large. In such cases the F value (14.20) can become small, signaling no difference between treatment means, when in fact such a difference exists.

In this section we present an experimental design referred to as a *randomized block design*. The purpose of this design is to control some of the extraneous sources of variation by removing such variation from the MSE term. This design tends to provide a more powerful hypothesis test in terms of the ability to detect differences between treatment means. To illustrate, let us consider a stress study for air traffic controllers.

EXAMPLE 14.2

A study directed at measuring the fatigue and stress on air traffic controllers has resulted in proposals for modification and redesign of the controller's work station. After consideration of several designs for the work station, three specific alternatives have been selected as having the best potential for reducing controller stress. The key question is: To what extent do the three alternatives differ in terms of their effect on controller stress? To answer this question we need to design an experiment that will provide measurements of air traffic controller stress under each alternative.

• • •

If a completely randomized design were used for the study in Example 14.2 random samples of controllers would be assigned to each work station alternative. However, it is believed that controllers differ substantially in terms of their ability to handle stressful situations. What is high stress to one controller might be only moderate or even low stress to another. Thus, when considering the within-group source of variation (estimated by MSE), we must realize that this variation includes both random error and error due to individual controller differences. In fact, for this study management expected controller variability to be a major contributor to the MSE term.

One way to separate the effect of the individual differences is to use the randomized block design. This design will identify the variability stemming from individual controller differences and remove it from the MSE term. The randomized block design calls for a single sample of controllers. Each controller in the sample is tested using each of the three work station alternatives. In experimental design terminology, the work station is the *factor of interest,* and the controllers are referred to as the *blocks. Blocking* is the process of using the same experimental unit (controller) for all treatments (work station alternatives) of the factor. For simplicity, we will refer to the work station alternatives as System A, System B, and System C.

To provide the necessary data, the three types of work stations were installed at the Cleveland Control Center in Oberlin, Ohio. Six controllers were selected at random and assigned to operate each of the systems. The order in which each controller (block) operated the systems (treatments) was selected at random. (This is called a randomized block design because, within each block, the treatments are randomly assigned.) A follow-up interview and a medical examination of each controller participating in the study provided a measure of the stress for each controller on each system. The data are shown in Table 14.4.

A summary of the stress data collected is shown in Table 14.5. In this table we

TABLE 14.4

A Randomized Block Design for the Air Traffic Controllers Stress Test

		Block 1 (Controller 1)	Block 2 (Controller 2)	Block 3 (Controller 3)	Block 4 (Controller 4)	Block 5 (Controller 5)	Block 6 (Controller 6)
Treatments	System A	15	14	10	13	16	13
	System B	15	14	11	12	13	13
	System C	18	14	15	17	16	13

Stress
Value

TABLE 14.5

Summary of Stress Data for the Air Traffic Controllers Stress Test

		Controller 1	Controller 2	Controller 3	Controller 4	Controller 5	Controller 6	Row or Treatment Totals	Treatment Means ($\bar{x}_i$)
				Blocks					
	System A	15	14	10	13	16	13	81	13.5
Treatments	System B	15	14	11	12	13	13	78	13
	System C	18	14	15	17	16	13	93	15.5
Column or block totals		48	42	36	42	45	39	252 ← Overall Sum $\sum_i \sum_j x_{ij}$	

$$\text{Sum of squares } \sum_i \sum_j x_{ij}^2 = (15)^2 + (14)^2 + (10)^2 + \cdots + (16)^2 + (13)^2 = 3598$$

have included column totals (blocks) and row totals (treatments) as well as some other sums that will be helpful in making the sum of squares computations for the ANOVA procedure. Since lower stress values are viewed as better, the sample data available would seem to favor System B with its mean stress rating of 13. However, the usual question remains: Do the sample results justify the conclusion that the mean stress levels for the three systems differ? That is, are the differences statistically significant? An analysis of variance computation similar to the one performed for the completely randomized design can be used to answer this statistical question.

The Analysis of Variance Procedure for a Randomized Block Design

The ANOVA procedure for the randomized block design requires us to partition the sum of squares total (SST) into three sources: sum of squares between treatments, sum of squares due to blocks, and sum of squares due to error. The formula for this partitioning is as follows.

$$\underset{\substack{\uparrow \\ \text{Sum of Squares} \\ \text{Total}}}{\text{SST}} = \underset{\substack{\uparrow \\ \text{Treatments}}}{\text{SSTR}} + \underset{\substack{\uparrow \\ \text{Blocks}}}{\text{SSB}} + \underset{\substack{\uparrow \\ \text{Error}}}{\text{SSE}} \qquad (14.21)$$

This partitioning of the sum of squares is summarized in the ANOVA table for the randomized block design as shown in Table 14.6. The notation used in this table is as follows.

$$k = \text{the number of treatments,}$$

$$b = \text{the number of blocks,}$$

$$n_T = \text{the total sample size } n_T = kb.$$

TABLE 14.6

ANOVA Table for the Randomized Block Design with k Treatments and b Blocks

Source of Variation	Sum of Squares	Degrees of Freedom	Mean Square	F
Between treatments	SSTR	$k - 1$	$\text{MSTR} = \dfrac{\text{SSTR}}{k - 1}$	$\dfrac{\text{MSTR}}{\text{MSE}}$
Between blocks	SSB	$b - 1$	$\text{MSB} = \dfrac{\text{SSB}}{b - 1}$	
Error	SSE	$(k - 1)(b - 1)$	$\text{MSE} = \dfrac{\text{SSE}}{(k - 1)(b - 1)}$	
Total	SST	$n_T - 1$		

Note that the ANOVA table in Table 14.6 also shows how the $n_T - 1$ total degrees of freedom are partitioned such that $k - 1$ go to treatments, $b - 1$ go to blocks, and $(k - 1)(b - 1)$ go to the error term. The mean square column shows the sum of squares divided by the degrees of freedom, and $F = \text{MSTR}/\text{MSE}$ is the F ratio used to test for a significant difference among the treatment means. The primary contribution of the randomized block design is that by blocking, we have removed the controller differences from the MSE term and obtained a more powerful test for the stress differences in the three work station alternatives.

Computations and Conclusions

As we saw in the completely randomized design, the primary difficulty was computing the appropriate sum of squares values. The steps for easing this computation for a randomized block design are shown below.

Computational Steps for Computing Sums of Squares for a Randomized Block Design

In addition to k, b, and n_T as previously defined, we use the following notation:

$$x_{ij} = \text{value of the observation under treatment } i \text{ in block } j$$

$$T_{i.} = \text{the total of all observations in treatment } i$$

$$T_{.j} = \text{the total of all observations in block } j$$

$$T = \text{the total of all observations}$$

$$\bar{x}_{i.} = \text{sample mean of the } i\text{th treatment}$$

$$\bar{x}_{.j} = \text{sample mean for the } j\text{th block}$$

$$\bar{\bar{x}} = \text{overall sample mean}$$

STEP 1 Compute $T_{i.} = \sum_j x_{ij}$, $T_{.j} = \sum_i x_{ij}$ and $T = \sum_i \sum_j x_{ij}$.

STEP 2 Compute $\sum_i \sum_j x_{ij}^2$.

STEP 3 Compute the total sum of squares (SST).

$$\text{SST} = \sum_i \sum_j (x_{ij} - \bar{\bar{x}})^2 = \sum_i \sum_j x_{ij}^2 - \frac{T^2}{n_T} \qquad (14.22)$$

STEP 4   Compute the sum of squares between treatments (SSTR):

$$\text{SSTR} = \sum_i b(\bar{x}_{i.} - \bar{\bar{x}})^2 = \frac{\sum_i T_{i.}^2}{b} - \frac{T^2}{n_T} \qquad (14.23)$$

STEP 5   Compute the sum of squares due to blocks (SSB).

$$\text{SSB} = \sum_j k(\bar{x}_{.j} - \bar{\bar{x}})^2 = \frac{\sum_j T_{.j}^2}{k} - \frac{T^2}{n_T} \qquad (14.24)$$

STEP 6   Compute the sum of squares due to error (SSE).

$$\text{SSE} = \text{SST} - \text{SSTR} - \text{SSB} \qquad (14.25)$$

EXAMPLE 14.2 (continued)

For the air traffic controller data in Table 14.5, these steps lead to the following sums of squares.

STEP 1 $T_{1.} = 15 + 14 + \cdots + 13 = 81$, $T_{2.} = 78$, $T_{3.} = 93$
$T_{.1} = 15 + 15 + 18 = 48$, $T_{.2} = 42$, $T_{.3} = 36$, $T_{.4} = 42$, $T_{.5} = 45$,
$T_{.6} = 39$, and $T = 15 + 14 + \cdots + 16 + 13 = 252$

STEP 2 $\sum_i \sum_j x_{ij}^2 = 15^2 + 14^2 + \cdots + 16^2 + 13^2 = 3598$

STEP 3 $\text{SST} = 3598 - \dfrac{(252)^2}{18} = 70$

STEP 4 $\text{SSTR} = \dfrac{(81)^2 + (78)^2 + (93)^2}{6} - \dfrac{(252)^2}{18} = 21$

STEP 5 $\text{SSB} = \dfrac{(48)^2 + (42)^2 + \cdots + (39)^2}{3} - \dfrac{(252)^2}{18} = 30$

STEP 6 $\text{SSE} = 70 - 21 - 30 = 19$

These sums of squares divided by their degrees of freedom provide the corresponding mean square values shown in Table 14.7. The F ratio used to test for differences between treatment means is $\text{MSTR}/\text{MSE} = 10.5/1.9 = 5.53$. Checking the F values in Table 4 of Appendix B, we find that the critical F value at $\alpha = .05$ (2 numerator degrees of freedom and 10 denominator degrees of freedom) is 4.10. With

TABLE 14.7

ANOVA Table for the Air Traffic Controller Stress Test

Source of Variation	Sum of Squares	Degrees of Freedom	Mean Square	F
Between treatments	21	2	10.5	$\frac{10.5}{1.9} = 5.53*$
Between blocks	30	5	6.0	
Error	19	10	1.9	
Total	70	17		

*Significant at $\alpha = .05$; critical $F = 4.10$

$F = 5.53$, we reject the null hypothesis $H_0: \mu_1 = \mu_2 = \mu_3$ and conclude that the work station designs differ in terms of their mean stress effects on air traffic controllers.

• • •

Before leaving this section let us make some general comments about the randomized block design. The blocking as described in this section is referred to as a *complete* block design; the word *complete* indicates that all k treatments are applied in each block. Experimental designs employing blocking where some but not all treatments are applied to each block are referred to as *incomplete* block designs. A discussion of incomplete block designs is outside the scope of this text.

Note that in the air traffic controller stress test each controller in the study was required to use all three systems. While this guarantees a complete block design, in some cases blocking is carried out with "similar" experimental units in each block. For example, assume that in a pretest of air traffic controllers the population of controllers was divided into stress classifications ranging from extremely high-stress individuals to extremely low-stress individuals. The blocking could still have been accomplished by randomly selecting three controllers from each of the stress classifications to participate in the study. Each block would then be formed from three controllers in the same stress class. The randomized aspect of the block design would be conducted by randomly assigning the three controllers in each block to the three systems.

EXERCISES

15. An automobile dealer conducted a test to determine if the time needed to complete a minor engine tune-up depends upon whether a computerized engine analyzer or an electronic analyzer is used. Because tuneup time varies among compact, intermediate, and full-sized cars, the three types of cars were used as blocks in the experiment. The data obtained are shown.

		Compact	Car Type Intermediate	Full-size
Analyzer	Computerized Electronic	50 42	55 44	63 46
		↑ Time (minutes)		

Using $\alpha = .05$, test for any significant differences.

16. An important factor in selecting software for word processng and data base management systems is the time required to learn how to use a particular system. In order to evaluate three file management systems, a firm designed a test involving five different word processing operators. Since operator variability was believed to be a significant factor, each of the five operators was trained on each of the three file management systems. The data obtained are shown.

	Operator				
	1	2	3	4	5
System A	16	19	14	13	18
System B	16	17	13	12	17
System C	24	22	19	18	22
	↑ Time (hours)				

Using $\alpha = .05$, test to see if there is any difference in training time for the three systems.

17. The following data were obtained for a randomized block design involving five treatments and three blocks: SST = 430, SSTR = 310, SSB = 85. Set up the ANOVA table and test for any significant differences. Use $\alpha = .05$.

18. Five different auditing procedures were compared with respect to total audit time. To control for possible variation due to the person conducting the audit, four accountants were selected randomly and treated as blocks in the experiment. The following values were obtained using the ANOVA procedure: SST = 100, SSTR = 45, SSB = 36. Using $\alpha = .05$, test to see if there is any significant difference in total audit time stemming from the auditing procedure used.

19. An experiment has been conducted for four treatments using eight blocks. Complete the analysis of variance table shown on page 492.

Source	Sum of Squares	Degrees of Freedom	Mean Square	F
Between Treatments	900			
Blocks	400			
Error				
Total	1800			

Using $\alpha = .05$, test for any significant differences.

20. Consider the following experimental results of a randomized block design. Make the calculations necessary to set up the analysis of variance table, and using $\alpha = .05$, test for any significant differences.

			Blocks			
		1	2	3	4	5
	A	10	12	18	20	8
Treatment	B	9	6	15	18	7
	C	8	5	14	18	8

14.5 INFERENCES ON INDIVIDUAL TREATMENT MEANS

Suppose that in carrying out an analysis of variance procedure we reject the null hypothesis and conclude that the treatment means are not all the same. Sometimes we may be satisfied with this conclusion, but in other cases we will want to go a step further and determine where the differences occur. The purpose of this section is to show how to conduct statistical comparisons between pairs of treatment means.

EXAMPLE 14.2 (continued)

Recall that in the air traffic controllers stress test, we rejected the null hypothesis and concluded the three work station systems differed in terms of their effects on controller stress. In this case the followup question is: We believe that the systems differ; where do the differences occur?

Let us show the details of a procedure that could be used to test for the equality of two treatment means. For example, let us test to see if there is a significant difference between the means of Systems A and B. From the data in Table 14.5 we found the treatment means to be $\bar{x}_1 = 13.5$ for System A and $\bar{x}_2 = 13$ for System B; thus System B shows the better (lower) stress level. But is the sample information sufficient to justify the conclusion that a difference in stress level exists between System A and System B?

• • •

A simple test for a difference between two treatment means is based on the t distribution (as presented for the two population case in Chapter 11). The mean square error (MSE) is used as an unbiased estimate of the population variance σ^2. A confidence interval estimate of the difference in the two population means (μ_1 and μ_2) is given by the following expression.

Interval Estimate for $\mu_1 - \mu_2$

$$\bar{x}_1 - \bar{x}_2 \pm t_{\alpha/2} \sqrt{MSE \left(\frac{1}{n_1} + \frac{1}{n_2} \right)} \tag{14.26}$$

where

$\bar{x}_1$ = sample mean for the first treatment

$\bar{x}_2$ = sample mean for the second treatment

$t_{\alpha/2}$ = t value for the test, where the number of degrees of freedom for the t statistic is given by the degrees of freedom for error (see ANOVA table "Degrees of Freedom" column)

MSE = mean square error (see ANOVA table)

n_1 = sample size for the first treatment

n_2 = sample size for the second treatment

If the confidence interval in (14.26) includes the value 0, we have to conclude that there is no significant difference between the treatment means. In this case we cannot reject the hypothesis that no difference exists between the treatment means. However, if the confidence interval does not include the value 0, we conclude that there is a difference.

EXAMPLE 14.2 (continued)

Let us return to the air traffic controllers stress test and compare the stress levels for System A and System B at the .05 level of significance. The following data are needed:

$$\bar{x}_1 = 13.5 \quad \text{(System A)}, \quad \bar{x}_2 = 13 \quad \text{(System B)}$$

With $\alpha = .05$, $t_{.025} = 2.228$ (note: 10 degrees of freedom for error in ANOVA Table 14.7)

$MSE = 1.9$ (Table 14.7)

$n_1 = n_2 = 6$ (sample size for each treatment is 6, since the study involved 6 blocks)

Using these data and (14.26) we have

$$13.5 - 13 \pm 2.228\sqrt{1.9(\tfrac{1}{6} + \tfrac{1}{6})} = .5 \pm 2.228(.7958)$$
$$= .5 \pm 1.77$$

Thus a 95% confidence interval for the difference between system A and system B is given by -1.27 to 2.27. Since this interval includes 0, we are unable to reject the hypothesis that the systems provide the same level of stress.

•   •   •

At this point, analysis of variance has told us that the stress levels for the three systems are not all the same. However, the above result tells us that Systems A and B appear to have the same level of stress. With $\bar{x}_3 = 15.5$, System C appears to have the highest level of stress. Thus we should feel comfortable in concluding that the difference in population means is due to System C.

A word of caution is needed at this point. The above test of an individual difference should be applied only if we reject the null hypothesis of equal population means. That is, whereas it may appear natural to use (14.26) to compare all possible pairs of treatment means (System A versus System B, System A versus System C, System B versus System C), statistical problems can occur with this sequential approach. If a null hypothesis is true (that is, the two population means are equal) and a test is conducted using (14.26) with a type I error probability of $\alpha = .10$, there is a probability of .10 of rejecting the null hypothesis when it is really true; hence the probability of making a correct decision is .90. If two tests are conducted in this manner, the probability that a correct decision is made on both tests is $(.9)(.9) = .81$. Thus the probability that *at least* one of the tests would result in rejecting a true null hypothesis is $1 - .81 = .19$.* Thus we see that the total type I error probability using a sequential testing procedure to test two hypotheses at the .10 significance level is .19 and not .10. For three tests at the .10 level of significance the type I error probability is $1 - (.90)(.90)(.90) = 1 - .729 = .271$. The probability of making a type I error increases rapidly as the number of multiple tests increases.

---

*This assumes that the two tests are independent and hence the joint probability of the two events can be obtained simply by multiplying the individual probabilities. In fact the two tests are not independent (MSE is used in each test), and hence the probability of error may be even greater than shown.

A simple procedure to adjust (approximately) for this increasing probability of making the type I error is to reduce the α level for each separate test. The aim is to reduce to a satisfactory level the α level for all tests taken together. With an α level of significance desired and m tests to be made, we would use α/m as the probability of making a type I error on any one test. Then the overall probability of making a type I error on any one of the tests will approximately equal α.

Because of the difficulty of the type I error increasing when making multiple tests of difference between individual means, a variety of specialized tests have been developed. Often these tests carry the name of the developer. The better-known tests for multiple comparisons include the Duncan multiple range test, the Newman-Keuls test, Tukey's test, and Scheffe's method. References in the bibliography provide details for these methods. It is recommended that these tests be considered whenever multiple comparisons among treatment means are expected to be a major concern in the study.

EXERCISES

21. Refer to Exercise 1. Use the procedure described in this section to test for the equality of the mean for manufacturers 1 and 3. Use $\alpha = .05$. What conclusion can you make after carrying out this test?

22. In Exercise 2, does it make sense to use the procedure described in this section to test for the equality of the mean drying time for paints 1 and 4? Explain.

23. Refer to Exercise 4. Use the procedure described in this section to test for the equality of the mean mileage for automobiles A and B. What general conclusion can you make after carrying out this test? Use $\alpha = .05$.

24. Refer to Exercise 5. Use the procedure described in this section to test for the equality of brands Y and Z. What general conclusions can you make after carrying out this test? Use $\alpha = .05$.

Summary

In this chapter we showed how a completely randomized design and a randomized block design can be employed with an analysis of variance procedure to lead to conclusions about differences between the means of several populations or treatments. Specifically, the completely randomized design and the randomized block design were used to draw conclusions about differences between the means for k levels of a single factor. The primary purpose of the blocking in the randomized block design is to remove extraneous sources of variation from the error term. This blocking results in a better estimate of the error variance and a better test to determine whether or not the treatment means of the single factor differed significantly.

Although these two experimental designs required different formulas and computations, we showed that the basis for the statistical test is the development of two independent estimates of the population variance σ^2. One estimate, MSTR, is based upon the variation between the treatments. This value is an unbiased estimate of σ^2

only if the means $\mu_1, \mu_2, \ldots, \mu_k$ are all equal. A second estimate, MSE, is based upon the variation of the observations within each sample. MSE always is an unbiased estimate of σ^2. By computing $F = \text{MSTR}/\text{MSE}$ and using the F distribution, we developed a decision rule for determining whether to accept or reject the hypothesis that the treatment means are equal. In both the experimental designs considered, the partitioning of the sum of squares and degrees of freedom into their various sources enabled us to compute the appropriate values for making the analysis of variance calculations and tests.

Whenever an analysis of variance conclusion results in the rejection of the equal-means hypothesis, we may want to consider testing for a difference between the individual treatment means. We showed how the t distribution test could be used to compare two treatment means. However, we warned against indiscriminate use of this testing procedure because of the increasing probability of making a type I error. By simultaneously making several tests for individual differences, the probability increases of erroneously claiming that a difference exists. Several specialized tests are available if multiple comparisons of the treatment means are to be considered.

Statistics in Practice

CEREAL TASTE TEST

A major manufacturer of breakfast cereals for children continually strives for formula improvements that offer better-tasting products. In order to help evaluate the taste characteristics of potential new cereal formulations, the company hired Burke Marketing Services of Cincinnati, Ohio. Burke took on the task of designing an experiment for the cereal taste-test evaluation.

The four key ingredients, or factors, that were thought to enhance taste were:

1. Ratio of wheat to corn in the cereal flake (2 levels)
2. Types of sweetness (3 types)
3. Flavor bits (present or absent)
4. Cooking time (short, long)

Using all combinations of the four ingredients would lead to $2 \times 3 \times 2 \times 2 = 24$ different cereal formulations. If the 24 cereal formulations were tested independently, the sample sizes would have to be large in order to measure the effects of the various ingredients on taste perception. In order to use fewer respondents, thus saving time and money, an experimental design, called a *fractional factorial design*, was recommended. Only 9 of the 24 cereal formulations were used in the study, but they were selected in such a way that the best possible measure of the independent effect of the four ingredients could be obtained.

Then, 108 children were selected to taste-test the various cereal formulations.

Each child taste-tested a preselected group of three different formulations. The order of tasting was randomized for each child. The data collected reflected the child's preference for the taste of the cereal.

The statistical technique of analysis of variance was used to analyze the sample data obtained from the 108 children. Using a significance test for each factor, the research conclusions reached were as follows:

a. The flake composition was an influential factor in the taste evaluation.
b. The sweetener type was an influential factor in the taste evaluation.
c. The flavor bit detracted from the taste rating for the cereal.
d. The cooking time was not a significant factor.

Given the analysis of variance conclusions, Burke's experimental procedure provided the company with excellent insights into how to improve the taste of its cereal products. In particular, the company began further studies with the flake composition and sweetener factors in order to develop better-tasting cereals.

Glossary

Replication—The number of times each experimental condition is repeated in an experiment. It is the sample size associated with each treatment combination.

Analysis of variance (ANOVA)—A statistical procedure for determining whether or not the means of several populations are equal.

Factor—Another word for the variable of interest in an ANOVA procedure.

Treatment—Different levels of a factor.

Experimental units—The objects of interest in the experiment.

Completely randomized design—An experimental design where the treatments are randomly assigned to the experimental units.

Mean square—The sum of squares divided by its corresponding degrees of freedom. This quantity is used in the F ratio to determine if significant differences in means exist or not.

ANOVA table—A table used to summarize the analysis of variance computations and results. It contains columns showing the source of variation, the degrees of freedom, the sum of squares, the mean squares, and the F values.

Partitioning—The process of allocating the total sum of squares and degrees of freedom into the various components.

Randomized block design—An experimental design employing blocking. The treatments are randomly assigned within each block.

Blocking—The process of using the same or similar experimental units for all treatments. The purpose of blocking is to remove a source of variation from the error term and hence provide a more powerful test for a difference in population or treatment means.

Key Formulas

Completely Randomized Designs

The Sum of Squares About the Mean

$$\text{SST} = \sum_i \sum_j (\overline{x}_{ij} - \overline{\overline{x}})^2 = \sum_i \sum_j x_{ij}{}^2 - \frac{T^2}{n_T} \qquad (14.17)$$

The Sum of Squares Due to Treatments

$$\text{SSTR} = \sum_i n_i(\overline{x}_i - \overline{\overline{x}})^2 = \sum_i \frac{T_i{}^2}{n_i} - \frac{T^2}{n_T} \qquad (14.18)$$

The Sum of Squares Due to Error

$$\text{SSE} = \text{SST} - \text{SSTR} \qquad (14.19)$$

The *F* Value

$$F = \frac{\text{MSTR}}{\text{MSE}} \qquad (14.20)$$

Randomized Block Designs

The Total Sum of Squares

$$\text{SST} = \sum_i \sum_j (x_{ij} - \overline{\overline{x}})^2 = \sum_i \sum_j x_{ij}^2 - \frac{T^2}{n_T} \qquad (14.22)$$

The Sum of Squares Between Treatments

$$\text{SSTR} = \sum_i b(\overline{x}_{i.} - \overline{\overline{x}})^2 = \frac{\sum_i T_{i.}^2}{b} - \frac{T^2}{n_T} \qquad (14.23)$$

The Sum of Squares Due to Blocks

$$\text{SSB} = \sum_j k(\overline{x}_{.j} - \overline{\overline{x}})^2 = \frac{\sum_j T_{.j}^2}{k} - \frac{T^2}{n_T} \qquad (14.24)$$

The Sum of Squares Due to Error

$$SSE = SST - SSTR - SSB \qquad (14.25)$$

Review Quiz

TRUE/FALSE

1. In a completely randomized design, the treatments must be randomly assigned to the experimental units.
2. In an experimental design involving three treatments, adding a replication requires adding three experimental units.
3. The analysis of variance procedure can be used to test the null hypothesis that k population means are equal.
4. The only assumption required for the ANOVA test on the difference of k population means is that the response variable is normally distributed.
5. In computing the between-treatments estimate of the population variance, we cannot assume that the null hypothesis is true.
6. The degrees of freedom associated with the sum of squares between treatments is the same as the number of treatments.
7. The within-treatments estimate of the population variance is called the mean square due to error.
8. The sum of squares due to error has a number of degrees of freedom equal to the total sample size minus the number of treatments minus 1.
9. In a completely randomized design the degrees of freedom associated with the total sum of squares is the sum of the degrees of freedom for SSTR and the degrees of freedom for SSE.
10. An unbalanced experimental design is one in which the number of experimental units exceeds the number of treatments.
11. Whenever a randomized block design is used, the F test should not be used.
12. If the null hypothesis that k population means are equal is rejected, we can then use a chi-square test to determine which individual means differ.

MULTIPLE CHOICE

Use the following information to answer questions 13–17. A statistics professor wishes to know the effect of class format on student learning, as measured by improvement on examination scores from the beginning to the end of the semester. The five class formats to be studied reflect different emphases on homework problems and computer exercises. Sixty students are randomly selected for this study; 12 students are randomly assigned to each class format.

13. In this example, the term *factor* is illustrated by
    a. the change in exam scores
    b. the class formats
    c. the different amounts of homework and computer work in the different class formats
    d. the 12 students in the sample from each class format
    e. the 60 students in the random sample

14. The term *treatment* is illustrated by
    a. the change in exam scores
    b. the class formats
    c. the different amounts of homework and computer work in the different class formats
    d. the 12 students in the sample from each class format
    e. the 60 students in the random sample

15. The term *replication* is illustrated by
    a. the change in exam scores
    b. the class formats
    c. the different amounts of homework and computer work in the different class formats
    d. the 12 students in the sample from each class format
    e. the 60 students in the random sample

16. The term *response* is illustrated by
    a. the change in exam scores
    b. the class formats
    c. the different amounts of homework and computer work in the different class formats
    d. the 12 students in the sample from each class format
    e. the 60 students in the random sample

17. This example best reflects which experimental design?
    a. randomized block design
    b. completely randomized design
    c. factorial experiments
    d. individual treatment means

18. What is the sampling distribution used in the test for the equality of more than two population means?
    a. normal distribution
    b. t distribution
    c. chi-square distribution
    d. F distribution

19. Which of the following is a required condition for using an ANOVA procedure on data from several populations?
    a. The data is obtained from independently selected samples.
    b. The populations are all normally distributed.
    c. The populations have the same variance.
    d. All of the above are necessary conditions.

20. An ANOVA procedure is used for data that was obtained from three sample groups, each comprised of four observations. The degrees of freedom for the

critical value of F is

a. 11
b. 2, 9
c. 2, 11
d. 3, 11

Supplementary Exercises

25. A simple random sample of the asking price ($1000s) of four houses currently for sale in each of two residential areas resulted in the following data.

Area 1	Area 2
92	90
89	102
98	96
105	88

   a. Use the procedure developed in Chapter 11 and test if the mean asking price is the same in both areas. Use $\alpha = .05$.
   b. Use the ANOVA procedure to test if the mean asking price is the same. Compare your analysis with (a). Use $\alpha = .05$.

26. Suppose that in Exercise 25 data were collected for another residential area. The asking prices for the simple random sample from the third area were as follows: $81,000, $86,000, $75,000, and $90,000. Are the mean asking prices for all three areas the same? Use $\alpha = .05$.

27. An analysis of the number of units sold by ten salespersons in each of four sales territories resulted in the following data.

	Sales Territory			
	1	2	3	4
Number of salespersons	10	10	10	10
Average number sold ($\bar{x}$)	130	120	132	114
Sample variance (s^2)	72	64	69	67

   Test at the $\alpha = .05$ level if there is any significant difference in the mean number of units sold in the four sales territories.

28. Suppose that in Exercise 27 the number of salespersons in each territory was as follows: $n_1 = 10$, $n_2 = 12$, $n_3 = 10$, and $n_4 = 15$. Using the same data for $\bar{x}$ and s^2 as given in Exercise 27, test at the $\alpha = .05$ level if there is any significant difference in the mean number of units sold in the four sales territories.

29. Consider an analysis of variance for the three-group data shown.

Group 1	Group 2	Group 3
54	63	73
60	60	71
57	59	77
55	59	64
69	69	70

  a. What are the hypotheses that will be tested by the analysis of variance procedure?
  b. What assumptions are made?
  c. Calculate the sum of squares.
  d. Show the analysis of variance table.

30. Three different assembly methods have been proposed for a new product. In order to determine which assembly method results in the greatest number of parts produced per hour, 30 workers were randomly selected and assigned to use one of the proposed methods. The number of units produced by each worker is given.

Method A	97	73	93	100	73	91	100	86	92	95
Method B	93	100	93	55	77	91	85	73	90	83
Method C	99	94	87	66	59	75	84	72	88	86

Use these data and test to see if the mean number of parts produced with each method is the same. Use $\alpha = .05$.

31. In order to test to see if there is any significant difference in the mean number of units produced per week by each of three production methods, the following data were collected.

Method 1	Method 2	Method 3
58	52	48
64	63	57
55	65	59
66	58	47
67	62	49

At the $\alpha = .05$ level of significance is there any difference in the means for the three methods?

32. Pappashales Restaurant is considering introducing a new specialty sandwich. For a determination of the effect of sandwich price on sales, the new sandwich was test-marketed at three prices in selected company restaurants. The following data, in terms of the number of sandwiches sold per day, were obtained.

$1.49	$1.79	$1.99
925	910	860
850	845	935
930	905	820
955	860	845

At the $\alpha = .05$ level of significance is there any difference in the mean number of sandwiches sold per day for the three prices? What should management of Pappashales do?

33. Hargreaves Automotive Parts, Inc. would like to compare the mileage for four different types of brake linings. Thirty linings of each type were produced and placed on a fleet of rental cars. The number of miles that each brake lining lasted until it no longer met the required federal safety standard was recorded, and an average value was computed for each type of lining. The following data were obtained.

	Sample Size	Sample Mean	Standard Deviation
Type A	30	32,000	1,450
Type B	30	27,500	1,525
Type C	30	34,200	1,650
Type D	30	30,300	1,400

Use these data and test to see if the corresponding population means are equal. Use = .05.

34. In a study of the educational skills of students in various school environments, samples of students were taken from inner-city schools, suburban schools, and rural schools. Test scores on reading skills of the students are shown below. Using a .05 level of significance, test for a significant difference between the mean reading scores of the three groups.

School System	Mean Reading Scores									
Inner-city	93	62	99	84	67	77	79	76	78	95
Suburban	72	79	75	65	62	69	93	80	84	81
Rural	94	92	95	78	67	87	95	90	95	97

35. A manufacturer of batteries for electronic toys and calculators is considering three new battery designs. An attempt was made to determine if the mean lifetime in hours is the same for each of the three designs. The following battery lifetime data were collected.

Design A	Design B	Design C
78	112	115
98	99	101
88	101	100
96	116	120

Test to see if the population means are equal. Use $\alpha = .05$.

36. At the end of each quarter, college students submit course evaluations to university administrators. An overall evaluation of each course an instructor teaches is then computed. Currently, four faculty are being considered for a teacher-of-the-year award. The overall course evaluation rating for each course taught by each instructor is shown below.

Instructor	Course Evaluation Rating								
Black	88	80	79	68	96	69			
Jennings	87	97	82	85	99	99	85	94	
Swanson	88	76	68	82	85	82	84	83	81
Wilson	80	85	56	71	89	87			

    a. Do the data support the conclusion that student evaluations indicate a difference in teaching abilities among the four candidates? Use a .05 level of significance.

    b. If a significant difference exists, use a test on individual treatment means to select the higher-rated instructors.

37. Refer to Exercise 35. Use the procedure described in Section 14.5 to test for the equality of the mean for Design A and Design B. What conclusion can you make after carrying out the above test? Use $\alpha = .05$.

38. In Exercise 32, would it make sense to do a test on individual treatment means as described in Section 14.5? Explain.

39. A research firm tests the miles per gallon obtained with three brands of gasoline. Because of different gasoline performance characteristics in different brands of automobiles, five brands of automobiles are selected and treated as blocks in the experiment. That is, each brand of automobile is tested with each type of gasoline. The results of the experiment are shown below:

		Blocks: Automobiles				
		A	B	C	D	E
	I	18	24	30	22	20
Gasoline brands	II	21	26	29	25	23
	III	20	27	34	24	24

Miles per gallon

With $\alpha = .05$, is there a significant difference in the mean miles per gallon characteristics of the three brands of gasoline?

40. Analyze the experimental data provided in Exercise 39 using the ANOVA procedure for completely randomized designs. Compare your findings with those obtained in Exercise 39. What is the advantage of attempting to remove the block effect?

41. The following data were obtained for a randomized block design involving three treatments and four blocks: SST = 148, SSTR = 84, SSB = 50. Set up the ANOVA table and test for any significant differences. Use $\alpha = .05$.

42. Three different road-repair compounds were tested at four different highway locations. At each location, three sections of the road were repaired, with each section using one of the three compounds. Data were then collected on the number of days of traffic usage before additional repair was required. These data are as follows.

		Location			
		1	2	3	4
	A	99	73	85	103
Compound	B	82	72	85	97
	C	81	79	82	86

With $\alpha = .01$ test to see if there is a significant difference in the compounds.

43. Four types of fertilizers were studied in an attempt to determine the effect of the fertilizers on the yield of corn crops. Five locations (blocks) were selected for test purposes. Each location represented a different type of soil condition that might be present for the corn. The four fertilizers were tested at each location and the yield data collected. Using the data shown and a .01 level of significance, determine if there is evidence that different fertilizers will provide a different corn crop yield.

		Location				
		1	2	3	4	5
	A	21	24	26	22	27
	B	20	22	23	21	24
Fertilizers	C	16	23	22	21	23
	D	15	20	22	21	22

Computer Exercise

As part of a long-term study of individuals 65 years of age or older, sociologists and physicians at Upstate Medical Center recently conducted an experiment to study the relationship between geographic location and depression levels. A sample of 60 individuals was selected; 20 of the individuals were lifetime residents of Florida, 20 were lifetime residents of New York, and 20 were lifetime residents of North Carolina. To account for any possible effects that might be due to the health status of the individual, 50% of the subjects that were sampled from each state had a chronic health condition (arthritis, hypertension, hearing loss, or heart ailment), and 50% did not have a chronic health condition.

Each individual who participated in the study was given a standardized test in order to measure depression. Higher test scores indicate higher levels of depression. The data are shown. A code of 1 is used to indicate that the subject had a chronic health condition and a code of 2 is used to indicate that the subject did not have a chronic health condition.

Florida		New York		North Carolina	
Score	Condition	Score	Condition	Score	Condition
7	2	9	2	14	1
10	1	19	1	14	1
12	1	10	1	20	1
19	1	11	2	15	1
11	1	8	2	6	2
2	2	17	1	5	2
4	2	15	1	12	2
13	1	11	1	8	2
14	1	9	2	12	1
17	1	6	2	9	2
8	2	7	2	16	1
5	2	12	1	12	1
11	1	20	1	7	2
18	1	10	2	13	1
7	2	13	1	6	2
4	2	10	2	15	1
3	2	12	1	4	2
4	2	15	1	6	2
1	2	4	2	6	2
9	1	10	2	10	1

QUESTIONS

1. Use numerical and graphical measures to summarize the data.
2. Ignoring the data on health condition, does there appear to be a significant difference in depression levels for residents of the three states?
3. Considering health condition as a block, does there appear to be a significant difference in depression levels for residents of the three states? Explain.

CHAPTER 15

Linear Regression and Correlation

What You Will Learn in This Chapter

- how to use regression analysis to develop an equation that estimates how two variables are related

- what the least squares method is

- how to compute and interpret the coefficient of determination

- how to use the t and F distributions to test for significant relationships between variables

- how to use a regression equation for estimation and prediction

- how to compute and interpret the sample correlation coefficient

Contents

Statistics in the News

STATISTICS INDICATE BENGALS WILL BEAT STEELERS SUNDAY

Professional football teams record vast amounts of data concerning running plays, passing plays, running yardage, passing yardage, number of interceptions, number of fumbles, and so on. But what does all this statistical data tell the coaches, and what—if anything—does it tell them about how the team can improve its chances of winning the game against the next opponent?

Bud Goode of Los Angeles is a statistician who began computer studies of professional football teams 15 years ago. Currently, ten National Football League teams use his computerized statistical summaries to analyze past football games and indicate ways in which a team may be able to exploit weaknesses of its next opponent. Each week he evaluates key statistics involving the Cincinnati Bengals and their next opponent.

As part of the statistical analysis prior to Sunday's game with the Pittsburgh Steelers, data on the number of interceptions thrown by the Steelers over the past 16 games were put into the computer. Using a method known as linear regression, the computer developed an equation that predicted the number of interceptions the Steelers would throw in their next game. The slope of the linear regression equation was positive, which indicated the Steelers were showing a trend, or tendency, to increase the number of interceptions they would throw during a game. A similar analysis of the number of interceptions made by the Cincinnati Bengals defense over the past 16 games

A strong pass rush by the defensive line will be necessary for the Bengals to intercept Pittsburgh's quarterback.

showed a positive slope, indicating the Bengals were showing a trend, or tendency, to increase their ability to intercept their opponent's passes.

Statistical analysis of data for the entire league shows that each pass interception contributes an average of 5 points to a team's score. Thus the statistical summaries of interception data for the Steelers and Bengals indicate trends favoring the Bengals in next Sunday's game. The extra interception being predicted for the Bengals should provide the winning margin, provided the two teams play according to statistical form. And thus, on Sunday at least, the statistics and the computer are favoring the Bengals to beat the Steelers.

Based on "Bengals vs. Steelers," *Cincinnati Enquirer* (September 19, 1982). For the record, the regulation game ended in a 20–20 tie, with the Steelers eventually winning 26–20 in overtime. The Cincinnati Bengals threw three interceptions during the game; the Pittsburgh Steelers, none.

In Chapter 4 we showed how correlation and regression analyses could be used to investigate the relationship between two variables. Recall that correlation analysis is a procedure for determining the extent to which two variables are linearly related, whereas regression analysis is a technique that can be used to develop a mathematical equation showing how the variables are related. The "Statistics in the News" article describes the use of regression analysis to predict the number of pass interceptions by a football team. Making predictions is one of the most common uses of regression analysis. In this chapter we extend our study of regression and correlation analyses to include the consideration of statistical inference.

We caution the reader before beginning this chapter that neither regression nor correlation analysis can be interpreted as establishing cause-effect relationships. Regression and correlation analyses can indicate only how or to what extent the variables are associated with each other. Any conclusions about a cause-and-effect relationship must be based on the judgment of the analyst.

15.1 AN INTRODUCTION TO REGRESSION ANALYSIS

In regression terminology, the variable which is being predicted by the mathematical equation is called the *dependent* variable. The variable being used to predict the value of the dependent variable is called the *independent* variable. Common statistical notation is to use y to denote the dependent variable and x to denote the independent variable.

EXAMPLE 15.1

New Hampshire Gas and Electric would like to develop an equation that can be used to predict the daily demand for electricity using data on the average daily temperature. The utility's desire to predict demand for electricity would suggest making electric demand the dependent variable for the analysis and temperature the independent variable. Thus, y = electric demand and x = temperature.

• • •

EXAMPLE 15.2

The instructor in a freshman computer science course is interested in the relationship between time spent using the computer system (x) and the final exam score (y). Data

TABLE 15.1

Data for Example 15.2

x = Hours Using Computer System	y = Final Exam Score
45	40
30	35
90	75
60	65
105	90
65	50
90	90
80	80
55	45
75	65

collected from a sample of 10 students who took the course last quarter are presented in Table 15.1.

Figure 15.1 shows graphically the data presented in Table 15.1. The number of hours spent using the computer system is shown on the horizontal axis, with the final exam score on the vertical axis. A graph such as this is known as a *scatter diagram*. The usual practice is to plot the independent variable on the horizontal axis and the dependent variable on the vertical axis. The advantage of a scatter diagram is that it provides an overview of the data and enables us to draw preliminary conclusions about a possible relationship between the variables.

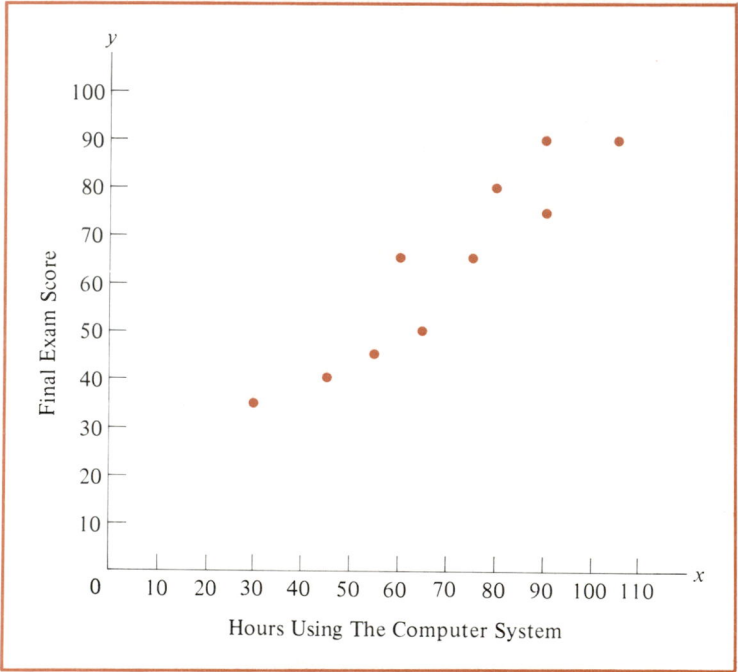

FIGURE 15.1
Scatter Diagram for Example 15.2

What preliminary conclusions can we draw from Figure 15.1? It appears that fewer hours spent using the computer system are associated with lower final exam scores and that more hours spent using the computer system are associated with higher final exam scores. It also appears that the relationship between the two variables can be approximated by a straight line. In Figure 15.2 we have drawn a straight line through the data that appears to provide a good linear approximation of the relationship between the variables.

• • •

Least Squares Method

In Chapter 4 we referred to the straight line fitted to the data involving two variables as the estimated regression line. Recall that, in general, the equation for the estimated regression line is written as follows.

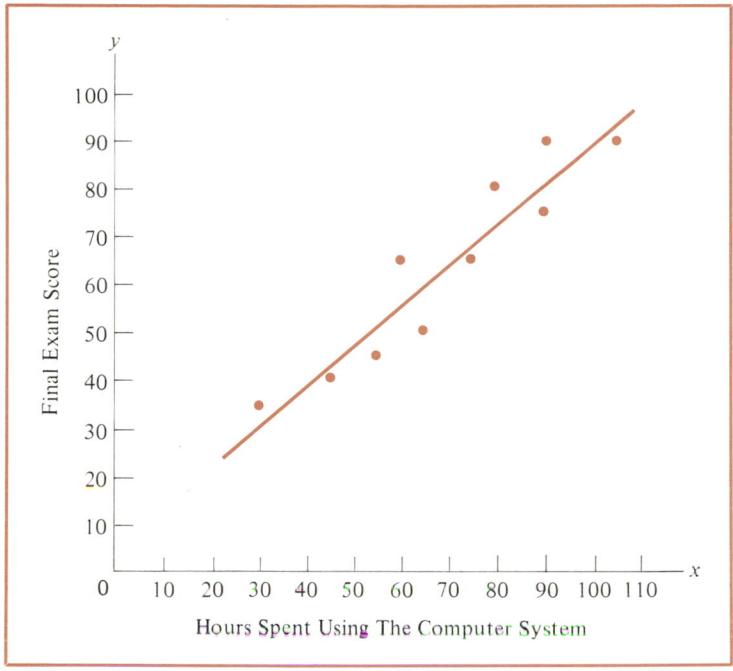

FIGURE 15.2
Straight-Line Approximation for Example 15.2

Estimated Regression Line

$$\hat{y} = b_0 + b_1 x \tag{15.1}$$

where

b_0 = y-intercept of the estimated regression line

b_1 = slope of the estimated regression line

Application of the *least squares method* provides the values of b_0 and b_1 that make the sum of squares of the differences between the observed values of the dependent variable (y) and the estimated values of the dependent variable ($\hat{y}$) a minimum. In other words, the least squares method is a procedure for determining the values of b_0 and b_1 (and, hence, $\hat{y}$) such that these values minimize

$$\Sigma(y_i - \hat{y}_i)^2 \tag{15.2}$$

Figure 15.3 illustrates the least squares objective.

Using differential calculus, it can be shown that the values of b_0 and b_1 that minimize (15.2) can be found by using (15.3) and (15.4).

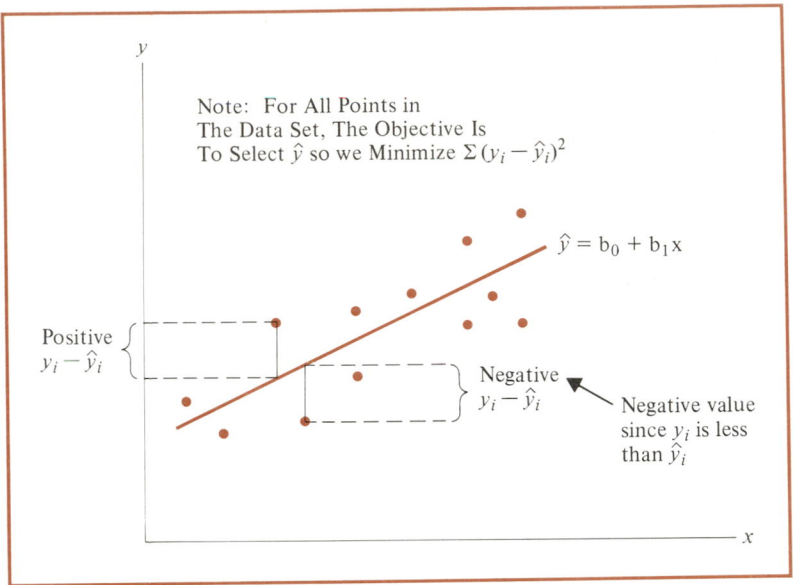

FIGURE 15.3
Least Squares Objective

Slope and y-Intercept for the Estimated Regression Line

$$b_1 = \frac{\Sigma x_i y_i - (\Sigma x_i \Sigma y_i)/n}{\Sigma x_i^2 - (\Sigma x_i)^2/n} \qquad (15.3)$$

$$b_0 = \bar{y} - b_1 \bar{x} \qquad (15.4)$$

EXAMPLE 15.2 (continued)

The calculations shown below are used to compute the values of b_1 and b_0 using (15.3) and (15.4).

x_i	y_i	$x_i y_i$	x_i^2
45	40	1,800	2,025
30	35	1,050	900
90	75	6,750	8,100
60	65	3,900	3,600
105	90	9,450	11,025
65	50	3,250	4,225
90	90	8,100	8,100
80	80	6,400	6,400
55	45	2,475	3,025
75	65	4,875	5,625
695	635	48,050	53,025

Now, following the recommended practice introduced in Chapter 4 of computing the values of b_0 and b_1 to at least four decimal places, we obtain

$$b_1 = \frac{48{,}050 - (695)(635)/10}{53{,}025 - (695)^2/10}$$

$$= \frac{3917.5}{4722.5} = .8295$$

$$b_0 = 63.5 - .8295(69.5)$$

$$= 5.8498$$

Rounding the values for b_0 and b_1 to two decimal places, we have $\hat{y} = 5.85 + .83x$ as the estimated regression line.

$$\bullet \quad \bullet \quad \bullet$$

In this chapter we are concerned with testing the statistical significance of the regression relationship equation, using the estimated regression line to predict the value of y, and developing interval estimates of y.

EXERCISES

1. The following data were collected regarding the weight (pounds) of women swimmers and their height (inches).

Height	68	64	62	65	66
Weight	132	108	102	115	128

   a. Develop a scatter diagram for these data with height on the horizontal axis.
   b. What does the scatter diagram developed in (a) indicate about the relationship between the two variables?
   c. Try to approximate the relationship between height and weight by drawing a straight line through the data.
   d. Develop the estimated regression line by computing the values of b_0 and b_1 using (15.3) and (15.4).
   e. If a swimmer's height is 63 inches, what would you estimate her weight to be?

2. Given are five observations taken for two variables, x and y.

Observation	x_i	y_i
1	2	25
2	3	25
3	5	20
4	1	30
5	8	16

    a. Develop a scatter diagram for these data.

    b. Use the method of least squares to compute an estimated regression line for the data.

3. The following data were collected regarding the monthly starting salaries and the grade point averages (GPA) for undergraduate students who had obtained a degree in political science.

GPA	Monthly Salary
2.6	$1100
3.4	1400
3.6	1800
3.2	1300
3.5	1600
2.9	1200

    a. Develop a scatter diagram for these data with GPA on the horizontal axis.

    b. What does the scatter diagram developed in (a) indicate about the relationship between the two variables?

    c. Draw a straight line through the data to approximate a linear relationship between GPA and salary.

    d. Use the least squares method to develop the estimated regression line.

    e. Predict the monthly starting salary for a student with a 3.0 GPA and for a student with a 3.5 GPA.

4. Eddie's Restaurants collected the following data on the relationship between advertising and sales at a sample of five restaurants.

Advertising Expenditures ($1000s)	Sales ($1000s)
1.0	19.0
4.0	44.0
6.0	40.0
10.0	52.0
14.0	53.0

Develop a scatter diagram for these data with advertising expenditures on the horizontal axis and sales on the vertical axis. Does there appear to be a linear relationship?

5. Shown are some data that a sales manager has collected concerning annual sales and years of experience.

Salesperson	Years of Experience	Annual Sales ($1000s)
1	1	80
2	3	97
3	4	92
4	4	102
5	6	103
6	8	111
7	10	119
8	10	123
9	11	117
10	13	136

   a. Develop a scatter diagram for these data with years of experience on the horizontal axis.
   b. Use the method of least squares to compute an estimated regression line for the relationship between years of experience and annual sales.

6. Tyler Realty collected the following data regarding the selling price of new homes and the size of the homes measured in terms of square footage of living space.

Square Footage	Selling Price
2500	$124,000
2400	$108,000
1800	$ 92,000
3000	$146,000
2300	$110,000

   a. Develop a scatter diagram for these data with square footage on the horizontal axis.
   b. Try to approximate the relationship between square footage and selling price by drawing a straight line through the data.
   c. Does there appear to be a linear relationship?
   d. Develop an estimated regression line using the least squares method.
   e. Predict the selling price for a home with 2700 square feet.

7. The owner of a local grocery store varied the price of a product for six consecutive weeks. The following data show the price per unit and the number sold that week.

Price	Number Sold
$.60	220
$.62	200
$.58	280
$.60	250
$.64	190
$.62	240

   a. Develop a scatter diagram for the data with price on the horizontal axis.
   b. What does the scatter diagram developed in (a) indicate about the relationship between price and the number of units sold?
   c. Develop an estimated regression line that can be used to predict the number of units sold, given the price.
   d. Predict the number of units sold at a price of $.63.
   e. Use the estimated regression line to predict the number of units sold at a price of $.64. How closely does this predicted value compare to the number of units the grocer actually sold at a price of $.64?

8. A university medical center has developed a test designed to measure a patient's stress level. The test is designed so that higher scores on the test correspond to higher levels of stress. As part of a research study, the blood pressure (low reading) of patients who took the test was recorded. The following results were obtained.

Stress Test Score	Blood Pressure
53	70
94	91
64	78
73	78
82	85
90	84

   a. Develop a scatter diagram for these data with stress test score on the horizontal axis. Does a linear relationship between the two variables appear to be appropriate?
   b. Develop the estimated least squares line for these data.
   c. Estimate an individual's blood pressure if he or she scored 85 on the stress test.

15.2 ASSUMPTIONS IN REGRESSION ANALYSIS

In Example 15.2 we found the least squares estimated regression line $\hat{y} = 5.85 + .83x$. The slope ($b_1 = .83$) and y-intercept ($b_0 = 5.85$) for this line were computed by applying (15.3) and (15.4) to the sample data. Recall from Chapter 9 that when data were available for just one variable, such as y, our objective was to use the value of the sample statistic (e.g., $\bar{y}$) to make inferences about the corresponding population parameter (e.g., μ). Our objective in this chapter is very similar, but we want to take advantage of the relationship between y and x to make better inferences.

If the scatter diagram indicates that the relationship between x and y can be approximated by a straight line, we make the assumption that for the population, the mean value of y, denoted by $E(y)$, is linearly related to x. Using the Greek letter β, we can write the *regression line,* which describes the relationship in the population, as follows.

$$E(y) = \beta_0 + \beta_1 x \qquad (15.5)$$

where

$$E(y) = \text{mean value of } y \text{ for a given value of } x$$

$$\beta_0 = y\text{-intercept of the regression line}$$

$$\beta_1 = \text{slope of the regression line}$$

In (15.6), $E(y)$ represents the average of all possible values of y that could occur at a particular value of x. Also, β_0 and β_1 are referred to as *population parameters* and are estimated using the sample values b_0 and b_1, respectively. Thus $\hat{y} = b_0 + b_1 x$

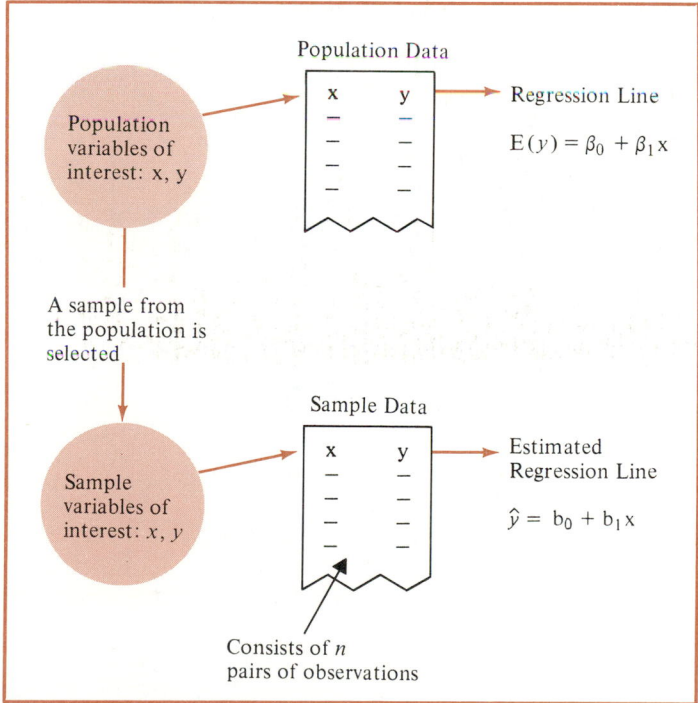

FIGURE 15.4
Estimating the Population Regression Line Using Sample Data

becomes an estimate of $E(y) = \beta_0 + \beta_1 x$. Hence we referred to $\hat{y} = b_0 + b_1 x$ as the *estimated regression line* and $E(y) = \beta_0 + \beta_1 x$ as the regression line. Figure 15.4 summarizes these concepts.

Based upon the concept of a regression line, we can now state the usual assumptions made in regression analysis.

Assumptions in Regression Analysis

1. For a particular value of x, the values of y are normally distributed about the regression line $E(y) = \beta_0 + \beta_1 x$.
2. The variance of y, denoted by σ^2, is the same for each value of x.
3. The values of y are independent.

EXAMPLE 15.2 (continued)

Let us return to the example involving the relationship between time spent using the computer system (x) and the final exam score (y). Consider students who have spent 80 hours on the computer system ($x = 80$); when $x = 80$, the value of $E(y)$ in (15.5) represents the average final exam score that would occur for these students. Figure 15.5 provides a graphical interpretation of the assumptions in regression analysis and their implications. As shown in Figure 15.5, the value of $E(y)$ changes according to the

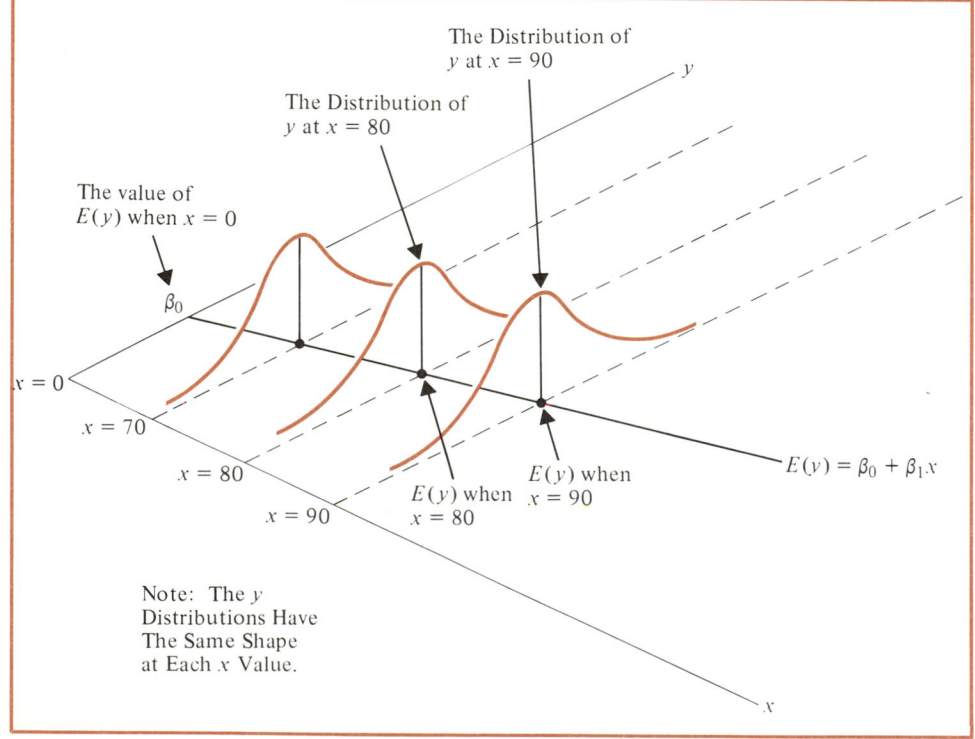

FIGURE 15.5
Illustration of Regression Assumptions

specific value of x considered. Note that the probability distribution of y about the regression line forms a normal distribution, each with the same shape and hence the same variance.

• • •

15.3 THE COEFFICIENT OF DETERMINATION

We have introduced the least squares method as a technique for finding b_0 and b_1 by minimizing the sum of squares of the differences between the observed values of the dependent variable (y) and the predicted values of the dependent variable ($\hat{y}$). Note that the differences between y and $\hat{y}$ for any data point or observation actually represent the errors in using $\hat{y}$ to estimate y. That is, for any observation the error is $y - \hat{y}$. This difference is also referred to as a *residual*. Thus the resulting sum of squares is referred to as the *sum of squares due to error*, or the *residual sum of squares*. We use SSE to represent this quantity.

Sum of Squares Due to Error

$$\text{SSE} = \Sigma(y_i - \hat{y}_i)^2 \tag{15.6}$$

EXAMPLE 15.2 (continued)

The calculations required to compute SSE for the data presented in Example 15.2 are shown.

x_i = Hours Using the Computer System	y_i = Final Exam Score	$\hat{y}_i = 5.85 + .83x$	$y_i - \hat{y}_i$	$(y_i - \hat{y}_i)^2$
45	40	43.20	-3.20	10.24
30	35	30.75	4.25	18.06
90	75	80.55	-5.55	30.80
60	65	55.65	9.35	87.42
105	90	93.00	-3.00	9.00
65	50	59.80	-9.80	96.04
90	90	80.55	9.45	89.30
80	80	72.25	7.75	60.06
55	45	51.50	-6.50	42.25
75	65	68.10	-3.10	9.61
				452.78

SSE

Thus SSE = 452.78 is a measure of the error in using the estimated regression line $\hat{y} = 5.85 + .83x$ to predict y.

• • •

Now suppose we were asked to develop an estimate of the final exam score for Example 15.2 without using the number of hours spent on the computer system. We could not use the estimated regression line and would have to use the value of the sample mean, $\bar{y} = 635/10 = 63.5$, as the best estimate of the final exam score. The error in using $\bar{y}$ to predict y is given by $y - \bar{y}$. The corresponding total sum of squares, denoted by SST, is often referred to as the *sum of squares about the mean*.

Total Sum of Squares

$$SST = \Sigma(y_i - \bar{y})^2 \qquad (15.7)$$

EXAMPLE 15.2 (continued)

The calculations required to compute SST are shown.

y_i = Final Exam Score	$y_i - \bar{y} = y_i - 63.5$	$(y_i - \bar{y})^2$
40	−23.5	552.25
35	−28.5	812.25
75	11.5	132.25
65	1.5	2.25
90	26.5	702.25
50	−13.5	182.25
90	26.5	702.25
80	16.5	272.25
45	−18.5	342.25
65	1.5	2.25
		3702.50
		↗ SST

Thus SST = 3702.50.

• • •

Consider the following question: Is $\hat{y}$ a better predictor of y, or is it really no better than using $\bar{y}$? For example 15.2 it should be clear that $\hat{y}$ is a better predictor since SST = 3702.50 is much larger than SSE = 452.78. To see why consider Figure 15.6, where we show the estimated regression line $\hat{y} = 5.85 + .83x$ and the line corresponding to $y = \bar{y} = 63.5$. Note in general that the points cluster more closely around the estimated regression line than they do about the line $\bar{y} = 63.5$. We can think of SST as a measure of how well the observations cluster about the $\bar{y}$ line, and SSE as a measure of how well the observations cluster about the $\hat{y}$ line.

To measure how much the estimated regression line deviates from the line $y = \bar{y}$

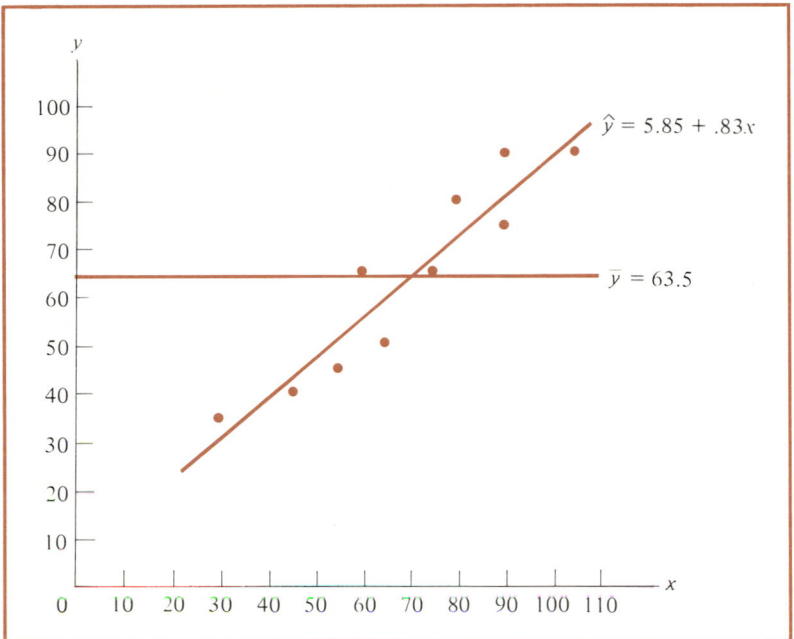

FIGURE 15.6
Deviations About the Estimated Regression Line and the Line $y = \bar{y}$ for Example 15.2

another sum of squares is computed. This sum of squares is commonly called the *sum of squares due to regression,* denoted by SSR. The sum of squares due to regression can be written as follows.

Sum of Squares Due to Regression

$$SSR = \Sigma(\hat{y}_i - \bar{y})^2 \qquad (15.8)$$

Because of difficulties with rounding errors the above formula is rarely used to compute SSR. (A computational formula is given at the end of the section.) From the discussion above we should expect that SSE, SST, and SSR are related. The relationship among SSE, SST, and SSR is as shown.

Relationship Among SST, SSR, and SSE

$$SST = SSR + SSE \qquad (15.9)$$

where

SST = total sum of squares about the mean

SSR = sum of squares due to regression

SSE = sum of squares due to error

EXAMPLE 15.2 (continued)

Recall that for Example 15.2, we found SSE = 452.78 and SST = 3702.50. Using (15.9), we can then conclude that the portion of the total sum of squares explained by the regression relationship is

$$\text{SSR} = \text{SST} - \text{SSE} = 3702.50 - 452.78 = 3249.72$$

• • •

Now let us see how these sums of squares can be used to provide a measure of the strength of the regression relationship. We would have the strongest possible relationship if every observation happened to lie on the least squares fitted line—the line would pass through each point, and we would have SSE = 0. Hence, for a perfect fit SSR must equal SST, and thus the ratio SSR/SST = 1. On the other hand, a poorer fit to the observed data results in a larger SSE. Since SST = SSR + SSE, however, the largest SSE (and *hence* worst fit) occurs when SSR = 0. In this case the estimated regression line does not help predict y. Thus the worst possible fit yields the ratio SSR/SST = 0.

If we were to use the ratio SSR/SST to evaluate the strength of the regression relationship, we would have a measure that could take on values between 0 and 1, with values closer to 1 implying a stronger relationship. The fraction SSR/SST is called the *coefficient of determination* and is represented by the symbol r^2.

Coefficient of Determination

$$r^2 = \frac{\text{SSR}}{\text{SST}}$$ (15.10)

To interpret r^2 better, we note that SSR, the difference between SST and SSE, really measures the portion of SST that is explained by the estimated regression line. Thus we can think of r^2 as

$$r^2 = \frac{\text{Sum of Squares Explained by Regression}}{\text{Total Sum of Squares (Before Regression)}}$$

When it is expressed as a percentage, r^2 can be interpreted as the percentage of the total sum of squares (SST) that can be explained using the estimated regression line. Statisticians often use r^2 as a measure of the goodness-of-fit of the estimated regression line.

EXAMPLE 15.2 (continued)

The value of the coefficient of determination for Example 15.2 is

$$r^2 = \frac{\text{SSR}}{\text{SST}} = \frac{3249.72}{3702.50} = .88$$

Thus we conclude that the estimated regression line has accounted for 88% of the total sum of squares. We should be very pleased with such a large r^2 value.

$$\bullet \quad \bullet \quad \bullet$$

Computational Efficiencies

When using a calculator to compute the value of the coefficient of determination, computational efficiencies can be realized by computing SSR and SST using equations (15.11) and (15.12).

Sum of Squares Due to Regression—Computational Formula

$$SSR = \frac{[\Sigma x_i y_i - (\Sigma x_i \Sigma y_i)/n]^2}{\Sigma x_i^2 - (\Sigma x_i)^2/n} \tag{15.11}$$

Total Sum of Squares—Computational Formula

$$SST = \Sigma y_i^2 - \frac{(\Sigma y_i)^2}{n} \tag{15.12}$$

EXAMPLE 15.2 (continued)

The following calculations show the computation of r^2 using (15.11) and (15.12).

x_i	y_i	$x_i y_i$	x_i^2	y_i^2
45	40	1,800	2,025	1,600
30	35	1,050	900	1,225
90	75	6,750	8,100	5,625
60	65	3,900	3,600	4,225
105	90	9,450	11,025	8,100
65	50	3,250	4,225	2,500
90	90	8,100	8,100	8,100
80	80	6,400	6,400	6,400
55	45	2,475	3,025	2,025
75	65	4,875	5,625	4,225
695	635	48,050	53,025	44,025

Using (15.11) we obtain

$$SSR = \frac{[48{,}050 - (695)(635)/10]^2}{53{,}025 - (695)^2/10} = \frac{(3917.5)^2}{4722.5} = 3249.72$$

Using (15.12) we also obtain

$$SST = 44,025 - \frac{(635)^2}{10} = 3702.50$$

Thus,

$$r^2 = \frac{SSR}{SST} = \frac{3249.72}{3702.50} = .88$$

Note that this is the same value that we obtained previously.

• • •

EXERCISES

9. Refer again to the data in Exercise 1.
   a. Compute SSE, SST, and SSR using (15.7), (15.8), and (15.9).
   b. Recompute SSR and SST using (15.11) and (15.12). Do you get the same results as in (a)?
   c. Compute the coefficient of determination, r^2. Comment on the strength of the relationship.

10. Refer again to the data in Exercise 2.
    a. Compute SSR and SST using (15.11) and (15.12).
    b. What percentage of the total sum of squares is accounted for by the estimated regression line?

11. Refer again to the data in Exercise 3.
    a. Compute SSE, SST, and SSR using (15.7), (15.8), and (15.9).
    b. Recompute SSR and SST using (15.11) and (15.12). Do you get the same results as in (a)?
    c. Compute the coefficient of determination, r^2. Comment on the strength of the relationship.

12. Refer again to the data in Exercise 4.
    a. Compute SSR and SST using (15.11) and (15.12).
    b. What percentage of the total sum of squares is accounted for by the fitted regression line?

13. A medical laboratory estimates the amount of protein in liver samples through the use of a regression model. A spectrometer emitting light shines through a substance containing the sample, and the amount of light absorbed is used to estimate the amount of protein in the sample. A new regression formula is developed daily because of differing amounts of dye in the solution. On one day six samples with known protein concentrations gave the following absorbence readings.

Absorbence Reading (x)	Milligrams of Protein (y)
.509	0
.756	20
1.020	40
1.400	80
1.570	100
1.790	127

a. Use these data to develop an estimated regression line relating the light absorbence reading to milligrams of protein present in the sample.

b. Compute r^2. Would you feel comfortable using this regression model to estimate the amount of protein in a sample?

c. In a sample just received the light absorbence reading was 0.941. Estimate the amount of protein in the sample.

14. The data from Exercise 2 are repeated here.

Observation	x_i	y_i
1	2	25
2	3	25
3	5	20
4	1	30
5	8	16

a. Compute $\bar{x}$ and $\bar{y}$.

b. Substitute the values of $\bar{x}$ and $\bar{y}$ for x and $\hat{y}$ in the estimated regression equation. Do these values satisfy the equation?

c. Will the least squares line always pass through the point corresponding to $(\bar{x}, \bar{y})$? Why or why not?

15.4 TESTING FOR SIGNIFICANCE

In the previous section we saw how the coefficient of determination (r^2) could be used as a measure of the strength of the regression relationship. Larger values of r^2 indicate a stronger relationship; smaller values indicate a weaker one. However, this measure does not allow us to conclude that the relationship is or is not statistically significant. In order to draw conclusions concerning statistical significance, we must take the sample size into consideration. In this section we show how to conduct significance tests that allow us to draw conclusions about the significance of a regression relationship.

An Estimate of σ^2

In the previous section we used the sum of squares due to error, SSE, as a measure of the variability of the actual observations about the estimated regression line. This quantity is also used to develop an estimate of σ^2, the population variance of the y values about the regression line. Recall from our earlier definition of sample variance that we divided the sum of the squared deviations about the sample mean by $n - 1$ to obtain an unbiased estimate of the population variance. We used $n - 1$ instead of n because 1 degree of freedom was lost when the sample mean was used to compute the sum of the squared deviations about the mean. In other words, 1 degree of freedom was lost because one parameter used in computing the sum of squares, the population mean, had to be estimated from the sample data. In regression analysis we must estimate the parameters β_0 and β_1 to compute SSE; that is, we use the estimates b_0 and b_1—obtained from the sample data—to compute the sum of squares due to error. For this reason, 2 degrees of freedom are lost; hence we must divide SSE by $n - 2$ to obtain an unbiased estimate of σ^2. The estimate obtained is called the *mean square due to error*, denoted by MSE.

Estimate of σ^2

$$\text{MSE} = \frac{\text{SSE}}{n - 2} \tag{15.13}$$

EXAMPLE 15.2 (continued)

Recall that in Example 15.2, SSE = 452.78. Thus,

$$\text{MSE} = \frac{\text{SSE}}{n - 2} = \frac{452.78}{8} = 56.60$$

$$\bullet \quad \bullet \quad \bullet$$

In the discussion that follows, we use MSE as an unbiased estimate of σ^2 in tests for the significance of the regression line.

t Test

Recall that the underlying regression equation is assumed to be $E(y) = \beta_0 + \beta_1 x$. If there really exists a relationship of this form between x and y, β_1 would have to differ from 0. Thus a conclusion regarding the significance of the regression relationship can be tested using the following hypotheses:

$$H_0: \beta_1 = 0$$

$$H_1: \beta_1 \neq 0$$

Before presenting the t test, we need to consider the properties of b_1, the least squares estimator of β_1. First, let us consider what would have happened if we had used a different random sample for the same regression study. For example, suppose that in Example 15.2 we had used the final exam records of 10 different students. A regression analysis of this new data, or sample, probably would result in an estimated regression line similar to our previous estimated regression line, $\hat{y} = 5.85 + .83x$. However, it is doubtful that we would obtain exactly the same values for b_0 and b_1. Thus b_0 and b_1 are themselves variables whose values depend upon the data items (the values of x_i and y_i) included in the sample. Recall the discussion of sampling distributions in Chapter 8; b_0 and b_1 must have their own sampling distributions. The properties of the sampling distribution for b_1 are defined as follows.

Sampling Distribution of b_1

Distribution Form:	Normal
Mean:	$E(b_1) = \beta_1$
Variance:	$\sigma_{b_1}^2 = \sigma^2 \dfrac{1}{\Sigma x_i^2 - (\Sigma x_i)^2/n}$

Since we do not know the value of σ^2, we develop an estimate of $\sigma_{b_1}^2$, denoted by $s_{b_1}^2$, by first estimating σ^2 with MSE. Thus we obtain the following.

Estimate of Variance of b_1

$$s_{b_1}^2 = \text{MSE} \frac{1}{\Sigma x_i^2 - (\Sigma x_i)^2/n} \tag{15.14}$$

EXAMPLE 15.2 (continued)

For Example 15.2, MSE = 56.60. Thus

$$s_{b_1}^2 = 56.60 \frac{1}{53,025 - (695)^2/10}$$

$$= 56.60 \frac{1}{4722.5}$$

$$= .0120$$

Hence

$$s_{b_1} = \sqrt{.0120} = .1095$$

· · ·

The t test regarding β_1 is based on the fact that

$$\frac{b_1 - \beta_1}{s_{b_1}}$$

has a t distribution with $n - 2$ degrees of freedom. If the null hypothesis is true, then $\beta_1 = 0$ and b_1/s_{b_1} has a t distribution with $n - 2$ degrees of freedom. Using b_1/s_{b_1} as a test statistic, we use the following decision rule to test $H_0: \beta_1 = 0$ versus $H_1: \beta_1 \neq 0$.

$$\text{Accept } H_0 \text{ if } -t_{\alpha/2} \leq \frac{b_1}{s_{b_1}} \leq t_{\alpha/2}$$

Reject otherwise

where $t_{\alpha/2}$ has $n - 2$ degrees of freedom and where α is the level of significance for the test.

EXAMPLE 15.2 (continued)

Recall that $b_1 = .83$ and $s_{b_1} = .1095$. Thus we have $b_1/s_{b_1} = .83/.1095 = 7.58$. From Table 2 of Appendix B we find that the t value corresponding to $\alpha = .01$ and $n - 2 = 8$ degrees of freedom is $t_{.005} = 3.355$. Since $b_1/s_{b_1} = 7.58 > 3.355$, we reject H_0 and conclude at the .01 level of significance that β_1 is not equal to zero. Thus we conclude that there is a significant relationship between time spent on the computer system and final exam score.

• • •

F Test

The t test has been used to test the null hypothesis $H_0: \beta_1 = 0$. An F test can also be used. In regression analysis with only one independent variable the t test and the F test yield the same results. But with more than one independent variable, only the F test can be used to test for a significant relationship between a dependent variable and a set of independent variables. Here we will introduce the F test and show that it leads to the same conclusion as the t test. In Chapter 16 we illustrate the use of the F test for multiple regression analysis.

The hypotheses we are testing are the same as before.

$$H_0: \beta_1 = 0$$

$$H_1: \beta_1 \neq 0$$

The logic behind the use of the F test for determining whether or not the relationship between x and y is statistically significant is based upon our being able to develop two independent estimates of σ^2. We have just seen that MSE provides an estimate of σ^2. If the null hypothesis $H_0: \beta_1 = 0$ is true, the mean square due to regression (denoted MSR) provides another *independent* estimate of σ^2.

To compute MSR we first note that for any sum of squares the mean square is the sum of squares divided by its degrees of freedom. The number of degrees of freedom for the sum of squares due to regression, SSR, is always equal to the number of independent variables. Since in this chapter we are concerned only with models involving one independent variable, the number of regression degrees of freedom is 1. Using DF as an abbreviation for degrees of freedom, we can write

$$\begin{array}{c}\text{Mean Square} \\ \text{Due to Regression}\end{array} = \text{MSR} = \frac{\text{SSR}}{\text{Regression DF}}$$

$$(15.15)$$

$$\text{MSR} = \frac{\text{SSR}}{\text{Number of Independent Variables}}.$$

The F test concerning the significance of the regression relationship is based on the following F statistic.

$$F = \frac{\text{MSR}}{\text{MSE}} \qquad (15.16)$$

If the null hypothesis ($H_0: \beta_1 = 0$) is true, then the value that we obtain when computing this ratio would be a value that could be obtained if we were randomly sampling values from an F distribution with 1 degree of freedom in the numerator and $n - 2$ degrees of freedom in the denominator. In Table 4 of Appendix B, we provide critical values for the F distribution. For example, this table shows that for an F distribution with 1 degree of freedom in the numerator and 8 degrees of freedom in the denominator, only 1% of the values are as large or larger than the table value of 11.26. Thus if this were the appropriate F distribution for a regression problem, we would compare the value of $F = \text{MSR}/\text{MSE}$ to 11.26; if F were greater than 11.26, we would have sufficient statistical evidence to reject the null hypothesis at the .01 level of significance.

EXAMPLE 15.2 (continued)

Recall that for this problem $n = 10$, MSE = 56.6, and MSR = SSR/1 = 3249.72. Thus

$$F = \frac{\text{MSR}}{\text{MSE}} = \frac{3249.72}{56.60} = 57.42$$

Since the numerator in this ratio, MSR, has 1 degree of freedom and the denominator, MSE, has $n - 2 = 10 - 2 = 8$ degrees of freedom, we refer to an F distribution with 1 degree of freedom in the numerator and 8 degrees of freedom in the denominator. As mentioned previously, the critical value at the .01 level of significance is 11.26. Thus we reject the null hypothesis that $\beta_1 = 0$ and conclude that there is a statistically significant relationship.

$$\bullet \quad \bullet \quad \bullet$$

EXERCISES

15. The data from Exercise 2 are repeated here.

Observation	x_i	y_i
1	2	25
2	3	25
3	5	20
4	1	30
5	8	16

   a. Compute an estimate of σ^2.

   b. Compute an estimate of the variance of b_1.

   c. Use the t test to test the hypotheses

$$H_0: \beta_1 = 0$$

$$H_1: \beta_1 \neq 0$$

at the .05 level of significance.

   d. Use the F test to test the hypotheses in (c) at the $\alpha = .05$ level of significance.

16. Given are five observations collected in a regression study on two variables.

Observation	x_i	y_i
1	2	2
2	4	3
3	5	2
4	7	6
5	8	4

   a. Develop the estimated regression equation for these data.

   b. Use the t test to test the hypotheses

$$H_0: \beta_1 = 0$$

$$H_1: \beta_1 \neq 0$$

at the $\alpha = .05$ level of significance.

   c. Use the F test to test the hypotheses in (b) at the $\alpha = .05$ level of significance.

17. Refer to Exercise 3, where an estimated regression line relating GPA to monthly starting salaries was developed. SSR and SST were computed in Exercise 11.

   a. Compute an estimate of σ^2.

   b. Compute $s_{b_1}^2$ and s_{b_1}.

   c. Use the t test to determine if GPA and salary are related at the $\alpha = .05$ level of significance.

   d. Use the F test to determine if GPA and salary are related at the $\alpha = .05$ level of significance.

18. Refer to Exercise 6, where an estimated regression line relating square footage to selling prices of new homes was developed. Test whether or not selling price and square footage are related at the $\alpha = .01$ level of significance.

19. Refer to Exercise 7, where an estimated regression line relating the price of a product and the number of units sold was developed. Test whether or not price and the number of units sold are related at the $\alpha = .05$ level of significance.

20. Refer to Exercise 5, where an estimated regression line relating years of experience and annual sales was developed. At the $\alpha = .05$ level of significance, determine whether or not annual sales and years of experience are related.

21. Refer to Exercise 13, where an estimated regression line relating light absorbence readings and milligrams of protein present in a liver sample was developed. Test whether or not the absorbence readings and amount of protein present are related at the $\alpha = .01$ level of significance.

15.5 USING THE ESTIMATED REGRESSION LINE FOR INTERVAL ESTIMATES

For Example 15.2 we concluded that the final exam grade (y) and the number of hours spent on the computer system (x) were related. Moreover, the estimated regression line $\hat{y} = 5.85 + .83x$ appears to describe adequately the relationship between x and y. Now we can begin to develop estimates of y for a given value of x. First let us consider point estimates.

EXAMPLE 15.2 (continued)

Suppose we want to develop a point estimate for the mean final exam score of *all* students who spend 85 hours on the computer system. The point estimate of $E(y)$ in this case is equal to $\hat{y} = 5.85 + .83x = 5.85 + .83(85) = 76.40$.

   Next, let us consider developing a prediction of the final exam score for David Edmunds, *one* particular student who has spent 85 hours on the computer system. The prediction of his final exam score is also given by $\hat{y} = 5.85 + .83x = 5.85 + .83(85) = 76.40$. Thus we see that the point estimate for the mean value of y for a given value of x is the same as the prediction of an individual value of y for a given value of x.

• • •

   There are two types of interval estimates to consider. The first is a *confidence interval* estimate of the mean value of y for a particular value of x, and the second type is a *prediction interval* estimate of an individual value of y corresponding to a given value of x. Thus confidence intervals are associated with mean values of y and prediction intervals are associated with individual values of y.

   Let $E(y_p)$ denote the mean value of y corresponding to a particular value of x, x_p, and $\hat{y}_p$ the predicted value of y corresponding to x_p. The following formulas show how confidence and prediction intervals are computed.

Confidence Interval for $E(y_p)$

$$\hat{y}_p \pm t_{\alpha/2} \sqrt{MSE\left[\frac{1}{n} + \frac{(x_p - \bar{x})^2}{\Sigma x_i^2 - (\Sigma x_i)^2/n}\right]} \qquad (15.17)$$

where

$$x_p = \text{a particular value of } x$$
$$y_p = \text{the value of } y \text{ corresponding to } x_p$$
$$\hat{y}_p = \text{the estimate of } E(y_p) = b_0 + b_1 x_p$$
$$E(y_p) = \text{the mean value of } y \text{ at } x_p$$

Prediction Interval for y_p

$$\hat{y}_p \pm t_{\alpha/2} \sqrt{MSE\left[1 + \frac{1}{n} + \frac{(x_p - \bar{x})^2}{\Sigma x_i^2 - (\Sigma x_i)^2/n}\right]} \qquad (15.18)$$

EXAMPLE 15.2 (continued)

Let us develop a 95% confidence interval for the mean value of y when $x = 85$. Using (15.17) with $x_p = 85$ and $\hat{y}_p = 5.85 + .83(85) = 76.40$, we obtain

$$76.40 \pm t_{\alpha/2} \sqrt{56.6\left[\frac{1}{10} + \frac{(85 - 69.5)^2}{53,025 - (695)^2/10}\right]}$$

We need to find the t value from Table 2 of Appendix B corresponding to $n - 2 = 10 - 2 = 8$ degrees of freedom and $\alpha = .05$. Doing so, we find $t_{.025} = 2.306$. Hence the resulting interval is

$$76.40 \pm (2.306) \sqrt{56.6 \, [.15087]} = 76.40 \pm 6.74$$

Thus, we obtain 69.66 to 83.14 as the 95% confidence interval for the mean final exam score when $x = 85$.

Similarly, using (15.18) we obtain the following 95% prediction interval for an individual value of y at $x = 85$.

$$76.40 \pm (2.306) \sqrt{56.6\left[1 + \frac{1}{10} + \frac{(85 - 69.5)^2}{53,025 - (695)^2/10}\right]}$$

$$= 76.40 \pm (2.306) \sqrt{56.6 \, [1.15087]} = 76.40 \pm 18.61$$

Thus 57.79 to 95.01 is the 95% prediction interval for the final exam score for one individual student who has spent 85 hours on the computer system. Note that this prediction interval is wider than the confidence interval for the mean final exam score. This difference simply reflects the fact that we are able to predict the mean final exam score with more precision than we can the final exam score for any one particular individual.

• • •

EXERCISES

22. As an extension of Exercise 3, develop a 95% confidence interval for estimating the mean starting salary for students with 3.0 GPA.

23. As an extension of Exercise 6, develop a 90% confidence interval for predicting the mean selling price for homes with 2,200 square feet of living space.

24. As an extension of Exercise 3, develop a 95% confidence interval for estimating the starting salary of Joe Heller, who has a GPA of 3.0.

25. As an extension of Exercise 6, develop a 95% confidence interval for the selling price of a home on Highland Terrace with 2,800 square feet.

26. State in your own words why a smaller confidence interval is obtained when a mean value is predicted than when an individual value is predicted.

27. A study conducted by the Department of Transportation regarding driving speed and mileage for mid-size automobiles resulted in the following data.

Driving Speed	Mileage
30	28
50	25
40	25
55	23
30	30
25	32
60	21
25	35
50	26
55	25

a. Determine the estimated regression line that relates mileage to the driving speed.

b. At the $\alpha = .05$ level of significance, determine whether or not mileage and driving speed are related.

c. Did the estimated regression line provide a good fit to the data?

d. Develop a 95% confidence interval for estimating the mean mileage for cars that are driven at 50 miles per hour.

e. If we were interested in one specific car that was driven at 50 miles per hour, how would our estimate of mileage change as compared to the estimate developed in (d)?

15.6 COMPUTER SOLUTION OF REGRESSION PROBLEMS

As we have seen, the computational aspects associated with regression analysis can be quite time-consuming. In this section we discuss how the computational burden can be simplified by using a computer software package. The general procedure followed in using computer packages is for the user to input the data (x and y values for the sample) together with some instructions concerning the types of analyses that are required. The software package performs the analyses and prints the results in an output report. Before discussing the details of this approach, we discuss the use of the ANOVA table as a device for summarizing the calculations performed in regression analysis. The ANOVA table is an important component of the output report produced by most software packages for regression analysis.

The ANOVA Table

In Chapter 14 we saw how the ANOVA table could provide a convenient summary of the computations for analysis of variance. In regression analysis a similar table can be developed. Table 15.2 shows the general form of the ANOVA table for two-variable regression studies. It can be seen that the relationship that holds among the sum of squares (that is, SST = SSR + SSE) also holds for the degrees of freedom. That is:

$$\text{Total DF} = \text{Regression DF} + \text{Error DF}$$

TABLE 15.2

General Form of the ANOVA Table for Two-Variable Regression Analysis

Source of Variation	Sum of Squares	Degrees of Freedom	Mean Square
Regression	SSR	1	$MSR = \dfrac{SSR}{1}$
Error	SSE	$n - 2$	$MSE = \dfrac{SSE}{(n - 2)}$
Total (about the mean)	SST	$n - 1$	

Computer Output

In Figure 15.7 we show the computer output for Example 15.2 using Minitab. The interpretation of the output is as follows.

1. The computer package prints the estimated regression line as $Y = 5.85 + 0.830X$. Note that Y represents the estimated value of the dependent variable (which we denoted by $\hat{y}$).

2. A table is printed that shows the value of b_0 and b_1, the standard deviation of each coefficient, and the t value obtained by dividing each coefficient value by its standard deviation. Thus to test $H_0: \beta_1 = 0$ versus $H_1: \beta_1 \neq 0$, we could compare 7.58 (the corresponding T-RATIO) to the table value. This is the procedure we described in the last part of Section 15.4.

3. The computer package prints the estimate of σ, $s = 7.523$ (which we referred to as MSE = 7.523), as well as information regarding the strength of the relationship. Note that R-SQUARED = 87.8 PERCENT is the coefficient of determination (multiplied by 100), which we denoted by r^2.

4. The ANOVA table (as described earlier in this section) is printed below the heading "ANALYSIS OF VARIANCE." Note that using this table, the F value equal to MSR/MSE could easily be computed in order to provide an alternate test of $H_0: \beta_1 = 0$ versus $H_1: \beta_1 \neq 0$.

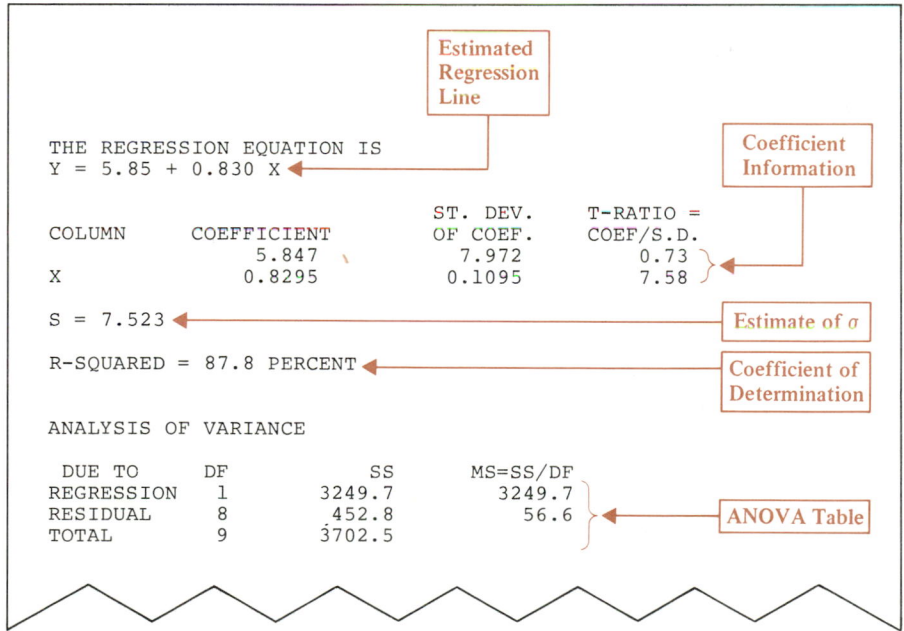

FIGURE 15.7
Computer Solution Using the Minitab Statistical Computing System

As you can see, the computer output from Minitab is fairly easy to interpret given our current background in regression analysis. When large data sets are involved, computer packages provide the only practical means for solving regression problems.

EXERCISES

28. The commercial division of a real estate firm is conducting a regression analysis of the relationship between x, annual gross rents ($1000s), and y, selling price ($1000s) for apartment buildings. Data have been collected on a number of properties recently sold, and the accompanying output has been obtained in a computer run.
   a. How many apartment buildings were in the sample?
   b. Write the estimated regression line.
   c. What is the value of s_{b_1}?
   d. Test the significance of the relationship at an $\alpha = .05$ level of significance.
   e. Estimate the selling price of an apartment building with gross annual rents of $50,000.

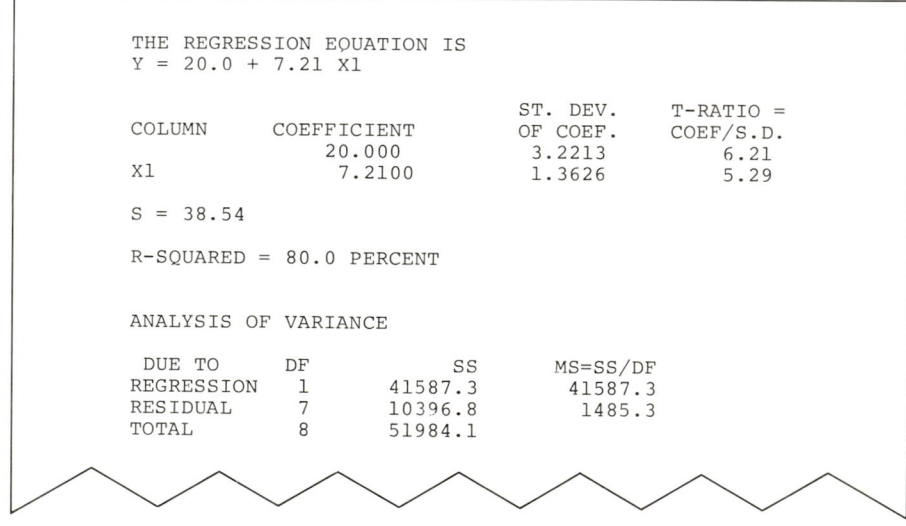

```
          THE REGRESSION EQUATION IS
          Y = 20.0 + 7.21 X1

                                        ST. DEV.      T-RATIO =
          COLUMN       COEFFICIENT      OF COEF.      COEF/S.D.
                           20.000         3.2213          6.21
          X1                7.2100        1.3626          5.29

          S = 38.54

          R-SQUARED = 80.0 PERCENT


          ANALYSIS OF VARIANCE

             DUE TO      DF          SS      MS=SS/DF
          REGRESSION     1       41587.3      41587.3
          RESIDUAL       7       10396.8       1485.3
          TOTAL          8       51984.1
```

29. Shown is a portion of the computer output (page 537 top) for a regression analysis relating maintenance expense (dollars per month) to usage (hours per week) of a particular brand of computer terminal.
   a. Write the estimated regression line.
   b. Test to see if monthly maintenance expense is related to usage at the .01 level of significance.
   c. Use the estimated regression line to predict monthly maintenance expense for any terminal that is used 25 hours per week.

```
           THE REGRESSION EQUATION IS
           Y = 6.1092 + .8951 X1

                                        ST. DEV.      T-RATIO =
           COLUMN       COEFFICIENT     OF COEF.      COEF/S.D.
                          6.1092          0.9361          6.53
           X1             0.8951          0.1490          6.01

           S = 6.61

           R-SQUARED = 81.9 PERCENT


           ANALYSIS OF VARIANCE

             DUE TO      DF            SS       MS=SS/DF
           REGRESSION     1        1575.76       1575.76
           RESIDUAL       8         349.14         43.64
           TOTAL          9        1924.90
```

30. A regression model relating x, number of salespersons at a branch office, to y, annual sales at the office ($1,000's), has been developed. Shown is the computer output from a regression analysis of the data.
    a. Write the estimated regression line.
    b. How many branch offices were involved in the study?
    c. Test the significance of the relationship at an $\alpha = .05$ level of significance.
    d. Predict the annual sales at the Memphis branch office. This branch has 12 salespersons.

```
           THE REGRESSION EQUATION IS
           Y = 80.0 + 50.00 X1

                                        ST. DEV.      T-RATIO =
           COLUMN       COEFFICIENT     OF COEF.      COEF/S.D.
                          80.0           11.333          7.06
           X1             50.0            5.482          9.12

           S = 9.06

           R-SQUARED = 74.8 PERCENT


           ANALYSIS OF VARIANCE

             DUE TO      DF            SS       MS=SS/DF
           REGRESSION     1        6828.6        6828.6
           RESIDUAL      28        2298.8          82.1
           TOTAL         29        9127.4
```

15.7 CORRELATION ANALYSIS

There are some situations in which we are not as concerned with the equation that relates two variables as in measuring the extent to which the two variables are related. In such cases a statistical technique referred to as correlation analysis can be used to determine the strength of the relationship between the two variables.*

Recall from Chapter 4 that the sample correlation coefficient r is defined such that values of r are between -1 and $+1$. A value of $+1$ indicates that x and y are perfectly related in a positive linear sense. That is, all the points in the scatter diagram lie on a straight line that has a positive slope. A value of -1 indicates that x and y are perfectly related in a negative sense. That is, all the points in the scatter diagram lie on a straight line that has a negative slope. Values of the sample correlation coefficient close to zero indicate that x and y are not linearly related.

The formula for computing the *sample correlation coefficient* that was presented in Chapter 4 is shown below.

Sample Correlation Coefficient

$$r = \frac{\Sigma x_i y_i - (\Sigma x_i \Sigma y_i)/n}{\sqrt{\Sigma x_i^2 - (\Sigma x_i)^2/n}\,\sqrt{\Sigma y_i^2 - (\Sigma y_i)^2/n}} \tag{15.19}$$

EXAMPLE 15.2 (continued)

Using (15.19), the computations required to compute the sample correlation coefficient are as follows:

x_i	y_i	$x_i y_i$	x_i^2	y_i^2
45	40	1,800	2,025	1,600
30	35	1,050	900	1,225
90	75	6,750	8,100	5,625
60	65	3,900	3,600	4,225
105	90	9,450	11,025	8,100
65	50	3,250	4,225	2,500
90	90	8,100	8,100	8,100
80	80	6,400	6,400	6,400
55	45	2,475	3,025	2,025
75	65	4,875	5,625	4,225
Totals 695	635	48,050	53,025	44,025

*In correlation analysis it is assumed that x and y are both random variables and are both normally distributed.

$$r = \frac{48{,}050 - (695)(635)/10}{\sqrt{53{,}025 - (695)^2/10}\,\sqrt{44{,}025 - (635)^2/10}}$$

$$= \frac{3917.5}{\sqrt{(4722.5)}\,\sqrt{(3702.5)}} = \frac{3917.5}{4181.5136} = .94$$

The sample correlation coefficient of $r = +.94$ shows that there is a strong positive linear relationship between x and y.

• • •

The sample correlation coefficient can also be computed as a by-product of a regression analysis. Let us see how.

Determining the Sample Correlation Coefficient From the Regression Analysis Results

In this discussion we assume that the least squares estimated regression line is $\hat{y} = b_0 + b_1 x$. In such cases the sample correlation coefficient can be computed using one of the following formulas.

Sample Correlation Coefficient Using Regression Analysis Output

$$r = \pm \sqrt{\text{Coefficient of Determination}} \qquad (15.20)$$

$$r = b_1 \left(\frac{s_x}{s_y}\right) \qquad (15.21)$$

where

$$s_x = \sqrt{\frac{\Sigma(x_i - \bar{x})^2}{(n-1)}} \qquad \text{(Sample Standard Deviation for } x\text{)}$$

$$s_y = \sqrt{\frac{\Sigma(y_i - \bar{y})^2}{(n-1)}} \qquad \text{(Sample Standard Deviation for } y\text{)}$$

With (15.20) it is not clear what the proper sign of the correlation coefficient should be. However, from (15.21) we see that the sign of the sample correlation coefficient must be the same as the sign of b_1, the slope of the estimated regression line.

EXAMPLE 15.2 (continued)

Recall that for Example 15.2 $r^2 = .88$. Hence

$$r = \sqrt{.88}$$

$$= +.94$$

Note that the sign of the sample correlation coefficient in this example is positive, since the slope ($b_1 = .83$) is positive and indicates a positive relationship between the variables. In addition, note that the value of r computed directly from the regression analysis is the same as the value of r we computed earlier using Equation (15.19).

$$\bullet \quad \bullet \quad \bullet$$

A Test for Significant Correlation

The sample correlation coefficient is a point estimator of the population correlation coefficient. Using ρ (rho) to denote the population correlation coefficient, a statistical test for the significance of a linear relationship between x and y can be performed by testing the following hypotheses:

$$H_0: \rho = 0$$

$$H_1: \rho \neq 0$$

These hypotheses are equivalent to the hypotheses regarding the significance of β_1, the slope of the regression line. Recall that the appropriate hypotheses in that case are

$$H_0: \beta_1 = 0$$

$$H_1: \beta_1 \neq 0$$

Thus if we reject the null hypothesis $H_0: \beta_1 = 0$ (see Section 15.4), we can reject the null hypothesis $H_0: \rho = 0$ and conclude that $\rho \neq 0$. This implies x and y are linearly related.

An alternate test for significant correlation can be performed by computing the following test statistic.

$$r \sqrt{\frac{n - 2}{1 - r^2}} \tag{15.22}$$

If $H_0: \rho = 0$ is true, then the value of (15.22) has a t distribution with $n - 2$ degrees of freedom. For $\alpha = .05$ and $n - 2 = 10 - 2 = 8$ degrees of freedom, we see that the appropriate t value from Table 2 of Appendix B is 2.306. Thus if the value of (15.22) exceeds 2.306 or is less than -2.306, we must reject the null hypothesis, $H_0: \rho = 0$.

EXAMPLE 15.2 (continued)

For Example 15.2 the value of (15.22) is

$$.94 \sqrt{\frac{8}{1 - .88}} = 7.68$$

Since 7.68 exceeds the t value of 2.306, we reject $H_0: \rho = 0$ and hence conclude that the population correlation coefficient is not 0.

$$\bullet \quad \bullet \quad \bullet$$

EXERCISES

31. Refer to Exercise 28 and compute the sample correlation coefficient between annual gross rent and selling price of apartment buildings.

32. The following estimated regression line has been developed to estimate the relationship between x, the number of units produced per week, and y, the total weekly cost of production ($):

$$\hat{y} = 60 + 3.2x$$

The standard deviation of weekly production is 10 units, and the standard deviation of weekly cost is $35.00. Compute the sample correlation coefficient r.

33. Eight observations on two random variables are given.

x_i	y_i
2	11
9	4
6	6
8	5
4	9
7	4
5	9
6	7

a. Compute r.
b. Test the hypotheses

$$H_0: \rho = 0$$

$$H_1: \rho \neq 0$$

at the $\alpha = .01$ level of significance.

34. Use the data in Exercise 3 to compute the sample correlation coefficient between GPA and salary.

35. The data presented in Exercise 27 are shown.

Speed	30	50	40	55	30	25	60	25	50	55
Mileage	28	25	25	23	30	32	21	35	26	25

    a.   Compute the sample correlation coefficient for these data.

    b.   Test the hypotheses

$$H_0: \rho = 0$$

$$H_1: \rho \neq 0$$

at the $\alpha = .01$ level of significance.

Summary

In this chapter we studied the topics of regression and correlation analysis. We discussed how regression analysis can be used to develop an equation showing how two variables are related and how correlation analysis can be used to determine the strength of the relationship between the variables.

Before concluding our discussion, however, we would like to emphasize a potential misinterpretation of these studies. In the introduction to this chapter we indicated that it is important to realize that regression and correlation analyses can indicate only how or to what extent the variables are associated with each other. These techniques cannot be interpreted directly as showing cause and effect relationships. For instance, suppose that a regression study of sales volumes and the salesperson's family size indicated that larger family sizes are associated with higher sales volumes. Although these variables are related, it is doubtful that just by increasing family size we can cause an increase in sales volumes. On the other hand, it may seem reasonable that a relationship between advertising expenditures and sales volume is truly a cause and effect relationship. That is, it seems reasonable to suppose that increasing advertising expenditures could cause an increase in sales volume. Thus we have two associations or relationships—family size versus sales volume and advertising expenditures versus sales volume. One of these probably is not causal, the other probably is. In both cases the regression analysis has indicated only an association. Any conclusions about a cause and effect relationship must be based on the judgment of the analyst.

Statistics in Practice

DETERMINING OPTIMUM FEED FOR POULTRY

Monsanto Company, the nation's fourth-largest chemical company, utilizes regression analysis in many facets of its operations. The following example, taken from product applications research in the Nutrition Chemicals Division, illustrates the use of regression analysis in marketing a methionine supplement for use in poultry, swine, and cattle feeds.

Poultry growers work with very high volumes and low profit margins. They have invested large amounts in the accurate definition of nutritional requirements for poultry, since optimal feed compositions result in more rapid growth and in a higher final body weight for a given feed intake. Feed efficiency, which relates gain in body weight to amount of feed consumed, is monitored closely over growth cycles.

A study was conducted by Monsanto in order to determine the relationship between the percentage of supplemental methionine used in feed and the body weight of the poultry. Using the data collected in this study, regression analysis was used to develop the following estimated regression line.

$$\hat{y} = 0.21 + 0.42x$$

where

$\hat{y}$ = estimated body weight in kilograms

x = percentage of supplemental methionine used in the feed

The coefficient of determination, r^2, was .78, indicating a reasonably good fit for the data.

This regression analysis was helpful to Monsanto in enabling the design of feeds that would provide the best possible weight gain for poultry per feed dollar expended. This application is typical of the role that regression analysis has played at Monsanto as the company works to develop quality products for its customers.

Glossary

Scatter diagram—A graph of the data involving two variables. The independent variable appears on the horizontal axis and the dependent variable appears on the vertical axis.

Least squares method—The approach used to develop an estimated regression equation that minimizes the sum of squares of the vertical distances from the points to the least squares fitted line.

Regression line—The mathematical equation relating the independent variable to the expected value of the dependent variable; that is, $E(y) = \beta_0 + \beta_1 x$.

Estimated regression line—The estimate of the regression line obtained by the least squares method; that is, $\hat{y} = b_0 + b_1 x$.

Residual—The difference between the actual value of the dependent variable and the value predicted using the estimated regression line; i.e., $y_i - \hat{y}_i$.

Coefficient of determination (r^2)—A measure of the variation explained by the estimated regression line. It is a measure of how well the estimated regression line fits the data.

Sample correlation coefficient (r)—A statistical measure of the linear association between two variables.

Key Formulas

Estimated Regression Line

$$\hat{y} = b_0 + b_1 x \tag{15.1}$$

Least Squares Objective

$$\min \Sigma(y_i - \hat{y}_i)^2 \tag{15.2}$$

Slope and *y*-intercept for the Estimated Regression Line

$$b_1 = \frac{\Sigma x_i y_i - (\Sigma x_i \Sigma y_i)/n}{\Sigma x_i^2 - (\Sigma x_i)^2/n} \tag{15.3}$$

$$b_0 = \bar{y} - b_1 \bar{x} \tag{15.4}$$

Regression Line

$$E(y) = \beta_0 + \beta_1 x \tag{15.5}$$

Sum of Squares Due to Error

$$SSE = \Sigma(y_i - \hat{y}_i)^2 \tag{15.6}$$

Total Sum of Squares

$$SST = \Sigma(y_i - \bar{y})^2 \tag{15.7}$$

Sum of Squares Due to Regression

$$SSR = \Sigma(\hat{y}_i - \bar{y})^2 \tag{15.8}$$

Relationship Among SST, SSR, and SSE

$$SST = SSR + SSE \tag{15.9}$$

Coefficient of Determination

$$r^2 = \frac{SSR}{SST} \tag{15.10}$$

Sum of Squares Due to Regression—Computational Formula

$$SSR = \frac{[\Sigma x_i y_i - (\Sigma x_i \Sigma y_i)/n]^2}{\Sigma x_i^2 - (\Sigma x_i)^2/n} \tag{15.11}$$

Total Sum of Squares—Computational Formula

$$\text{SST} = \Sigma y_i^2 - (\Sigma y_i)^2/n \tag{15.12}$$

Estimate of $\overset{2}{\sigma}$

$$\text{MSE} = \frac{\text{SSE}}{n-2} \tag{15.13}$$

Estimated Variance of b_1

$$s_{b_1}^2 = \text{MSE}\left(\frac{1}{\Sigma x_i^2 - (\Sigma x_i)^2/n}\right) \tag{15.14}$$

Mean Square Due to Regression

$$\text{MSR} = \frac{\text{SSR}}{\text{Regression DF}} = \frac{\text{SSR}}{\text{Number of Independent Variables}} \tag{15.15}$$

The F Statistic

$$F = \frac{\text{MSR}}{\text{MSE}} \tag{15.16}$$

Confidence Interval for $E(y_p)$

$$\hat{y}_p \pm t_{\alpha/2}\sqrt{\text{MSE}\left[\frac{1}{n} + \frac{(x_p - \bar{x})^2}{\Sigma x_i^2 - (\Sigma x_i)^2/n}\right]} \tag{15.17}$$

Prediction Interval for y_p

$$\hat{y}_p \pm t_{\alpha/2}\sqrt{\text{MSE}\left(1 + \frac{1}{n} + \frac{(x_p - \bar{x})^2}{\Sigma x_i^2 - (\Sigma x_i)^2/n}\right)} \tag{15.18}$$

Sample Correlation Coefficient

$$r = \frac{\Sigma x_i y_i - (\Sigma x_i \Sigma y_i)/n}{\sqrt{\Sigma x_i^2 - (\Sigma x_i)^2/n}\sqrt{\Sigma y_i^2 - (\Sigma y_i)^2/n}} \tag{15.19}$$

Sample Correlation Coefficient Using Regression Analysis Output

$$r = \pm\sqrt{\text{Coefficient of Determination}} \tag{15.20}$$

or

$$r = b_1\left(\frac{s_x}{s_y}\right) \tag{15.21}$$

Test Statistic for an Alternate Test for Significant Correlation

$$r \sqrt{\frac{n-2}{1-r^2}} \qquad\qquad (15.22)$$

Review Quiz

TRUE/FALSE

1. The least squares method is used to determine an estimated regression line that minimizes the squared deviations of the data values from the line.
2. The least squares method is applicable only in situations where the estimated regression line has a positive slope.
3. If the slope of the estimated regression line is positive, the correlation coefficient must be negative.
4. The slope of the estimated regression line (b_1) is a sample statistic, since, like other sample statistics, it is computed from the sample observations.
5. The coefficient of determination is the square root of the correlation coefficient.
6. The sum of squares due to regression (SSR) plus the sum of squares due to error (SSE) must equal the total sum of squares (SST).
7. A t test can be used to test whether or not there is a significant regression relationship.
8. The sampling distribution of b_1 is normal if the usual regression assumptions are satisfied.
9. An interval estimate for a particular value of the dependent variable yields a smaller interval than an interval estimate of the expected value of the dependent variable.
10. Minitab cannot be used for the computer solution of regression problems.
11. If two variables are perfectly linearly related, the sample correlation coefficient must equal -1 or 1.
12. The residual is the difference between the actual value of a dependent variable and the value predicted by the estimated regression line.

MULTIPLE CHOICE

13. If two variables x and y have a significant linear relationship, then
     a. there may or may not be any causal relationship between x and y
     b. x causes y to happen
     c. y causes x to happen
     d. answers (b) and (c) are both correct
14. For the estimated regression line $\hat{y} = 3 - 10x$, the correlation coefficient r
     a. equals 0
     b. is less than 0
     c. is greater than 0

15. If the correlation coefficient for two variables is $-.9$, the coefficient of determination is
   a. .9
   b. $-.81$
   c. .81

16. If a data set has SST = 200 and SSE = 150, then the coefficient of determination is
   a. .25
   b. .50
   c. .75
   d. 50

17. Compared to the prediction interval for a particular value of y, the confidence interval for a mean value of y will be
   a. narrower
   b. wider
   c. not enough information is given

18. A sample correlation coefficient is calculated from 15 pairs of x and y observations. The t distribution used to determine whether or not this coefficient is statistically significant will have how many degrees of freedom?
   a. 13
   b. 14
   c. 15
   d. 28

19. Which of the following is correct?
   a. SST = SSR $-$ SSE
   b. SSE = SSR $+$ SSE
   c. SSR = SSE $+$ SST
   d. SSE = SST $-$ SSR

20. The coefficient of determination is calculated as
   a. SST/SSE
   b. SSR/SST
   c. SSR/SSE
   d. SSE/SSR

21. Which of the following is an appropriate test statistic to test the null hypothesis that there is no linear relationship between x and y?
   a. SSR/SST
   b. MSE
   c. MSR/MSE
   d. MSE/MST

22. Which of the following points are *always* on the estimated linear regression line?
   a. $x = \bar{x}, y = \bar{y}$
   b. $x = 0, y = 0$
   c. $x = 1, y = 0$
   d. $x = 0, y = 1$

Supplementary Exercises

36. What is the difference between regression analysis and correlation analysis?

37. Does a high value of r^2 imply that two variables are causally related? Explain.

38. In your own words, explain the difference between an interval estimate of the mean value of y for a given x and an interval estimate for an individual value of y for a given x.

39. How do we measure how closely the actual data points are to the estimated regression line? That is, how do we measure the goodness-of-fit of the regression line?

40. What is the purpose of testing whether or not $\beta_1 = 0$?

41. In a manufacturing process the assembly line speed (feet per minute) was thought to affect the number of defective parts found during the inspection process. To test this theory, management devised a situation where the same batch of parts was inspected visually at a variety of line speeds. The following data were collected.

Line Speed	Number of Defective Parts Found
20	21
20	19
40	15
30	16
60	14
40	17

a. Develop the estimated regression line that relates line speed to the number of defective parts found.

b. At the $\alpha = .05$ level of significance determine whether or not line speed and number of defective parts found are related.

c. Did the estimated regression line provide a good fit to the data?

d. Develop a 95% confidence interval to predict the mean number of defective parts for a line speed of 50 feet per minute.

42. The PJH&D Company is in the process of deciding whether or not to purchase a maintenance contract for its new word processing system. They feel that maintenance expense should be related to usage and have collected the following information on weekly usage (hours) and annual maintenance expense.

a. Develop the estimated regression line that relates annual maintenance expense, in hundreds of dollars, to weekly usage.

b. Test the significance of the relationship in (a) at the $\alpha = .05$ level of significance.

c. PJH&D expects to operate the word processor 30 hours per week. Develop a 95% prediction interval for the company's annual maintenance expense.

d. If the maintenance contract costs $3000 per year, would you recommend purchasing it? Why or why not?

Weekly Usage (Hours)	Annual Maintenance Expense ($100s)
13	17.0
10	22.0
20	30.0
28	37.0
32	47.0
17	30.5
24	32.5
31	39.0
40	51.5
38	40.0

43. A sociologist was hired by a large city hospital to investigate the relationship between the number of unauthorized days that an employee is absent per year and the distance (miles) between home and work for employees. A sample of ten employees was chosen, and the following data were collected.

Distance to Work (miles)	Number of Days Absent
1	8
3	5
4	8
6	7
8	6
10	3
12	5
14	2
14	4
18	2

a. Develop a scatter diagram for the above data. Does a linear relationship appear reasonable? Explain.

b. Develop the least squares estimated regression line.

c. Is there a significant relationship between the two variables? Use $\alpha = .05$.

d. Did the estimated regression line provide a good fit? Explain.

e. Use the estimated regression line developed in (b) to develop a 95% confidence interval estimate of the expected number of days absent for employees living 5 miles from the company.

44. The owner of a chain of fast-food restaurants would like to investigate the relationship between the daily sales volume of a company restaurant and the number of competitor restaurants within a 1-mile radius of the firm's restaurant. The following data have been collected.

Number of Competitors Within 1 Mile	Sales ($)
1	3,600
1	3,300
2	3,100
3	2,900
3	2,700
4	2,500
5	2,300
5	2,000

a. Develop the least squares estimated regression line that relates daily sales volume to the number of competitor restaurants within a 1-mile radius.
b. Is there a significant relationship between the two variables? Use $\alpha = .05$.
c. Did the estimated regression line provide a good fit? Explain.
d. Use the estimated regression line developed in (a) to develop a 95% interval estimate of the daily sales volume for a particular company restaurant that has four competitors within a 1-mile radius.

45. The regional transit authority for a major metropolitan area would like to determine if there is any relationship between the age of a bus and the annual maintenance cost. A sample of 10 buses resulted in the following data.

Age of Bus (years)	Maintenance Cost ($)
1	350
2	370
2	480
2	520
2	590
3	550
4	750
4	800
5	790
5	950

a. Compute the sample correlation coefficient for the above data.
b. Using the sample correlation coefficient, test to see if the two variables are significantly related. Use $\alpha = .10$.

46. Reconsider the regional transit authority problem presented in Exercise 45.
a. Develop the least squares estimated regression line.
b. Test to see if the two variables are significantly related at $\alpha = .05$.
c. Did the least squares line provide a good fit to the observed data? Explain.
d. Develop a 90% confidence interval estimate of the maintenance cost for a specific bus that is 4 years old.

47. A psychology professor at Givens College is interested in the relation between time spent studying and total points earned in the course. Data collected on 10 students who took the course last quarter are given below.

Time Spent Studying	Total Points Earned
45	40
30	35
90	75
60	65
105	90
65	50
90	90
80	80
55	45
75	65

   a. Compute the sample correlation coefficient for the above data.
   b. Use the sample correlation coefficient and test to see if there is a significant relationship at the $\alpha = .05$ level.

48. Reconsider the Givens College data in Exercise 47.
   a. Develop an estimated regression line relating total points earned to hours spent studying.
   b. Test the significance of the model at the $\alpha = .05$ level.
   c. Predict the total points earned by Mark Sweeney. He spent 95 hours studying.
   d. Develop a 90% confidence interval for the total points earned by Mark Sweeney.

49. A sociologist collected the following data regarding the ages of wives and husbands when they were married.

Wife's age	19	42	28	25	36
Husband's age	20	32	31	24	33

   a. Develop the least squares estimated regression line.
   b. Test to see if the two variables (wife's age and husband's age) are related at the $\alpha = .05$ level of significance. Discuss the result.

Computer Exercise

As part of a study of transportation safety, the Department of Transportation collected data on the number of fatal accidents and the percentage of licensed drivers under the age of 21. The values for the two variables for a 1-year period were obtained from a sample of 42 cities. In order to account for differences in the population sizes of the various cities, the measure for the number of fatal accidents was defined in terms of the number of fatal accidents per 1000 licenses. The data are given.

Percent Under 21	Number of Fatal Accidents per 1000 Licenses	Percent Under 21	Number of Fatal Accidents per 1000 Licenses
13	2.962	17	4.100
12	0.708	8	2.190
8	0.885	16	3.623
12	1.652	15	2.623
11	2.091	9	0.835
17	2.627	8	0.820
18	3.830	14	2.890
8	0.368	8	1.267
13	1.142	15	3.224
8	0.645	10	1.014
9	1.028	10	0.493
16	2.801	14	1.443
12	1.405	18	3.614
9	1.433	10	1.926
10	0.039	14	1.643
9	0.338	16	2.943
11	1.849	12	1.913
12	2.246	15	2.814
14	2.855	13	2.634
14	2.352	9	0.926
11	1.294	17	3.256

QUESTIONS

1. Develop numerical and graphical measures to summarize the data.
2. Does there appear to be a relationship between the number of fatal accidents per 1000 licenses and the percentage of drivers under the age of 21? Explain.
3. What conclusions and recommendations can you derive from your analysis?

CHAPTER 16

Multiple Regression

What You Will Learn in This Chapter

- what multiple regression analysis is

- the important role computer packages play in performing multiple regression analysis

- how to use the t and F distributions to test for significant relationships in multiple regression analysis

Contents

Statistics in the News

FATHER VERSUS MOTHER IN CUSTODY FOR CHILDREN OF DIVORCE

Mothers have traditionally been awarded custody of children in divorce cases, due to the presumption by the courts that mothers were uniquely suited to raising children and also because of the powerful influence of our economic system, which has favored male employment. Thus children tend to go with the mother, whereas the father works to provide the financial support. However, the changing roles of both men and women and the various interrelated social changes have recently caused courts to consider fathers as custodians of children in divorce cases.

While there are many variables, or factors, that need to be evaluated in making a custody decision, one consideration is that of providing the best environment for the children's academic achievement. Researchers Frederick Shilling and Patrick Lynch have investigated the effect that custody with the father rather than the mother has on the academic achievement of eighth-grade children. Data for the Shilling-Lynch study were obtained from a sample of approximately 3000 single-parent children in the eighth grade. In the sample, children living with fathers numbered 550, whereas 2610 children lived with the mothers.

Academic performance of the children was measured on reading, mathematics, and a composite of these two measures. Other data available included a measure of socioeconomic status based on parental occupation and education, sex of the child, sex of the single parent, residence (urban, suburban, rural) and a measure of the student's

Jeffrey Grosscup

The courts have historically favored mother-custody for children of divorce.

perception of parental interest in school. Using a method known as multiple linear regression, the researchers developed an equation that used the socioeconomic status, sex of child, sex of parent, residence, and parental interest variables to predict academic performance scores. In terms of the sex of the single parent, the researchers found that verbal, mathematical, and overall academic achievement scores were better for children living with their mothers, other things being equal. The study can be construed as an encouragement to courts to continue assigning young children of broken families to the mothers instead of fathers, especially when academic achievement is an important factor in determining the best placement for the child.

Based on "Father Versus Mother Custody and Academic Achievement of Eighth Grade Children," by Frederick Shilling and Patrick D. Lynch, *Journal of Research and Development in Education* 18 (1985).

In Chapter 15 we discussed how regression analysis can be used to develop a mathematical equation representing the relationship between two variables. Recall that the variable being predicted or explained by the mathematical equation is called the *dependent variable*; the variable being used to predict or explain the value of the dependent variable is called the *independent variable*. In this chapter we continue our study of regression analysis by considering situations that involve two or more independent variables. Regression analysis involving two or more independent variables is called *multiple regression analysis*. "Statistics in the News" reported on a multiple regression analysis using several independent variables to predict, or explain, the dependent variable of academic performance.

16.1 AN INTRODUCTION TO MULTIPLE REGRESSION ANALYSIS

To provide an introduction to multiple regression analysis, consider a simple example for which the technique of multiple regression analysis can be applied.

EXAMPLE 16.1

Researchers studied the relationship of beer consumption, state alcohol policies and motor vehicle regulations to the number of fatal automobile accidents.* One objective of the study was to develop an equation that could be used to predict the number of fatal accidents given the driving age (percent of drivers under 21) and the number of outlets selling alcohol for on-premises consumption. The number of fatal accidents would be the dependent variable, and the driving age and the number of outlets would be the independent variables.

• • •

In Example 16.1 we see that there are two independent variables. Thus to develop a notation that can be used to refer to these two independent variables, let

$$x_1 = \text{Driving Age}$$

and

$$x_2 = \text{Number of Outlets}$$

The advantage of this notation is that with multiple regression problems involving more than two independent variables, we can continue to refer to each independent variable as x with an appropriate subscript. As we did in Chapter 15, we refer to the dependent variable with the letter y. Thus in Example 16.1 we denote the dependent variable as y = number of fatal accidents.

EXAMPLE 16.2

Butler Trucking Company is an independent trucking company located in southern California. A major portion of Butler's business involves deliveries throughout its local area. In an effort to develop better work schedules, the owner would like to develop an equation that could be used to help predict daily travel time for a truck. It is believed that the two most important predictors of daily travel time are the number of miles

*This study is described in more detail in "Statistics in Practice" at the end of this chapter.

traveled and the number of deliveries made. Data collected for 10 days of operation are shown in Table 16.1.

TABLE 16.1

Data for the Butler Trucking Company Example

Day	x_1 = Miles Traveled	x_2 = Number of Deliveries	y = Travel Time (Hours)
1	100	4	9.3
2	50	3	4.8
3	100	4	8.9
4	100	2	5.8
5	50	2	4.2
6	80	1	6.8
7	75	3	6.6
8	80	2	5.9
9	90	3	7.6
10	90	2	6.1

• • •

In Example 16.2, suppose we believe that daily travel time (y) is related to both miles traveled (x_1) and number of deliveries (x_2) by the following equation.

$$E(y) = \beta_0 + \beta_1 x_1 + \beta_2 x_2 \tag{16.1}$$

Equation (16.1) is called a *multiple regression equation*. $E(y)$ represents the average of all possible values of y that could occur at a given value of x_1 and a given value of x_2; β_0, β_1, and β_2 are referred to as *population parameters*. Note that if $\beta_2 = 0$, then x_2 is not related to y, and hence the multiple regression equation reduces to the one-independent-variable equation discussed in Chapter 15; that is, $E(y) = \beta_0 + \beta_1 x_1$. The multiple regression equation of (16.1) can be extended to the case of more independent variables by simply adding more terms, such as $\beta_3 x_3$, $\beta_4 x_4$, and so on. Equation (16.2) shows the general case for a situation involving p independent variables:

$$E(y) = \beta_0 + \beta_1 x_1 + \beta_2 x_2 + \cdots + \beta_p x_p \tag{16.2}$$

Note that if β_3, β_4, $\cdots$, β_p all equal zero, (16.2) reduces to the two-independent-variable multiple regression equation of (16.1).

To obtain more insight into the form of the relationship given by (16.1), refer to Figure 16.1. The graph of this regression equation is a plane in three-dimensional space. In Figure 16.1 we show such a graph with x_1 and x_2 on the horizontal axis and y on the vertical axis. The specific data point shown in Figure 16.1 is for the case where $x_1 = x_1^*$ and $x_2 = x_2^*$.

The same assumptions that we made for the one-independent-variable case in Chapter 15 also apply in multiple regression analysis. To simplify the discussion, we

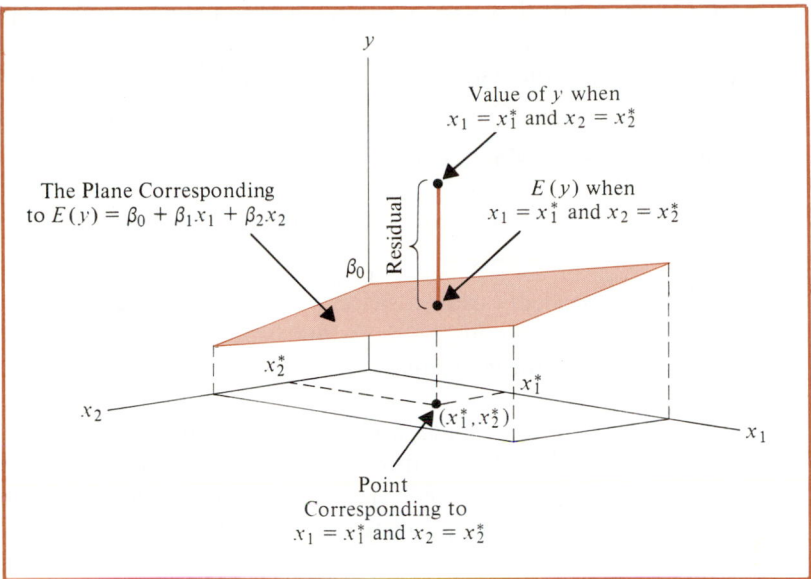

FIGURE 16.1
Graph of the Regression Equation for Multiple
Regression Analysis Involving Two Independent Variables

state these assumptions for the two-independent-variable case, recognizing that the extension to the general case of p independent variables can be easily made.

Assumptions in Multiple Regression Analysis

1. For a specified value of x_1 and x_2, the values of y are normally distributed about the regression equation

$$E(y) = \beta_0 + \beta_1 x_1 + \beta_2 x_2$$

2. The variance of y, denoted by σ^2, is the same for all values of x_1 and x_2.
3. The values of y are independent.

EXAMPLE 16.2 (continued)

Let us return to the example involving the relationship between the travel time for a truck and the two independent variables, miles traveled and number of deliveries. Consider any truck that travels 50 miles ($x_1 = 50$) and makes two deliveries ($x_2 = 2$). When $x_1 = 50$ and $x_2 = 2$, the value for $E(y)$ in (16.1) represents the average travel time that would occur for all trucks that travel 50 miles and make two deliveries. For these values of x_1 and x_2, the values of y are assumed to be normally distributed with a mean of $\beta_0 + \beta_1(50) + \beta_2(2)$ and variance of σ^2; moreover, the values of travel time are assumed to be independent of one another.

• • •

In the previous chapter the least squares method was used to develop estimates of β_0 and β_1 for the one-independent-variable case. In multiple regression analysis, the least squares method is used in an analogous manner to develop estimates of the parameters $\beta_0, \beta_1, \beta_2, \cdots, \beta_p$. These estimates are denoted by $b_0, b_1, b_2, \cdots, b_p$; thus the corresponding estimated regression equation is written as follows.

Estimated Regression Equation

$$\hat{y} = b_0 + b_1x_1 + b_2x_2 + \cdots + b_px_p \qquad (16.3)$$

In the two independent variables case, an estimate of the regression equation $E(y) = \beta_0 + \beta_1x_1 + \beta_2x_2$ is given by $\hat{y} = b_0 + b_1x_1 + b_2x_2$. Note that b_0 is the least squares estimate of β_0, b_1 is the least squares estimate of β_1, and b_2 is the least squares estimate of β_2.

At this point you should begin to see the similarity between the concepts of multiple regression analysis and those of the previous chapter. We have just extended the concepts of regression analysis involving one independent variable to the case involving two or more independent variables. In the next section we discuss how to develop the estimated regression equation for the data presented in Example 16.2.

16.2 DEVELOPING THE ESTIMATED REGRESSION EQUATION

In Chapter 15 we presented formulas for estimating β_0 and β_1 for the regression line given by $E(y) = \beta_0 + \beta_1x_1$. In the general multiple regression case, the usual presentation of formulas for computing the estimates of $\beta_0, \beta_1, \beta_2$, etc., involves the use of matrix algebra and is beyond the scope of this text. However, for the special case of two independent variables, we can show what is involved.

Recall from Chapter 15 that the least squares method chooses the values of the estimates of the population parameters that minimize the sum of squares of the differences between the observed values of the dependent variable and the estimated values. For the case of two independent variables, the least squares method is a procedure for determining the values of b_0, b_1, and b_2 that minimize

$$\Sigma (y_i - \hat{y}_i)^2$$

Note that $\hat{y}_i = b_0 + b_1x_{1i} + b_2x_{2i}$

where

$x_{1i} = i$th value of x_1

$x_{2i} = i$th value of x_2

$\hat{y}_i = $ predicted value of y when $x_1 = x_{1i}$ and $x_2 = x_{2i}$

Thus we can write the least squares criterion as follows.

$$\min \Sigma(y_i - b_0 - b_1x_{1i} - b_2x_{2i})^2 \qquad (16.4)$$

It can be shown, using calculus, that the values of b_0, b_1, and b_2 that minimize (16.4) must satisfy the three equations below, called the *normal equations*.

Normal Equations—Two Independent Variables

$$nb_0 + (\Sigma x_{1i})b_1 + (\Sigma x_{2i})b_2 = \Sigma y_i \tag{16.5}$$

$$(\Sigma x_{1i})b_0 + (\Sigma x_{1i}^2)b_1 + (\Sigma x_{1i} x_{2i})b_2 = \Sigma x_{1i}y_i \tag{16.6}$$

$$(\Sigma x_{2i})b_0 + (\Sigma x_{1i} x_{2i})b_1 + (\Sigma x_{2i}^2)b_2 = \Sigma x_{2i}y_i \tag{16.7}$$

EXAMPLE 16.2 (continued)

In order to apply the normal equations for the Butler Trucking problem we must first find the values of the coefficients of b_0, b_1, and b_2 for these equations, as well as the right-hand side values. The necessary data for Butler Trucking are contained in Table 16.2.

Using the information in Table 16.2 we can substitute into the normal equations (16.5) to (16.7) to obtain the following equations for the Butler trucking problem.

$$10b_0 + 815b_1 + 26b_2 = 66.0 \tag{16.8}$$

$$815b_0 + 69{,}625b_1 + 2165b_2 = 5594.0 \tag{16.9}$$

$$26b_0 + 2165b_1 + 76b_2 = 180.6 \tag{16.10}$$

TABLE 16.2

Calculation of Coefficients for Normal Equations

y_i	x_{1i}	x_{2i}	x_{1i}^2	x_{2i}^2	$x_{1i}x_{2i}$	$x_{1i}y_i$	$x_{2i}y_i$
9.3	100	4	10,000	16	400	930	37.2
4.8	50	3	2,500	9	150	240	14.4
8.9	100	4	10,000	16	400	890	35.6
5.8	100	2	10,000	4	200	580	11.6
4.2	50	2	2,500	4	100	210	8.4
6.8	80	1	6,400	1	80	544	6.8
6.6	75	3	5,625	9	225	495	19.8
5.9	80	2	6,400	4	160	472	11.8
7.6	90	3	8,100	9	270	684	22.8
6.1	90	2	8,100	4	180	549	12.2
66.0	815	26	69,625	76	2165	5594	180.6

Since the least squares estimates must satisfy these three equations simultaneously, in order to obtain values for b_0, b_1, and b_2 we have to solve this system of three simultaneous linear equations in three variables. The solution* is given by $b_0 = .0367$, $b_1 = .0562$, and $b_2 = .7639$. Thus the estimated regression equation for Butler Trucking is

$$\hat{y} = .0367 + .0562x_1 + .7639x_2 \tag{16.11}$$

• • •

Note on Interpretation of Coefficients

Before continuing with our discussion of multiple regression, let us consider a modification of Example 16.2 in which we use only one independent variable, miles traveled, to predict the value of the dependent variable.

EXAMPLE 16.2 (continued)

Using the data in Table 16.1 we can use (15.3) and (15.4) to develop estimates of b_0 and b_1 for the estimated regression line $\hat{y} = b_0 + b_1x_1$; these estimates are $b_0 = 1.1314$ and $b_1 = .0671$. Thus the estimated regression line is $\hat{y} = 1.1314 + .0671x_1$.

One observation can be made concerning the relationship between the estimated regression equation with only miles traveled as an independent variable, and the one that includes the number of deliveries as a second independent variable. The value of b_1 is not the same in both cases. In the case where we use only one independent variable, we interpret $b_1 = .0671$ as the amount of change in y for a 1-unit change in the independent variable. In the estimate multiple regression equation, this interpretation must be modified somewhat. That is, we interpret each regression coefficient as follows: $b_1 = .0562$ represents the change in y corresponding to a 1-unit change in the independent variable y_1 when the other independent variable y_2 is held constant.

• • •

In general, in multiple regression analysis, we interpret each regression coefficient, b_i, as the change in y corresponding to a one unit change in x_i when all other independent variables are held constant.

Computer Solution

It can be shown that for multiple regression problems involving p independent variables, there are $p + 1$ normal equations that must be solved simultaneously for the estimated coefficients $b_0, b_1, b_2, \cdots, b_p$. The computational effort involved requires more sophisticated solution procedures than we have used in the solution of (16.5) to (16.7). Fortunately, computer software packages can be used to obtain these solutions with very little effort on the part of the user.

In Figure 16.2 we show the computer output from the Minitab computer package

---

*In the chapter appendix we show how the solution is obtained.

```
        THE REGRESSION EQUATION IS
        Y = 0.0367 + 0.0562 X1 + 0.764 X2

                                   ST. DEV.      T-RATIO =
        COLUMN        COEFFICIENT   OF COEF.      COEF/S.D.
        —                  0.037      1.326           0.03
        X1   MILES      0.05616      0.01564          3.59
        X2   DELIVERIES 0.7639       0.3053           2.50

        S = 0.8494

        R-SQUARED = 79.0 PERCENT
        R-SQUARED = 72.9 PERCENT, ADJUSTED FOR D.F.


        ANALYSIS OF VARIANCE

         DUE TO      DF          SS       MS=SS/DF
        REGRESSION    2      18.9499        9.4749
        RESIDUAL      7       5.0501        0.7214
        TOTAL         9      24.0000
```

FIGURE 16.2
Output Data for Multiple Regression Problem Solution Using Minitab

for the version of the Butler Trucking problem involving the two independent variables: miles traveled and number of deliveries. Note that in the column labeled "COEFFICIENT," the values are the same as we obtained (except for rounding) for b_0, b_1, and b_2. We discuss the remainder of the computer output in the following sections.

EXERCISES

1. The admissions' officer for Clearwater College developed the following estimated regression equation relating final college GPA to student's SAT mathematics scores and their high-school GPA.

$$\hat{y} = -1.41 + .0235x_1 + .00486x_2$$

where

x_1 = high-school grade-point average

x_2 = SAT mathematics score

y = Final college grade-point average

Interpret the coefficients in this estimated regression equation.

2. The personnel director for Electronics Associates developed the following estimated regression equation relating an employee's score on a job satisfaction

test to his or her length of service and wage rate.

$$\hat{y} = 14.4 - 8.69x_1 + 13.5x_2$$

where

$$x_1 = \text{length of service (years)}$$

$$x_2 = \text{wage rate (dollars)}$$

$$y = \text{job satisfaction test score (higher scores} \\ \text{indicate more job satisfaction)}$$

Interpret the coefficients in this estimated regression equation.

3. A chemist has developed the following estimated regression equation relating the yield of a process (y) to the temperature (x_1) and the pressure (x_2).

$$\hat{y} = 69 + 1.17x_1 - 3.97x_2$$

where temperature is measured in degrees Fahrenheit and pressure in pounds per square inch. Interpret the coefficients in this estimated regression equation.

4. A shoe store has developed the following estimated regression equation relating sales to inventory investment and advertising expenditures.

$$\hat{y} = 25 + 10x_1 + 8x_2$$

where

$$x_1 = \text{inventory investment (\$1000s)}$$

$$x_2 = \text{advertising expenditures (\$1000s)}$$

$$y = \text{sales (\$1000s)}$$

Interpret the coefficients (b_1 and b_2) in this estimated regression equation.

16.3 TESTING FOR A SIGNIFICANT RELATIONSHIP

In discussing tests for determining whether or not there is a significant relationship among the independent variables and the dependent variable, we initially consider a multiple regression situation involving two independent variables. We assume that we have an underlying regression equation of the form

$$E(y) = \beta_0 + \beta_1x_1 + \beta_2x_2$$

Therefore, the appropriate test for determining whether or not there is a significant relationship among x_1, x_2, and y is as follows.

$$H_0: \beta_1 = \beta_2 = 0$$

$$H_1: \text{At Least One of the Two Coefficients}$$
$$\text{Is Not Zero}$$

If we reject H_0, we conclude that there is a significant relationship among x_1, x_2, and y.

F Test

In Chapter 15 we showed how the ANOVA table could provide a summary of the computational aspects associated with regression analysis. In multiple regression situations, we will see that the ANOVA table also provides the same type of summary. For example, in Table 16.3 we show the general form of the ANOVA table for a multiple regression situation involving two independent variables.

The test concerning the significance of the relationship in multiple regression analysis is based upon the ratio of the two mean squares in the ANOVA table. That is, if we let

$$\text{MSR} = \frac{\text{SSR}}{2}$$

and

$$\text{MSE} = \frac{\text{SSE}}{(n - 3)}$$

then the appropriate test statistic is given by

$$F = \frac{\text{MSR}}{\text{MSE}} \tag{16.12}$$

TABLE 16.3

General Form of the ANOVA Table for a Multiple Regression Situation Involving Two Independent Variables

Source of Variation	Sum of Squares	Degrees of Freedom	Mean Square
Regression	SSR	2	$\text{MSR} = \frac{\text{SSR}}{2}$
Error	SSE	$n - 3$	$\text{MSE} = \frac{\text{SSE}}{n - 3}$
Total (about the mean)	SST	$n - 1$	

If the null hypothesis is true (i.e., $\beta_1 = \beta_2 = 0$), then the value that we obtain when computing this ratio should appear to be a value that would be obtained if we were randomly sampling values from an F distribution with 2 numerator degrees of freedom and $n - 3$ denominator degrees of freedom. In Table 4 of Appendix B, we provide critical values for the F distribution. For example, this table shows that for an F distribution with 2 numerator degrees of freedom and 7 denominator degrees of freedom, only 5% of the values obtained would be as large as or larger than the table value of 4.74. Thus if this were the appropriate F distribution for a multiple regression problem, we would compare the value of $F = $ MSR/MSE to 4.74; if this value exceeded 4.74, we would have sufficient statistical evidence to reject the null hypothesis at the 5% level of significance.

EXAMPLE 16.2 (continued)

In Figure 16.2 we showed the Minitab computer output for the Butler Trucking multiple regression problem. For this problem MSR = 9.4749 and MSE = .7214. Thus

$$F = \frac{\text{MSR}}{\text{MSE}} = \frac{9.4749}{.7214} = 13.134$$

Since the numerator in this ratio, MSR, has 2 degrees of freedom associated with it and the denominator, MSE, has 7 degrees of freedom, we refer to an F distribution with 2 numerator degrees of freedom and 7 denominator degrees of freedom. As we showed earlier, the critical value at the 5% level of significance in this case is 4.74. Thus we must reject the null hypothesis that $\beta_1 = \beta_2 = 0$ and conclude that there is a statistically significant relationship at the 5% level of significance.

· · ·

The General Anova Table and F Test

Now that we know how the F test can be applied for a multiple regression situation involving two independent variables, let us generalize our test to the case involving p independent variables.

The appropriate hypotheses to test to determine if there is a significant relationship are as follows:

$$H_0: \beta_1 = \beta_2 = \cdots = \beta_p = 0$$

$$H_1: \text{At Least One of the } p \text{ Coefficients}$$
$$\text{Is Not Equal to Zero}$$

Again, if we reject H_0, we can conclude that there is a significant relationship.

The general form of the ANOVA table for the multiple regression case involving p independent variables is shown in Table 16.4. The only change from the two-independent-variable case is in the degrees of freedom corresponding to SSR and SSE. Here the sum of squares due to regression has p degrees of freedom corresponding to the p independent variables, and the degrees of freedom corresponding to error has $n - $

TABLE 16.4

ANOVA Table for Multiple Regression Analysis Involving p Independent Variables

Source	Sum of Squares	Degrees of Freedom	Mean Square
Regression	SSR	p	$\text{MSR} = \dfrac{\text{SSR}}{p}$
Error	SSE	$n - p - 1$	$\text{MSE} = \dfrac{\text{SSE}}{n - p - 1}$
Total (about the mean)	SST	$n - 1$	

$p - 1$ degrees of freedom. When looking up the critical value from the F distribution table, the numerator degrees of freedom are p and the denominator degrees of freedom are $n - p - 1$.

t Test for Significance of Individual Parameters

If, after using the F test, we conclude that the multiple regression relationship is significant (that is, we conclude that at least one of the β_i is not equal to 0), it is often of interest to conduct tests to see which of the individual parameters, β_i, are significant. The t test that we introduced in Chapter 15 for the case of one independent variable can also be used to test for the significance of the individual parameters in multiple regression analysis.

The hypothesis test we wish to make is the same for the coefficient of each independent variable. It is stated as follows.

$$H_0: \beta_i = 0$$

$$H_1: \beta_i \neq 0$$

In Chapter 15 we learned how to conduct such a test for the case where there is only one independent variable. The hypotheses were:

$$H_0: \beta_1 = 0$$

$$H_1: \beta_1 \neq 0$$

To test these hypotheses we computed the sample statistic b_1/s_{b_1}, where b_1 is the least squares estimate of β_1 and s_{b_1} is the estimate of the standard deviation of the sampling distribution of b_1. We learned that the sampling distribution of b_1/s_{b_1} follows a t distribution with $n - 2$ degrees of freedom. Thus to conduct the hypothesis test, we chose a value of α, found a rejection region determined by $t_{\alpha/2}$, computed b_1/s_{b_1} and rejected H_0 if

$$\frac{b_1}{s_{b_1}} > t_{\alpha/2}$$

or

$$\frac{b_1}{s_{b_1}} < -t_{\alpha/2}$$

The procedure for testing the individual parameters in the multiple regression case is essentially the same. The only differences are in the number of degrees of freedom for the appropriate t distribution in the formula for computing s_{b_i}. The number of degrees of freedom is the same as for the sum of squares due to error. Thus we use $n - p - 1$ degrees of freedom, where p is the number of independent variables. (Note that for the case of one independent variable, this reduces to the $n - 2$ degrees of freedom used in Chapter 15.) The formula for s_{b_i} is more involved, and we do not present it here; however, s_{b_i} is calculated and printed by most computer software packages for multiple regression analysis.

EXAMPLE 16.2 (continued)

Let us return to the Butler Trucking problem to test the significance of the parameters β_1 and β_2. In the Minitab printout (Figure 16.2), the values of b_1, b_2, s_{b_1}, and s_{b_2} are given as

$$b_1 = .05616 \qquad s_{b_1} = .01564$$

$$b_2 = .7639 \qquad s_{b_2} = .3053$$

Therefore, for the parameters β_1 and β_2, we obtain

$$\frac{b_1}{s_{b_1}} = \frac{.05616}{.01564} = 3.59$$

$$\frac{b_2}{s_{b_2}} = \frac{.7639}{.3053} = 2.50$$

Note that both of these values were provided by the Minitab output of Figure 16.2 under the column labeled "T-RATIO = COEF/S.D." Using $\alpha = .05$ and $10 - 2 - 1 = 7$ degrees of freedom, we can find the appropriate $t_{\alpha/2}$ value for our hypothesis tests in Table 2 of Appendix B. We obtain

$$t_{.025} = 2.365$$

Now, since $b_1/s_{b_1} = 3.59 > 2.365$, we reject the hypothesis that $\beta_1 = 0$. Furthermore, since $b_2/s_{b_2} = 2.50 > 2.365$, we reject the hypothesis that $\beta_2 = 0$.

$$\bullet \quad \bullet \quad \bullet$$

Cautionary Note

The test that we have just discussed should be used only if we believe that the assumptions presented in Section 16.1 are appropriate. These assumptions regarding the regression equation are important because statisticians have used these assumptions as the basis for developing the tests introduced in this section. A technique referred to as residual analysis plays an important part in determining if the model assumptions appear to be satisfied and what actions can be taken to correct any observed problems. A discussion of this type of analysis is beyond the scope of this text. Thus in the following discussion, we assume that the assumptions presented in Section 16.1 are satisfied.

EXERCISES

5. Shown below is a partial Minitab computer output from a regression analysis.

```
        THE REGRESSION EQUATION IS
        Y = 8.103 + 7.602 X1 + 3.111 X2

                                    ST. DEV.      T-RATIO =
        COLUMN·      COEFFICIENT    OF COEF.      COEF/S.D.
        —               8.103         2.667       _____
        X1              7.602         2.105       _____
        X2              3.111         0.613       _____

        S = 3.35

        R-SQUARED = 92.3 PERCENT
        R-SQUARED = _____ PERCENT, ADJUSTED FOR D.F.


        ANALYSIS OF VARIANCE

          DUE TO      DF            SS         MS=SS/DF
        REGRESSION    __           1612        _____
        RESIDUAL      12           ____        _____
        TOTAL         __           ____        _____
```

a. Compute the appropriate t ratios.
b. Test for the significance of β_1 and β_2 at $\alpha = .05$.
c. Compute the entries in the DF, SS, and MS = SS/DF columns.

6. In Exercise 4 the following estimated regression equation for relating sales to inventory investment and advertising expenditures was given.

$$\hat{y} = 25 + 10x_1 + 8x_2$$

The data used to develop the model came from a survey of 10 stores. In addition to the estimated regression equation, it was found as a result of a computer run that SST = 16,000 and SSR = 12,000.

a. Compute SSE, MSE, and MSR.

b. Use an F test and an $\alpha = .05$ level of significance to determine if there is a significant relationship among the variables.

7. Shown is a partial Minitab computer printout for a multiple regression problem involving two independent variables.

```
        THE REGRESSION EQUATION IS
        Y = 11.61 + 2.16 X1 + 4.80 X2

                                    ST. DEV.    T-RATIO =
        COLUMN      COEFFICIENT     OF COEF.    COEF/S.D.
        —             11.61           3.07      _____
        X1             2.16           0.69      _____
        X2             4.80           1.03      _____

        ANALYSIS OF VARIANCE

          DUE TO    DF          SS      MS=SS/DF
        REGRESSION  __        90.3      _____
        RESIDUAL    12      _____    _____
        TOTAL       __       108.6      _____
```

a. Find the appropriate values for regression and total degrees of freedom.

b. Find SSE, MSR, MSE, and the t ratios for the coefficients.

c. Compute F and test at the $\alpha = .01$ level whether a significant relationship exists or not.

d. Use the t test and $\alpha = .05$ to test $H_0: \beta_1 = 0$ and $H_0: \beta_2 = 0$.

8. The following estimated regression equation involving three independent variables has been developed:

$$\hat{y} = 18.31 + 8.12x_1 + 17.9x_2 - 3.6x_3$$

Computer output indicates that $s_{b_1} = 2.1$, $s_{b_2} = 9.72$, and $s_{b_3} = .71$. There were 15 observations in the study. Test each hypothesis.

a. $H_0: \beta_1 = 0$ at $\alpha = .05$

b. $H_0: \beta_2 = 0$ at $\alpha = .05$

c. $H_0: \beta_3 = 0$ at $\alpha = .05$

16.4 DETERMINING THE STRENGTH OF THE RELATIONSHIP

At this point we have concluded for the Butler Trucking problem, that the estimated regression equation $\hat{y} = .0367 + .0562x_1 + .7639x_2$ is statistically significant. That is, there is a significant relationship among x_1, x_2, and y, and so the estimated regression equation might be useful in predicting y. Now we would like to compute a measure of the strength of the relationship, or the goodness-of-fit of the regression equation to the data.

In Chapter 15 we used the coefficient of determination (r^2) to evaluate the strength of the regression relationship. Recall that r^2 is computed as

$$r^2 = \frac{SSR}{SST}$$

In multiple regression analysis, we compute a similar quantity called the *multiple coefficient of determination*.

Multiple Coefficient of Determination

$$R^2 = \frac{SSR}{SST} \tag{16.13}$$

When multiplied by 100, the multiple coefficient of determination represents the percentage of variability in y that is explained by the estimated regression equation.

EXAMPLE 16.2 (continued)

In the case of Butler Trucking, we find that

$$R^2 = \frac{18.9499}{24.0000} = .7896$$

Therefore, 78.96% of the variability in y is explained by the relationship with miles traveled and number of deliveries. For the Butler Trucking problem, it can be shown that the regression model with only miles traveled (x_1) as the independent variable has an r^2 value of .60. Therefore, with the addition of the second independent variable, number of deliveries (x_2), the percentage of variability explained has increased from 60% to 78.96%.

• • •

In general, it is always true that R^2 will increase as more independent variables are added to the regression equation because adding variables to the equation causes the prediction errors to be smaller, hence reducing SSE. Since SST = SSR + SSE, when SSE gets smaller SSR must get larger, causing R^2 = SSR/SST to increase.

Many analysts recommend adjusting R^2 for the number of independent variables, in order to avoid overestimating the impact on the amount of explained variability, of adding an independent variable. This *adjusted multiple coefficient of determination* is computed as follows.

Adjusted Multiple Coefficient of Determination

$$R_a^2 = 1 - (1 - R^2) \frac{n - 1}{n - p - 1} \qquad (16.14)$$

EXAMPLE 16.2 (continued)

For the Butler Trucking problem, we obtain

$$R_a^2 = 1 - (1 - .7896) \frac{10 - 1}{10 - 2 - 1}$$

$$= 1 - (.2104)(1.2857)$$

$$= .7295$$

Note that both values of the coefficient of determination are provided by the Minitab output shown in Figure 16.2.

• • •

EXERCISES

9. Refer to the computer output from a regression run shown in Exercise 7.
   a. Compute R^2.
   b. Compute R_a^2.
   c. Does the model appear to explain a large amount of the variability in the data?

10. Refer to Exercise 6.
    a. For the estimated regression equation given, compute R^2.
    b. Compute R_a^2.
    c. Does the model appear to explain a large amount of the variability in the data?

11. Refer to Exercise 5.
    a. Compute R_a^2.
    b. Does the model appear to explain a large amount of the variability in the data?

12. In a regression analysis involving 18 observations and four independent variables, it was determined that SSR = 18,051.63 and SSE = 1,014.3.
    a. Determine R^2 and R_a^2.
    b. Test the significance of the relationship at the $\alpha = .01$ level of significance.

13. In a regression analysis involving 30 observations, the following estimated regression equation was obtained.

$$\hat{y} = 17.6 + 3.8x_1 - 2.3x_2 + 7.6x_3 + 2.7x_4.$$

For this model SST = 1805 and SSR = 1760.

a. Compute R^2.
b. Compute R_a^2.
c. At $\alpha = .05$ test the significance of the relationship among the variables.

16.5 ESTIMATION AND PREDICTION

Estimating the mean value of y and predicting an individual value of y in multiple regression is similar to that for the case of regression analysis involving one independent variable. First, recall that in Chapter 15 we showed that the point estimate of the expected value of y for a given value of x was the same as the prediction of an individual value of y. In both cases we used $\hat{y} = b_0 + b_1x$ as the point estimate. In multiple regression we use the same procedure. That is, we substitute the given values of $x_1, x_2, \ldots, x_p$ into the estimated regression equation and use the corresponding value of $\hat{y}$ as the point estimate.

EXAMPLE 16.2 (continued)

Suppose that for the Butler Trucking problem, we wanted to use the estimated regression equation involving x_1 (miles traveled) and x_2 (number of deliveries) to do the following:

1. Estimate the mean value of travel time for all trucks that travel 50 miles and make two deliveries.
2. Predict the travel time for one specific truck that travels 50 miles and makes two deliveries.

Using the estimated regression equation $\hat{y} = .0367 + .0562x_1 + .7639x_2$ with $x_1 = 50$ and $x_2 = 2$, we obtain the following value of $\hat{y}$.

$$\hat{y} = .0367 + .0562(50) + .7639(2) = 4.3745.$$

Hence the point estimate of travel time in both cases is approximately 4.4 hours.

• • •

To develop interval estimates for the mean value of y and for an individual value of y, we use a procedure similar to that for the case of regression analysis involving one independent variable. The formulas required, however, are beyond the scope of the text. Nevertheless, computer packages for multiple regression analysis will often provide confidence intervals once the values of $x_1, x_2, \ldots, x_p$ are specified by the user.

EXAMPLE 16.2 (continued)

In Table 16.5 we show 95% confidence interval estimates for the Butler Trucking problem for selected values of x_1 and x_2. Note that the prediction interval for an individual estimate of y is wider than the interval estimate for the expected value of y. This simply reflects the fact that for given values of x_1 and x_2, we can predict the mean travel time for all trucks with more precision than we can the travel time for one specific truck.

TABLE 16.5

95% Confidence Interval and 95% Prediction Interval Estimates for the Butler Trucking Problem

Value of x_1	Value of x_2	Expected Value of y Lower Limit	Upper Limit	Individual Value of y Lower Limit	Upper Limit
50	2	3.0841	5.6649	1.9869	6.7621
50	3	3.7127	6.5642	2.6750	7.6018
80	1	3.9907	6.7097	2.9006	7.6926
80	2	5.2984	6.8226	3.9120	8.2091
100	2	6.0774	8.2916	4.8908	9.4782
100	4	7.4853	9.9394	6.3584	11.0662

EXERCISES

14. Reconsider Exercise 1, where the estimated regression equation relating the college GPA (y) to the high-school GPA (x_1) and the SAT mathematics score (x_2) was

$$\hat{y} = -1.41 + .0235x_1 + .00486x_2$$

Estimate the final college GPA for a student who has a high-school average of 84 and a score of 540 on the SAT mathematics test.

15. Reconsider Exercise 2, where the estimated regression equation developed was

$$\hat{y} = 14.4 - 8.69x_1 + 13.5x_2$$

where

$$x_1 = \text{length of service (years)}$$

$$x_2 = \text{wage rate (dollars)}$$

$$y = \text{job satisfaction test score (higher scores indicate more job satisfaction)}$$

Develop an estimate of the job satisfaction test score for an employee that has 4 years of service and makes $6.50 per hour.

16. In Exercise 3 we described a situation where a chemist developed the following estimated regression equation relating the yield of a process (y) to the temperature (x_1) and the pressure (x_2).

$$\hat{y} = 69 + 1.17x_1 - 3.97x_2$$

where temperature is measured in degrees Fahrenheit and pressure in pounds per square inch. Estimate the yield of the process corresponding to a temperature of 80° and a pressure of 16 pounds per square inch.

17. Recall Exercise 4, where we described a situation in which a shoe store had developed the following estimated regression equation relating sales ($1000s) to inventory investment ($1000s) and advertising expenditure ($1000s):

$$\hat{y} = 25 + 10x_1 + 8x_2$$

Estimate sales if there is a $15,000 investment in inventory and an advertising budget of $10,000.

Summary

In this chapter we showed how extensions of the concepts of linear regression can be used to develop an estimated regression equation for predicting y that involves two or more independent variables. We noted that the interpretation of the coefficients had to be modified somewhat for this case. For instance, we interpreted b_i as an estimate of the change in the dependent variable y that would result from a 1-unit change in independent variable x_i when the other independent variables are held constant.

A key part of any multiple regression study is the use of a computer software package for carrying out the computational work. Many excellent packages exist and, after a short learning period, can be used to develop the estimated regression equation and conduct the appropriate significance tests. We illustrated the use of such a statistical package for the Butler Trucking problem.

Statistics in Practice

RESEARCH FOR DRIVING SAFETY

Research studies encouraged by agencies such as the U.S. Department of Transportation and the National Safety Council are often directed at obtaining a better understanding of the causes of motor vehicle accidents and fatalities. Such research has been an aid to legislators and other policy makers who seek to maximize safety on the nation's streets and highways. Many of the studies on motor vehicle accidents and fatalities are statistical in nature and employ techniques such as multiple regression.

A study by Israel Colon and Henry Cutter investigated the factors related to motor vehicle fatalities based on data available from the United States Department of Transportation Fatal Accident Reporting System. In particular, data from all 50 states and the District of Columbia were examined in order to determine the relationship between alcohol availability and fatal motor vehicle accidents. Two multiple regression equations were developed: one with the number of fatal accidents as the dependent variable and one with the number of motor vehicle fatalities (potentially more than one per accident) as the dependent variable. In both cases, the seven independent or predictor variables considered were:

1. driving age (percent of drivers under 21)
2. beverage-purchase age
3. average beer consumption
4. number of outlets per million population selling alcohol for on-premise consumption
5. percentage of metropolitan residents
6. percentage of male drivers
7. mileage per driver per year

Results of the two regression analyses were similar, with the coefficient of determination slightly over .75 for each. Significance testing on the individual regression coefficients pointed to driving age, average beer consumption, and the number of on-premise consumption outlets as the important independent variables. The researchers concluded their study with interpretations that suggested possible strategies for reducing motor vehicle fatalities.

"The Relationship of Beer Consumption and State Alcohol and Motor Vehicle Policies to Fatal Accidents," by Israel Colon and Henry S. G. Cutter, *Journal of Safety Research* 14 (1983).

Glossary

Multiple regression analysis—Regression analysis involving two or more independent variables.

Normal equations—The equations that must be solved to compute $b_0, b_1, b_2, \ldots, b_p$ the least squares estimates of $\beta_0, \beta_1, \beta_2, \ldots \beta_p$

Multiple coefficient of determination (R^2)—When multiplied by 100, R^2 represents the percentage of variability in y that is explained by the estimated regression equation.

Adjusted multiple coefficient of determination (R_a^2)—The value of R^2 adjusted for the number of independent variables.

Key Formulas

Multiple Regression Equation Involving p Independent Variables

$$E(y) = \beta_0 + \beta_1 x_1 + \cdots + \beta_p x_p \tag{16.2}$$

Estimated Regression Equation

$$\hat{y} = b_0 + b_1 x_1 + \cdots + b_p x_p \tag{16.3}$$

The F Statistic

$$F = \frac{\text{MSR}}{\text{MSE}} \tag{16.12}$$

Multiple Coefficient of Determination

$$R^2 = \frac{\text{SSR}}{\text{SST}} \tag{16.13}$$

Adjusted Multiple Coefficient of Determination

$$R_a^2 = 1 - (1 - R^2)\left(\frac{n-1}{n-p-1}\right) \tag{16.14}$$

Review Quiz

TRUE/FALSE

1. In multiple regression analysis, there are two or more dependent variables.
2. The least squares method cannot be used to develop the coefficients of the estimated regression equation for multiple regression analysis.
3. The coefficient of x_1 in an estimated regression equation will be the same no matter how many independent variables are involved.
4. The computer solution of a multiple regression problem provides all the coefficients for the estimated regression equation.

5. An F test can be used to test for a significant relationship in multiple regression analysis.
6. The number of degrees of freedom for the error sum of squares in multiple regression analysis decreases as the number of independent variables increases.
7. The number of degrees of freedom in the t test for the significance of individual parameters in multiple regression analysis is always $n - 3$.
8. The adjusted multiple coefficient of determination will always be smaller than the unadjusted multiple coefficient of determination.

MULTIPLE CHOICE

For questions 9–12, consider a regression model in which two independent variables, x_1 and x_2, are used to explain the dependent variable, y.

9. In the test of the hypotheses $H_0: \beta_1 = \beta_2 = 0$ and H_1: either β_1 or β_2 or both $\neq 0$, the test statistic MSR/MSE has a sampling distribution that is the
   a. t distribution
   b. F distribution
   c. normal distribution
   d. chi-square distribution
10. The degrees of freedom for the distribution in question 9 are
   a. $n - 2$
   b. $n - 3$
   c. 3 and $n - 2$
   d. 2 and $n - 3$
11. Which sampling distribution is used to test $H_0: \beta_2 = 0$, $H_1: \beta_2 \neq 0$?
   a. t distribution
   b. F distribution
   c. normal distribution
   d. chi-square distribution
12. The degrees of freedom for the distribution in question 11 are
   a. $n - 2$
   b. $n - 3$
   c. 3 and $n - 2$
   d. 2 and $n - 3$

Supplementary Exercises

18. Recall that in Exercise 1, the admissions' officer for Clearwater College developed the following estimated regression equation relating final college GPA to student's SAT mathematics scores and their high-school GPA.

$$\hat{y} = -1.41 + .0235x_1 + .00486x_2$$

where

$$x_1 = \text{high-school grade-point average}$$

$$x_2 = \text{SAT mathematics score}$$

$$y = \text{final college grade-point average}$$

A portion of the Minitab computer output is shown.

```
THE REGRESSION EQUATION IS
Y = - 1.41 + 0.0235 X1 + 0.00486 X2

                               ST. DEV.     T-RATIO =
COLUMN        COEFFICIENT      OF COEF.     COEF/S.D.
                -1.4053         0.4848        -2.90
X1             0.023467        0.008666      _____
X2             _____        0.001077      _____

S = 0.1298


R-SQUARED = _____ PERCENT
R-SQUARED = _____ PERCENT, ADJUSTED FOR D.F.


ANALYSIS OF VARIANCE

   DUE TO       DF            SS        MS=SS/DF
REGRESSION      __         1.76209     _____
RESIDUAL        __         _____    _____
TOTAL           9          1.88000
```

a. Complete the missing entries in this output.
b. Compute F and test at the $\alpha = .05$ level whether a significant relationship exists or not.
c. Did the estimated regression equation provide a good fit to the data? Explain.
d. Use the t test and $\alpha = .05$ to test $H_0: \beta_1 = 0$ and $H_0: \beta_2 = 0$.

19. Recall that in Exercise 2 the personnel director for Electronics Associates developed the following estimated regression equation relating an employee's score on a job satisfaction test to their length of service and wage rate.

$$\hat{y} = 14.4 - 8.69x_1 + 13.5x_2$$

where

$$x_1 = \text{length of service (years)}$$

$$x_2 = \text{wage rate (dollars)}$$

$$y = \text{job satisfaction test score (higher scores indicate more job satisfaction)}$$

A portion of the Minitab computer output is shown:

```
THE REGRESSION EQUATION IS
Y = 14.4 - 8.69 X1 + 13.5 X2

                                 ST. DEV.     T-RATIO =
COLUMN        COEFFICIENT        OF COEF.     COEF/S.D.
                 14.448            8.191         1.76
X1             _____            1.555        _____
X2              13.517            2.085        _____

S = 3.773

R-SQUARED = _____ PERCENT
R-SQUARED = _____ PERCENT, ADJUSTED FOR D.F.


ANALYSIS OF VARIANCE

   DUE TO     DF            SS        MS=SS/DF
REGRESSION    2          _____     _____
RESIDUAL      __          71.17      _____
TOTAL         7          720.00
```

a. Complete the missing entries in this output.
b. Compute F and test at the $\alpha = .05$ level whether a significant relationship exists or not.
c. Did the estimated regression equation provide a good fit to the data? Explain.
d. Use the t test and $\alpha = .05$ to test $H_0: \beta_1 = 0$ and $H_0: \beta_2 = 0$.

20. Recall that in Exercise 3 a chemist has developed the following estimated

```
THE REGRESSION EQUATION IS
Y = 69.0 + 1.17 X1 - 3.97 X2

                                 ST. DEV.     T-RATIO =
COLUMN        COEFFICIENT        OF COEF.     COEF/S.D.
                 69.02            29.96         2.30
X1              1.1741           0.3396        _____
X2             -3.9667           0.9174        _____

S = 7.035

R-SQUARED = _____ PERCENT
R-SQUARED = _____ PERCENT, ADJUSTED FOR D.F.


ANALYSIS OF VARIANCE

   DUE TO     DF            SS        MS=SS/DF
REGRESSION    __         1333.60     _____
RESIDUAL      __          346.40     _____
TOTAL         9          _____
```

regression equation relating the yield of a process (y) to the temperature (x_1) and the pressure (x_2).

$$\hat{y} = 69 + 1.17x_1 - 3.97x_2$$

where temperature is measured in degrees Fahrenheit and pressure in pounds per square inch. A portion of the Minitab computer output is shown at the bottom of page 578.

a. Complete the missing entries in this output.
b. Compute F and test at the $\alpha = .05$ level whether a significant relationship exists or not.
c. Did the estimated regression equation provide a good fit to the data? Explain.
d. Use the t test and $\alpha = .05$ to test $H_0: \beta_1 = 0$ and $H_0: \beta_2 = 0$.

21. Bauman Construction Company makes bids on a variety of projects. In an effort to estimate the bid to be made by one of its competitors, Bauman has obtained data on 15 previous bids and developed the following estimated regression equation.

$$\hat{y} = 80 + 45x_1 - 3x_2$$

where

$$y = \text{competitor's bid (\$1000s)}$$

$$x_1 = \text{square feet (1000s)}$$

$$x_2 = \text{local index of construction activity}$$

a. Estimate the competitor's bid on a project involving 50,000 square feet and an index of construction activity of 70.
b. If SSR = 19,780 and SST = 21,533, test at $\alpha = .01$ the significance of the relationship.

22. The following estimated regression equation was developed for a model involving two independent variables.

$$\hat{y} = 40.7 + 8.63x_1 + 2.71x_2$$

After dropping x_2 from the model, the least squares method was used again to obtain an estimated regression equation involving only x_1 as an independent variable.

$$\hat{y} = 42.0 + 9.01x_1$$

Give an interpretation of the coefficient of x_1 in each model.

23. The owner of TAI Movie Theaters, Inc. would like to investigate the effect of television advertising on weekly gross revenue for special promotion films. The following historical data were developed.

a. Using these data, develop an estimated regression equation relating weekly gross revenue to television advertising expenditure.

Weekly Gross Revenue ($1000s)	Television Advertising ($1000s)
96	5.0
90	2.0
95	4.0
92	2.5
95	3.0
94	3.5
94	2.5
94	3.0

b. Estimate the weekly gross revenue in a week in which $3500 is spent on television advertising.

24. As an extension of Exercise 23, consider the possibility of incorporating the effect of newspaper advertising as well as television advertising on weekly gross revenue. The following data were developed from historical records.

Weekly Gross Revenue ($1000s)	Newspaper Advertising ($1000s)	Television Advertising ($1000s)
96	1.5	5.0
90	2.0	2.0
95	1.5	4.0
92	2.5	2.5
95	3.3	3.0
94	2.3	3.5
94	4.2	2.5
94	2.5	3.0

Let

$$x_1 = \text{newspaper advertising (\$1000s)}$$

$$x_2 = \text{television advertising (\$1000s)}$$

$$y = \text{weekly gross revenue (\$1000s)}$$

Shown on page 581 is the Minitab computer output for these data.

a. Complete the missing entries in this table.
b. Compute F and test at the $\alpha = .05$ level whether a significant relationship exists or not.
c. Did the estimated regression equation provide a good fit to the data? Explain.
d. Use the t test and $\alpha = .05$ to test $H_0: \beta_1 = 0$ and $H_0: \beta_2 = 0$.
e. Estimate the weekly gross revenue for a week in which $3500 is spent on television advertising and $1500 is spent on newspaper advertising.

```
THE REGRESSION EQUATION IS
Y = 83.2 + 1.30 X1 + 2.29 X2

                                 ST. DEV.     T-RATIO =
  COLUMN        COEFFICIENT       OF COEF.    COEF/S.D.
                   83.230          1.574        52.88
  X1               1.3010          0.3207      _____
  X2               2.2902          0.3041      _____

  S = 0.6426

  R-SQUARED = _____ PERCENT
  R-SQUARED = _____ PERCENT, ADJUSTED FOR D.F.


  ANALYSIS OF VARIANCE

    DUE TO      DF          SS       MS=SS/DF
  REGRESSION   ___       23.435     _____
  RESIDUAL     ___       _____    _____
  TOTAL        ___       25.500
```

f. Is the coefficient for television advertising expenditures the same as the coefficient obtained in Exercise 2? Interpret this coefficient in each case.

25. Heller Company believes that the quantity of lawnmowers sold depends on the price of its mower and the price of a competitor's mower. Let

$$y = \text{quantity sold (1000s)}$$

$$x_1 = \text{price of competitor's mower (dollars)}$$

$$x_2 = \text{price of Heller's mower (dollars)}$$

Management would like an estimated regression equation that relates quantity sold to the price of the Heller mower and the competitor's mower. The following data are available concerning prices in 10 different cities.

Competitor's Price (x_1)	Heller's Price (x_2)	Quantity Sold (y)
120	100	102
140	110	100
190	90	120
130	150	77
155	210	46
175	150	93
125	250	26
145	270	69
180	300	65
150	250	85

The Minitab computer output for these data is shown on page 582.

```
THE REGRESSION EQUATION IS
Y = 66.5 + 0.414 X1 - 0.270 X2

                                  ST. DEV.    T-RATIO =
  COLUMN        COEFFICIENT       OF COEF.    COEF/S.D.
                   66.52            41.88        1.59
  X1               0.4139          0.2604       _____
  X2              -0.26978         0.08091      _____

  S = 18.74

  R-SQUARED = _____ PERCENT
  R-SQUARED = _____ PERCENT, ADJUSTED FOR D.F.


  ANALYSIS OF VARIANCE

     DUE TO      DF           SS        MS=SS/DF
  REGRESSION     __         _____      _____
  RESIDUAL       __         2457.3      _____
  TOTAL          __         7076.1
```

a. Complete the missing entries in this table.
b. Compute F and test at the $\alpha = .05$ level whether a significant relationship exists or not.
c. Did the estimated regression equation provide a good fit to the data? Explain.
d. Use the t test and $\alpha = .05$ to test $H_0: \beta_1 = 0$ and $H_0: \beta_2 = 0$.
e. Predict the quantity sold in a city where Heller prices its mower at $160 and the competitor prices its mower at $170.

26. In a regression analysis involving 27 observations the following estimated regression equation was developed.

$$\hat{y} = 16.3 + 2.3x_1 + 12.1x_2 - 5.8x_3$$

Also, the following standard errors were obtained:

$$s_{b_1} = .53 \qquad s_{b_2} = 8.15 \qquad s_{b_3} = 1.30$$

At an $\alpha = .05$ level of significance conduct the following hypothesis tests.

a. $H_0: \beta_1 = 0$ versus $H_1: \beta_1 \neq 0$
b. $H_0: \beta_2 = 0$ versus $H_1: \beta_2 \neq 0$
c. $H_0: \beta_3 = 0$ versus $H_1: \beta_3 \neq 0$

Computer Exercise

A government agency conducted a study to determine the relationship between income, household size, and the amount charged in the last 12 months on credit cards.

The objective of this study was to develop an equation that could be used to predict the amount charged on credit cards given the individual's income and household size. The following data were obtained.

Income ($1000s)	Household Size	Amount Charged	Income ($1000s)	Household Size	Amount Charged
33	3	2780	29	2	2491
32	2	2748	26	3	2300
35	1	3070	43	3	3690
34	2	2831	35	4	2935
35	2	3047	35	2	2946
37	1	3061	35	4	2670
38	2	3390	39	1	3485
35	2	3091	30	2	2722
41	2	3415	29	3	2620
47	1	4107	36	1	3126
42	1	3429	32	3	2726
33	5	2521	41	1	3486
38	1	3493	35	5	2720
38	3	3104	36	2	3172
34	3	2752	33	2	2769
30	4	2401	31	1	2804
27	2	2276	37	1	3170
34	2	2865	36	3	2923
29	3	2578	27	2	2270
37	1	3173	36	2	3053
38	5	3091	25	2	2266
33	3	2686	41	6	3187
38	5	2921	35	2	2947
28	4	2250	40	5	3224
35	5	2865	35	2	3105

QUESTIONS

1. Use numerical and graphical measures to summarize these data.
2. Does there appear to be any relationship between the amount charged and the individual's income? Between the amount charged and the household size?
3. Develop an estimated regression equation that can be used to predict the amount charged.

APPENDIX TO CHAPTER 16: Solving the Normal Equations for the Butler Trucking Company

In the chapter we developed an estimated regression equation for Butler Trucking involving two independent variables, miles traveled and number of deliveries. Substituting the data for this problem (see Section 16.2) into (16.5) to (16.7) provided the following normal equations:

$$10b_0 + 815b_1 + 26b_2 = 66.0, \tag{16.8}$$

$$815b_0 + 69{,}625b_1 + 2{,}165b_2 = 5{,}594.0, \tag{16.9}$$

$$26b_0 + 2{,}165b_1 + 76b_2 = 180.6. \tag{16.10}$$

By multiplying (16.8) by 81.5 and subtracting the result from (16.9) we can eliminate b_0 and obtain an equation involving b_1 and b_2 only:

$$
\begin{array}{r}
815b_0 + 69{,}625.0b_1 + 2{,}165b_2 = 5{,}594.0 \\
-815b_0 - 66{,}422.5b_1 - 2{,}119b_2 = -5{,}379.0 \\
\hline
3{,}202.5b_1 + 46b_2 = 215.0.
\end{array} \tag{16A.1}
$$

Now multiply (16.8) by 2.6 and subtract the result from (16.10). This manipulation yields a second equation involving b_1 and b_2 only:

$$
\begin{array}{r}
26b_0 + 2{,}165b_1 + 76.0b_2 = 180.6 \\
-26b_0 - 2{,}119b_1 - 67.6b_2 = -171.6 \\
\hline
46b_1 + 8.4b_2 = 9.0.
\end{array} \tag{16A.2}
$$

With equations (16A.1) and (16A.2) we can solve simultaneously for b_1 and b_2. Multiplying (16A.2) by 46/8.4 and subtracting the result from (16A.1) gives us an equation involving b_1 only:

$$
\begin{array}{r}
3{,}202.5000b_1 + 46b_2 = 215.0000 \\
-251.9048b_1 - 46b_2 = -49.2857 \\
\hline
2{,}950.5952b_1 = 165.7143.
\end{array} \tag{16A.3}
$$

Using (16A.3) to solve for b_1 we get

$$b_1 = \frac{165.7143}{2{,}950.5952} = .056163.$$

Using this value for b_1 we can substitute into (16A.2) to solve for b_2:

$$46(.056163) + 8.4b_2 = 9,$$

$$2.583498 + 8.4b_2 = 9,$$

$$8.4b_2 = 6.416502,$$

$$b_2 = .7638693.$$

Now we can substitute the values obtained for b_1 and b_2 into (16.8), thus obtaining b_0:

$$10b_0 + 815(.056163) + 26(.7638693) = 66.0,$$
$$10b_0 + 45.772845 \quad\ + 19.860602 \quad\ = 66.0,$$
$$10b_0 = .366553,$$
$$b_0 = .0366553.$$

Rounding to four significant digits, we obtain the following estimated regression equation for Butler Trucking:

$$\hat{y} = .0367 + .0562x_1 + .7639x_2. \tag{16A.4}$$

CHAPTER 17

Nonparametric Methods

What You Will Learn in This Chapter

- when nonparametric methods are applicable

- the advantages of nonparametric methods

- how to use the Wilcoxon and Mann-Whitney tests for differences between two populations

- how to use and interpret the sign test

- how to compute the Spearman rank correlation coefficient

Contents

Statistics in the News

PHYSICAL CONTACT IN THE FAMILY

Physical contact is an important means of connecting people with one another. The touch, the handshake, the hug, the embrace, and the kiss reflect different levels of physical intimacy and commitment. Physical contact is a nonverbal means of interpersonal bonding and an important aspect of family activity. Children require the physical contact of parents, especially during infancy, if they are to thrive. Although physical contact is an important aspect of close relationships, it has been analyzed infrequently by social scientists because of its delicacy and related measurement difficulties.

A recent study by Oscar Grusky, Phillip Bonacich, and Mark Peyrot attempts to measure and test hypotheses about physical contact within the family unit. Using 48 two-parent, two-child families, the researchers asked each family to pose for a family photograph. In this setting, the researchers observed who touched whom and the frequency of the physical contacts.

The researchers hypothesized that higher-status persons in the family unit would touch those of lower status more often than vice-versa. The touches were expected to come from father to mother, from parent to child, and/or from older child to younger child. Using a technique from nonparametric statistics known as the sign test, the researchers found support for this hypothesis. They noted that fathers were significantly more likely to touch mothers than vice-versa and that parents were more likely to touch children than vice-versa.

The researchers found that children were more likely to be touched than parents and that mothers were significantly more likely to be touched than fathers. Younger children were also more likely to be touched than older children. Fathers were touched less than any other member of the family and experienced the least physical contact of any family member.

Michael Hayman, Stock, Boston

Physical contact among father, mother and children helps provide a close family relationship.

The researchers conclude by noting that family physical contact is a complex multidimensional phenomenon. On one level it signifies status and power differential between members; on another level it reflects warmth, closeness, support and expressiveness. Clearly touching and physical contact are important aspects in the health and well-being of a family unit. Further research should contribute to a better understanding of physical contact in family relationships.

Based on "Physical Contact in the Family," by Oscar Grusky, Phillip Bonacich, and Mark Peyrot, *Journal of Marriage and the Family* (August 1984).

In this chapter we introduce several statistical procedures that do not require knowledge of the form of the probability distribution from which the measurements come. The methods of statistical inference we shall study here are called *nonparametric methods*. Since nonparametric methods do not require assumptions about the form of the population probability distribution, they are often referred to as *distribution-free methods*. "Statistics in the News" described the use of a nonparametric statistical method known as the sign test.

17.1 WILCOXON RANK-SUM TEST

In this section we present a nonparametric statistical test to determine if there are any differences between two populations. This nonparametric test is based upon two independent random samples, one from each population. Recall that in Chapter 11 we conducted a test to determine if there were a difference between the means of two populations. The hypotheses tested were

$$H_0: \mu_1 = \mu_2$$

$$H_1: \mu_1 \neq \mu_2$$

In the small-sample case, the statistical method for this test was based on two assumptions.

1. Both populations are normally distributed.
2. The population variances are equal.

The nonparameteric method we use in this section does not require either of these two assumptions.*

A nonparametric alternative for analyzing data from independent samples from two populations was developed by Wilcoxon and is referred to as the *Wilcoxon rank-sum test*. Instead of testing for the differences between the means (parameters) of the two populations, the Wilcoxon rank-sum procedure is designed to test whether or not the two populations are the same. The hypotheses for the Wilcoxon rank-sum test are

$$H_0: \text{The Two Populations Are Identical}$$

$$H_1: \text{The Two Populations Are Not Identical}$$

We use the following example to demonstrate the methodology of the Wilcoxon rank-sum test.

---

*The only assumption made about the population for the nonparametric methods is that the data values are continuous; the purpose of this assumption is to make ties theoretically impossible. In practice, however, we usually have a few ties because of rounding or the fact that the continuity assumption is not strictly satisfied.

EXAMPLE 17.1

The Third National Bank has two branch offices. Data collected from two independent simple random samples, one from each branch, are shown in Table 17.1. Do the data indicate that the populations of checking account balances at the two branch banks are or are not identical?

The first step in the rank-sum procedure is to rank the *combined* data from the lowest to the highest values. Using the combined set of 22 observations shown in Table 17.1, we find the lowest data value of $750 (sixth item of sample 2) and assign to it a rank of 1. Continuing the ranking, we have the following.

Account Balance	Item	Assigned Rank
$ 750	6th of sample 2	1
$ 800	5th of sample 2	2
$ 805	7th of sample 1	3
$ 850	2nd of sample 2	4
.	.	.
.	.	.
.	.	.
$1195	4th of sample 1	21
$1200	3rd of sample 1	22

In ranking the combined data, we may find that two or more data values are the same. In this case, these same values are given the *average* ranking of their positions in the combined data set. This situation of *ties* occurs with the ranking of the 22 account

TABLE 17.1

Account Balances for Two Branches of the Third National Bank

Branch 1		Branch 2	
Sampled Account	Account Balance ($)	Sampled Account	Account Balance ($)
1	1095	1	885
2	955	2	850
3	1200	3	915
4	1195	4	950
5	925	5	800
6	950	6	750
7	805	7	865
8	945	8	1000
9	875	9	1050
10	1055	10	935
11	1025		
12	975		

balances from the two branch banks. For example, the balance of $945 (eighth item of sample one) will be assigned the rank of 11. However, the next two values in the data set are tied with values of $950 (see the sixth item of sample 1 and the fourth item of sample 2). Since these two values will be considered for assigned ranks of 12 and 13, they are both given the assigned rank of 12.5. At the next highest data value of $955, we continue the ranking process by assigning $955 the rank of 14. Table 17.2 shows the entire data set with the assigned rank of each observation.

TABLE 17.2

Combined Ranking of the Data in the Two Samples from the Third National Bank

	Branch 1				Branch 2	
Sampled Account	Account Balance ($)	Rank		Sampled Account	Account Balance ($)	Rank
1	1095	20		1	885	7
2	955	14		2	850	4
3	1200	22		3	915	8
4	1195	21		4	950	12.5
5	925	9		5	800	2
6	950	12.5		6	750	1
7	805	3		7	865	5
8	945	11		8	1000	16
9	875	6		9	1050	18
10	1055	19		10	935	10
11	1025	17			Sum of ranks	83.5
12	975	15				
	Sum of ranks	169.5				

The next step in the Wilcoxon rank-sum test is to sum the ranks for each sample. These sums are shown in Table 17.2. The test procedure can be based upon the sum of the ranks for either sample. In the following discussion, we use the sum of the ranks for the sample from Branch 1. We denote this sum by the symbol W; for this example, $W = 169.5$.

• • •

Let us investigate for a moment the properties of the sum of ranks. First of all, what is the smallest value that W could assume in Example 17.1? Clearly, the smallest value for W corresponds to the case where the 12 sample accounts from Branch 1 have ranks 1, 2, 3, . . . , 12. The sum of the ranks in this case is $1 + 2 + \cdot \cdot \cdot + 12 = 78$. Thus the smallest possible value for W is 78.

Next, let us consider the maximum possible value for W. Clearly, the maximum value for W occurs if the ranks of the 12 sampled items are 11, 12, . . . , 22. Note that this occurs if the ranks for the ten items in sample 2 are 1, 2, . . . , 10. Thus the maximum value for W is equal to $11 + 12 + \cdot \cdot \cdot + 22 = 198$. What we have shown, then, is that for Example 17.1 W must take on a value between 78 and 198.

Note that values of W close to 78 imply that Branch 1 has smaller account balances. Values of W close to 198 imply that Branch 1 has higher account balances.

Thus if the null hypothesis of two identical populations is true, we would expect the value of W to be somewhere close to the average of the above two values, or $(78 + 198)/2 = 138$. Using this insight and the fact that the actual sum of the ranks for sample 1 is $W = 169.5$, we might be tempted to conclude that Branch 1 is carrying the higher account balances. However, when we realize that the data shown are for relatively small sample sizes, we may not be ready to draw a general conclusion about the differences in the two populations. The Wilcoxon rank-sum test provides a statistical procedure for determining if there is sufficient evidence to conclude that the populations are in fact different.

Under the assumption that the two populations are identical, it can be shown that with two independent samples of size n_1 and n_2, the sampling distribution of W is as follows.

Sampling Distribution of W for Identical Populations

Mean: $$\mu_W = \frac{1}{2} n_1(n_1 + n_2 + 1) \qquad (17.1)$$

Standard Deviation: $$\sigma_W = \sqrt{\frac{1}{12} n_1 n_2(n_1 + n_2 + 1)} \qquad (17.2)$$

Distribution Form: Approximately Normal Provided $n_1 \geq 10$ and $n_2 \geq 10$

EXAMPLE 17.1 (continued)

For Branch 1, we have

$$\mu_W = \frac{1}{2} 12(12 + 10 + 1) = 138$$

$$\sigma_W = \sqrt{\frac{1}{12} 12(10)(12 + 10 + 1)} = 15.17$$

The sampling distribution of W is shown in Figure 17.1. Following the usual hypothesis-testing procedure, we compute the standardized test statistic z to determine if the observed value of W appears to be from the sampling distribution of Figure 17.1. If W appears to be from this distribution, we will accept the hypothesis that the two populations providing the independent random samples are identical. However, if W does not appear to be from this distribution, we will reject the null hypothesis and conclude that the populations are not identical. Computing the test statistic we have

$$z = \frac{W - \mu_W}{\sigma_W} = \frac{169.5 - 138}{15.17} = 2.08$$

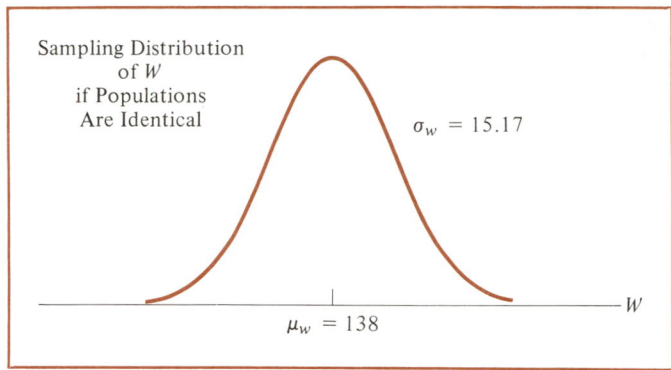

FIGURE 17.1
Sampling Distribution of *W* for the Third National Bank Example

At an $\alpha = .05$ level of significance, we know that, in order to accept H_0, z should be between -1.96 and $+1.96$. Since $z = 2.08$ is not in this interval, we reject H_0. Thus we conclude that the two populations are not identical. That is, the populations of account balances at the two branches are not the same.

$$\bullet \quad \bullet \quad \bullet$$

In summary, the Wilcoxon rank-sum test follows the steps outlined below in order to determine if two independent random samples are selected from identical populations.

1. Rank the combined sample observations from lowest to highest, with tied values being assigned the average of the tied rankings.
2. Compute W, the sum of the ranks for the first sample.
3. Make the test for significant differences between the two populations by using the observed value of W and comparing it to the sampling distribution of W for identical populations (see equations (17.1) and (17.2)). The value of the standardized test statistic z will provide the basis for accepting or rejecting the null hypothesis of identical populations.

Exceptions to this procedure arise if either sample size is less than 10 or if numerous ties occur in the ranking of the combined data. In the first case, the small sample size invalidates the use of the normal approximation of the sampling distribution of W. However, a table of exact values for W in the small-sample-size case is available in most books on nonparametric methods. This table can be used to determine if a significant difference exists between the two populations. If numerous tied ranks are observed, a slight modification in the formula for σ_W is required. The necessary modification can be found in most advanced nonparametric reference texts.

Mann-Whitney Test

A test equivalent to the Wilcoxon rank-sum test was proposed by Mann and Whitney and is referred to as the *Mann-Whitney test*. Instead of working directly with the sum of the ranks to carry out the test under the assumption of identical populations, Mann and Whitney proposed the use of a test statistic U whose value is related to W by the following expression.

$$U = (\text{Largest Possible Value of } W) - W$$

$$= \left(n_1 n_2 + \frac{n_1(n_1 + 1)}{2}\right) - W \tag{17.3}$$

EXAMPLE 17.1 (continued)

For the Third National Bank example, the largest possible value of W was 198. Thus the value of U would be

$$U = 198 - 169.5 = 28.5$$

It can be shown that if the two populations are identical, the sampling distribution of U can be approximated for large n_1 and n_2 (that is, $n_1, n_2 \geq 10$) by a normal distribution with mean

$$\mu_U = \frac{1}{2} n_1 n_2 \tag{17.4}$$

and standard deviation

$$\sigma_U = \sqrt{\tfrac{1}{12} n_1 n_2 (n_1 + n_2 + 1)}. \tag{17.5}$$

Let us compare these results with those for the normal approximation to W in the Wilcoxon rank-sum test. We see that only the means differ ($\mu_W \neq \mu_U$ but $\sigma_W = \sigma_U$). For the Third National Bank example, $\mu_U = 60$ and $\sigma_U = 15.17$. Thus at the $\alpha = .05$ level of significance, the decision rule will be

Accept H_0 (identical populations) if $-1.96 \leq z \leq 1.96$

Reject H_0 otherwise

where

$$z = \frac{U - \mu_U}{U_U}$$

Since

$$z = \frac{28.5 - 60}{15.17} = -2.08$$

is less than -1.96 we reject H_0.

Thus we see that the Mann-Whitney test procedure leads to the same conclusion as the Wilcoxon rank-sum test. Hence either test can be used, depending upon the preference of the user.

• • •

The nonparametric tests discussed in this section are used to determine whether or not two populations are identical. The statistical tests, such as the t test described in Chapter 11, test the equality of two population means. When we reject the hypothesis that the means are equal, we conclude that the populations differ only in their means. When we reject the hypothesis that the populations are identical using one of the methods of this section, we cannot state how they differ. The populations could have different means, different variances, and/or different forms. Nonetheless, if we had assumed that the populations were the same in every way except for the means, a rejection of H_0 using a nonparametric method would have implied that the means differed. The major advantage of the nonparametric methods, however, is that they do not require any assumptions about the form of the probability distribution from which the measurements come.

EXERCISES

1. Mileage performance tests were conducted for two models of automobiles. Twelve automobiles of each model were randomly selected and a miles-per-gallon rating for each model was developed based upon 1000 miles of highway driving. The data are shown.

	Model 1		Model 2
Automobile	Miles per Gallon	Automobile	Miles per Gallon
1	20.6	1	21.3
2	19.9	2	17.6
3	18.6	3	17.4
4	18.9	4	18.5
5	18.8	5	19.7
6	20.2	6	21.1
7	21.0	7	17.3
8	20.5	8	18.8
9	19.8	9	17.8
10	19.8	10	16.9
11	19.2	11	18.0
12	20.5	12	20.1

Use $\alpha = .10$ and test for a significant difference in the populations of miles-per-gallon ratings for the two models.

2. The number of hours per day of television viewing was recorded for a sample of children and a sample of adults. Using an $\alpha = .10$ level of significance, does there appear to be a difference in television viewing times for children and adults?

Children	Adults
3.5	2.0
4.5	5.5
1.5	3.0
2.5	2.0
3.0	0.5
3.0	1.0
2.5	0.0
4.5	2.0
4.0	3.0
2.0	1.5
5.0	
4.0	

3. The following data from police records show the number of daily crime reports from a sample of days during the winter months and a sample of days during the summer months. Using a .05 level of significance, determine if there is a significant difference between the number of crime reports in winter and summer months.

Winter	Summer
18	28
20	18
15	24
16	32
21	18
20	29
12	23
16	38
19	28
20	18

4. A certain brand of microwave oven was priced at 10 stores in Dallas and 13 stores in San Antonio. The data are as follows.

Dallas	San Antonio
445	460
489	451
405	435
485	479
439	475
449	445
436	429
420	434
430	410
405	422
	425
	459
	430

Use a .05 level of significance and test whether or not prices for the microwave oven are the same in the two cities.

5. Independent random samples of faculty at two colleges were taken, and annual salaries ($1000s) were recorded. Use a .10 level of significance to test if there is a difference between faculty salaries at the two colleges.

College 1	College 2
36	22
18	16
22	19
15	15
19	12
27	24
42	25
48	19
31	14
29	18
33	

17.2 WILCOXON SIGNED-RANK TEST

In the previous section we described a nonparametric statistical test that could be used to determine if two independent random samples were selected from identical populations. An important alternative to the two-independent-samples procedure occurs whenever we pair or match items such that each observation from one sample has a corresponding matched observation from the other sample. In some cases the same individual or same unit generates the observations or measurements for *both* *populations* and therefore provides the matched or paired observations. In other cases two separate, but similar, individuals or units are identified. One of the individuals generates a measurement for one population, and the second individual generates a corresponding measurement for the other population.

In Chapter 11 we described a statistical test of differences between population means based on matched samples. That test is often referred to as the *paired difference test*. The methodology of the paired difference test requires us to measure the observed *difference* for each pair of observations. Then, under the assumption that the population of differences between the pairs is normally distributed, a *t* test can be used to test the null hypothesis of no difference between population means. If some questions exist concerning the appropriateness of the assumption of normally distributed differences or if it is possible only to rank-order the differences from most similar to most dissimilar, a nonparametric statistical method becomes desirable. Wilcoxon provided the methodology for the nonparametric analysis of the paired differences or matched-sample analysis. The resulting test is referred to as the *Wilcoxon signed-rank test*.

EXAMPLE 17.2

A manufacturing firm is attempting to determine if a difference in task-completion time exists for two production methods. A sample of 11 workers was selected, and each worker completed a production task using each of the two production methods. The production method that each worker performed first was selected randomly. Thus each worker in the sample provided a pair of observations, as shown in Table 17.3. A positive difference in task completion times indicates that method 1 requires more time and a negative difference in times indicates that method 2 requires more time. Do the data indicate that the methods are significantly different in terms of task-completion times?

TABLE 17.3

Production Task Completion Times (minutes)

Worker	Method 1	Method 2	Difference
1	10.2	9.5	.7
2	9.6	9.8	−.2
3	9.2	8.8	.4
4	10.6	10.1	.5
5	9.9	10.3	−.4
6	10.2	9.3	.9
7	10.6	10.5	.1
8	10.0	10.0	.0
9	11.2	10.6	.6
10	10.7	10.2	.5
11	10.6	9.8	.8

The question raised is whether or not the two methods provide differences in task-completion times. In effect, we have two populations of task-completion times, one population associated with each method. The hypotheses that will be tested are

$$H_0: \text{The Populations Are Identical}$$

$$H_1: \text{The Populations Are Not Identical}$$

Certainly, if H_0 cannot be rejected, we will conclude that the task-completion times are similar for the two methods. However, if H_0 can be rejected, we will conclude that the two methods differ in terms of task-completion times.

• • •

The first step of the Wilcoxon signed-rank test requires that we rank the *absolute value* of the differences in the two methods. To do this we first discard any differences of zero and then rank the remaining absolute differences from lowest to highest. Tied differences are assigned average rank values, as in the Wilcoxon rank-sum test. The ranking of the absolute values of differences for Example 17.2 is shown in Table 17.4. Note that the difference of 0 for worker 8 is discarded from the rankings; then the smallest absolute difference of .1 is assigned the rank of 1. This ranking of absolute differences continues with the largest absolute difference of .9 assigned the rank of 10. The absolute differences of .4 for workers 3 and 5 are assigned the average rank of 3.5, while the absolute differences of .5 for workers 4 and 10 are assigned the average rank of 5.5

Once the ranks of the absolute differences have been determined, the ranks are given the sign of the original difference in the data. For example, the .1 difference, which was assigned the rank of 1, is given the value of $+1$ because the observed

TABLE 17.4

Ranking of Absolute Differences for the Production Task Completion Time Example

Worker	Difference	Absolute Value of Difference	Rank
1	.7	.7	8
2	−.2	.2	2
3	.4	.4	3.5
4	.5	.5	5.5
5	−.4	.4	3.5
6	.9	.9	10
7	.1	.1	1
8	0	0	—
9	.6	.6	7
10	.5	.5	5.5
11	.8	.8	9

difference between the two methods was positive for this individual. The .2 difference, which was assigned the rank of 2, is given the value of -2 because the observed difference between the two methods was negative for this individual. The complete list of signed ranks, together with their sum, is shown in Table 17.5.

Let us return to the original hypothesis of identical task-completion times for the two methods. If the populations representing task-completion times for each of the two methods are identical, we would expect the positive ranks and the negative ranks to cancel each other, so that the sum of the signed rank values would be approximately 0. Thus the test for significance under the Wilcoxon signed-rank test involves determining whether or not the computed sum of signed ranks ($+44$ in Example 17.2) is significantly different than 0.

Let T denote the sum of the signed-rank values in a Wilcoxon signed-rank test. It can be shown that if the two populations are identical and that the number of matched pairs of data is 10 or more, the sampling distribution of T can be approximated as follows.

Sampling Distribution of T for Identical Populations

$$\text{Mean:} \quad \mu_T = 0 \tag{17.6}$$

$$\text{Standard Deviation:} \quad \sigma_T = \sqrt{\frac{n(n+1)(2n+1)}{6}} \tag{17.7}$$

Distribution Form: Approximately Normal Provided $n \geq 10$

TABLE 17.5

Signed Ranks for the Production Task Completion Time Example

Worker	Difference	Rank of Absolute Difference	Signed Rank
1	.7	8	+8
2	−.2	2	−2
3	.4	3.5	+3.5
4	.5	5.5	+5.5
5	−.4	3.5	−3.5
6	.9	10	+10
7	.1	1	+1
8	.0	—	—
9	.6	7	+7
10	.5	5.5	+5.5
11	.8	9	+9
		Sum of signed ranks	+44

EXAMPLE 17.2 (continued)

For this example, we have $n = 10$, since we discarded the observation with the difference of 0 (worker 8). Thus using (17.7), we have

$$\sigma_T = \sqrt{\frac{10(11)(21)}{6}} = 19.62$$

The sampling distribution of T under the assumption of identical populations is shown in Figure 17.2.

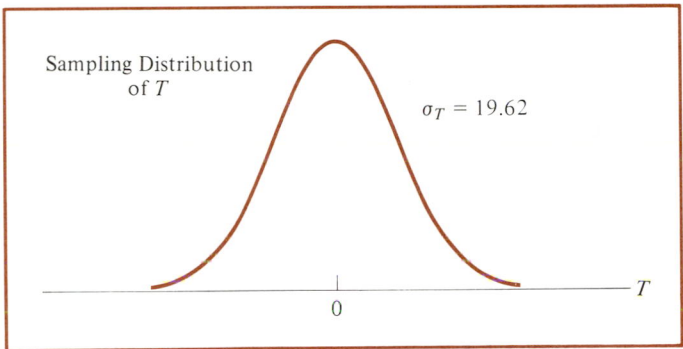

Sampling Distribution of T

$\sigma_T = 19.62$

T

0

FIGURE 17.2
Sampling Distribution of the Wilcoxon *T* Value for the Production Task Completion Time Example

The value of the standarized test statistic z is as follows.

$$z = \frac{T - \mu_T}{\sigma_T} = \frac{44 - 0}{19.62} = 2.24$$

Testing the null hypothesis of no difference using a level of significance of $\alpha = .05$, the acceptance region will be from $z = -1.96$ to $z = +1.96$. With the value of $z = 2.24$, we reject H_0 and conclude that the two populations are not identical and that the methods differ in terms of task-completion times. Although we have determined that a difference between populations exists, the Wilcoxon signed-rank test does not enable us to conclude in what ways the populations differ. However, the fact that method 2 showed the shorter completion times for 8 of the 11 workers would lead us to conclude that differences between the two populations indicate method 2 to be the better production method.

• • •

EXERCISES

6. A sample of 10 individuals was used in a study to test the effects of a relaxant on the time required to fall asleep for male adults. Data for 10 subjects showing the

number of minutes required to fall asleep with and without the relaxant are given. Use a .05 level of significance to determine if the relaxant reduces the time required to fall asleep. What is your conclusion?

Subject	Without Relaxant	With Relaxant
1	15	10
2	12	10
3	22	12
4	8	11
5	10	9
6	7	5
7	8	10
8	10	7
9	14	11
10	9	6

7. A test is conducted of two overnight mail-delivery services. Two samples of identical deliveries are set up such that both delivery services are notified of the need for a delivery at the same time. The number of hours required to make the delivery is recorded for each service. Do the data shown suggest a difference in the delivery times for the two services? Use a .05 level of significance for the test.

Delivery	Service 1	Service 2
1	24.5	28.0
2	26.0	25.5
3	28.0	32.0
4	21.0	20.0
5	18.0	19.5
6	36.0	28.0
7	25.0	29.0
8	21.0	22.0
9	24.0	23.5
10	26.0	29.5
11	31.0	30.0

8. The midterm and final examination scores for a sample of students in a German class are given. Do the data indicate that student performances differ on the two examinations? Use a .10 level of significance.

Student	Midterm	Final
1	72	81
2	55	62
3	60	62
4	80	77
5	75	92
6	81	84
7	90	94
8	73	78
9	71	70
10	78	87

9. Harding Investors, Inc. provides a 6-week training program for newly hired management trainees. As part of the normal evaluation procedure, the firm gives each trainee a pretest and post-test. Use a one-tailed test with $\alpha = .05$ and analyze the following data as part of the evaluation of the firm's management training program. What is your conclusion?

Trainee	Pretest Score	Post-test Score
1	45	65
2	60	70
3	65	63
4	60	67
5	52	60
6	62	58
7	57	70
8	70	65
9	72	80

10. Eight test-market cities were selected as part of a market research study designed to evaluate the effectiveness of a particular advertising campaign. The sales dollars for each city were recorded for the week prior to the promotional program. Then the campaign was conducted for two weeks, with new sales data collected for the week immediately following the campaign. The resulting data with sales in thousands of dollars are shown.

City	Pre-campaign Sales	Post-campaign Sales
Dayton	100	105
Cincinnati	120	140
Columbus	95	90
Cleveland	140	130
Indianapolis	80	82
Louisville	65	55
St. Louis	90	105
Pittsburgh	140	152

Use $\alpha = .05$. What conclusion would you draw concerning the value of the advertising program?

17.3 THE SIGN TEST FOR PAIRED COMPARISONS

The paired-differences or matched-sample approach is a good experimental design for identifying differences between two populations. If quantitative measures, such as the task-completion times of Example 17.2, are available, then the nonparametric Wilcoxon signed-rank test described in the previous section is applicable.

In some situations we may want to use a matched-sample design to test for differences between two populations but find it is not possible to obtain the quantitative measures for each population required by the Wilcoxon signed-rank test. For instance, common applications involve using a sample of individuals to provide ratings for two items, such as two political candidates, two brands of a product, two cities, and so on. If each individual can quantitatively rate each item, the Wilcoxon signed-rank test can be conducted. However, in many cases, the individuals will be able to state the higher-rated or more-preferred item but will be unable to provide a quantitative measure of the difference in preference. In this situation, a nonparametric method, called the *sign test,* can be used to determine if the preferences for two items are identical.

EXAMPLE 17.3

Mueller Beverage Products of Milwaukee, Wisconsin, has conducted a market research study designed to determine if there is a consumer preference for Mueller's Old Brew Beer over the individual consumer's usual beer. Each individual participating in the test was provided with a glass of his or her usual beer and a glass of Mueller's Old Brew. The two glasses were not labeled, and thus the individuals had no way of knowing beforehand which of the two glasses was Mueller's Old Brew and which was the individual's usual brand. The glass that each individual tasted first was randomly selected. After tasting the beer in each glass, the individuals were asked to indicate their *preferred* beer. The test results from a sample of 24 individuals are shown in

Table 17.6. If an individual selected Mueller's Old Brew as the preferred beer, a plus sign was recorded. On the other hand, if the individual stated a preference for his or her usual brand, a minus sign was recorded. Do the data for the 24 individuals indicate a significant difference in the preferences for the beers?

• • •

The sign test can be used to determine if there is a preference for one of two items. The null hypothesis being tested is that there is no preference, or that the proportion preferring each brand is $p = .50$. In effect, the sign test is comparing the observed proportion of plus signs to the hypothesized proportion $p = .50$ under the assumption of no preference. The hypotheses being tested are as follows.

$$H_0: \text{No preference exists}$$

$$H_1: \text{A preference exists}$$

TABLE 17.6

Sign Test Data Collection for the Mueller Beer Study

Individual	Brand Preferred	Value Recorded
1	Old Brew	+
2	Old Brew	+
3	Usual Brand	−
4	Old Brew	+
5	Usual Brand	−
6	Old Brew	+
7	Usual Brand	−
8	Old Brew	+
9	Old Brew	+
10	Usual Brand	−
11	Old Brew	+
12	Usual Brand	−
13	Usual Brand	−
14	Usual Brand	−
15	Old Brew	+
16	Usual Brand	−
17	Old Brew	+
18	Old Brew	+
19	Old Brew	+
20	Usual Brand	−
21	Old Brew	+
22	Old Brew	+
23	Usual Brand	−
24	Old Brew	+

Under the assumption of no preference, the number of plus signs in the sample follows a binomial probability distribution with $p = .50$. In Chapter 7 we stated that the normal distribution was a good approximation of the binomial distribution if both $np \geq 5$ and $n(1 - p) \geq 5$. Since we assume $p = .50$ in this test, the normal distribution may be used to approximate the binomial distribution with samples of size 10 or more. Using this normal approximation, the sampling distribution for the number of plus signs is as follows.

Sampling Distribution of the Number of Plus Signs when No Preference Exists

$$\text{Mean:} \quad \mu = .50n \tag{17.8}$$

$$\text{Standard Deviation:} \quad \sigma = \sqrt{.25n} \tag{17.9}$$

Distribution Form: Approximately Normal Provided $n \geq 10$

EXAMPLE 17.3 (continued)

With $n = 24$, we have $\mu = .50(24) = 12$ and $\sigma = \sqrt{.25(24)} = 2.45$. Thus the sampling distribution of the number of plus signs for the Mueller's Old Brew Beer is as shown in Figure 17.3.

Refering to Table 17.6, we see that 14 of the 24 preferences were described by a plus sign. Using the sampling distribution of Figure 17.3, the standardized test statistic associated with the 14 plus signs is

$$z = \frac{14 - 12}{2.45} = .82$$

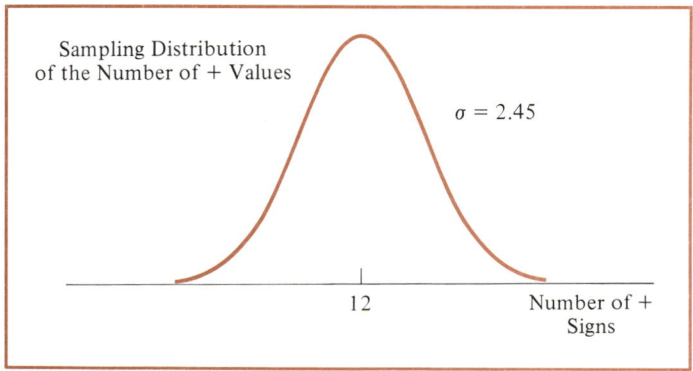

FIGURE 17.3
Sampling Distribution for the Number of Plus Ratings for Mueller's Old Brew Beer

With $\alpha = .05$, we will reject H_0 if z is not between -1.96 and $+1.96$. With $z = .82$ the sign test shows we cannot reject H_0. Hence there is insufficient evidence to conclude that a consumer preference exists between Mueller's Old Brew and the individual's usual brand. While this is the conclusion supported by the sign test, it actually may be a positive indicator for the Mueller product. That is, the fact that we were not able to detect a significant difference between Mueller's Old Brew and an individual's usual brand is indicative of a possibility that individuals may switch to Old Brew on the basis of it being similar in quality to their own individual brand.

• • •

In Example 17.3, all 24 individuals in the study were able to express a preference. In some situations, however, one or more of the individuals in a study may not be able to state a preference. In such cases the individual's response can be removed from the study and the analysis conducted with a smaller sample size.

EXAMPLE 17.4

A poll taken during a recent presidential election campaign asked 200 registered voters to rate the Democratic and Republican candidates in terms of best overall foreign policy. Results of the poll showed 72 rated the Democratic candidate higher, 103 rated the Republican candidate higher, and 25 indicated no difference between the candidates. Does the poll indicate that there is a significant difference between public perception of the foreign policies of the two candidates?

Using the sign test, we see that $n = 200 - 25 = 175$ individuals were able to indicate the candidate they believed had the best overall foreign policy. Using (17.5) and (17.6), we find the sampling distribution of the number of plus signs has the following properties.

$$\mu = .50n = .50(175) = 87.5$$

$$\sigma = .25n = \sqrt{.25(175)} = 6.6$$

In addition, with $n = 175$ we can assume that the distribution is approximately normal.

Using the number of times the Democratic candidate received the higher foreign policy rating as the number of plus signs (72), we have the following value of the standardized test statistic

$$z = \frac{72 - 87.5}{6.6} = -2.35$$

Note that $z = -2.35$ is less than -1.96. Thus the hypothesis of no difference in foreign policy for the two candidates should be rejected at the .05 level of significance. Based on this study the Republican candidate is perceived to have the higher-rated foreign policy.

• • •

Hypothesis Tests About a Median

In earlier chapters we described statistical tests that can be used to make inferences about population means. We now want to show how the sign test can be used to conduct hypothesis tests about the value of a population median. Recall that the median splits a population such that 50% of the values fall at the median or above and 50% fall at the median or below. We can apply the sign test to conduct a hypothesis test about the value of a median by using a plus sign whenever the data value in the sample is above the median and a minus sign whenever the data value in the sample is below the median. Any data value exactly equal to the hypothesized value of the median should be discarded. The computations for the sign test are done in exactly the same manner as shown in Examples 17.3 and 17.4

EXAMPLE 17.5

The following hypothesis test is being conducted on the median price of new homes in St. Louis, Missouri.

$$H_0: \text{Median} = \$65,000$$

$$H_1: \text{Median} \neq \$65,000$$

From a sample of 62 sales of new homes, 37 had prices above $65,000, 23 had prices below $65,000, and 2 had prices of exactly $65,000. Use the sign test to draw a conclusion about the median price of the new homes.

Using (17.5) and (17.6) for the $n = 60$ homes with prices different than $65,000, we have

$$\mu = .50(60) = 30$$

$$\sigma = \sqrt{.25(60)} = 3.87$$

Using 37 as the number of plus signs, the standardized test statistic becomes

$$z = \frac{37 - 30}{3.87} = 1.81$$

With a level of significance of $\alpha = .05$, we cannot reject H_0 if z is between -1.96 and $+1.96$. Since $z = 1.81$ we are unable to reject H_0 at the .05 level of significance and must conclude that, based on this data, the assumption that the median selling price of a new home is $65,000 cannot be rejected.

$$\bullet \quad \bullet \quad \bullet$$

EXERCISES

11. The following data show the preferences indicated by 10 individuals in taste tests involving two brands of coffee.

Individual	Brand A Versus Brand B	Individual	Brand A Versus Brand B
1	+	6	+
2	+	7	−
3	+	8	+
4	−	9	−
5	+	10	+

With $\alpha = .10$, test for a significant difference in the preferences for the two brands. A plus indicates a preference for brand A over brand B.

12. In a television preference poll a sample of 180 individuals was asked to state a preference for one of the two shows aired at the same time on Friday evenings. "Big Town Detective" was favored by 100, 65 favored "The Friday Variety Special," and 15 were unable to state a preference for one over the other. Is there evidence of a significant difference in the preferences for the two shows? Use $\alpha = .05$ for the test.

13. Menu planning at the Hampshire House Restaurant involves the question of customer preferences for steak and seafood. A sample of 250 customers was asked to state a preference for the two menu items. A preference for steak was stated by 140, and 110 stated a preference for seafood. Use $\alpha = .05$ and test for a difference in the preference for the two menu items.

14. The nationwide median hourly wage for a particular labor group is $9.50 per hour. A sample of 200 individuals in this labor group was taken in a city. 134 individuals had a wage rate less than $9.50 per hour; 54 individuals had a wage rate greater than $9.50 per hour; 12 individuals had a wage rate of $9.50. Test the null hypothesis that the median hourly wage in this city is the same as the nationwide median hourly wage. Use a .02 level of significance.

15. In a sample of 150 college basketball games, it was found that the home team won 98 games. Test to see if this data supports the position that there is a home-team advantage in college basketball.

16. From court house records, it is found that in 120 divorce cases, the filing for divorce was initiated by the wife 82 times. Test for a difference in divorce filings between husbands and wives. Using a .01 level of significance, what is your conclusion?

17.4 RANK CORRELATION

Correlation was introduced in Chapters 4 and 15 as a measure of the linear association between two variables for which quantitative data are available. In this section we consider measures of association between two variables when only rank-order data are available. The *Spearman rank-correlation coefficient, r_s,* has been developed for this purpose.

The formula for the Spearman rank-correlation coefficient is as follows.

Spearman Rank-Correlation Coefficient

$$r_s = 1 - \frac{6 \Sigma d_i^2}{n(n^2 - 1)} \qquad (17.10)$$

where

n = the number of items or individuals being ranked

x_i = the rank of item *i* with respect to one variable

y_i = the rank of item *i* with respect to a second variable

$d_i = x_i - y_i$

EXAMPLE 17.6

A company wants to determine if individuals who were expected at the time of employment to be better salespersons actually turn out to have better sales records. To investigate this question, the vice-president in charge of personnel has carefully reviewed the original job interview summaries, academic records, and letters of recommendation for 10 current members of the firm's sales force. Based on the review of this information, the vice president ranked the 10 individuals in terms of their potential for success, basing the assessment solely upon the information available at the

TABLE 17.7

Sales Potential and Actual 2-Year Sales in Units for 10 Salespersons

Salesperson	Ranking of Potential	2-Year Sales (units)	Ranking According to 2-Year Sales
A	2	400	1
B	4	360	3
C	7	300	5
D	1	295	6
E	6	280	7
F	3	350	4
G	10	200	10
H	9	260	8
I	8	220	9
J	5	385	2

time of employment. Then a list was obtained of the number of units sold by each salesperson over the first 2 years. Based on actual sales performance, a second ranking of the 10 salespersons was carried out. Table 17.7 shows the relevant data and the two rankings. The statistical question involves determining whether or not there is agreement between the ranking of potential at the time of employment and the ranking based upon the actual sales performance over the first 2 years. Let us compute the Spearman rank-correlation coefficient for the data in Table 17.7. The computations for the rank-correlation coefficient are summarized in Table 17.8. Here we see that the rank-correlation coefficient is a positive .73. The Spearman rank-correlation coefficient ranges from -1.0 to $+1.0$, with an interpretation similar to the sample correlation coefficient in that positive values near 1.0 indicate a strong association between the rankings; as one rank increased, the other rank increased. On the other hand, rank correlations near -1.0 indicate a strong negative association in the ranks (as one rank increases, the other rank decreases). The value $r_s = .73$ indicates a positive correlation between potential and actual performance. Individuals ranked high on potential tend to rank high on performance.

• • •

A Test for Significant Rank Correlation

At this point, we have seen how sample results can be used to compute the sample rank-correlation coefficient. As with many other statistical procedures, we may wish to use the sample results to make an inference about the population rank correlation, ρ_s, between two variables. In Example 17.6, the population rank-correlation coefficient

TABLE 17.8

Computation of the Spearman Rank-Correlation Coefficient for Sales Potential and Sales Performance

Salesperson	x_i = Ranking of Potential	y_i = Ranking of Sales Performance	$d_i = x_i - y_i$	d_i^2
A	2	1	1	1
B	4	3	1	1
C	7	5	2	4
D	1	6	-5	25
E	6	7	-1	1
F	3	4	-1	1
G	10	10	0	0
H	9	8	1	1
I	8	9	-1	1
J	5	2	3	9
				$\Sigma d_i^2 = 44$

$$r_s = 1 - \frac{6\Sigma d_i^2}{n(n^2-1)} = 1 - \frac{6(44)}{10(100-1)} = .73$$

could be obtained by making the rank-correlation coefficient computations for all members of the sales force. However, we would like to avoid all this data collection and make an inference about the population rank-correlation based on the sample rank-correlation coefficient, r_s. To make this inference, we must test the following hypotheses:

$$H_0: \rho_s = 0$$

$$H_1: \rho_s \neq 0$$

Under the null hypothesis of no rank correlation ($\rho_s = 0$), the rankings are independent and the sampling distribution of r_s is as follows.

Sampling Distribution of r_s When the Rankings are Independent*

Mean: $\qquad\qquad\qquad \mu_{r_s} = 0$ $\qquad\qquad$ (17.11)

Standard Deviation: $\quad \sigma_{r_s} = \sqrt{\dfrac{1}{(n-1)}}$ $\qquad$ (17.12)

Form: Approximately Normal Provided $n \geq 10$.

EXAMPLE 17.6 (continued)

The sample rank-correlation coefficient for sales potential and sales performance in Example 17.6 was $r_s = .73$. Use this value to test for a significant rank correlation.

From (17.11) we have $\mu_{r_s} = 0$, and from (17.12) we have $\sigma_{r_s} = \sqrt{1/(10-1)} = .33$. Using the standardized test statistic, we have

$$z = \frac{r_s - \mu_{r_s}}{\sigma_{r_s}} = \frac{.73 - 0}{.33} = 2.21$$

Using a level of significance of $\alpha = .05$, we see that the null hypothesis of no correlation must be rejected if $z < -1.96$ or if $z > 1.96$. Since $z = 2.21 > 1.96$, we reject the hypothesis of no rank correlation. Thus we can conclude that a significant rank correlation exists between sales potential and sales performance for the population.

• • •

---

*The sampling distribution of r_s is based on the assumption that the two rankings are independent; that is, the rank-correlation for the population is zero.

EXERCISES

17. Consider the following two sets of rankings for six items.

	Case One			Case Two	
Item	First Ranking	Second Ranking	Item	First Ranking	Second Ranking
A	1	1	A	1	6
B	2	2	B	2	5
C	3	3	C	3	4
D	4	4	D	4	3
E	5	5	E	5	2
F	6	6	F	6	1

Note that in the first case the rankings are identical, while in the second case the rankings are exactly opposite. What value should you expect for the Spearman rank-correlation coefficient for each of these cases? Explain. Calculate the rank-correlation coefficient for each case.

18. In the baseball draft eight players are ranked by a scout in terms of speed and then in terms of power hitting.

Player	Speed Ranking	Power-Hitting Ranking
A	1	8
B	2	5
C	3	6
D	4	7
E	5	2
F	6	3
G	7	4
H	8	1

Use the Spearman rank-correlation coefficient to measure the association between speed and power. Use $\alpha = .05$ and test for the significance of this correlation coefficient.

19. A high-school mathematics teacher and an English teacher ranked 10 honor roll students on ability in the respective subjects. The rankings are as follows.

Student	Mathematics Ranking	English Ranking
1	4	7
2	1	5
3	9	2
4	3	6
5	10	4
6	8	1
7	5	8
8	7	3
9	2	9
10	6	10

    a. What is the Spearman rank correlation coefficient? Comment on your interpretation of the value of the rank correlation.

    b. Test for a significant rank correlation using a .05 level of significance.

20. At a national diving meet, the final rankings of the top 10 divers in the 1-meter and 3-meter diving competitions are as follows.

Diver	1 Meter	3 Meter
A	2	3
B	5	7
C	1	1
D	8	6
E	10	9
F	9	8
G	7	10
H	4	4
I	3	2
J	6	5

    a. What is the Spearman rank correlation coefficient for the ordering of the divers in the two competitions?

    b. Use a .05 level of significance to test for a significant rank correlation.

21. In a poll of men and women television viewers, preferences for the top 10 shows led to the following rankings. Is there a relationship between the rankings

provided for the two groups? Use $\alpha = .10$.

Television Show	Ranking by Men	Ranking by Women
1	1	5
2	5	10
3	8	6
4	7	4
5	2	7
6	3	2
7	10	9
8	4	8
9	6	1
10	9	3

22. A student organization surveyed both recent graduates and current students in an attempt to obtain information on the quality of teaching at a particular university. An analysis of the responses provided the following rankings for 10 professors on the basis of teaching ability.

Professor	Ranking by Current Students	Ranking by Recent Graduates
1	4	6
2	6	8
3	8	5
4	3	1
5	1	2
6	2	3
7	5	7
8	10	9
9	7	4
10	9	10

Do the rankings given by the current students agree with the rankings given by the recent graduates? Use $\alpha = .10$ and test for a significant rank correlation.

Summary

In this chapter we discussed statistical methods that do not require assumptions about the population probability distribution. These methods are called nonparametric or distribution-free methods. There are two major advantages in working with nonparametric methods. First, as we have just noted, fewer assumptions are required; hence the methods can be applied to a wider range of problems. Second, many of the nonparametric methods require only rank-order or preference data.

The Wilcoxon rank-sum test provides a nonparametric procedure for identifying differences in two populations whenever two independent samples are selected. The Wilcoxon signed-rank test provides a similar methodology for cases when the samples involve matched pairs. The sign test provides a nonparametric procedure for identifying differences in two populations when the only data available are preferences between matched items. This test does not even require rank-order data. The Spearman rank-correlation coefficient provides a measure of association between two ranked sets of items.

Statistics in Practice

DETERMINING A NEW OFFICE LOCATION

West Shell Realtors is a real estate firm located in southwestern Ohio, southeastern Indiana, and northern Kentucky. In addition to monthly statistical summaries for ongoing operations, the company uses statistical considerations to guide company plans and strategies. One area in which statistical analysis has played an important part for West Shell involves the determination of new sales locations. That is, each time the company addresses the question of the best place to locate a new office, a variety of data must be analyzed to assist in evaluating and comparing alternative office location sites.

In one such instance the company had identified two areas as prime candidates for a new office location: Clifton and Roselawn. The statistical issues involved determining in what ways the two areas were alike and in what ways they differed. Although a variety of factors were considered in comparing the two areas, let us focus on one factor: the selling price of homes in the two areas.

The actual sales prices of units sold over a period of time was viewed as a sample of sales for the area. There were few data on the number of units sold in both areas, and a normal distribution assumption for selling prices was judged to be inappropriate. Thus nonparametric statistical methods were considered as ways to analyze the available data.

A sample of 25 home sales in the Clifton area showed a mean selling price of $55,250; a sample of 18 home sales in the Roselawn area showed a mean selling price of $50,375. The Wilcoxon rank-sum test was used to determine whether or not the selling prices in the two areas appeared to be identical or not. Using a .05 level of significance, the critical values for the Wilcoxon rank-sum test are 470 and 630. The value of W for the total sample of 43 recorded sales in the two areas is 595.2. In this case the nonparametric test leads to the acceptance of the hypothesis that the two populations are identical. The selection basis for the location of the new office should now focus on criteria other than unit selling price, since the areas are believed to be similar on this factor.

Glossary

Nonparametric methods—A collection of statistical methods that generally requires very few, if any, assumptions about the population distribution. These methods can be applied when only rank-order or preference data are available.

Distribution-free methods—Another name for nonparametric statistical methods suggested by the lack of assumptions required concerning the population distribution.

Wilcoxon rank-sum test—A nonparametric statistical test for identifying differences between two populations based on the analysis of two independent samples.

Mann-Whitney test—A nonparametric statistical test for identifying differences between two populations based on the analysis of two independent samples. It is equivalent to the Wilcoxon rank-sum test.

Wilcoxon signed-rank test—A nonparametric statistical test for identifying differences between two populations based on the analysis of two matched or paired samples.

Sign test—A nonparametric statistical test for identifying differences between two populations based on the analysis of two matched or paired samples. Preference data are all that is required.

Spearman rank-correlation coefficient—A correlation measure based on rank-order data for two variables. It provides a measure of monotonicity of the relationship between the two variables.

Key Formulas

Wilcoxon Rank-Sum Test

$$\mu_w = \frac{1}{2} n_1(n_1 + n_2 + 1) \tag{17.1}$$

$$\sigma_w = \sqrt{\frac{1}{12} n_1 n_2 (n_1 + n_2 + 1)} \tag{17.2}$$

Mann-Whitney Test

$$U = (\text{largest possible value of } W) - W$$

$$= \left(n_1 n_2 + \frac{n_1(n_1 + 1)}{2} \right) - W \tag{17.3}$$

$$\mu_U = \frac{1}{2} n_1 n_2 \tag{17.4}$$

$$\sigma_U = \sqrt{\frac{1}{12} n_1 n_2 (n_1 + n_2 + 1)} \tag{17.5}$$

Wilcoxon Signed-Rank Test

$$\mu_T = 0 \tag{17.6}$$

$$\sigma_T = \sqrt{\frac{n(n + 1)(2n + 1)}{6}} \tag{17.7}$$

Spearman Rank Correlation Coefficient

$$r_s = 1 - \frac{6\Sigma d_i^2}{n(n^2 - 1)} \tag{17.10}$$

Review Quiz

TRUE/FALSE

1. Nonparametric methods are often referred to as distribution-free methods.
2. Nonparametric statistical methods are not applicable to qualitative data.
3. With the Wilcoxon rank-sum test we must assume the populations sampled from are normal.
4. If ties occur in ranking the data from the two samples, the Wilcoxon rank-sum test cannot be used.
5. With the Wilcoxon rank-sum test, we need to know only the two sample sizes in order to compute the standard deviation of the rank sum.
6. With the Wilcoxon rank-sum test, we need to know only the two sample sizes in order to compute the expected value of the rank sum.
7. The Wilcoxon signed-rank test uses paired differences to test whether or not two populations are identical.
8. With the Wilcoxon signed-rank test, the expected value of the sum of signed ranks is equal to the sample size.
9. The sign test can be used to test whether or not individuals prefer one item over another.
10. The Spearman rank correlation coefficient can be computed only if quantitative data is available for the variables.

MULTIPLE CHOICE

11. A nonparametric test for the equivalence of two populations would be used instead of a parametric test for the equivalence of the population parameters if
    a. the samples are very large
    b. the samples are not independent
    c. no information about the populations is available
    d. the parametric test is always used in this situation
12. A variable assumes the following values.

$$10, 12, 15, 15, 16, 18$$

What rank is assigned to 15 in a rank-sum test?

a.   3.5 for both values of 15

b.   3 for one 15 and 4 for the other 15

c.   only one value of 15 should be used and given the rank 3

d.   no rank needed because ties should be omitted from the rankings

13.   For a Wilcoxon rank-sum test, the first and second samples have sizes 15 and 24, respectively. The expected value of W, is

a.   180

b.   300

c.   0

d.   not enough information given

14.   In a Wilcoxon signed-rank test, the two samples each have $n = 17$. The expected value of T is

a.   0

b.   145

c.   298

d.   300

15.   A municipal transit system has collected data on which of two seating configurations is preferred by the passengers on its buses. In a sign test to determine if one seating arrangement is significantly preferred, the null hypothesis would be

a. $H_0: \mu = 0$

b. $H_0: \mu = .5$

c. $H_0: p = 0$

d. $H_0: p = .5$

Supplementary Exercises

23.   Starting salaries were recorded for 10 recent graduates at each of two community colleges. Use $\alpha = .10$ and test for the differences in the starting salaries from the two colleges.

| **Eastern College** | | **Western College** | |
Student	Monthly Salary ($)	Student	Monthly Salary ($)
1	890	1	1000
2	950	2	1020
3	1200	3	1140
4	1150	4	1000
5	1300	5	975
6	1350	6	925
7	990	7	900
8	1050	8	1025
9	1400	9	1075
10	1450	10	930

24. Independent random samples of houses in two different neighborhoods of a large city have been collected and data on assessed valuation recorded.

Neighborhood 1	Neighborhood 2
18,000	16,500
16,000	20,500
12,000	23,000
20,000	17,500
19,000	22,000
17,000	21,000
16,500	21,500
19,000	19,500
15,500	17,000
16,000	23,500
17,500	21,000
18,000	22,000

Test at the $\alpha = .05$ level of significance whether or not assessed values are the same in each neighborhood.

25. The following data show product weights for items produced on two production lines. Test for a difference between the product weights for the two lines. Use $\alpha = .10$.

Production Line 1	Production Line 2
13.6	13.7
13.8	14.1
14.0	14.2
13.9	14.0
13.4	14.6
13.2	13.5
13.3	14.4
13.6	14.8
12.9	14.5
14.4	14.3
	15.0
	14.9

26. Twelve homemakers were asked to estimate the retail selling price of two models of refrigerators. The estimates of selling price provided by the homemakers are shown.

Homemaker	Model 1	Model 2
1	$650	$ 900
2	760	720
3	740	690
4	700	850
5	590	920
6	620	800
7	700	890
8	690	920
9	900	1000
10	500	690
11	610	700
12	720	700

Use these data and test at the .05 level of significance to determine if there is a difference in the homemaker's perception of selling price for the two models.

27. A study was designed to evaluate the weight-gain potential of a new poultry feed. A sample of 12 chickens was used in a 6-week study. The weight of each chicken was recorded before and after the 6-week test period. The difference between the before and after weights of each chicken are as follows: 1.5, 1.2, −0.2, 0.0, .5, .7, .8, 1.0, 0.0, .6, 0.2, −.01. A negative value indicates a weight loss during the test period, whereas 0.0 indicates no weight change over the period. Use a .05 level of significance to determine if the new feed appears to provide a weight gain for the chickens.

28. In a soft-drink taste test, 48 individuals stated a preference for one of two well-known brands. Results showed 28 favoring brand A, 16 favoring brand B, and 4 undecided. Use the sign test with $\alpha = .10$ and determine whether or not there is a significant difference in the preferences for the two brands of soft-drinks.

29. Use the sign test and perform the statistical analysis that will help us determine whether or not the task-completion times for two production methods differ. The data are as follows.

Worker	Method 1 (minutes)	Method 2 (minutes)
1	10.2	9.5
2	9.6	9.8
3	9.2	8.8
4	10.6	10.1
5	9.9	10.3
6	10.2	9.3
7	10.6	10.5
8	10.0	10.0
9	11.2	10.6
10	10.7	10.2
11	10.6	9.8

Use a plus if the difference is positive and a minus if the difference is negative. Test with $\alpha = .05$.

30. Mayfield Products, Inc. has collected data on preferences of 12 individuals concerning cleaning power of two brands of detergent. The individuals and their preferences are shown below. A plus indicates a preference for brand A.

Individual	Brand A Versus Brand B	Individual	Brand A Versus Brand B
1	−	7	−
2	+	8	+
3	+	9	+
4	+	10	−
5	−	11	+
6	+	12	+

With $\alpha = .10$, test for a significant difference in the preference for the two brands.

31. The following data are from a sample of 22 items: 14, 25, 28, 18, 26, 32, 41, 19, 26, 15, 20, 26, 27, 19, 22, 31, 38, 21, 19, 25, 26, and 30. Test the hypothesis that the population has a median of 20. Using a .10 level of significance, what is your conclusion?

32. A group of investment analysts ranked 12 companies, first with respect to book value and then with respect to growth potential.

Company	Ranking of Book Value	Ranking of Growth Potential
1	12	2
2	2	9
3	8	6
4	1	11
5	9	4
6	7	5
7	3	12
8	11	1
9	4	7
10	5	10
11	6	8
12	10	3

For these data does a relationship exist between the companies' book values and growth potentials? Use $\alpha = .05$.

33. Two individuals provided the following preference rankings of seven soft-drinks. For $\alpha = .05$, is there a significant rank correlation for the two individuals?

Soft-Drink	Ranking by Individual 1	Ranking by Individual 2
A	1	3
B	3	2
C	5	5
D	6	7
E	7	6
F	4	1
G	2	4

34. A sample of 15 students obtained the following rankings on midterm and final examinations in a statistics course.

Midterm Rank	Final Rank
1	4
2	7
3	1
4	3
5	8
6	2
7	5
8	12
9	6
10	9
11	14
12	15
13	11
14	10
15	13

Compute the Spearman rank-correlation coefficient for the data and test for a significant correlation with $\alpha = .10$.

Appendixes

A.
References and Bibliography

General

Freund, J. E., and R. E. Walpole, *Mathematical Statistics,* 3rd ed., Englewood Cliffs, N.J., Prentice-Hall, 1980.

Hogg, R. V., and A. T. Craig, *Introduction to Mathematical Statistics,* 4th ed., New York, Macmillan, 1978.

Mood, A. M., F. A. Graybill, and D. C. Boes, *Introduction to the Theory of Statistics,* 3rd ed., New York, McGraw-Hill, 1974.

Neter, J., W. Wasserman, and G. W. Whitmore, *Applied Statistics,* 2nd ed., Boston, Allyn & Bacon, 1982.

Exploratory Data Analysis

Hartwig, F., and B. E. Dearing, *Exploratory Data Analysis,* Beverly Hills, Sage Publications, 1979

Tukey, J. W., *Exploratory Data Analysis,* Menlo Park, California, Addison-Wesley, 1977

Vellerman, P. F., and D. C. Hoaglin, *Applications, Basics, and Computing of Exploratory Data Analysis,* Boston, Duxbury Press, 1981

Probability

Feller, W., *An Introduction to Probability Theory and Its Applications,* Vol. I, 3rd ed., New York, John Wiley & Sons, 1968.

Feller, W., *An Introduction to Probability Theory and Its Applications,* Vol. II, 2nd ed., New York, John Wiley & Sons, 1971.

Hoel, P. G., S. C. Port, and C. J. Stone, *Introduction to Probability Theory,* Boston, Houghton Mifflin, 1971.

Parzen, E., *Modern Probability Theory and Its Applications,* New York, John Wiley & Sons, Inc., 1960.

Zehna, P. W., *Probability Distributions and Statistics,* Boston, Allyn & Bacon, 1970.

Sampling Methods

Cochran, W. G., *Sampling Techniques,* 3rd ed., New York, John Wiley & Sons, 1977.

Kish, L., *Survey Sampling,* New York, John Wiley & Sons, 1965.

Scheaffer, R. L., W. Mendenhall, and L. Ott, *Elementary Survey Sampling,* 2nd ed., North Scituate, Mass., Duxbury Press, 1979.

Williams, B., *A Sampler on Sampling,* New York, John Wiley & Sons, 1978.

Analysis of Variance and Experimental Design

Anderson, V. L., and R. A. McLean, *Design of Experiments: A Realistic Approach,* New York, Marcel Dekker, 1974.

Cochran, W. G., and G. M. Cox, *Experimental Designs,* 2nd ed., New York, John Wiley & Sons, 1957.

Mendenhall, W., *Introduction to Linear Models and the Design and Analysis of Experiments,* Belmont, Calif., Duxbury Press, 1968.

Montgomery, D. C., *Design and Analysis of Experiments,* 2nd ed., New York, John Wiley & Sons, 1983

Regression Analysis

Draper, N. R., and H. Smith, *Applied Regression Analysis,* 2nd ed., New York, John Wiley & Sons, 1981.

Kleinbaum, D. G., and L. L. Kupper, *Applied Regression Analysis and Other Multivariable Methods,* North Scituate, Mass., Duxbury Press, 1978.

Mosteller, F., and J. W. Tukey, *Data Analysis and Regression: A Second Course in Statistics,* Reading, Mass., Addison-Wesley, 1977.

Neter, J., W. Wasserman and M. Kutner, *Applied Linear Statistical Models,* 2nd. ed. Homewood, Ill., Richard D. Irwin, 1985.

Neter, J., W. Wasserman, and M. Kutner, *Applied Linear Regression Models,* Homewood, Ill., Richard D. Irwin, 1983

Nonparametric Methods

Conover, W. J., *Practical Nonparametric Statistics,* 2nd ed., New York, John Wiley & Sons, 1980.

Gibbons, J. D., I. Olkin, and M. Sobel, *Selecting and Ordering Populations: A New Statistical Methodology,* New York, John Wiley & Sons, 1977.

Mosteller, F., and R. E. K. Rourke, *Sturdy Statistics,* Reading, Mass., Addison-Wesley, 1973.

Siegel, S., *Nonparametric Statistics for the Behavioral Sciences,* New York, McGraw-Hill, 1956.

B.
Tables

TABLE 1

Standard Normal Distribution

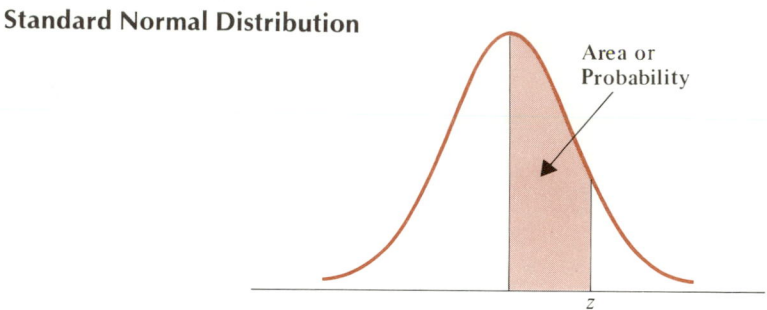

Area or Probability

Entries in the table give the area under the curve between the mean and z standard deviations above the mean. For example, for $z = 1.25$ the area under the curve between the mean and z is .3944.

z	.00	.01	.02	.03	.04	.05	.06	.07	.08	.09
.0	.0000	.0040	.0080	.0120	.0160	.0199	.0239	.0279	.0319	.0359
.1	.0398	.0438	.0478	.0517	.0557	.0596	.0636	.0675	.0714	.0753
.2	.0793	.0832	.0871	.0910	.0948	.0987	.1026	.1064	.1103	.1141
.3	.1179	.1217	.1255	.1293	.1331	.1368	.1406	.1443	.1480	.1517
.4	.1554	.1591	.1628	.1664	.1700	.1736	.1772	.1808	.1844	.1879
.5	.1915	.1950	.1985	.2019	.2054	.2088	.2123	.2157	.2190	.2224
.6	.2257	.2291	.2324	.2357	.2389	.2422	.2454	.2486	.2518	.2549
.7	.2580	.2612	.2642	.2673	.2704	.2734	.2764	.2794	.2823	.2852
.8	.2881	.2910	.2939	.2967	.2995	.3023	.3051	.3078	.3106	.3133
.9	.3159	.3186	.3212	.3238	.3264	.3289	.3315	.3340	.3365	.3389
1.0	.3413	.3438	.3461	.3485	.3508	.3531	.3554	.3577	.3599	.3621
1.1	.3643	.3665	.3686	.3708	.3729	.3749	.3770	.3790	.3810	.3830
1.2	.3849	.3869	.3888	.3907	.3925	.3944	.3962	.3980	.3997	.4015
1.3	.4032	.4049	.4066	.4082	.4099	.4115	.4131	.4147	.4162	.4177
1.4	.4192	.4207	.4222	.4236	.4251	.4265	.4279	.4292	.4306	.4319
1.5	.4332	.4345	.4357	.4370	.4382	.4394	.4406	.4418	.4429	.4441
1.6	.4452	.4463	.4474	.4484	.4495	.4505	.4515	.4525	.4535	.4545
1.7	.4554	.4564	.4573	.4582	.4591	.4599	.4608	.4616	.4625	.4633
1.8	.4641	.4649	.4656	.4664	.4671	.4678	.4686	.4693	.4699	.4706
1.9	.4713	.4719	.4726	.4732	.4738	.4744	.4750	.4756	.4761	.4767
2.0	.4772	.4778	.4783	.4788	.4793	.4798	.4803	.4808	.4812	.4817
2.1	.4821	.4826	.4830	.4834	.4838	.4842	.4846	.4850	.4854	.4857
2.2	.4861	.4864	.4868	.4871	.4875	.4878	.4881	.4884	.4887	.4890
2.3	.4893	.4896	.4898	.4901	.4904	.4906	.4909	.4911	.4913	.4916
2.4	.4918	.4920	.4922	.4925	.4927	.4929	.4931	.4932	.4934	.4936
2.5	.4938	.4940	.4941	.4943	.4945	.4946	.4948	.4949	.4951	.4952
2.6	.4953	.4955	.4956	.4957	.4959	.4960	.4961	.4962	.4963	.4964
2.7	.4965	.4966	.4967	.4968	.4969	.4970	.4971	.4972	.4973	.4974
2.8	.4974	.4975	.4976	.4977	.4977	.4978	.4979	.4979	.4980	.4981
2.9	.4981	.4982	.4982	.4983	.4984	.4984	.4985	.4985	.4986	.4986
3.0	.4986	.4987	.4987	.4988	.4988	.4989	.4989	.4989	.4990	.4990

TABLE 2

t Distribution

Entries in the table give t_α values, where α is the area or probability in the upper tail of the t distribution. For example, with 10 degrees of freedom and a .05 area in the upper tail, $t_{.05} = 1.812$.

Degrees of Freedom	Area in Upper Tail				
	.10	.05	.025	.01	.005
1	3.078	6.314	12.706	31.821	63.657
2	1.886	2.920	4.303	6.965	9.925
3	1.638	2.353	3.182	4.541	5.841
4	1.533	2.132	2.776	3.747	4.604
5	1.476	2.015	2.571	3.365	4.032
6	1.440	1.943	2.447	3.143	3.707
7	1.415	1.895	2.365	2.998	3.499
8	1.397	1.860	2.306	2.896	3.355
9	1.383	1.833	2.262	2.821	3.250
10	1.372	1.812	2.228	2.764	3.169
11	1.363	1.796	2.201	2.718	3.106
12	1.356	1.782	2.179	2.681	3.055
13	1.350	1.771	2.160	2.650	3.012
14	1.345	1.761	2.145	2.624	2.977
15	1.341	1.753	2.131	2.602	2.947
16	1.337	1.746	2.120	2.583	2.921
17	1.333	1.740	2.110	2.567	2.898
18	1.330	1.734	2.101	2.552	2.878
19	1.328	1.729	2.093	2.539	2.861
20	1.325	1.725	2.086	2.528	2.845
21	1.323	1.721	2.080	2.518	2.831
22	1.321	1.717	2.074	2.508	2.819
23	1.319	1.714	2.069	2.500	2.807
24	1.318	1.711	2.064	2.492	2.797
25	1.316	1.708	2.060	2.485	2.787
26	1.315	1.706	2.056	2.479	2.779
27	1.314	1.703	2.052	2.473	2.771
28	1.313	1.701	2.048	2.467	2.763
29	1.311	1.699	2.045	2.462	2.756
30	1.310	1.697	2.042	2.457	2.750
40	1.303	1.684	2.021	2.423	2.704
60	1.296	1.671	2.000	2.390	2.660
120	1.289	1.658	1.980	2.358	2.617
α	1.282	1.645	1.960	2.326	2.576

TABLE 3

Chi-Square Distribution

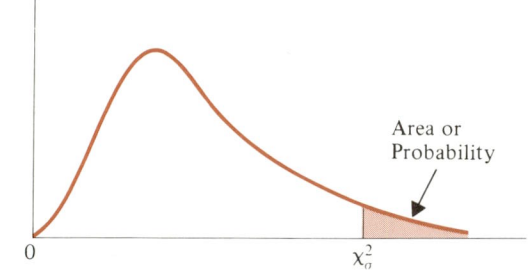

Area or
Probability

0 χ^2_α

Entries in the table give χ^2_α values, where α is the area or probability in the upper tail of the chi-square distribution. For example, with 10 degrees of freedom and a .01 area in the upper tail, $\chi^2_{.01} = 23.2093$.

Degrees of Freedom	Area in Upper Tail									
	.995	.99	.975	.95	.90	.10	.05	.025	.01	.005
1	$392{,}704 \times 10^{-10}$	$157{,}088 \times 10^{-9}$	$982{,}069 \times 10^{-9}$	$393{,}214 \times 10^{-8}$	.0157908	2.70554	3.84146	5.02389	6.63490	7.87944
2	.0100251	.0201007	.0506356	.1025878	.210720	4.60517	5.99147	7.37776	9.21034	10.5966
3	.0717212	.114832	.215795	.351846	.584375	6.25139	7.81473	9.34840	11.3449	12.8381
4	.206990	.297110	.484419	.710721	1.063623	7.77944	9.48773	11.1433	13.2767	14.8602
5	.411740	.554300	.831211	1.145476	1.61031	9.23635	11.0705	12.8325	15.0863	16.7496
6	.675727	.872085	1.237347	1.63539	2.20413	10.6446	12.5916	14.4494	16.8119	18.5476
7	.989265	1.239043	1.68987	2.16735	2.83311	12.0170	14.0671	16.0128	18.4753	20.2777
8	1.344419	1.646482	2.17973	2.73264	3.48954	13.3616	15.5073	17.5346	20.0902	21.9550
9	1.734926	2.087912	2.70039	3.32511	4.16816	14.6837	16.9190	19.0228	21.6660	23.5893
10	2.15585	2.55821	3.24697	3.94030	4.86518	15.9871	18.3070	20.4831	23.2093	25.1882
11	2.60321	3.05347	3.81575	4.57481	5.57779	17.2750	19.6751	21.9200	24.7250	26.7569
12	3.07382	3.57056	4.40379	5.22603	6.30380	18.5494	21.0261	23.3367	26.2170	28.2995
13	3.56503	4.10691	5.00874	5.89186	7.04150	19.8119	22.3621	24.7356	27.6883	29.8194
14	4.07468	4.66043	5.62872	6.57063	7.78953	21.0642	23.6848	26.1190	29.1413	31.3193
15	4.60094	5.22935	6.26214	7.26094	8.54675	22.3072	24.9958	27.4884	30.5779	32.8013
16	5.14224	5.81221	6.90766	7.96164	9.31223	23.5418	26.2962	28.8454	31.9999	34.2672
17	5.69724	6.40776	7.56418	8.67176	10.0852	24.7690	27.5871	30.1910	33.4087	35.7185
18	6.26481	7.01491	8.23075	9.39046	10.8649	25.9894	28.8693	31.5264	34.8053	37.1564
19	6.84398	7.63273	8.90655	10.1170	11.6509	27.2036	30.1435	32.8523	36.1908	38.5822

20	7.43386	8.26040	9.59083	10.8508	12.4426	28.4120	31.4104	34.1696	37.5662	39.9968
21	8.03366	8.89720	10.28293	11.5913	13.2396	29.6151	32.6705	35.4789	38.9321	41.4010
22	8.64272	9.54249	10.9823	12.3380	14.0415	30.8133	33.9244	36.7807	40.2894	42.7958
23	9.26042	10.19567	11.6885	13.0905	14.8479	32.0069	35.1725	38.0757	41.6384	44.1813
24	9.88623	10.8564	12.4011	13.8484	15.6587	33.1963	36.4151	39.3641	42.9798	45.5585
25	10.5197	11.5240	13.1197	14.6114	16.4734	34.3816	37.6525	40.6465	44.3141	46.9278
26	11.1603	12.1981	13.8439	15.3791	17.2919	35.5631	38.8852	41.9232	45.6417	48.2899
27	11.8076	12.8786	14.5733	16.1513	18.1138	36.7412	40.1133	43.1944	46.9630	49.6449
28	12.4613	13.5648	15.3079	16.9279	18.9392	37.9159	41.3372	44.4607	48.2782	50.9933
29	13.1211	14.2565	16.0471	17.7083	19.7677	39.0875	42.5569	45.7222	49.5879	52.3356
30	13.7867	14.9535	16.7908	18.4926	20.5992	40.2560	43.7729	46.9792	50.8922	53.6720
40	20.7065	22.1643	24.4331	26.5093	29.0505	51.8050	55.7585	59.3417	63.6907	66.7659
50	27.9907	29.7067	32.3574	34.7642	37.6886	63.1671	67.5048	71.4202	76.1539	79.4900
60	35.5346	37.4848	40.4817	43.1879	46.4589	74.3970	79.0819	83.2976	88.3794	91.9517
70	43.2752	45.4418	48.7576	51.7393	55.3290	85.5271	90.5312	95.0231	100.425	104.215
80	51.1720	53.5400	57.1532	60.3915	64.2778	96.5782	101.879	106.629	112.329	116.321
90	59.1963	61.7541	65.6466	69.1260	73.2912	107.565	113.145	118.136	124.116	128.299
100	67.3276	70.0648	74.2219	77.9295	82.3581	118.498	124.342	129.561	135.807	140.169

TABLE 4

F Distribution

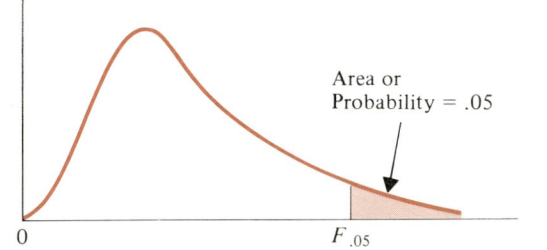

Entries in the table give F_α values, where α is the area or probability in the upper tail of the F distribution. For example, with 12 numerator degrees of freedom, 15 denominator degrees of freedom, and a .05 area in the upper tail, $F_{.05} = 2.48$.

Table of $F_{.05}$ Values

Denominator Degrees of Freedom	Numerator Degrees of Freedom																			
	1	2	3	4	5	6	7	8	9	10	12	15	20	24	30	40	60	120	∞
1	161.4	199.5	215.7	224.6	230.2	234.0	236.8	238.9	240.5	241.9	243.9	245.9	248.0	249.1	250.1	251.1	252.2	253.3	254.3
2	18.51	19.00	19.16	19.25	19.30	19.33	19.35	19.37	19.38	19.40	19.41	19.43	19.45	19.45	19.46	19.47	19.48	19.49	19.50
3	10.13	9.55	9.28	9.12	9.01	8.94	8.89	8.85	8.81	8.79	8.74	8.70	8.66	8.64	8.62	8.59	8.57	8.55	8.53
4	7.71	6.94	6.59	6.39	6.26	6.16	6.09	6.04	6.00	5.96	5.91	5.86	5.80	5.77	5.75	5.72	5.69	5.66	5.63
5	6.61	5.79	5.41	5.19	5.05	4.95	4.88	4.82	4.77	4.74	4.68	4.62	4.56	4.53	4.50	4.46	4.43	4.40	4.36
6	5.99	5.14	4.76	4.53	4.39	4.28	4.21	4.15	4.10	4.06	4.00	3.94	3.87	3.84	3.81	3.77	3.74	3.70	3.67
7	5.59	4.74	4.35	4.12	3.97	3.87	3.79	3.73	3.68	3.64	3.57	3.51	3.44	3.41	3.38	3.34	3.30	3.27	3.23
8	5.32	4.46	4.07	3.84	3.69	3.58	3.50	3.44	3.39	3.35	3.28	3.22	3.15	3.12	3.08	3.04	3.01	2.97	2.93
9	5.12	4.26	3.86	3.63	3.48	3.37	3.29	3.23	3.18	3.14	3.07	3.01	2.94	2.90	2.86	2.83	2.79	2.75	2.71
10	4.96	4.10	3.71	3.48	3.33	3.22	3.14	3.07	3.02	2.98	2.91	2.85	2.77	2.74	2.70	2.66	2.62	2.58	2.54
11	4.84	3.98	3.59	3.36	3.20	3.09	3.01	2.95	2.90	2.85	2.79	2.72	2.65	2.61	2.57	2.53	2.49	2.45	2.40
12	4.75	3.89	3.49	3.26	3.11	3.00	2.91	2.85	2.80	2.75	2.69	2.62	2.54	2.51	2.47	2.43	2.38	2.34	2.30
13	4.67	3.81	3.41	3.18	3.03	2.92	2.83	2.77	2.71	2.67	2.60	2.53	2.46	2.42	2.38	2.34	2.30	2.25	2.21
14	4.60	3.74	3.34	3.11	2.96	2.85	2.76	2.70	2.65	2.60	2.53	2.46	2.39	2.35	2.31	2.27	2.22	2.18	2.13

15	4.54	3.68	3.29	3.06	2.90	2.79	2.71	2.64	2.59	2.54	2.48	2.40	2.33	2.29	2.25	2.20	2.16	2.11	2.07
16	4.49	3.63	3.24	3.01	2.85	2.74	2.66	2.59	2.54	2.49	2.42	2.35	2.28	2.24	2.19	2.15	2.11	2.06	2.01
17	4.45	3.59	3.20	2.96	2.81	2.70	2.61	2.55	2.49	2.45	2.38	2.31	2.23	2.19	2.15	2.10	2.06	2.01	1.96
18	4.41	3.55	3.16	2.93	2.77	2.66	2.58	2.51	2.46	2.41	2.34	2.27	2.19	2.15	2.11	2.06	2.02	1.97	1.92
19	4.38	3.52	3.13	2.90	2.74	2.63	2.54	2.48	2.42	2.38	2.31	2.23	2.16	2.11	2.07	2.03	1.98	1.93	1.88
20	4.35	3.49	3.10	2.87	2.71	2.60	2.51	2.45	2.39	2.35	2.28	2.20	2.12	2.08	2.04	1.99	1.95	1.90	1.84
21	4.32	3.47	3.07	2.84	2.68	2.57	2.49	2.42	2.37	2.32	2.25	2.18	2.10	2.05	2.01	1.96	1.92	1.87	1.81
22	4.30	3.44	3.05	2.82	2.66	2.55	2.46	2.40	2.34	2.30	2.23	2.15	2.07	2.03	1.98	1.94	1.89	1.84	1.78
23	4.28	3.42	3.03	2.80	2.64	2.53	2.44	2.37	2.32	2.27	2.20	2.13	2.05	2.01	1.96	1.91	1.86	1.81	1.76
24	4.26	3.40	3.01	2.78	2.62	2.51	2.42	2.36	2.30	2.25	2.18	2.11	2.03	1.98	1.94	1.89	1.84	1.79	1.73
25	4.24	3.39	2.99	2.76	2.60	2.49	2.40	2.34	2.28	2.24	2.16	2.09	2.01	1.96	1.92	1.87	1.82	1.77	1.71
26	4.23	3.37	2.98	2.74	2.59	2.47	2.39	2.32	2.27	2.22	2.15	2.07	1.99	1.95	1.90	1.85	1.80	1.75	1.69
27	4.21	3.35	2.96	2.73	2.57	2.46	2.37	2.31	2.25	2.20	2.13	2.06	1.97	1.93	1.88	1.84	1.79	1.73	1.67
28	4.20	3.34	2.95	2.71	2.56	2.45	2.36	2.29	2.24	2.19	2.12	2.04	1.96	1.91	1.87	1.82	1.77	1.71	1.65
29	4.18	3.33	2.93	2.70	2.55	2.43	2.35	2.28	2.22	2.18	2.10	2.03	1.94	1.90	1.85	1.81	1.75	1.70	1.64
30	4.17	3.32	2.92	2.69	2.53	2.42	2.33	2.27	2.21	2.16	2.09	2.01	1.93	1.89	1.84	1.79	1.74	1.68	1.62
40	4.08	3.23	2.84	2.61	2.45	2.34	2.25	2.18	2.12	2.08	2.00	1.92	1.84	1.79	1.74	1.69	1.64	1.58	1.51
60	4.00	3.15	2.76	2.53	2.37	2.25	2.17	2.10	2.04	1.99	1.92	1.84	1.75	1.70	1.65	1.59	1.53	1.47	1.39
120	3.92	3.07	2.68	2.45	2.29	2.17	2.09	2.02	1.96	1.91	1.83	1.75	1.66	1.61	1.55	1.50	1.43	1.35	1.25
∞	3.84	3.00	2.60	2.37	2.21	2.10	2.01	1.94	1.88	1.83	1.75	1.67	1.57	1.52	1.46	1.39	1.32	1.22	1.00

Table of $F_{.025}$ Values

Denominator Degrees of Freedom	Numerator Degrees of Freedom																			
	1	2	3	4	5	6	7	8	9	10	12	15	20	24	30	40	60	120	∞
1	647.8	799.5	864.2	899.6	921.8	937.1	948.2	956.7	963.3	968.6	976.7	984.9	993.1	997.2	1,001	1,006	1,010	1,014	1,018
2	38.51	39.00	39.17	39.25	39.30	39.33	39.36	39.37	39.39	39.40	39.41	39.43	39.45	39.46	39.46	39.47	39.48	39.49	39.50
3	17.44	16.04	15.44	15.10	14.88	14.73	14.62	14.54	14.47	14.42	14.34	14.25	14.17	14.12	14.08	14.04	13.99	13.95	13.90
4	12.22	10.65	9.98	9.60	9.36	9.20	9.07	8.98	8.90	8.84	8.75	8.66	8.56	8.51	8.46	8.41	8.36	8.31	8.26
5	10.01	8.43	7.76	7.39	7.15	6.98	6.85	6.76	6.68	6.62	6.52	6.43	6.33	6.28	6.23	6.18	6.12	6.07	6.02
6	8.81	7.26	6.60	6.23	5.99	5.82	5.70	5.60	5.52	5.46	5.37	5.27	5.17	5.12	5.07	5.01	4.96	4.90	4.85
7	8.07	6.54	5.89	5.52	5.29	5.21	4.99	4.90	4.82	4.76	4.67	4.57	4.47	4.42	4.36	4.31	4.25	4.20	4.14
8	7.57	6.06	5.42	5.05	4.82	4.65	4.53	4.43	4.36	4.30	4.20	4.10	4.00	3.95	3.89	3.84	3.78	3.73	3.67
9	7.21	5.71	5.08	4.72	4.48	4.32	4.20	4.10	4.03	3.96	3.87	3.77	3.67	3.61	3.56	3.51	3.45	3.39	3.33
10	6.94	5.46	4.83	4.47	4.24	4.07	3.95	3.85	3.78	3.72	3.62	3.52	3.42	3.37	3.31	3.26	3.20	3.14	3.08
11	6.72	5.26	4.63	4.28	4.04	3.88	3.76	3.66	3.59	3.53	3.43	3.33	3.23	3.17	3.12	3.06	3.00	2.94	2.88
12	6.55	5.10	4.47	4.12	3.89	3.73	3.61	3.51	3.44	3.37	3.28	3.18	3.07	3.02	2.96	2.91	2.85	2.79	2.72
13	6.41	4.97	4.35	4.00	3.77	3.60	3.48	3.39	3.31	3.25	3.15	3.05	2.95	2.89	2.84	2.78	2.72	2.66	2.60
14	6.30	4.86	4.24	3.89	3.66	3.50	3.38	3.29	3.21	3.15	3.05	2.95	2.84	2.79	2.73	2.67	2.61	2.55	2.49
15	6.20	4.77	4.15	3.80	3.58	3.41	3.29	3.20	3.12	3.06	2.96	2.86	2.76	2.70	2.64	2.59	2.52	2.46	2.40
16	6.12	4.69	4.08	3.73	3.50	3.34	3.22	3.12	3.05	2.99	2.89	2.79	2.68	2.63	2.57	2.51	2.45	2.38	2.32
17	6.04	4.62	4.01	3.66	3.44	3.28	3.16	3.06	2.98	2.92	2.82	2.72	2.62	2.56	2.50	2.44	2.38	2.32	2.25
18	5.98	4.56	3.95	3.61	3.38	3.22	3.10	3.01	2.93	2.87	2.77	2.67	2.56	2.50	2.44	2.38	2.32	2.26	2.19
19	5.92	4.51	3.90	3.56	3.33	3.17	3.05	2.96	2.88	2.82	2.72	2.62	2.51	2.45	2.39	2.33	2.27	2.20	2.13
20	5.87	4.46	3.86	3.51	3.29	3.13	3.01	2.91	2.84	2.77	2.68	2.57	2.46	2.41	2.35	2.29	2.22	2.16	2.09
21	5.83	4.42	3.82	3.48	3.25	3.09	2.97	2.87	2.80	2.73	2.64	2.53	2.42	2.37	2.31	2.25	2.18	2.11	2.04
22	5.79	4.38	3.78	3.44	3.22	3.05	2.93	2.84	2.76	2.70	2.60	2.50	2.39	2.33	2.27	2.21	2.14	2.08	2.00
23	5.75	4.35	3.75	3.41	3.18	3.02	2.90	2.81	2.73	2.67	2.57	2.47	2.36	2.30	2.24	2.18	2.11	2.04	1.97
24	5.72	4.32	3.72	3.38	3.15	2.99	2.87	2.78	2.70	2.64	2.54	2.44	2.33	2.27	2.21	2.15	2.08	2.01	1.94
25	5.69	4.29	3.69	3.35	3.13	2.97	2.85	2.75	2.68	2.61	2.51	2.41	2.30	2.24	2.18	2.12	2.05	1.98	1.91
26	5.66	4.27	3.67	3.33	3.10	2.94	2.82	2.73	2.65	2.59	2.49	2.39	2.28	2.22	2.16	2.09	2.03	1.95	1.88
27	5.63	4.24	3.65	3.31	3.08	2.92	2.80	2.71	2.63	2.57	2.47	2.36	2.25	2.19	2.13	2.07	2.00	1.93	1.85
28	5.61	4.22	3.63	3.29	3.06	2.90	2.78	2.69	2.61	2.55	2.45	2.34	2.23	2.17	2.11	2.05	1.98	1.91	1.83
29	5.59	4.20	3.61	3.27	3.04	2.88	2.76	2.67	2.59	2.53	2.43	2.32	2.21	2.15	2.09	2.03	1.96	1.89	1.81
30	5.57	4.18	3.59	3.25	3.03	2.87	2.75	2.65	2.57	2.51	2.41	2.31	2.20	2.14	2.07	2.01	1.94	1.87	1.79
40	5.42	4.05	3.46	3.13	2.90	2.74	2.62	2.53	2.45	2.39	2.29	2.18	2.07	2.01	1.94	1.88	1.80	1.72	1.64
60	5.29	3.93	3.34	3.01	2.79	2.63	2.51	2.41	2.33	2.27	2.17	2.06	1.94	1.88	1.82	1.74	1.67	1.58	1.48
120	5.15	3.80	3.23	2.89	2.67	2.52	2.39	2.30	2.22	2.16	2.05	1.94	1.82	1.76	1.69	1.61	1.53	1.43	1.31
∞	5.02	3.69	3.12	2.79	2.57	2.41	2.29	2.19	2.11	2.05	1.94	1.83	1.71	1.64	1.57	1.48	1.39	1.27	1.00

Table of $F_{.01}$ Values

Denominator Degrees of Freedom	Numerator Degrees of Freedom																			
	1	2	3	4	5	6	7	8	9	10	12	15	20	24	30	40	60	120	∞
1	4,052	4,999.5	5,403	5,625	5,764	5,859	5,928	5,982	6,022	6,056	6,106	6,157	6,209	6,235	6,261	6,287	6,313	6,339	6,366
2	98.50	99.00	99.17	99.25	99.30	99.33	99.36	99.37	99.39	99.40	99.42	99.43	99.45	99.46	99.47	99.47	99.48	99.49	99.50
3	34.12	30.82	29.46	28.71	28.24	27.91	27.67	27.49	27.35	27.23	27.05	26.87	26.69	26.60	26.50	26.41	26.32	26.22	26.13
4	21.20	18.00	16.69	15.98	15.52	15.21	14.98	14.80	14.66	14.55	14.37	14.20	14.02	13.93	13.84	13.75	13.65	13.56	13.46
5	16.26	13.27	12.06	11.39	10.97	10.67	10.46	10.29	10.16	10.05	9.89	9.72	9.55	9.47	9.38	9.29	9.20	9.11	9.06
6	13.75	10.92	9.78	9.15	8.75	8.47	8.26	8.10	7.98	7.87	7.72	7.56	7.40	7.31	7.23	7.14	7.06	6.97	6.88
7	12.25	9.55	8.45	7.85	7.46	7.19	6.99	6.84	6.72	6.62	6.47	6.31	6.16	6.07	5.99	5.91	5.82	5.74	5.65
8	11.26	8.65	7.59	7.01	6.63	6.37	6.18	6.03	5.91	5.81	5.67	5.52	5.36	5.28	5.20	5.12	5.03	4.95	4.86
9	10.56	8.02	6.99	6.42	6.06	5.80	5.61	5.47	5.35	5.26	5.11	4.96	4.81	4.73	4.65	4.57	4.48	4.40	4.31
10	10.04	7.56	6.55	5.99	5.64	5.39	5.20	5.06	4.94	4.85	4.71	4.56	4.41	4.33	4.25	4.17	4.08	4.00	3.91
11	9.65	7.21	6.22	5.67	5.32	5.07	4.89	4.74	4.63	4.54	4.40	4.25	4.10	4.02	3.94	3.86	3.78	3.69	3.60
12	9.33	6.93	5.95	5.41	5.06	4.82	4.64	4.50	4.39	4.30	4.16	4.01	3.86	3.78	3.70	3.62	3.54	3.45	3.36
13	9.07	6.70	5.74	5.21	4.86	4.62	4.44	4.30	4.19	4.10	3.96	3.82	3.66	3.59	3.51	3.43	3.34	3.25	3.17
14	8.86	6.51	5.56	5.04	4.69	4.46	4.28	4.14	4.03	3.94	3.80	3.66	3.51	3.43	3.35	3.27	3.18	3.09	3.00
15	8.68	6.36	5.42	4.89	4.56	4.32	4.14	4.00	3.89	3.80	3.67	3.52	3.37	3.29	3.21	3.13	3.05	2.96	2.87
16	8.53	6.23	5.29	4.77	4.44	4.20	4.03	3.89	3.78	3.69	3.55	3.41	3.26	3.18	3.10	3.02	2.93	2.84	2.75
17	8.40	6.11	5.18	4.67	4.34	4.10	3.93	3.79	3.68	3.59	3.46	3.31	3.16	3.08	3.00	2.92	2.83	2.75	2.65
18	8.29	6.01	5.09	4.58	4.25	4.01	3.84	3.71	3.60	3.51	3.37	3.23	3.08	3.00	2.92	2.84	2.75	2.66	2.57
19	8.18	5.93	5.01	4.50	4.17	3.94	3.77	3.63	3.52	3.43	3.30	3.15	3.00	2.92	2.84	2.76	2.67	2.58	2.49
20	8.10	5.85	4.94	4.43	4.10	3.87	3.70	3.56	3.46	3.37	3.23	3.09	2.94	2.86	2.78	2.69	2.61	2.52	2.42
21	8.02	5.78	4.87	4.37	4.04	3.81	3.64	3.51	3.40	3.31	3.17	3.03	2.88	2.80	2.72	2.64	2.55	2.46	2.36
22	7.95	5.72	4.82	4.31	3.99	3.76	3.59	3.45	3.35	3.26	3.12	2.98	2.83	2.75	2.67	2.58	2.50	2.40	2.31
23	7.88	5.66	4.76	4.26	3.94	3.71	3.54	3.41	3.30	3.21	3.07	2.93	2.78	2.70	2.62	2.54	2.45	2.35	2.26
24	7.82	5.61	4.72	4.22	3.90	3.67	3.50	3.36	3.26	3.17	3.03	2.89	2.74	2.66	2.58	2.49	2.40	2.31	2.21
25	7.77	5.57	4.68	4.18	3.85	3.63	3.46	3.32	3.22	3.13	2.99	2.85	2.70	2.62	2.54	2.45	2.36	2.27	2.17
26	7.72	5.53	4.64	4.14	3.82	3.59	3.42	3.29	3.18	3.09	2.96	2.81	2.66	2.58	2.50	2.42	2.33	2.23	2.13
27	7.68	5.49	4.60	4.11	3.78	3.56	3.39	3.26	3.15	3.06	2.93	2.78	2.63	2.55	2.47	2.38	2.29	2.20	2.10
28	7.64	5.45	4.57	4.07	3.75	3.53	3.36	3.23	3.12	3.03	2.90	2.75	2.60	2.52	2.44	2.35	2.26	2.17	2.06
29	7.60	5.42	4.54	4.04	3.73	3.50	3.33	3.20	3.09	3.00	2.87	2.73	2.57	2.49	2.41	2.33	2.23	2.14	2.03
30	7.56	5.39	4.51	4.02	3.70	3.47	3.30	3.17	3.07	2.98	2.84	2.70	2.55	2.47	2.39	2.30	2.21	2.11	2.01
40	7.31	5.18	4.31	3.83	3.51	3.29	3.12	2.99	2.89	2.80	2.66	2.52	2.37	2.29	2.20	2.11	2.02	1.92	1.80
60	7.08	4.98	4.13	3.65	3.34	3.12	2.95	2.82	2.72	2.63	2.50	2.35	2.20	2.12	2.03	1.94	1.84	1.73	1.60
120	6.85	4.79	3.95	3.48	3.17	2.96	2.79	2.66	2.56	2.47	2.34	2.19	2.03	1.95	1.86	1.76	1.66	1.53	1.38
∞	6.63	4.61	3.78	3.32	3.02	2.80	2.64	2.51	2.41	2.32	2.18	2.04	1.88	1.79	1.70	1.59	1.47	1.32	1.00

TABLE 5

Binomial Probabilities

Entries in the table give the probability of x successes in n trials of a binomial experiment, where p is the probability of a success on one trial. For example, with six trials and $p = .40$, the probability of two successes is .3110.

n	x	.05	.10	.15	.20	.25	.30	.35	.40	.45	.50
1	0	.9500	.9000	.8500	.8000	.7500	.7000	.6500	.6000	.5500	.5000
	1	.0500	.1000	.1500	.2000	.2500	.3000	.3500	.4000	.4500	.5000
2	0	.9025	.8100	.7225	.6400	.5625	.4900	.4225	.3600	.3025	.2500
	1	.0950	.1800	.2550	.3200	.3750	.4200	.4550	.4800	.4950	.5000
	2	.0025	.0100	.0225	.0400	.0625	.0900	.1225	.1600	.2025	.2500
3	0	.8574	.7290	.6141	.5120	.4219	.3430	.2746	.2160	.1664	.1250
	1	.1354	.2430	.3251	.3840	.4219	.4410	.4436	.4320	.4084	.3750
	2	.0071	.0270	.0574	.0960	.1406	.1890	.2389	.2880	.3341	.3750
	3	.0001	.0010	.0034	.0080	.0156	.0270	.0429	.0640	.0911	.1250
4	0	.8145	.6561	.5220	.4096	.3164	.2401	.1785	.1296	.0915	.0625
	1	.1715	.2916	.3685	.4096	.4219	.4116	.3845	.3456	.2995	.2500
	2	.0135	.0486	.0975	.1536	.2109	.2646	.3105	.3456	.3675	.3750
	3	.0005	.0036	.0115	.0256	.0469	.0756	.1115	.1536	.2005	.2500
	4	.0000	.0001	.0005	.0016	.0039	.0081	.0150	.0256	.0410	.0625
5	0	.7738	.5905	.4437	.3277	.2373	.1681	.1160	.0778	.0503	.0312
	1	.2036	.3280	.3915	.4096	.3955	.3602	.3124	.2592	.2059	.1562
	2	.0214	.0729	.1382	.2048	.2637	.3087	.3364	.3456	.3369	.3125
	3	.0011	.0081	.0244	.0512	.0879	.1323	.1811	.2304	.2757	.3125
	4	.0000	.0004	.0022	.0064	.0146	.0284	.0488	.0768	.1128	.1562
	5	.0000	.0000	.0001	.0003	.0010	.0024	.0053	.0102	.0185	.0312
6	0	.7351	.5314	.3771	.2621	.1780	.1176	.0754	.0467	.0277	.0156
	1	.2321	.3543	.3993	.3932	.3560	.3025	.2437	.1866	.1359	.0938
	2	.0305	.0984	.1762	.2458	.2966	.3241	.3280	.3110	.2780	.2344
	3	.0021	.0146	.0415	.0819	.1318	.1852	.2355	.2765	.3032	.3125
	4	.0001	.0012	.0055	.0154	.0330	.0595	.0951	.1382	.1861	.2344
	5	.0000	.0001	.0004	.0015	.0044	.0102	.0205	0.369	.0609	.0938
	6	.0000	.0000	.0000	.0001	.0002	.0007	.0018	.0041	.0083	.0156
7	0	.6983	.4783	.3206	.2097	.1335	.0824	.0490	.0280	.0152	.0078
	1	.2573	.3720	.3960	.3670	.3115	.2471	.1848	.1306	.0872	.0547
	2	.0406	.1240	.2097	.2753	.3115	.3177	.2985	.2613	.2140	.1641
	3	.0036	.0230	.0617	.1147	.1730	.2269	.2679	.2903	.2918	.2734
	4	.0002	.0026	.0109	.0287	.0577	.0972	.1442	.1935	.2388	.2734
	5	.0000	.0002	.0012	.0043	.0115	.0250	.0466	.0774	.1172	.1641
	6	.0000	.0000	.0001	.0004	.0013	.0036	.0084	.0172	.0320	.0547
	7	.0000	.0000	.0000	.0000	.0001	.0002	.0006	.0016	.0037	.0078

This table is reproduced by permission from R. S. Burington and D. C. May, *Handbook of Probability and Statistics with Tables*. New York: McGraw-Hill Book Company, 1970.

TABLE 5

(*Continued*)

n	x	.05	.10	.15	.20	p .25	.30	.35	.40	.45	.50
8	0	.6634	.4305	.2725	.1678	.1001	.0576	.0319	.0168	.0084	.0039
	1	.2793	.3826	.3847	.3355	.2670	.1977	.1373	.0896	.0548	.0312
	2	.0515	.1488	.2376	.2936	.3115	.2965	.2587	.2090	.1569	.1094
	3	.0054	.0331	.0839	.1468	.2076	.2541	.2786	.2787	.2568	.2188
	4	.0004	.0046	.0185	.0459	.0865	.1361	.1875	.2322	.2627	.2734
	5	.0000	.0004	.0026	.0092	.0231	.0467	.0808	.1239	.1719	.2188
	6	.0000	.0000	.0002	.0011	.0038	.0100	.0217	.0413	.0703	.1094
	7	.0000	.0000	.0000	.0001	.0004	.0012	.0033	.0079	.0164	.0312
	8	.0000	.0000	.0000	.0000	.0000	.0001	.0002	.0007	.0017	.0039
9	0	.6302	.3874	.2316	.1342	.0751	.0404	.0207	.0101	.0046	.0020
	1	.2985	.3874	.3679	.3020	.2253	.1556	.1004	.0605	.0339	.0176
	2	.0629	.1722	.2597	.3020	.3003	.2668	.2162	.1612	.1110	.0703
	3	.0077	.0446	.1069	.1762	.2336	.2668	.2716	.2508	.2119	.1641
	4	.0006	.0074	.0283	.0661	.1168	.1715	.2194	.2508	.2600	.2461
	5	.0000	.0008	.0050	.0165	.0389	.0735	.1181	.1672	.2128	.2461
	6	.0000	.0001	.0006	.0028	.0087	.0210	.0424	.0743	.1160	.1641
	7	.0000	.0000	.0000	.0003	.0012	.0039	.0098	.0212	.0407	.0703
	8	.0000	.0000	.0000	.0000	.0001	.0004	.0013	.0035	.0083	.0176
	9	.0000	.0000	.0000	.0000	.0000	.0000	.0001	.0003	.0008	.0020
10	0	.5987	.3487	.1969	.1074	.0563	.0282	.0135	.0060	.0025	.0010
	1	.3151	.3874	.3474	.2684	.1877	.1211	.0725	.0403	.0207	.0098
	2	.0746	.1937	.2759	.3020	.2816	.2335	.1757	.1209	.0763	.0439
	3	.0105	.0574	.1298	.2013	.2503	.2668	.2522	.2150	.1665	.1172
	4	.0010	.0112	.0401	.0881	.1460	.2001	.2377	.2508	.2384	.2051
	5	.0001	.0015	.0085	.0264	.0584	.1029	.1536	.2007	.2340	.2461
	6	.0000	.0001	.0012	.0055	.0162	.0368	.0689	.1115	.1596	.2051
	7	.0000	.0000	.0001	.0008	.0031	.0090	.0212	.0425	.0746	.1172
	8	.0000	.0000	.0000	.0001	.0004	.0014	.0043	.0106	.0229	.0439
	9	.0000	.0000	.0000	.0000	.0000	.0001	.0005	.0016	.0042	.0098
	10	.0000	.0000	.0000	.0000	.0000	.0000	.0000	.0001	.0003	.0010
11	0	.5688	.3138	.1673	.0859	.0422	.0198	.0088	.0036	.0014	.0005
	1	.3293	.3835	.3248	.2362	.1549	.0932	.0518	.0266	.0125	.0054
	2	.0867	.2131	.2866	.2953	.2581	.1998	.1395	.0887	.0513	.0269
	3	.0137	.0710	.1517	.2215	.2581	.2568	.2254	.1774	.1259	.0806
	4	.0014	.0158	.0536	.1107	.1721	.2201	.2428	.2365	.2060	.1611
	5	.0001	.0025	.0132	.0388	.0803	.1321	.1830	.2207	.2360	.2256
	6	.0000	.0003	.0023	.0097	.0268	.0566	.0985	.1471	.1931	.2256
	7	.0000	.0000	.0003	.0017	.0064	.0173	.0379	.0701	.1128	.1611
	8	.0000	.0000	.0000	.0002	.0011	.0037	.0102	.0234	.0462	.0806
	9	.0000	.0000	.0000	.0000	.0001	.0005	.0018	.0052	.0126	.0269
	10	.0000	.0000	.0000	.0000	.0000	.0000	.0002	.0007	.0021	.0054
	11	.0000	.0000	.0000	.0000	.0000	.0000	.0000	.0000	.0002	.0005

TABLE 5

(*Continued*)

n	x	.05	.10	.15	.20	p .25	.30	.35	.40	.45	.50
12	0	.5404	.2824	.1422	.0687	.0317	.0138	.0057	.0022	.0008	.0002
	1	.3413	.3766	.3012	.2062	.1267	.0712	.0368	.0174	.0075	.0029
	2	.0988	.2301	.2924	.2835	.2323	.1678	.1088	.0639	.0339	.0161
	3	.0173	.0853	.1720	.2362	.2581	.2397	.1954	.1419	.0923	.0537
	4	.0021	.0213	.0683	.1329	.1936	.2311	.2367	.2128	.1700	.1208
	5	.0002	.0038	.0193	.0532	.1032	.1585	.2039	.2270	.2225	.1934
	6	.0000	.0005	.0040	.0155	.0401	.0792	.1281	.1766	.2124	.2256
	7	.0000	.0000	.0006	.0033	.0115	.0291	.0591	.1009	.1489	.1934
	8	.0000	.0000	.0001	.0005	.0024	.0078	.0199	.0420	.0762	.1208
	9	.0000	.0000	.0000	.0001	.0004	.0015	.0048	.0125	.0277	.0537
	10	.0000	.0000	.0000	.0000	.0000	.0002	.0008	.0025	.0068	.0161
	11	.0000	.0000	.0000	.0000	.0000	.0000	.0001	.0003	.0010	.0029
	12	.0000	.0000	.0000	.0000	.0000	.0000	.0000	.0000	.0001	.0002
13	0	.5133	.2542	.1209	.0550	.0238	.0097	.0037	.0013	.0004	.0001
	1	.3512	.3672	.2774	.1787	.1029	.0540	.0259	.0113	.0045	.0016
	2	.1109	.2448	.2937	.2680	.2059	.1388	.0836	.0453	.0220	.0095
	3	.0214	.0997	.1900	.2457	.2517	.2181	.1651	.1107	.0660	.0349
	4	.0028	.0277	.0838	.1535	.2097	.2337	.2222	.1845	.1350	.0873
	5	.0003	.0055	.0266	.0691	.1258	.1803	.2154	.2214	.1989	.1571
	6	.0000	.0008	.0063	.0230	.0559	.1030	.1546	.1968	.2169	.2095
	7	.0000	.0001	.0011	.0058	.0186	.0442	.0833	.1312	.1775	.2095
	8	.0000	.0000	.0001	.0011	.0047	.0142	.0336	.0656	.1089	.1571
	9	.0000	.0000	.0000	.0001	.0009	.0034	.0101	.0243	.0495	.0873
	10	.0000	.0000	.0000	.0000	.0001	.0006	.0022	.0065	.0162	.0349
	11	.0000	.0000	.0000	.0000	.0000	.0001	.0003	.0012	.0036	.0095
	12	.0000	.0000	.0000	.0000	.0000	.0000	.0000	.0001	.0005	.0016
	13	.0000	.0000	.0000	.0000	.0000	.0000	.0000	.0000	.0000	.0001
14	0	.4877	.2288	.1028	.0440	.0178	.0068	.0024	.0008	.0002	.0001
	1	.3593	.3559	.2539	.1539	.0832	.0407	.0181	.0073	.0027	.0009
	2	.1229	.2570	.2912	.2501	.1802	.1134	.0634	.0317	.0141	.0056
	3	.0259	.1142	.2056	.2501	.2402	.1943	.1366	.0845	.0462	.0222
	4	.0037	.0349	.0998	.1720	.2202	.2290	.2022	.1549	.1040	.0611
	5	.0004	.0078	.0352	.0860	.1468	.1963	.2178	.2066	.1701	.1222
	6	.0000	.0013	.0093	.0322	.0734	.1262	.1759	.2066	.2088	.1833
	7	.0000	.0002	.0019	.0092	.0280	.0618	.1082	.1574	.1952	.2095
	8	.0000	.0000	.0003	.0020	.0082	.0232	.0510	.0918	.1398	.1833
	9	.0000	.0000	.0000	.0003	.0018	.0066	.0183	.0408	.0762	.1222
	10	.0000	.0000	.0000	.0000	.0003	.0014	.0049	.0136	.0312	.0611
	11	.0000	.0000	.0000	.0000	.0000	.0002	.0010	.0033	.0093	.0222
	12	.0000	.0000	.0000	.0000	.0000	.0000	.0001	.0005	.0019	.0056
	13	.0000	.0000	.0000	.0000	.0000	.0000	.0000	.0001	.0002	.0009
	14	.0000	.0000	.0000	.0000	.0000	.0000	.0000	.0000	.0000	.0001

TABLE 5

(*Continued*)

n	x	.05	.10	.15	.20	*p* .25	.30	.35	.40	.45	.50
15	0	.4633	.2059	.0874	.0352	.0134	.0047	.0016	.0005	.0001	.0000
	1	.3658	.3432	.2312	.1319	.0668	.0305	.0126	.0047	.0016	.0005
	2	.1348	.2669	.2856	.2309	.1559	.0916	.0476	.0219	.0090	.0032
	3	.0307	.1285	.2184	.2501	.2252	.1700	.1110	.0634	.0318	.0139
	4	.0049	.0428	.1156	.1876	.2252	.2186	.1792	.1268	.0780	.0417
	5	.0006	.0105	.0449	.1032	.1651	.2061	.2123	.1859	.1404	.0916
	6	.0000	.0019	.0132	.0430	.0917	.1472	.1906	.2066	.1914	.1527
	7	.0000	.0003	.0030	.0138	.0393	.0811	.1319	.1771	.2013	.1964
	8	.0000	.0000	.0005	.0035	.0131	.0348	.0710	.1181	.1647	.1964
	9	.0000	.0000	.0001	.0007	.0034	.0116	.0298	.0612	.1048	.1527
	10	.0000	.0000	.0000	.0001	.0007	.0030	.0096	.0245	.0515	.0916
	11	.0000	.0000	.0000	.0000	.0001	.0006	.0024	.0074	.0191	.0417
	12	.0000	.0000	.0000	.0000	.0000	.0001	.0004	.0016	.0052	.0139
	13	.0000	.0000	.0000	.0000	.0000	.0000	.0001	.0003	.0010	.0032
	14	.0000	.0000	.0000	.0000	.0000	.0000	.0000	.0000	.0001	.0005
	15	.0000	.0000	.0000	.0000	.0000	.0000	.0000	.0000	.0000	.0000
16	0	.4401	.1853	.0743	.0281	.0100	.0033	.0010	.0003	.0001	.0000
	1	.3706	.3294	.2097	.1126	.0535	.0228	.0087	.0030	.0009	.0002
	2	.1463	.2745	.2775	.2111	.1336	.0732	.0353	.0150	.0056	.0018
	3	.0359	.1423	.2285	.2463	.2079	.1465	.0888	.0468	.0215	.0085
	4	.0061	.0514	.1311	.2001	.2252	.2040	.1553	.1014	.0572	.0278
	5	.0008	.0137	.0555	.1201	.1802	.2099	.2008	.1623	.1123	.0667
	6	.0001	.0028	.0180	.0550	.1101	.1649	.1982	.1983	.1684	.1222
	7	.0000	.0004	.0045	.0197	.0524	.1010	.1524	.1889	.1969	.1746
	8	.0000	.0001	.0009	.0055	.0197	.0487	.0923	.1417	.1812	.1964
	9	.0000	.0000	.0001	.0012	.0058	.0185	.0442	.0840	.1318	.1746
	10	.0000	.0000	.0000	.0002	.0014	.0056	.0167	.0392	.0755	.1222
	11	.0000	.0000	.0000	.0000	.0002	.0013	.0049	.0142	.0337	.0667
	12	.0000	.0000	.0000	.0000	.0000	.0002	.0011	.0040	.0115	.0278
	13	.0000	.0000	.0000	.0000	.0000	.0000	.0002	.0008	.0029	.0085
	14	.0000	.0000	.0000	.0000	.0000	.0000	.0000	.0001	.0005	.0018
	15	.0000	.0000	.0000	.0000	.0000	.0000	.0000	.0000	.0001	.0002
	16	.0000	.0000	.0000	.0000	.0000	.0000	.0000	.0000	.0000	.0000
17	0	.4181	.1668	.0631	.0225	.0075	.0023	.0007	.0002	.0000	.0000
	1	.3741	.3150	.1893	.0957	.0426	.0169	.0060	.0019	.0005	.0001
	2	.1575	.2800	.2673	.1914	.1136	.0581	.0260	.0102	.0035	.0010
	3	.0415	.1556	.2359	.2393	.1893	.1245	.0701	.0341	.0144	.0052
	4	.0076	.0605	.1457	.2093	.2209	.1868	.1320	.0796	.0411	.0182
	5	.0010	.0175	.0668	.1361	.1914	.2081	.1849	.1379	.0875	.0472
	6	.0001	.0039	.0236	.0680	.1276	.1784	.1991	.1839	.1432	.0944
	7	.0000	.0007	.0065	.0267	.0668	.1201	.1685	.1927	.1841	.1484
	8	.0000	.0001	.0014	.0084	.0279	.0644	.1134	.1606	.1883	.1855
	9	.0000	.0000	.0003	.0021	.0093	.0276	.0611	.1070	.1540	.1855

TABLE 5

(*Continued*)

n	x	.05	.10	.15	.20	p .25	.30	.35	.40	.45	.50
17	10	.0000	.0000	.0000	.0004	.0025	.0095	.0263	.0571	.1008	.1484
	11	.0000	.0000	.0000	.0001	.0005	.0026	.0090	.0242	.0525	.0944
	12	.0000	.0000	.0000	.0000	.0001	.0006	.0024	.0081	.0215	.0472
	13	.0000	.0000	.0000	.0000	.0000	.0001	.0005	.0021	.0068	.0182
	14	.0000	.0000	.0000	.0000	.0000	.0000	.0001	.0004	.0016	.0052
	15	.0000	.0000	.0000	.0000	.0000	.0000	.0000	.0001	.0003	.0010
	16	.0000	.0000	.0000	.0000	.0000	.0000	.0000	.0000	.0000	.0001
	17	.0000	.0000	.0000	.0000	.0000	.0000	.0000	.0000	.0000	.0000
18	0	.3972	.1501	.0536	.0180	.0056	.0016	.0004	.0001	.0000	.0000
	1	.3763	.3002	.1704	.0811	.0338	.0126	.0042	.0012	.0003	.0001
	2	.1683	.2835	.2556	.1723	.0958	.0458	.0190	.0069	.0022	.0006
	3	.0473	.1680	.2406	.2297	.1704	.1046	.0547	.0246	.0095	.0031
	4	.0093	.0700	.1592	.2153	.2130	.1681	.1104	.0614	.0291	.0117
	5	.0014	.0218	.0787	.1507	.1988	.2017	.1664	.1146	.0666	.0327
	6	.0002	.0052	.0301	.0816	.1436	.1873	.1941	.1655	.1181	.0708
	7	.0000	.0010	.0091	.0350	.0820	.1376	.1792	.1892	.1657	.1214
	8	.0000	.0002	.0022	.0120	.0376	.0811	.1327	.1734	.1864	.1669
	9	.0000	.0000	.0004	.0033	.0139	.0386	.0794	.1284	.1694	.1855
	10	.0000	.0000	.0001	.0008	.0042	.0149	.0385	.0771	.1248	.1669
	11	.0000	.0000	.0000	.0001	.0010	.0046	.0151	.0374	.0742	.1214
	12	.0000	.0000	.0000	.0000	.0002	.0012	.0047	.0145	.0354	.0708
	13	.0000	.0000	.0000	.0000	.0000	.0002	.0012	.0045	.0134	.0327
	14	.0000	.0000	.0000	.0000	.0000	.0000	.0002	.0011	.0039	.0117
	15	.0000	.0000	.0000	.0000	.0000	.0000	.0000	.0002	.0009	.0031
	16	.0000	.0000	.0000	.0000	.0000	.0000	.0000	.0000	.0001	.0006
	17	.0000	.0000	.0000	.0000	.0000	.0000	.0000	.0000	.0000	.0001
	18	.0000	.0000	.0000	.0000	.0000	.0000	.0000	.0000	.0000	.0000
19	0	.3774	.1351	.0456	.0144	.0042	.0011	.0003	.0001	.0002	.0000
	1	.3774	.2852	.1529	.0685	.0268	.0093	.0029	.0008	.0002	.0000
	2	.1787	.2852	.2428	.1540	.0803	.0358	.0138	.0046	.0013	.0003
	3	.0533	.1796	.2428	.2182	.1517	.0869	.0422	.0175	.0062	.0018
	4	.0112	.0798	.1714	.2182	.2023	.1491	.0909	.0467	.0203	.0074
	5	.0018	.0266	.0907	.1636	.2023	.1916	.1468	.0933	.0497	.0222
	6	.0002	.0069	.0374	.0955	.1574	.1916	.1844	.1451	.0949	.0518
	7	.0000	.0014	.0122	.0443	.0974	.1525	.1844	.1797	.1443	.0961
	8	.0000	.0002	.0032	.0166	.0487	.0981	.1489	.1797	.1771	.1442
	9	.0000	.0000	.0007	.0051	.0198	.0514	.0980	.1464	.1771	.1762
	10	.0000	.0000	.0001	.0013	.0066	.0220	.0528	.0976	.1449	.1762
	11	.0000	.0000	.0000	.0003	.0018	.0077	.0233	.0532	.0970	.1442
	12	.0000	.0000	.0000	.0000	.0004	.0022	.0083	.0237	.0529	.0961
	13	.0000	.0000	.0000	.0000	.0001	.0005	.0024	.0085	.0233	.0518
	14	.0000	.0000	.0000	.0000	.0000	.0001	.0006	.0024	.0082	.0222

TABLE 5

(*Continued*)

n	x	.05	.10	.15	.20	p .25	.30	.35	.40	.45	.50
19	15	.0000	.0000	.0000	.0000	.0000	.0000	.0001	.0005	.0022	.0074
	16	.0000	.0000	.0000	.0000	.0000	.0000	.0000	.0001	.0005	.0018
	17	.0000	.0000	.0000	.0000	.0000	.0000	.0000	.0000	.0001	.0003
	18	.0000	.0000	.0000	.0000	.0000	.0000	.0000	.0000	.0000	.0000
	19	.0000	.0000	.0000	.0000	.0000	.0000	.0000	.0000	.0000	.0000
20	0	.3585	.1216	.0388	.0115	.0032	.0008	.0002	.0000	.0000	.0000
	1	.3774	.2702	.1368	.0576	.0211	.0068	.0020	.0005	.0001	.0000
	2	.1887	.2852	.2293	.1369	.0669	.0278	.0100	.0031	.0008	.0002
	3	.0596	.1901	.2428	.2054	.1339	.0716	.0323	.0123	.0040	.0011
	4	.0133	.0898	.1821	.2182	.1897	.1304	.0738	.0350	.0139	.0046
	5	.0022	.0319	.1028	.1746	.2023	.1789	.1272	.0746	.0365	.0148
	6	.0003	.0089	.0454	.1091	.1686	.1916	.1712	.1244	.0746	.0370
	7	.0000	.0020	.0160	.0545	.1124	.1643	.1844	.1659	.1221	.0739
	8	.0000	.0004	.0046	.0222	.0609	.1144	.1614	.1797	.1623	.1201
	9	.0000	.0001	.0011	.0074	.0271	.0654	.1158	.1597	.1771	.1602
	10	.0000	.0000	.0002	.0020	.0099	.0308	.0686	.1171	.1593	.1762
	11	.0000	.0000	.0000	.0005	.0030	.0120	.0336	.0710	.1185	.1602
	12	.0000	.0000	.0000	.0001	.0008	.0039	.0136	.0355	.0727	.1201
	13	.0000	.0000	.0000	.0000	.0002	.0010	.0045	.0146	.0366	.0739
	14	.0000	.0000	.0000	.0000	.0000	.0002	.0012	.0049	.0150	.0370
	15	.0000	.0000	.0000	.0000	.0000	.0000	.0003	.0013	.0049	.0148
	16	.0000	.0000	.0000	.0000	.0000	.0000	.0000	.0003	.0013	.0046
	17	.0000	.0000	.0000	.0000	.0000	.0000	.0000	.0000	.0002	.0011
	18	.0000	.0000	.0000	.0000	.0000	.0000	.0000	.0000	.0000	.0002
	19	.0000	.0000	.0000	.0000	.0000	.0000	.0000	.0000	.0000	.0000
	20	.0000	.0000	.0000	.0000	.0000	.0000	.0000	.0000	.0000	.0000

TABLE 6

Random Digits

63271	59986	71744	51102	15141	80714	58683	93108	13554	79945
88547	09896	95436	79115	08303	01041	20030	63754	08459	28364
55957	57243	83865	09911	19761	66535	40102	26646	60147	15702
46276	87453	44790	67122	45573	84358	21625	16999	13385	22782
55363	07449	34835	15290	76616	67191	12777	21861	68689	03263
69393	92785	49902	58447	42048	30378	87618	26933	40640	16281
13186	29431	88190	04588	38733	81290	89541	70290	40113	08243
17726	28652	56836	78351	47327	18518	92222	55201	27340	10493
36520	64465	05550	30157	82242	29520	69753	72602	23756	54935
81628	36100	39254	56835	37636	02421	98063	89641	64953	99337
84649	48968	75215	75498	49539	74240	03466	49292	36401	45525
63291	11618	12613	75055	43915	26488	41116	64531	56827	30825
70502	53225	03655	05915	37140	57051	48393	91322	25653	06543
06426	24771	59935	49801	11082	66762	94477	02494	88215	27191
20711	55609	29430	70165	45406	78484	31639	52009	18873	96927
41990	70538	77191	25860	55204	73417	83920	69468	74972	38712
72452	36618	76298	26678	89334	33938	95567	29380	75906	91807
37042	40318	57099	10528	09925	89773	41335	96244	29002	46453
53766	52875	15987	46962	67342	77592	57651	95508	80033	69828
90585	58955	53122	16025	84299	53310	67380	84249	25348	04332
32001	96293	37203	64516	51530	37069	40261	61374	05815	06714
62606	64324	46354	72157	67248	20135	49804	09226	64419	29457
10078	28073	85389	50324	14500	15562	64165	06125	71353	77669
91561	46145	24177	15294	10061	98124	75732	00815	83452	97355
13091	98112	53959	79607	52244	63303	10413	63839	74762	50289
73864	83014	72457	22682	03033	61714	88173	90835	00634	85169
66668	25467	48894	51043	02365	91726	09365	63167	95264	45643
84745	41042	29493	01836	09044	51926	43630	63470	76508	14194
48068	26805	94595	47907	13357	38412	33318	26098	82782	42851
54310	96175	97594	88616	42035	38093	36745	56702	40644	83514
14877	33095	10924	58013	61439	21882	42059	24177	58739	60170
78295	23179	02771	43464	59061	71411	05697	67194	30495	21157
67524	02865	39593	54278	04237	92441	26602	63835	38032	94770
58268	57219	68124	73455	83236	08710	04284	55005	84171	42596
97158	28672	50685	01181	24262	19427	52106	34308	73685	74246
04230	16831	69085	30802	65559	09205	71829	06489	85650	38707
94879	56606	30401	02602	57658	70091	54986	41394	60437	03195
71446	15232	66715	26385	91518	70566	02888	79941	39684	54315
32886	05644	79316	09819	00813	88407	17461	73925	53037	91904
62048	33711	25290	21526	02223	75947	66466	06232	10913	75336
84534	42351	21628	53669	81352	95152	08107	98814	72743	12849
84707	15885	84710	35866	06446	86311	32648	88141	73902	69981
19409	40868	64220	80861	13860	68493	52908	26374	63297	45052
57978	48015	25973	66777	45924	56144	24742	96702	88200	66162
57295	98298	11199	96510	75228	41600	47192	43267	35973	23152
94044	83785	93388	07833	38216	31413	70555	03023	54147	06647
30014	25879	71763	96679	90603	99396	74557	74224	18211	91637
07265	69563	64268	88802	72264	66540	01782	08396	19251	83613
84404	88642	30263	80310	11522	57810	27627	78376	36240	48952
21778	02085	27762	46097	43324	34354	09369	14966	10158	76089

C. Summation Notation

Summations

DEFINITION

$$\sum_{i=1}^{n} x_i = x_1 + x_2 + \cdots + x_n. \tag{C.1}$$

Example: $x_1 = 5$, $x_2 = 8$, $x_3 = 14$:

$$\begin{aligned}
\sum_{i=1}^{3} x_i &= x_1 + x_2 + x_3 \\
&= 5 + 8 + 14 \\
&= 27.
\end{aligned}$$

RESULT 1

For a constant c:

$$\sum_{i=1}^{n} c = \underbrace{(c + c + \cdots + c)}_{n \text{ times}} = nc. \tag{C.2}$$

645

Example: $c = 5$, $n = 10$:

$$\sum_{i=1}^{10} 5 = 10(5) = 50.$$

Example: $c = \bar{x}$:

$$\sum_{i=1}^{n} \bar{x} = n\bar{x}.$$

RESULT 2

$$\sum_{i=1}^{n} cx_i = cx_1 + cx_2 + \cdots + cx_n$$

$$= c(x_1 + x_2 + \cdots + x_n) = c\sum_{i=1}^{n} x_i. \qquad (C.3)$$

Example: $x_1 = 5$, $x_2 = 8$, $x_3 = 14$, $c = 2$:

$$\sum_{i=1}^{3} 2x_i = 2\sum_{i=1}^{3} x_i = 2(27) = 54.$$

RESULT 3

$$\sum_{i=1}^{n} (ax_i + by_i) = a\sum_{i=1}^{n} x_i + b\sum_{i=1}^{n} y_i. \qquad (C.4)$$

Example: $x_1 = 5$, $x_2 = 8$, $x_3 = 14$, $a = 2$, $y_1 = 7$, $y_2 = 3$, $y_3 = 8$, $b = 4$:

$$\sum_{i=1}^{3} (2x_i + 4y_i) = 2\sum_{i=1}^{3} x_i + 4\sum_{i=1}^{3} y_i$$

$$= 2(27) + 4(18)$$

$$= 54 + 72$$

$$= 126.$$

Double Summations

Consider the following data involving the variable x_{ij}, where i is the subscript denoting the row position and j is the subscript denoting the column position:

		Column		
		1	2	3
Row	1	$x_{11} = 10$	$x_{12} = 8$	$x_{13} = 6$
	2	$x_{21} = 7$	$x_{22} = 4$	$x_{23} = 12$

DEFINITION

$$\sum_{i=1}^{n} \sum_{j=1}^{m} x_{ij} = (x_{11} + x_{12} + \cdots + x_{1m}) + (x_{21} + x_{22} + \cdots + x_{2m})$$

$$+ (x_{31} + x_{32} + \cdots + x_{3m}) + \cdots + (x_{n1} + x_{n2} + \cdots + x_{nm}). \quad (C.5)$$

Example:

$$\sum_{i=1}^{2} \sum_{j=1}^{3} x_{ij} = x_{11} + x_{12} + x_{13} + x_{21} + x_{22} + x_{23}$$

$$= 10 + 8 + 6 + 7 + 4 + 12$$

$$= 47.$$

DEFINITION

$$\sum_{i=1}^{n} x_{ij} = x_{1j} + x_{2j} + \cdots + x_{nj}. \quad (C.6)$$

Example:

$$\sum_{i=1}^{2} x_{i2} = x_{12} + x_{22}$$

$$= 8 + 4$$

$$= 12.$$

Shorthand Notation

Sometimes when a summation is for all values of the subscript, we use the following shorthand notations:

$$\sum_{i=1}^{n} x_i = \sum_i x_i, \quad (C.7)$$

$$\sum_{i=1}^{n} x_i = \sum x_i, \quad (C.8)$$

$$\sum_{i=1}^{n} \sum_{j=1}^{m} x_{ij} = \sum_i \sum_j x_{ij}, \quad (C.9)$$

$$\sum_{i=1}^{n} x_{ij} = \sum_i x_{ij}. \quad (C.10)$$

Answers to Review Quizzes

Chapter 1

1.	T	5.	b	8.	b
2.	F	6.	b	9.	b
3.	F	7.	a	10.	b
4.	T				

Chapter 2

1.	F	8.	F	15.	b
2.	T	9.	F	16.	c
3.	F	10.	T	17.	d
4.	T	11.	T	18.	a
5.	F	12.	F	19.	c
6.	T	13.	T	20.	a
7.	F	14.	b		

Chapter 3

1.	F	7.	F	12.	b
2.	F	8.	F	13.	a
3.	T	9.	T	14.	d
4.	T	10.	T	15.	a
5.	F	11.	b	16.	b
6.	F				

Chapter 4

1.	F	7.	T	12.	d
2.	T	8.	F	13.	b
3.	T	9.	F	14.	a
4.	F	10.	T	15.	c
5.	F	11.	b	16.	d
6.	T				

Chapter 5

1.	T	8.	F	14.	c
2.	T	9.	T	15.	b
3.	F	10.	F	16.	d
4.	F	11.	b	17.	d
5.	F	12.	a	18.	a
6.	T	13.	a	19.	d
7.	T				

Chapter 6

1.	T	8.	F	15.	c
2.	F	9.	T	16.	a
3.	T	10.	F	17.	b
4.	F	11.	T	18.	c
5.	F	12.	F	19.	d
6.	T	13.	a	20.	c
7.	T	14.	b		

Chapter 7

1.	F	7.	T	12.	a
2.	F	8.	F	13.	a
3.	T	9.	T	14.	c
4.	T	10.	T	15.	b
5.	F	11.	b	16.	d
6.	T				

Chapter 8

1.	T	8.	F	15.	b
2.	T	9.	F	16.	d
3.	F	10.	F	17.	b
4.	T	11.	c	18.	c
5.	T	12.	c	19.	d
6.	F	13.	a	20.	b
7.	F	14.	b		

Chapter 9

1.	F	9.	F	16.	b	
2.	T	10.	T	17.	a	
3.	T	11.	T	18.	c	
4.	T	12.	F	19.	a	
5.	T	13.	F	20.	b	
6.	F	14.	c	21.	b	
7.	T	15.	b	22.	c	
8.	F					

Chapter 10

1.	F	5.	F	8.	c	
2.	F	6.	T	9.	d	
3.	T	7.	d	10.	b	
4.	T					

Chapter 11

1.	T	7.	T	12.	a	
2.	F	8.	T	13.	d	
3.	T	9.	F	14.	b	
4.	F	10.	T	15.	a	
5.	F	11.	d	16.	a	
6.	F					

Chapter 12

1.	F	6.	T	11.	b	
2.	F	7.	F	12.	b	
3.	T	8.	T	13.	c	
4.	F	9.	a	14.	d	
5.	T	10.	c	15.	b	

Chapter 13

1.	T	8.	T	15.	b	
2.	F	9.	T	16.	b	
3.	F	10.	F	17.	b	
4.	F	11.	a	18.	d	
5.	T	12.	c	19.	b	
6.	T	13.	c	20.	d	
7.	F	14.	a			

Chapter 14

1.	T	**8.**	F	**15.**	d
2.	T	**9.**	T	**16.**	a
3.	T	**10.**	F	**17.**	b
4.	F	**11.**	F	**18.**	d
5.	F	**12.**	F	**19.**	d
6.	F	**13.**	b	**20.**	b
7.	T	**14.**	c		

Chapter 15

1.	T	**9.**	F	**16.**	a
2.	F	**10.**	F	**17.**	a
3.	F	**11.**	T	**18.**	a
4.	T	**12.**	T	**19.**	d
5.	F	**13.**	a	**20.**	b
6.	T	**14.**	b	**21.**	c
7.	T	**15.**	c	**22.**	a
8.	T				

Chapter 16

1.	F	**5.**	T	**9.**	b
2.	F	**6.**	T	**10.**	d
3.	F	**7.**	F	**11.**	a
4.	T	**8.**	T	**12.**	b

Chapter 17

1.	T	**6.**	T	**11.**	c
2.	F	**7.**	T	**12.**	a
3.	F	**8.**	F	**13.**	b
4.	F	**9.**	T	**14.**	a
5.	T	**10.**	F	**15.**	d

Answers to Even-Numbered Problems

Chapter 1

2. **a.** all grocery stores in Montgomery on the day of interest
   b. $1.60
4. **a.** 250 households used in the study
   b. all households in Cedar Bluff
6. **a.** women whose mothers took the drug during pregnancy and women whose mothers did not take the drug during pregnancy
   b. 15.8 abnormalities per 1000
   c. 7.9 abnormalities per 1000
8. **a.** percentage of defective items, average filling weights, etc.
   b. time and cost considerations
10. **a.** acceptable statement
   b. incorrect generalization
   c. acceptable statement
   d. not a justifiable conclusion for the entire population
   e. not statistically supportable

Chapter 2

2. nominal

6. ordinal
8. interval
10. a & b

Party	Frequency	Relative Frequency
Democrat	10	.33
Republican	15	.50
Independent	5	.17
Totals	30	1.00

12. a & b

Time	Frequency	Relative Frequency
7:00	3	.15
7:30	4	.20
8:00	4	.20
8:30	7	.35
9:00	2	.10
Totals	20	1.00

14. a & b

Period	Frequency	Relative Frequency
24	3	.12
36	6	.24
48	12	.48
60	4	.16
Totals	25	1.00

16. a & b

Depression Score	Frequency	Relative Frequency
25–34	3	.12
35–44	1	.04
45–54	2	.08
55–64	6	.24
65–74	4	.16
75–84	6	.24
85–94	2	.08
95–104	1	.04
Totals	25	1.00

18.

Call Duration	Frequency	Relative Frequency
2–3.9	5	.25
4–5.9	9	.45
6–7.9	4	.20
8–9.9	0	.00
10–11.9	2	.10
Totals	20	1.00

20. a & b

Score	Frequency	Relative Frequency
140–149	2	.10
150–159	7	.35
160–169	3	.15
170–179	6	.30
180–189	1	.05
190–199	1	.05
Totals	20	1.00

c & d

Score	Cumulative Frequency	Cumulative Relative Frequency
less than or equal to 149	2	.10
less than or equal to 159	9	.45
less than or equal to 169	12	.60
less than or equal to 179	18	.90
less than or equal to 189	19	.95
less than or equal to 199	20	1.00

22. a

Monthly Sales	Frequency
80–89	4
90–99	4
100–109	3
110–119	1
Total	12

26. a & b

Amount	Frequency	Relative Frequency
50–59	2	.08
60–69	5	.20
70–79	5	.20
80–89	9	.36
90–99	2	.08
100–109	1	.04
110–119	1	.04
Totals	25	1.00

28.

5	7
6	3 4
7	2 5 9
8	0 3 4 9 9
9	1 6 7 8 9
10	2 5 8
11	3 5
12	1 2 4
13	6

30.

3	4 8 9
4	1 4 6 7 7 8 8 9
5	1 2 3 5 8 8
6	1 6 6 8
7	1 2 3 4 7
8	4 6 9
9	0

32. a & b

Attendance	Frequency	Relative Frequency
10–14	2	.10
15–19	7	.35
20–24	7	.35
25–29	3	.15
30–34	1	.05
Totals	20	1.00

34. a & b

Airline	Frequency	Relative Frequency
American Air	8	.200
East Coast Air	13	.325
Suncoast	8	.200
Great Western	11	.275
Totals	40	1.000

36.

Number	Frequency
0–4	1
5–9	3
10–14	5
15–19	4
20–24	2
Total	15

38. a

Amount	Frequency	Relative Frequency
25–34.99	2	.10
35–44.99	6	.30
45–54.99	4	.20
55–64.99	4	.20
65–74.99	2	.10
75–84.99	2	.10
Totals	20	1.00

42.

Sales	Relative Frequency
8–10	.077
11–13	.077
14–16	.077
17–19	.307
20–22	.308
23–25	.077
26–28	.077
Total	1.000

48. a & b

Points Scored	Frequency	Relative Frequency
60–64	3	.12
65–69	1	.04
70–74	6	.24
75–79	4	.16
80–84	4	.16
85–89	3	.12
90–94	3	.12
95–99	1	.04
Totals	25	1.00

50.

9	4 7 8
10	2 3 5
11	4 7 8 9
12	1 2 6 7
13	5 6 8
14	1 2 4

Chapter 3

2. **a.** 178

    b. 178

    c. do not report a mode

    d. 184

4. mean = 17.4; median = 18; mode = 18; 70th percentile = 18

6.

	mean	**median**	**mode**
city	15.58	15.9	15.3
country	18.92	18.7	18.6, 19.4

8. mean = 400; variance = 250; standard deviation = 15.81

10. range = 7; variance = 4.97; standard deviation = 2.23

12. s^2 = .002; line should not be shut down

14. **a.** at least .56 are between $100 and $400

    b. at least .84 are between $0 and 500

    c. $26–$474

16. mean = 74.02; variance = 158.78; standard deviation = 12.60

18. mean = 64.7; variance = 367.67; standard deviation = 19.17

20. 5-number summary to be used to develop the box-and-whisker plot is: 1. 98; 2. 106; 3. 118; 4. 126; 5. 151

22. 5-number summary to be used to develop the box-and-whisker plot is: 1. 41; 2. 61; 3. 73.5; 4. 81; 5. 97

24. **a.** 12.4

    b. 12.5

    c. 12.5

    d. 12.0

    e. 2

    f. 1

    g. .52

    h. .72

    i. 6%

26. s^2 = 98,489.310; s = 9,924.18; range = 39,500

28. mean = 6.1; median = 6; mode = 4; variance = 6.13; standard deviation = 2.48

30. mean = 45; median = 43.5; mode = 50; variance = 134.36; standard deviation = 11.59

32. Cola
34. **a.** at least 51%
  b. at least 88%
  c. 27.876 to 52.124 million shares
36. mean = 51.5; variance = 227.37; standard deviation = 15.08
38. 5-number summary to be used to develop the box-and-whisker plot is: 1. 42; 2. 62; 3. 79; 4. 89; 5. 98
40. 5-number summary to be used to develop the box-and-whisker plot is: 1. 400; 2. 624; 3. 836; 4. 999; 5. 1278

Chapter 4

2. **b.** There appears to be a negative linear relationship
  c. -60
  d. $r = -0.97$
4. **b.** yes
6. **b.** linearly related
  d. $\hat{y} = 45.379 + .167 x$
  e. 66.25
8. **a.** yes
  b. $\hat{y} = 31.2501 - 2.4643 x$
  c. 11.5357
10. **a.** $\hat{y} = 11.635 + .5455 x$
  b. 28

12.

		Democrat	Republican	Independent	Totals
Educa-tional Level	Did not complete high school	40 (16.0)	20 (8.0)	10 (4.0)	70 (28.0)
	High School degree	30 (12.0)	35 (14.0)	15 (6.0)	80 (32.0)
	College degree	30 (12.0)	45 (18.0)	25 (10.0)	100 (40.0)
	Totals	100 (40.0)	100 (40.0)	50 (20.0)	250 (100.0)

14.

		Grade					
		A	B	C	D	F	Totals
Instructor	Johnson	4 (4.4)	12 (13.3)	5 (5.6)	2 (2.2)	3 (3.3)	26 (28.9)
	Pray	2 (2.2)	14 (15.6)	8 (8.9)	1 (1.1)	5 (5.6)	30 (33.3)
	Evans	12 (13.3)	8 (8.9)	10 (11.1)	3 (3.3)	1 (1.1)	34 (37.8)
	Totals	18 (20.0)	34 (37.8)	23 (25.6)	6 (6.6)	9 (10.0)	90 (100.0)

16. **a.** -9.86
  b. $-.94$
18. **b.** $r = -.86$
20. **a.** $\hat{y} = 10.529 + .9534\,x$
  b. 39.131 or \$3913.10
22. **a.** $\hat{y} = 21.9999 + 13.1667\,x$
  b. 74.6667 or \$746.67

24.

		Side Effect		
		None	Some	Totals
Drug Level	A	25 (25.0)	5 (5.0)	30 (30.0)
	B	30 (30.0)	20 (20.0)	50 (50.0)
	C	5 (5.0)	15 (15.0)	20 (20.0)
	Totals	60 (60.0)	40 (40.0)	100 (100.0)

Chapter 5

2. **a.** 6
4. **a.** 4
6. 1024
8. four
10. **a.** 36
  c. six
12. probabilities sum to 1.05; therefore, revise

14. no; probabilities do not sum to 1
16. no; there are four possible outcomes
18. .20; data does not confirm equal consumer preference
20. **a.** no; probabilities sum to .85
  b. revise probabilities so they sum to 1.0
22. **a.** 36
  c. $\frac{1}{6}$
  d. $\frac{5}{18}$
  e. no
  f. classical
24. **a.** .84
  b. .20
  c. .24
26. **a.** .76
  b. .24
28. **a.** yes
  b. .58
  c. .12
30. **a.** yes
  b. .78
  c. .08
  d. yes; $P(A \cap B) = 0$; $P(A \cup B) = .86$
32. .72
34. **b.** .9
  c. .1
36. **a.** yes
  b. .65
38. **b.** 20% own a foreign car; 80% a U.S. car
  c. .15
  d. .95
  e. .17
  f. .75
  g. not independent
40. **b.** son has a higher probability of having attended college
  c. .72
  d. .40
  e. not independent
42. **a.** .045
  b. .45
  c. not independent
  d. .355
  e. no
44. **a.** .20
  b. .70
  c. .30
46. **a.** .50
  b. .75
  c. .65

48. .625

50. **a.** 4

c. $R = \{(S, S), (S, U)\}$; $M = \{(S, S), (U, S)\}$

d. $\{(S, S), (S, U), (U, S)\}$

e. $\{(S, S)\}$

f. no

52. **a.** $\frac{3}{8}$; $\frac{1}{4}$; $\frac{3}{8}$; $\frac{1}{2}$

b. $\frac{1}{8}$

c. $\frac{1}{2}$

d. $\frac{1}{2}$

e. $\frac{1}{3}$

f. 0

g. 0

h. no

54. **a.** no

b. no

c. .50

56. **a.** .12; .20

b. not independent

58. **a.** .50; .30; .20

b. no

c. .40

d. no

e. probability of buying the product increases to .40

60. **a.** .25

b. yes

c. no

62. **a.** .25; 40.; .10

b. .25

c. B and S are independent; program appears to have no effect

64. **a.** .20

b. .35

c. 65%

66. yes; P(sale | callback) = .21

68. 3.44%

Chapter 6

2. a & b.   Let N = no improvement
          M = moderate improvement
          S = strong improvement

Outcome	Value of Random Variable
(N,N)	0
(N,M)	1
(N,S)	1
(M,N)	1
(M,M)	2
(M,S)	2
(S,N)	1
(S,M)	2
(S,S)	2

4. **a.** 0, 1, 2, . . . , 20; discrete
   b. 0, 1, 2, . . . ; discrete
   c. 0, 1, 2, . . . ; discrete
   d. $0 \leq x \leq 8$; continuous
   e. $x \geq 0$; continuous

6. **a**

x	$f(x)$
22	.10
23	.40
24	.30
25	.20
Total	1.00

8. **b.** 2/10
   c. 7/10
10. **a.** yes
   b. .333
   c. .833
12. **a.** 2.45
   b. 2.0475; 1.43

14. a.

x	$f(x)$
-5	20/38
5	18/38
Total	38/38 = 1.00

    b. -0.26
    c. 24.94; 4.99
    d. -26
16. c. 3.5
    d. variance = 2.9166; standard deviation = 1.708
18. variance = 9475; standard deviation = 97.34
20. a. 3 trials
    b. 2; boy or girl
    c. .5
    d. independent trials
    e. number of girls; discrete; 0, 1, 2, or 3
24. c. 2
    d.

Number of Defects	Probability
0	.9409
1	.0582
2	.0009

26. $f(1) = .4116; f(2) = .2646; f(0) = .2401$
28. a. .90
    b. .99
    c. .999
    d. yes
30. a. .5905
    b. .1937
    c. .6082
32. $\mu = 212.5; \sigma^2 = 31.88$
34. a. 1.6
    b. 120
36. no; yes; no

38. a.

x	0	1	2	3	4
$f(x)$	.10	.40	.30	.15	.05

    b. $\mu = 1.65;\ \sigma^2 = 1.0275$

40. a.

x	9	10	11	12	13
$f(x)$	.30	.20	.25	.05	.20

    b. 10.65
    c. 2.1275
    d. Looks good; E(Profit) = 1.35 million

44. a. .2992
    b. .4618
    c. .5382

46. a. .1328
    b. .1413

48. $\mu = 10.4;\ \sigma^2 = 9.88$

Chapter 7

2. a.

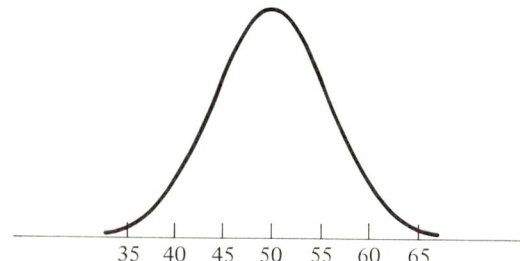

    b. .6826
    c. .9544

4. a.

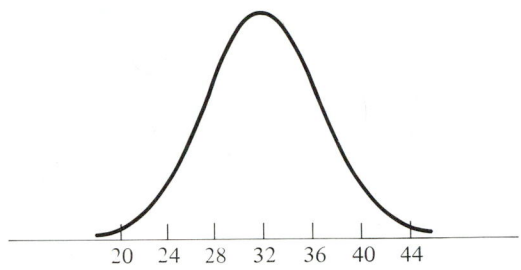

20 24 28 32 36 40 44

   b. .9544
   c. .9544 ÷ 2 = .4772
   d. .5000
   e. .1587

6. a. .3413
   b. .4332
   c. .4772
   d. .4938

8. a. .2967
   b. .4418
   c. .3300
   d. .5910
   e. .8849
   f. .2388

10. a. 1.96
   b. 0.61
   c. 1.12
   d. 0.44

12. a. 2.33
   b. 1.96
   c. 1.645
   d. 1.28

14. a. 4.75%
   b. .99%
   c. 79.38%
   d. $796.75

16. a. .0228
   b. .2857
   c. 9.522

18. a. .4592
   b. .0301
   c. 3976

20. a. .0377
   b. .0823
   c. .9108

22. a. .8336
   b. .0049

24. a. .8106
     b. .0838
     c. .6955
     c. .9500
26. a. .1587
     b. .0228
     c. .8185
28. .0062
30. a. .5899
     b. 29.04
32. a. .4706
     b. .0475
     c. 42,450
34. a. .1357
     b. 0.19%
     c. 3.14 or approximately 3 years
36. a. 54
     b. 4.98
     c. .1841
38. a. 5.16%
     b. 57.87%
     c. 99.55%
     d. approximately 0

Chapter 8

2. a. 10; ABC, ABD, ABE, ACD, ACE, ADE, BCD, BCE, BDE, CDE
    b. 1/10
4. It is easier due to the fact that we do not have to list all possible simple random samples.
6. a. 15; 1-2, 1-3, 1-4, 1-5, 1-6, 2-3, 2-4, 2-5, 2-6, 3-4, 3-5, 3-6, 4-5, 4-6, 5-6
    b. 7.75, 7.00, 9.00, 8.00, 8.25, 6.75, 8.75, 7.75, 8.00, 8.00, 7.00, 7.25, 9.00, 9.25, 8.25
    c.

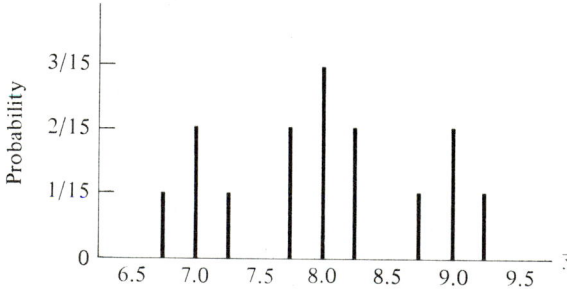

8. a.

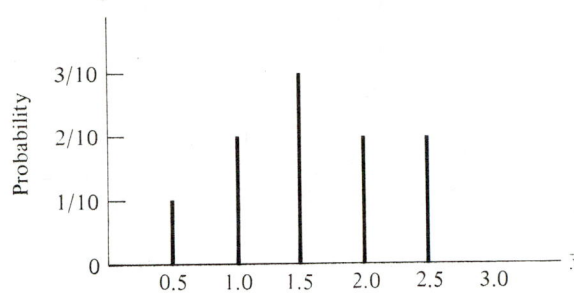

b.

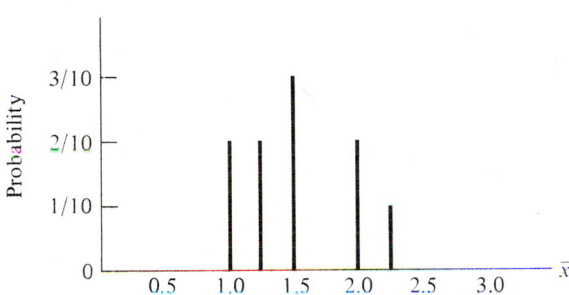

10. a. $\mu = 17; \sigma = 1.414$

b. AB, AC, AD, BC, BD, CD; 16, 17, 16, 18, 17, 18

c. $E(\bar{x}) = 17; \sigma_{\bar{x}} = .816$

d. $E(\bar{x}) = 17; \sigma_{\bar{x}} = .816$

12. a. $\mu = 185; \sigma = 45.72$

b. $E(\bar{x}) = 185; \sigma_{\bar{x}} = 23.28$

c. $E(\bar{x}) = 185; \sigma_{\bar{x}} = 18.67$

14. $E(\bar{x}) = 170; \sigma_{\bar{x}} = 4.43$

16. $n = 100$

18. a. 1.414

b. 1.414

c. 1.407

d. 1.343

e. $n/N > .05$ only in case d. For the other examples, $\sigma_{\bar{x}} \approx 1.41$ regardless of the population size.

20. Normal distribution with $E(\bar{x}) = 18$ and $\sigma_{\bar{x}} = .566$.

22. Normal distribution with $E(\bar{x}) = 16.5$ and $\sigma_{\bar{x}} = .18$.

24. a. Normal distribution with $E(\bar{x}) = 15.9$ and $\sigma_{\bar{x}} = .079$

b. .1020

26. a. Normal distribution with $E(\bar{x}) = 120$ and $\sigma_{\bar{x}} = 40/\sqrt{30} = 7.3$

b. .0031; .2482

28. a. The population is assumed to be normally distributed.

b. .9266

c. Take a larger sample with $n \geq 30$.

30. **a.** $N = 2000$; $\sigma_{\bar{x}} = 20.11$
    $N = 5000$; $\sigma_{\bar{x}} = 20.26$
    $N = 10,000$; $\sigma_{\bar{x}} = 20.31$
    b. $N = 2000$; .7850
    $N = 5000$; .7814
    $N = 10,000$; .7814
    c. yes, since all three probabilities are very similar.
32. **a.** .5036
    b. $n = 246$
34. **a.** A and B
    b. C and E
36. 364, 702, 782, 263, 281, 243, 493, 337, 525, 825.
38. **a.** 21; Identifying the seven airlines by A, B, C, D, E, F and G, we have samples of AB, AC, AD, AE, AF, AG, BC, BD, BE, BF, BG, CD, CE, CF, CG, DE, DF, DG, EF, EG and FG
    b. $1/21$
    c. United Airlines (B) appears in 6 samples.
    d. $6/21 = 2/7$
40. **a.** Normal distribution with $E(\bar{x}) = 14$ and $\sigma_{\bar{x}} = .566$
    b. Sample distribution is approximately normal
    c. .9616
    d. .6212
42. **a.** $n/N = .20$; yes
    b. 3.58
    c. .8384
44. .8764
46. 108, 290, 201, 292, 322, 009, 244, 249, 226, 125, (continue at top of next column) 147 and 113.
48. All three are convenience samples.

Chapter 9

2. 246.42 to 253.58
4. 12.86 to 14.34
6. $82.00 to $90.00
8. 33
10. **a.** 62
    b. 385
    c. 1537
12. **a.** 49
    b. 30 to 34
14. Type I error: Shutting down the machine when it is operating properly
    Type II error: Allowing the machine to operate when it is producing bad parts
16. **a.** $\mu = 400$
    $\mu \neq 400$
    b. $z = -4.38$; reject H_0.

18. **a.** $H_0 : \mu = 12.2$ Accept H_0 If $-2.33 \leq z \leq 2.33$.
        $H_1 : \mu \neq 12.2$ Reject H_0 Otherwise.
    b. $z = 1.06$; accept H_0
    c. .2892
20. $z = 1.77$; accept the president's claim.
    p value = .0384
22. $z = -2.32$; reject the manager's claim.
24. $z = 2.26$; reject the advertisement claim.
    p value = .0119
26. $z = -5.27$; reject H_0.
    p value is less than .001
28. **a.** 1.734
    b. -1.321
    c. 3.365
    d. -1.761 and 1.761
    e. -2.048 and 2.048
30. 7.26 to 9.74
32. 75.90 to 84.10; 75.03 to 84.97
34. **a.** 5.55
    b. 4.51 to 6.59
36. **a.** $t = 2.94$; reject H_0 and conclude $\mu > 55$
    b. approximately .011
38. $t = -3.20$; reject H_0 and conclude $\mu < 300$
40. 3248.55 to 3551.45
42. 128.78 to 134.10
44. 158
46. $H_0 : \mu = 220$
    $H_1 : \mu < 220$ Convert to the new line if H_0 is rejected.
48. $z = 2.37$; reject H_0
50. a. $H_0 : \mu = 72$
        $H_1 : \mu > 72$
    b. .0143
    c. Reject H_0 and conclude $\mu > 72$
52. 56.58 to 63.42
54. **a.** 55
    b. 49.33 to 60.67
    c. 46.61 to 63.39

Chapter 10

2. **a.** .775
    b. .5625
    c. .4375

4. **a.** 1234, 1235, 1236, 1245, 1246, 1256, 1345, 1346, 1356, 1456, 2345, 2346, 2356, 2456, and 3456.

   b.

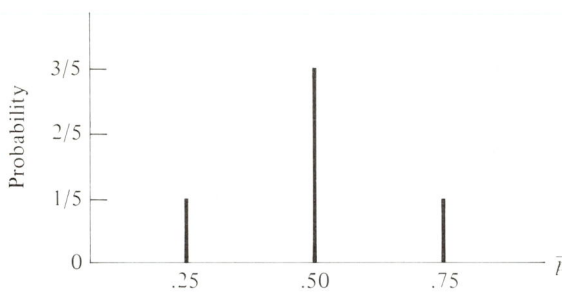

6. **a.** $E(\bar{p}) = .08$; $\sigma_{\bar{p}} = .0384$

   b. Normal distribution with $E(\bar{p}) = .08$ and $\sigma_{\bar{p}} = .0271$

   c. Normal distribution since both $np \geq 5$ and $n(1 - p) \geq 5$.

8. **a.** Normal distribution with $E(\bar{p}) = .30$ and $\sigma_{\bar{p}} = .0458$

   b. .9708

   c. .7242

10. **a.** no; $np = 16$, but $n(1 - p) = 4$.

   b. .9476

12. .4714

14. .691 to .843

16. .238 to .362

18. .028 to .062

20. 1399

22. **a.** .533

   b. .444 to .622

   c. no, since the interval is from .444 to .622

   d. 383

24. **a.** $z = 1.38$; Reject H_0.

   b. .0838; Reject H_0.

26. $z = -3.64$; p value $< .001$; Reject H_0.

28. $z = -2.07$; p value $= .0192$; Reject H_0.

30. $z = -1.40$; p value $= .0808$; Accept H_0.

32. .388 to .526

34. **a.** Normal distribution with $E(\bar{p}) = .15$ and $\sigma_{\bar{p}} = .0505$

   b. .4448

36. .9525

38. 4626

40. $z = 1.63$; p value $= .0516$; Accept H_0 and do not initiate the special offer.

42. $z = -1.66$; p value $= .0485$; Accept H_0.

Chapter 11

2. **a.** .30

   b. .142 to .458

   c. .112 to .488

4. 1354 to 2646

6. $z = 8.08$; p value is approximately 0; Reject H_0
8. $z = 3.19$; p value $< .002$; Reject H_0; System B preferred.
10. **a.** H_0: $\mu_1 - \mu_2 = 0$; H_1: $\mu_1 - \mu_2 > 0$.
    b. Normal distribution with $E(\bar{x}_1 - \bar{x}_2) = 0$ and $\sigma_{\bar{x}_1 - \bar{x}_2} = .56$
    c. Accept H_0 If $z \leq 1.645$; Reject H_0 Otherwise
    d. $z = 2.68$; Reject H_0 and go with supplier B.
12. $t = 2.29$; p value $= .0375$; Reject H_0.
14. $t = 2.33$; Accept H_0: $\mu_d = 0$.
16. .023 to .157
18. $z = -2.30$; p value $= .0214$; Reject H_0.
20. **a.** $z = 2.33$; Reject H_0.
    b. .019 to .221
22. 18.22 to 21.48; 17.55 to 22.15.
24. 45 for each city
26. $z = -5.14$; Reject H_0
28. $t = 1.37$; Accept H_0.
30. $-.171$ to $-.029$
32. $z = 1.42$; p value $= .1556$; Accept H_0.

Chapter 12

2. $3.95 \leq \sigma \leq 7.59$
4. $.02 \leq \sigma^2 \leq .12$
   $.14 \leq \sigma \leq .34$
6. $\chi^2 = 31.5 > \chi^2_{.10} = 29.6151$
   reject the claim
8. Do not reject the hypothesis that $\sigma^2 = 10$.
10. $s_1^2 = 14.44$, $s_2^2 = 27.04$, and $F = .534$
    With $F_{.05} = 3.79$ and $F_{.95} = .26$ we cannot reject.
12. Do not reject H_0: $\sigma_1^2 = \sigma_2^2$.
14. $F = 2.083 > F_{.05} = 1.94$.
    Conclude variances are not equal.
16. **a.** $s^2 = 900$
    b. $567.35 \leq \sigma^2 \leq 1689.72$
    c. $23.82 \leq \sigma \leq 41.11$
18. **a.** They do not support the historical variability.
    b. $.33 \leq \sigma^2 \leq 1.14$
    $.57 \leq \sigma \leq 1.07$
20. Do not conclude the variance is exceeding the standard.
22. Reject the null hypothesis that the variances are equal. Pooling is not appropriate.
24. $F_{.05} = 2.40$, $F_{.95} = .42$, $F = .43$. Do not conclude the departments differ.

Chapter 13

2. $\chi^2_{.10} = 4.61$ (2 degrees of freedom), $\chi^2 = 6.24$ conclude that there is a difference in color preferences.
4. $\chi^2_{.05} = 5.99$, $\chi^2 = 2.31$. Cannot conclude there is a preference.

6. $\chi^2_{.05} = 5.99$, $\chi^2 = 6.31$. Reject the null hypotheses that salesperson is independent of type of product.

8. Classifications are independent. $\chi^2_{.05} = 5.99$, $\chi^2 = 1.54$

10. Party affiliation is not independent of education level. $\chi^2_{.01} = 13.28$, $\chi^2 = 13.42$

12. We cannot conclude there is a consumer preference. $\chi^2_{.05} = 5.99$, $\chi^2 = 2.4375$

14. Shift and part quality are not independent. $\chi^2_{.05} = 5.99$, $\chi^2 = 8.11$

16. There is no significant difference. $\chi^2_{.01} = 13.28$, $\chi^2 = 6.32$.

18. Receiving a job offer is related to grade point average. $\chi^2_{.10} = 4.61$, $\chi^2 = 5.81$

20. Classifications are independent. $\chi^2_{.05} = 15.51$, $\chi^2 = 13.85$

Chapter 14

2. MSTR = 110, MSE = 43.25, F = 2.54, $F_{.05}$ = 3.24. There is no significant difference in drying times.

4. MSTR = 35, MSE = 2, F = 17.5, $F_{.05}$ = 3.89. There is a significant difference in mean miles per gallon.

6. MSTR = 35, MSE = 19.67, F = 1.78, $F_{.05}$ = 3.89. The effect of temperature on yield is not significant.

10. SSE = 180, F = 12.5

12. MSTR = 19.255, MSE = .9635, $F_{.05}$ = 3.10, F = 19.98. There is a significant difference.

14. MSTR = 808.93, MSE = 126, F = 6.42, $F_{.05}$ = 3.39. There is a significant difference.

16. MSTR = 51.67, MSE = .9167, F = 56.36, $F_{.05}$ = 4.46. There is a significant difference.

18. MSTR = 11.25, MSE = 1.5833, F = 7.11, $F_{.05}$ = 3.26. The total audit time for the procedures is not the same.

20. MSTR = 13.265, MSE = 2.01625, F = 6.58, $F_{.05}$ = 4.46. There is a significant difference.

22. No

24. Interval estimate: 5.2 to 20.8. Difference is significant.

26. MSTR = 196, MSE = 44, F = 4.45, $F_{.05}$ = 4.26. Mean asking prices are not the same.

28. MSTR = 873.47, MSE = 67.70, F = 12.9, $F_{.05}$ is between 2.76 and 2.84. Mean number of units sold is not the same.

30. MSTR = 210, MSE = 142.07, F = 1.48, $F_{.05}$ = 3.35. Difference is not significant.

32. MSTR = 2,633.34, MSE = 1850, F = 1.42, $F_{.05}$ = 4.26. No significant difference.

34. MSTR = 430, MSE = 106.89, F = 4.02, $F_{.05}$ = 3.5. Difference is not significant.

36. **a.** MSTR = 250.62, MSE = 79.6, F = 3.15, $F_{.05}$ = 2.99. There is a significant difference.

    b. Prof. Jennings is significantly higher than Prof. Swanson.

38. No.

40. It enables us to obtain significant results.

42. MSTR = 69.335, MSE = 41.555, F = 1.67, $F_{.01}$ = 10.92. There is no significant difference.

Chapter 15

2. **b.** $\hat{y} = 30.33 - 1.88x$
4. The relationship appears curvilinear.
6. **c.** yes
   d. $\hat{y} = 6.38 + 45.68x$
   e. \$129.716
8. **b.** $\hat{y} = 47.61 + 44x$
   c. 85.01
10. **a.** SSR = 108.47, SST = 114.8
    b. 94%
12. **a.** SSR = 558.47, SST = 757.2
    b. 73.75%
14. **a.** 3.8, 23.2
    b. yes
    c. yes
16. **a.** $\hat{y} = .75 + .51x$
    b. cannot reject H_0
    c. cannot reject H_0.
18. MSR = 1543.84 MSE = 32.05 F = 48.17 $F_{.01}$ = 34.12. Selling price and square footage are related.
20. MSR = 2272, MSE = 21.25, F = 106.92, $F_{.05}$ = 5.32. Reject H_0.
22. 1275.67 ± 151.63
24. 1275.67 ± 356.55
28. **a.** 9
    b. $\hat{y} = 20 + 7.21x$
    c. 1.3626
    d. Reject H_0.
    e. \$380,500
30. **a.** $\hat{y} = 80 + 50x$
    b. 30
    c. reject H_0
    d. \$680,000
32. $r = .91$
34. $r = .92$
42. **a.** $\hat{y} = 10.53 + .95x$
    b. reject H_0
    c. \$2865 to \$4941
    d. yes
44. **a.** $\hat{y} = 3766.67 - 322.22x$
    b. reject H_0
    c. $r^2 = .94$
    d. \$2115.98 to \$2839.60
46. **a.** $\hat{y} = 220 + 131.67x$
    b. reject H_0
    c. $r^2 = .87$
    d. \$595.72 to \$897.64
48. **a.** $\hat{y} = 5.85 + .83x$
    b. reject H_0

Chapter 16

6. **a.** SSE = 4000, MSE = 571.43, MSR = 6000
    b. Relationship is significant.
8. **a.** $t = 3.87$, reject H_0
    b. $t = 1.84$, do not reject H_0
    c. $t = -5.07$, reject H_0
10. **a.** $R^2 = .75$
    b. $R_a^2 = .68$
    c. It explains a reasonable amount but there is still a fair amount v plained—not bad.
12. **a.** $R^2 = .95$, $R_a^2 = .93$
    b. Relationship is significant.
14. 3.1884
16. 99.08
18. **b.** $F = 52.32$, $F_{.05} = 4.74$, Reject H_0
    c. $R_a^2 = .919$, good fit
    d. Reject H_0 in both cases.
20. **b.** $F = 13.47$, $F_{.05} = 4.74$, Reject H_0
    c. $R_a^2 = .735$, fair fit
    d. Reject H_0 in both cases
24. **b.** $F = 28.37$, $F_{.05} = 5.79$, Reject H_0
    c. $R_a^2 = .887$, good fit
    d. Reject H_0 in both cases
    e. $93,165
26. **a.** Reject H_0
    b. Do not reject H_0
    c. Reject H_0

Chapter 17

2. $w = 171$, $\mu_w = 138$, $\sigma_w = 15.17$. Critical values 113.05 and 162.95. Reject H_0.
4. $w = 116$, $\mu_w = 120$, $\sigma_w = 16.12$. Do not reject H_0.
6. $T = 36$, $\sigma_T = 19.62$. Reject H_0 using one-tailed test.
8. $T = -46$, $\sigma_T = 19.62$. Reject H_0.
10. $T = -13$, $\sigma_T = 14.3$. One-tailed test. Do not reject H_0.
12. $\mu = 82.5$, $\sigma = 6.4$. Reject H_0
14. $\mu = 94$, $\sigma = 6.86$. Reject H_0
16. $\mu = 60$, $\sigma = 5.48$. Reject H_0.
18. $r_s = -.81$, $\sigma_{r_s} = .38$. Reject H_0.
20. $r_s = .87$, $\sigma_{r_s} = .33$. Reject H_0
22. $r_s = .77$, $\sigma_{r_s} = .33$. Reject H_0
24. $\mu_w = 150$, $\sigma_w = 17.32$, $z = -2.97$. Reject H_0
26. $\sigma_T = 25.5$, $T = -66$, $z = -2.59$. Reject H_0
28. $\mu = 22$, $\sigma = 3.32$. Reject H_0
30. $\mu = 6$, $\sigma = 1.73$. Do not reject H_0
32. $r_s = -.92$ $\sigma_{r_s} = .30$. Reject H_0
34. $r_s = .76$ $\sigma_{r_s} = .267$. Reject H_0

Index

t Distribution

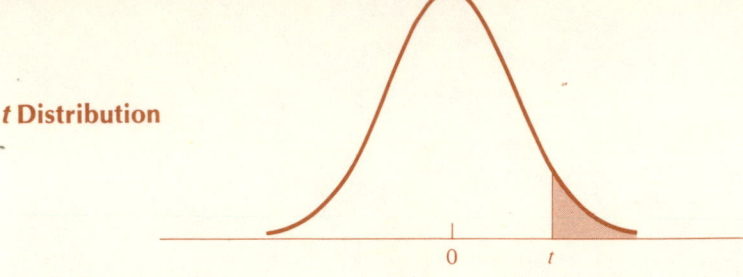

Entries in the table give t_α values, where α is the area or probability in the upper tail of the *t* distribution. For example, with 10 degrees of freedom and a .05 area in the upper tail, $t_{.05} = 1.812$.

Degrees of Freedom	Area in Upper Tail				
	.10	.05	.025	.01	.005
1	3.078	6.314	12.706	31.821	63.657
2	1.886	2.920	4.303	6.965	9.925
3	1.638	2.353	3.182	4.541	5.841
4	1.533	2.132	2.776	3.747	4.604
5	1.476	2.015	2.571	3.365	4.032
6	1.440	1.943	2.447	3.143	3.707
7	1.415	1.895	2.365	2.998	3.499
8	1.397	1.860	2.306	2.896	3.355
9	1.383	1.833	2.262	2.821	3.250
10	1.372	1.812	2.228	2.764	3.169
11	1.363	1.796	2.201	2.718	3.106
12	1.356	1.782	2.179	2.681	3.055
13	1.350	1.771	2.160	2.650	3.012
14	1.345	1.761	2.145	2.624	2.977
15	1.341	1.753	2.131	2.602	2.947
16	1.337	1.746	2.120	2.583	2.921
17	1.333	1.740	2.110	2.567	2.898
18	1.330	1.734	2.101	2.552	2.878
19	1.328	1.729	2.093	2.539	2.861
20	1.325	1.725	2.086	2.528	2.845
21	1.323	1.721	2.080	2.518	2.831
22	1.321	1.717	2.074	2.508	2.819
23	1.319	1.714	2.069	2.500	2.807
24	1.318	1.711	2.064	2.492	2.797
25	1.316	1.708	2.060	2.485	2.787
26	1.315	1.706	2.056	2.479	2.779
27	1.314	1.703	2.052	2.473	2.771
28	1.313	1.701	2.048	2.467	2.763
29	1.311	1.699	2.045	2.462	2.756
30	1.310	1.697	2.042	2.457	2.750
40	1.303	1.684	2.021	2.423	2.704
60	1.296	1.671	2.000	2.390	2.660
120	1.289	1.658	1.980	2.358	2.617
α	1.282	1.645	1.960	2.326	2.576

Reprinted by permission of Biometrika Trustees from Table 12, Percentage Points of the *t*-Distribution, 3rd Edition, 1966. E. S. Pearson and H. O. Hartley, *Biometrika Tables for Statisticians*, Vol. I.